Fundamentals of AUTOMATIC CONTROL

Fundamentals of

AUTOMATIC

JOHN WILEY & SONS, INC.

CONTROL

SOMESHWAR C. GUPTA

Southern Methodist University

LAWRENCE HASDORFF

Virginia Polytechnic Institute

New York · London · Sydney · Toronto

Also by Someshwar C. Gupta

TRANSFORM AND STATE VARIABLE METHODS IN LINEAR
SYSTEMS (1966)

Library of Congress Catalogue Card Number: 70-100322

SBN 471   33645   9

Printed in the United States of America

10   9   8   7   6   5   4   3

To
Kusum, Ruth, *Mona*, Debbie,
Rakesh, Henry, *Neena*, Karen

Special thanks are due
Professors E. I. Jury, T. J. Higgins,
M. E. Van Valkenburg, and J. C. Prabhakar

Preface

Within the last few years there has been an increasing emphasis on modern control theory based on the state variable approach. How has this affected the control curriculum at both the undergraduate and the graduate level?

Traditionally, the analysis and elementary synthesis of continuous-control systems by the use of frequency response and root locus methods have been taught at the undergraduate level. This was usually followed by a senior-level optional course in advanced synthesis. There also have been regular graduate courses in sampled-data and nonlinear control and, in some schools on random process control. A few textbooks have included a chapter of sampled data content, which could have been taught as an introduction to the subject at the late senior level. With the state variable methods and the modern optimal control, these traditional courses are undergoing a tremendous change. Many new courses now are coming into being. The traditional course on transform methods is being slowly changed to accommodate state variable methods, which may serve as the foundations of modern control theory. There are now regular courses on state-space techniques, optimal control, modern nonlinear control, and even on some very advanced topics such as distributed-parameter systems, computational aspects, Kalman filtering, and functional analysis applications.

This trend toward modern work has led many curriculums to neglect the traditional courses. For example, sampled-data control as a course is being discontinued in many schools. With so many topics to be covered, it has become desirable to collect the material that usually has been taught in traditional courses and to put it down in a systematic fashion so that it can be taught, selectively, in a regular senior-level course on a semester basis or, more thoroughly, in two quarterly courses. This material should cover the classical analysis and design of continuous and sampled-data systems, which still remain the most valid methods of design and still serve as an introduction to modern work.

With the above basic ideas in mind, we have developed this textbook. Both continuous- and sampled-data control systems are considered. The classical techniques are discussed, in detail, both for analysis and compensation. The state variable techniques are also discussed, giving the reader some basis for more modern work. Here again the analysis and synthesis techniques are given. Finally, an introduction is given to the integral square error compensation based on Wiener filtering theory; this gives the reader some idea about the criteria not based on trial and error used to measure the performance of the system. Thus we believe that this textbook gives all the basic classical material of traditional courses while laying the foundation for more modern work.

The material in its present form has been class-tested for a number of years and the present organization has been found successful. This material can be taught in a number of ways. If the student has had an exposure to transform and state variable methods (as is the case in a number of universities now) before taking control courses, selected material from all eleven chapters can be covered in one semester. Otherwise, the first six chapters can be covered in one quarter and the second quarter can be used for the last five chapters; on a one-semester basis, some of the topics have to be omitted. This book also may be used for traditional continuous-control and sampled-data control courses if they are taught separately. Ultimately this depends on the teacher and how he wants to use the text. We have, of course, tried to make the material to the point and the motivation at every point clear, thus providing a readable, up-to-date examination of the subject.

It should also be mentioned that there is considerable amount of numerical work associated with control theory and this is conveniently handled on the digital computer. We have solved many of the examples in the text through the use of the computer and we encourage the reader to use the computer, where appropriate, to solve the problems given at the end of each chapter. Many of the computer programs needed are easily developed; others can be obtained from computer program libraries associated with most computer centers.

Finally, we are grateful to many of our graduate students as well as the reviewers of the manuscript for helping us bring this book to publication.

S. C. Gupta
L. Hasdorff

Contents

4

MORE ON THE ANALYSIS OF CONTINUOUS-TIME SYSTEMS   121

5

DISCRETE-TIME SYSTEMS AND THEIR RESPONSE   163

6

ANALYSIS OF DISCRETE-TIME SYSTEMS   203

10

STATE VARIABLE FEEDBACK COMPENSATION   473

11

INTEGRAL-SQUARE ERROR COMPENSATION   523

SELECTED BIBLIOGRAPHY   573

INDEX   579

NOTATION

The reader will find the following explanation of notation useful in the understanding of this text.

$\cong$	approximately equal to
$\triangleq$	equal to by definition
$\forall$	for all
$1(t)$	unit step function
$\delta(t)$	delta function or impulse function
$|f(t)|$	absolute value of the function $f(t)$
$f^*(t)$	periodically sampled function $f(t)$ having values at sampling instants only
$g(t)$	impulse response of a system
$f_1(t) * f_2(t)$	convolution of functions $f_1(t)$ and $f_2(t)$
s	Laplace transform variable
$\mathscr{L}\{f(t)\} = F(s)$	unilateral Laplace transform of $f(t)$
$\mathscr{L}^{-1}$	inverse Laplace transform
$\mathscr{L}_B$	bilateral Laplace transform
p_i	a typical pole of a transfer function
z	Z-transform variable
z_i	a typical zero of a transfer function
$Z\{f(t)\} = F(z)$	Z-transform of $f(t)$
Z^{-1}	inverse Z-transform
Z_m	modified Z-transform
$u(t)$	input to open-loop system
$r(t)$	input to feedback system
$y(t), c(t)$	output of the system
$e(t)$	error signal

a	column vector; lower-case boldface letters are used to represent vectors in general
A	matrix; capital boldface letters are used to represent matrices in general; also used as system matrix
$\mathbf{A}^T$	transpose of matrix **A**
x	state vector
$\mathbf{x}_0$	inital value of the state vector
$\mathbf{x}(nT)$	state vector at sampling instants
b	associated input vector
c	output vector
$\mathbf{e}^{At}$	exponential matrix
T	sampling period
$\mathbf{A}_d$	system matrix based on the Discrete Model
$\mathbf{h}(T)$	$\int_0^T \mathbf{e}^{At}\mathbf{b}dt$
k	state variable feedback vector
ζ	damping factor
ω	angular frequency
ω_n	undamped natural frequency
ω_m	peak frequency
c_{step}	system response to a step input
c_{ramp}	system response to a ramp input
M_0	overshoot
t_s	settling time
τ	time constant, also a dummy variable
t_p	time to first peak of response-peak time
$GH(s)$	open-loop transfer function in the s-domain
$G(s), G_f(s)$	plant transfer function
$T(s)$	transfer function of the overall system in the s-domain
e_{ss}	steady-state value of error, general notation
$|G(j\omega)|$	magnitude of the transfer function
$\angle G(j\omega)$	phase angle of the transfer function
$\phi_n(\zeta)$	Mitrovic functions
K_p	position error constant
K_v	velocity error constant
K_a	acceleration error constant
$G_h(s)$	transfer function of the hold circuit

g_{oh}	impulse response of zero-order hold
g_{1h}	impulse response of first-order hold
$GH^*(j\omega)$	open-loop transfer function of the sampled-data system in the frequency domain
$GH(z)$	open-loop transfer function in the z-domain
$T(z)$	overall transfer function in the z-domain
v	frequency variable in the w-plane
Δ	delay factor
$T(z, m)$	overall transfer function using modified Z-transform
$G_c(s)$	transfer function of the compensator in the s-domain
$D(z)$	transfer function of the compensator in the z-domain
$I_{xy}(\tau)$	translation function of $x(\tau)$ and $y(\tau)$
$I_{xx}(\tau)$	autotranslation function of $x(\tau)$
$I_{xx}(s)$	bilateral Laplace transform of $I_{xx}(\tau)$
$I_{x^*x^*}(\tau)$	autotranslation function of $x^*(\tau)$
$I_{XX}(z)$	Z-transform of $I_{x^*x^*}(\tau)$
$I_{XX}(z, m)$	modified Z-transform of $I_{xx}(\tau)$
$\overline{x^2(t)}$	integral square value of $x(t)$
RHP	right-hand plane
RHS	right-hand side
LHP	left-hand plane
LHS	left-hand side
SDS	sampled-data systems
SVF	state variable feedback
ISV	integral square value
ISE	integral square error
ZOH	zero-order hold
FOH	first-order hold

Fundamentals of AUTOMATIC CONTROL

1

THE PROBLEM AND THE METHOD OF ATTACK

1.1   THE PLANT

Some Basic Assumptions

We begin the consideration of automatic control by discussing the "something" that is to be controlled. In the case of the engineer, this is most likely a physical system that is usually a collection of hardware. For our purposes, we shall call this thing to be controlled the *plant*. We could call it almost anything, but "plant" is a descriptive and convenient word that is frequently used in the literature; therefore, we shall use it.

There is a quantity—and, again, to the engineer this is usually a physical quantity—directly related to the plant upon which the specific interest of the controller centers. This quantity is called the *output* or, more specifically, the output of the plant. The term output has been chosen arbitrarily, but it is descriptive and conventional. The variable or the quantity by which the plant is actuated or driven is logically called the *input* (Fig. 1.1-1). A single output is the quantity to be controlled by means of a plant actuated by a single known input. We here assume the input can be any value chosen for convenience, whereas the output can be controlled only through the plant. It is tacitly assumed that the output is not directly coupled to the input. In general, the plants that the control engineer deals with have more than a single input and output. We consider only the single-input, single-output case, for it reduces the analysis problem to a tractable size while still illustrating the basic methods and tools available and what can be accomplished by the control engineer.

A very important, although seldom mentioned assumption is the observability of the output. That is, we assume the output may be detected and measured, and is available as a signal to be used by the controller.

To analyze the plant it is convenient to speak of the input and output signals. By the signal we mean the value of the given quantity as a function of time, i.e., the input signal is the value of the input quantity as a function of time; similarly, the output signal is the value of the output quantity as a function of time.

2

To get a better notion of a plant and the associated input and output, let us consider some examples of plants for which it might be desirable to design a control system.

Example 1.1-1   *The Electronic Amplifier*

Consider the case of a voltage amplifier where the plant is the amplifier itself. The voltage on the input terminals of the amplifier is the input to the plant, and the voltage on the output terminals is the corresponding output of the plant. The input signal is the voltage on the input terminals as a function of time, and the output signal is the voltage on the output terminals as a function of time.

Example 1.1-2   *The D-C Motor*

Consider the d-c motor itself as the plant. If we take the angular position of the shaft as the quantity we wish to control, then shaft position is the output. (We could also logically wish to control the angular velocity of the shaft, in which case this quantity would be considered the output.) The position of a d-c motor's shaft may be controlled by fixing the voltage applied to the field coils and varying the voltage applied to the terminals of the armature. If this method of control is chosen, then armature voltage is the input to the plant. The shaft position also may be controlled by fixing the voltage applied to the armature (at a nonzero value) and then varying the voltage applied to the field coils. If this method of control is chosen, then field coil voltage is the input to the plant; here the field coil voltage as a function of time is the input signal and shaft position as a function of time is the output signal.

We see from this example that a given plant can have more than one quantity that may be taken as the output and more than one quantity that may be taken as the input. If only single-input, single-output plants are considered, the assumption in the case of the output is that only one quantity is of interest in any specific case, and in the case of the input only one input quantity is subject to any variation while all other inputs are held at some constant value.

Example 1.1-3   *A Ship Steering System*

Here we consider a ship with its rudder and the mechanism for controlling the position of the rudder as the plant. The quantity we specifically wish to control is the heading of the ship, so heading is the output of the plant. The control by which the rudder is positioned (the helmsman's wheel) may be considered the input to the plant. The specific input quantity is the position of the rudder position control. Looking at the signals involved here, we say the heading of the ship as a function of time is the output signal. The position of the rudder position control as a function of time is the input signal.

The Linear Time-Invariant Plant

We again consider the single-input, single-output plant as schematically indicated in Fig. 1.1-1. We assume the input signal is some function $u(t)$ and corresponding to this input signal is the output signal $c(t)$. In other words, for the plant

$$u(t) \rightarrow c(t) \tag{1.1-1}$$

where the $\rightarrow$ indicates the operation that the plant performs upon the input signal to produce the output signal. Besides assuming that this is a single-input, single-output plant, additional mathematical assumptions on the behavior of the plant are required if the behavior of the plant is to be analyzed and a controller designed for it. These assumptions are that the plant is (1) *linear* and (2) *time-invariant*. By linearity of a given plant, we mean that if $u_1(t)$ and $u_2(t)$ are two input signals and

$$u_1(t) \rightarrow c_1(t)$$
$$u_2(t) \rightarrow c_2(t) \tag{1.1-2}$$

then the plant is said to be linear if

$$[au_1(t) + bu_2(t)] \rightarrow [ac_1(t) + bc_2(t)], \qquad -\infty < t < \infty \tag{1.1-3}$$

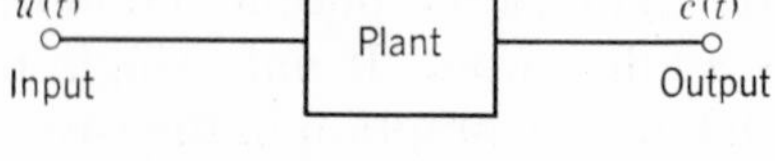

Figure 1.1-1   The plant.

where a and b are scalar constants. A system is said to be time invariant if

$$u(t) \rightarrow c(t), \qquad -\infty < t < \infty$$

implies that

$$u(t+\tau) \rightarrow c(t+\tau), -\infty < t < \infty, -\infty < \tau < \infty \qquad (1.1\text{-}4)$$

i.e., when the input is shifted by any amount τ, it produces the same output, now shifted by an amount τ.

The restrictions of linearity and time invariance as given in (1.1-3) and (1.1-4) may at first glance seem quite strong, which they are, since in the practical case no plant satisfies them exactly. However, there are many real plants which approximately satisfy these conditions over a considerable range of the magnitude of the input and output, and for any reasonable value of τ. And for an even larger number of plants, a linear time-invariant model that adequately approximates the responses of the plant over the useful ranges of the input and output may be found. Hence we can usually work within the restrictions which linearity and time invariance impose.

A linear time-invariant plant is important because for such a plant a considerable body of mathematical tools has been built; we can use these tools to analyze the response to any practically useful input. With the ability to analyze comes the ability to design the controllers, thus tailoring the response of the system to whatever is desired. The most commonly encountered linear time-invariant plants (and systems) are linear, lumped-parameter RLC electrical networks. A common class of linear time-invariant plants (those that we are mainly concerned with) are those whose input and output quantities satisfy a linear, constant-coefficient differential equation. The three examples we have seen are specific examples of plants that may be considered to be linear and time invariant over a considerable range of the input and output quantities.

1.2 THE CONTROL PROBLEM AND THE BASIC APPROACH

The Control Problem

What does the control engineer do with the linear, time-invariant, single-input, single-output plant? We consider here the simplest meaningful problem, a very common practical one: that of making the output of the plant [$c(t)$ in Fig. 1.1-1] track some given reference signal which we shall call $r(t)$. Ideally, then

$$c(t) \equiv r(t)$$

is what we wish to accomplish.

To illustrate the basic idea let us consider control problems that might be appropriate for the three plants considered in Section 1.1.

Example 1.2-1   *The Voltage Amplifier Used in a Loudspeaker System*

For the plant of Example 1.1-1, the voltage amplifier, we consider a public address system; the reference signal $[r(t)]$ is the electrical signal from the microphone into which someone is speaking. The voltage amplifier output terminals are connected to and driving a loudspeaker. The control problem is then to have the output voltage waveform $[c(t)]$ of the amplifier which is driving the loudspeaker exactly track the voltage waveform which is coming from the microphone $[r(t)]$. The voltage amplifier, the plant, has the job of making up the considerable difference between the microphone output power and the power required to drive the loudspeaker.

Example 1.2-2   *A Weathervane Repeater System*

Recall the plant of Example 1.1-2, the d-c motor. This motor may be used to drive a pointer on an indicator that will show the position of a weathervane, which may be located on a tower at some remote weather station. The reference signal $r(t)$ will then be the angular position of the weathervane, and the control problem is to make the angular position of the motor shaft [the output of the plant, $c(t)$], which will drive the pointer on the indicator, track the angular position of the weathervane. We assume some means of converting the weathervane's angular position into an electrical signal that may be applied to the input terminals of the d-c motor.

Example 1.2-3   *An Automatic Ship Steering System*

Here we consider the ship steering system of Example 1.1-3 and the control problem of designing a system that will make the ship automatically hold any desired heading. In this case the reference signal would most likely be the position of a dial on a control console. The problem then is to make the heading of the ship [the output of the plant, $c(t)$] track the position indicated on the dial on the control console, i.e., the reference signal $r(t)$.

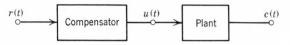

Figure 1.2-1   Open-loop compensation scheme.

The Basic Approach to the Design of the Controller

Now that we have this reference signal $r(t)$, we next consider the means by which we may make the output of the plant track $r(t)$. We may assume that if the reference signal $r(t)$ is applied to the input of the plant, the resulting output will not be satisfactory. One possibility is to add some sort of signal processor (let us call it a compensator) between the reference signal and the input of the plant, as diagrammed in Fig. 1.2-1. This is called open-loop compensation, for reasons that will be obvious shortly. The function of the compensator is to take the signal $r(t)$ and from it produce a $u(t)$; when applied to the plant, the $u(t)$ results in an output $c(t)$, which will be the reference signal $r(t)$ or an adequate approximation to it. The compensator has to act at least approximately like an inverse plant. Though this is a feasible approach, in practice it has not been found a desirable one, especially if the reference signal is not a fixed signal but a whole family of different signals.

A better approach than the open-loop scheme of Fig. 1.2-1 is to try to control with the difference between the output of the plant and the reference signal rather than with the reference signal itself. A possible arrangement for accomplishing this is shown in Fig. 1.2-2, where the device appearing at the input to the compensator takes the difference between the reference signal $r(t)$ and the plant output $c(t)$, i.e., $e(t) = r(t) - c(t)$. It may be seen that this is a natural tracking system if there is no signal inversion[1] between the input of the compensator and the output of the plant. If this condition is satisfied, the signal applied to the input to the compensator-plant combination

[1]A positive change in $e(t)$ produces a positive change in $c(t)$ and negative change in $e(t)$ produces a negative change in $c(t)$.

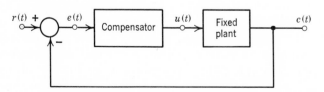

Figure 1.2-2   Closed-loop compensation scheme.

reduces the error signal $e(t)$, which is the difference between the reference and the plant output. And it should be noted, it does this for any reference signal $r(t)$ that we may apply. The role of the compensator is now simplified, for all that is required is that the signal not be inverted by the compensator-plant combination. To achieve this natural tracking system with the reduced compensator requirements, we have given up two things. First, the output of the plant can never be made to track the reference signal exactly, for no control signal is applied to the compensator (and hence the plant) unless there is a finite difference between the plant output and the reference. Second, there is the danger of instability, for if signal inversion should occur between the input to the compensator and the output of the plant, the control action will drive the output of the plant away from the reference signal instead of toward it. This must, of course, be avoided. In practice it has been possible to design compensators so that the error signal [$e(t)$ in Fig. 1.2-2] is always small [so that $r(t)$ is always very near $c(t)$] and also so that instability is avoided. Thus the compensation scheme of Fig. 1.2-2 is the preferred one.

When the output is fed back into the input (of the compensator-

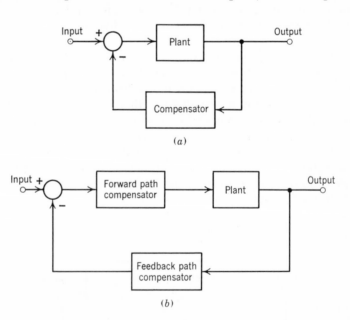

(a)

(b)

Figure 1.2-3 (a) Feedback path compensation. (b) Combined feedback and forward-path compensation.

plant), the system is called a *feedback system* (Fig. 1.2-2). Since the feedback signal is subtracted from the reference, it is more specifically a negative feedback system. The compensator seen in Fig. 1.2-2 is termed forward-path, unity-feedback compensation. The compensator may not appear in the forward path, for it may be in the feedback path as shown in Fig. 1.2-3a or in both as in Fig. 1.2-3b. In fact, there may be more than one feedback loop.

To illustrate controller design, we return to the three examples of control problems discussed earlier in this section. Consider first Example 1.2-1, the public address system. In this case we produce the error signal by providing some electrical means of subtracting the output voltage of the amplifier (or a portion thereof) from the voltage signal as received from the microphone. One applies the resulting error signal (through a compensator if necessary) to the input of the amplifier. If the output of the amplifier is more positive than the output voltage of the microphone, the signal applied to the input of the amplifier is such as to drive the output of the amplifier negative, i.e., in the direction that will bring the amplifier output voltage closer to the voltage at the output of the microphone. The reverse action occurs when the output voltage of the amplifier is more negative than the output voltage of the microphone. The output of the amplifier connected to the loudspeaker input terminals results in a very close replica of the microphone output voltage being impressed upon the input terminals of the system.

For Example 1.2-2 (the weathervane repeater system) means of detecting the difference between the angular position of the weathervane and the angular position of the d-c motor shaft (which is connected to the indicator on the repeater dial) must be found. This difference is the error signal. By applying the error signal through a compensator to the input terminals of the d-c motor, the motor shaft may be driven to the right when the indicator on the repeater dial is to the left of the weathervane position and to the left when the motor shaft is to the right of the weathervane position.

In Example 1.2-3 we have an automatic ship steering system. Here, by detecting the difference between the desired heading of the ship, indicated by the dial on the control console, and the actual heading of the ship, we get the error signal. The error signal through a suitable compensator can be made to apply an action to the rudder position control which will correct the heading of the ship to the left when the actual heading is to the right of the desired heading and to the right when the actual heading is to the left of the desired heading.

1.3   A SPECIAL CASE OF INTEREST: THE SAMPLED-DATA SYSTEM

In the control schemes discussed thus far, we have tacitly assumed that the signals within the system are available continuously to be fed into the various components of the system. Sometimes it is found that the control signals in a given system can be made to change only at discrete instances of time. Usually these instances of time are separated by some uniform interval of time. This type of discrete signal occurs quite naturally whenever digital equipment appears anywhere in the system. It also occurs quite naturally in the case of remote control systems in which a telemetry system appears in the control path. Usually the telemetry system must be shared among several different functions, so it is natural to "time-share" it in such a manner that it transmits control signals for a given system only at regularly spaced instances of time. These systems are in general called *sampled-data* or *discrete-time control systems*. We assume here that the fixed plant is of the same type as described in Section 1.1, i.e., single-input, single-output, linear, and time invariant. However, a sampler shall be assumed to appear at the input to the plant; this is shown in Fig. 1.3-1. The mathematical function of the sampler will be described in detail later; for now we assume that it is a switch which closes for an instant of time every T seconds.

For this case the problem considered will be the same as in the continuous-time case (the case in which the sampler does not appear). That is, it is desired to have the output $c(t)$ (see Fig. 1.3-1) track some specified input function of time $u(t)$. The method of accomplishing this is through the use of a closed-loop compensator. Basically this compensator will be assumed to appear as shown in Figs. 1.3-2a and 1.3-2b.

Compensation as shown in Fig. 1.3-2a, where samplers precede and follow the compensator, is called *discrete compensation*. It appears naturally where the fixed plant is flying about (e.g., in a satellite) and the controller is fixed (e.g., on the ground). Compensation as shown in Fig. 1.3-2b, where the compensator is placed between the input sampler and the fixed plant (no sampler after compensator), is termed *continuous compensation*. It appears naturally where the source providing the input signal $r(t)$ is flying about (e.g., in an

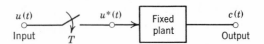

Figure 1.3-1   Fixed plant with sampler at input.

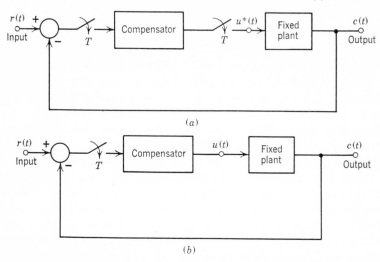

(a)

(b)

Figure 1.3-2 (a) Discrete closed-loop compensation of a sampled-data system. (b) Continuous closed-loop compensation of a sampled-data system.

aircraft) and the fixed plant is on the ground where no weight restriction applies. Both are closed-loop (feedback) systems and both present the problem of instability.

In summary, in the case of sampled-data or discrete-time systems, the control problem is to make the output $c(t)$ of a fixed plant track an input signal $r(t)$. The method of accomplishing this is by the design of closed-loop compensators which fit into the system in one of the two ways shown in Fig. 1.3-2. We shall also see that compensators may be designed by the same methods for the feedback path.

1.4 THE STATE VARIABLE APPROACH – A BROADENING OF THE BASIC ASSUMPTIONS

The Concept of State

Thus far we have considered only plants with a single input and a single output, where only the output was measurable. This reduces the plant to its simplest form, which has many applications. However, it is easy to see that a system designer may want to know much more about what is going on in the plant than just what the output is. Moreover, for purposes of design of a controller, it might be advantageous to measure and use quantities in the plant other than just the output. For instance, in the amplifier of Examples 1.1-1 and 1.2-1, it may be expected that not only is the voltage on the output terminal available to the designer who wishes to use the amplifier in a system, but the voltage on any node may be brought out for measurement and used as well. In the case of the d-c motor in Examples 1.1-2 and 1.2-2, besides just measuring shaft position, the designer may put a tachometer on the shaft and measure and thus make available for use the angular velocity of the shaft.

In general, what these quantities other than the output indicate is the dynamic state of the plant. The state of a system (or plant) may be defined as that minimum amount of information required to determine the future behavior of the system given the input. This definition is not too revealing, so let us try to clarify it with a few examples. A simple and commonly occurring example is the system (or plant) in which the input and output are related by an nth-order differential equation. Given the input, to determine a solution to this differential equation (and hence the behavior of the system) requires n linearly independent initial conditions. Thus this set of n initial conditions is the minimum amount of information required to specify the future behavior of the system, i.e., it can be said to specify the state of the system. In this case the state is specified by n independent quantities; these n quantities are called state variables. The convention is to term a system an nth-order system if n state variables are required to specify its state.

Similarly, we may consider simple RLC electrical networks. If the network has no voltage-source capacitor loops and no current-source inductor cut-sets, then to solve the network equations the initial values of the currents through the inductors and voltages across the capacitors are required. Thus for this type of network the currents through the inductors and the voltages across the capacitors may be said to specify the state of the network, i.e., to form a set of state variables.

Another example illustrating this concept of state is the position of a free body in space. To specify the position of the body for future time we need to know the present position and velocity of the body. To specify position and velocity in three-dimensional space, three position coordinates and three velocities in the direction of the position coordinates are required, a total of six quantities to specify the state, i.e., we have six state variables and a sixth-order system.

State Variable Feedback

If the state variables of a plant are available for measurement, and the usual case is that some, although perhaps not all, are; a method of controller design becomes possible using what is called state variable feedback. The basic idea here is as shown in Fig. 1.4-1, where it is seen that instead of feeding back the output of the plant, a weighted sum of the feedback variables is fed back and subtracted from the reference signal. In the block diagram in Fig. 1.4-1, the blocks in the feedback paths indicate a multiplication of the signal passing through by the constant factor k_i. The figure shows a compensator in series with the plant. With state variable feedback, the compensator is usually much simpler than with output feedback. Further, and not surprising in light of the much greater flexibility possible, a more stable and better responding system may be expected with a controller that uses state variable feedback. This will be seen later.

The main application of the state variable concept goes far beyond the design of controllers for linear, time-invariant, single-input, single-output systems. It has been found to provide a basis for the analysis and study of nonlinear, time-varying, multi-input and multi-output systems. The state variable matrix approach has also been found to be compatible with the use of the digital computer as a computational aid. These last-mentioned subjects are beyond the scope of this text and are mentioned only to indicate the range of possibilities inherent in the state variable approach.

1.5 THE METHOD OF ANALYSIS

The basic method of analysis used in this book is the transform or frequency-domain approach. All signals, which are functions of time, are transformed (by the $\mathscr{L}$-transform when time is continuous, by

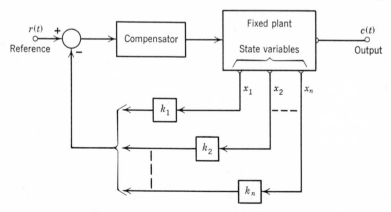

Figure 1.4-1 State variable feedback system.

the Z-transform, an extension of the $\mathscr{L}$-transform, when time is discrete) into functions of a complex variable. All operations are then accomplished in this complex-variable domain. In the past engineers have been very successful in applying this transform method to the design of controllers for linear time-invariant systems with single input and single output. This is primarily because for this case the transforms of the input and output are very simply related through what is called the transfer function. We consider the transfer function of such a system in some detail in Chapter 2; it is sufficient to note here that this makes the analysis of these systems, if not easy, at least tractable. These transform methods have hence been the basis by which the vast majority of presently existing control systems have been designed.

In this text we introduce the fundamentals of automatic control systems. Our chief concern is the linear, time-invariant, single-input, single-output system in both the continuous- and discrete-time cases. As mentioned earlier, the basic method of analysis is by transforms and the basic approach is through the consideration of the transfer function. This serves to show what may be achieved with automatic control systems and a few of the very basic methods of achieving these results. With this background in the classical, transfer-function approach to control systems, we then turn to a study of state variable methods for considering these same systems. Again, the method of analysis will be the use of the $\mathscr{L}$- and Z-transforms. Our study of the state variable methods will allow us to lay the basis for the discussion of compensation of these plants by state variable feedback.

Thus we attempt to lay a basis in classical control theory by considering the single-input, single-output, linear, time-invariant system by the now standard transfer-function method. Using this basis we then go on to an introduction of the newer state variable methods still using the $\mathscr{L}$-transform and Z-transform as the basic tools of analysis.

REFERENCES

1. Maxwell, J. C. "On Governors," *Proc. of the Royal Society of London*, **16**, 1868, in *Selected Papers on Mathematical Trends in Control Theory*, Dover, New York, 1964, pp. 270–283.
2. Bode, H. W. "Feedback–the History of an Idea," in *Selected Papers on Mathematical Trends in Control Theory*, Dover, New York, 1964, pp. 106–123.
3. Black, H. S. "Stabilized Feedback Amplifiers," *Bell System Tech. Journal*, 1934.
4. Minorsky, N. "Directional Stability and Automatically Steered Bodies," *Jour. of Amer. Soc. of Nav. Engrs.*, **34**, 1922, p. 280.

2

CONTINUOUS-TIME SYSTEMS AND THEIR RESPONSE

2.1 INTRODUCTION

Now that we have discussed the basic control problem and the importance of feedback, we turn to the elementary analysis procedures for the continuous-time control systems. One of the basic considerations in the analysis of control systems is their time response. This is usually the time behavior of the output of the system for a specified input. Here we introduce two methods of finding the time response of linear systems: the transfer function and the related impulse-response method. We shall develop these methods and use them to find the time response of some simple systems which are very basic in the analysis and synthesis of linear control systems. The responses of these simple systems illustrate the type of response that we shall later attempt to achieve for the more complex systems.

2.2 THE IMPULSE-RESPONSE AND TRANSFER FUNCTION OF A LINEAR TIME-INVARIANT SYSTEM

Two Tools: The Laplace Transform and the Impulse Function

I. THE LAPLACE TRANSFORM

A basic mathematical tool for the analysis and synthesis of systems whose input and output satisfy a linear, constant-coefficient differential equation is the Laplace transform (the $\mathscr{L}$-transform). The transform itself is defined by

$$\mathscr{L}\{f(t)\} = \int_0^\infty f(t)\, e^{-st}\, dt = F(s) \tag{2.2-1}$$

We take a function $f(\cdot)$ of t (t is usually time in the applications of the transform) and transform to a function $F(\cdot)$ of the Laplace variable s. Whereas t is a real variable, the variable s is complex and hence $F(s)$ is a function of a complex variable. In general we write

$$s = \sigma + j\omega \tag{2.2-2}$$

From the defining equation for the $\mathscr{L}$-transform it is seen that the transform applies to the function $f(t)$ only for $t \geq 0$. In applications of the transform it is assumed, though it may not be stated explicitly, that

$$f(t) = 0, \qquad t < 0 \qquad (2.2\text{-}3)$$

If we wish to apply this transform to a function that is nonzero for $t < 0$, the bilateral Laplace transform ($\mathscr{L}_B$-transform) is used.

In order for a function to have a $\mathscr{L}$-transform, it is necessary that the integral in the defining equation (2.2-1) converge for some value or values of s. Logically, then, we next ask what functions the $\mathscr{L}$-transform exists for. The answer to this is that the $\mathscr{L}$-transform exists for functions of exponential type, i.e., functions which satisfy

$$|f(t)| \leq K e^{\sigma^+ t}, \qquad t \geq 0, \quad K > 0, \quad \sigma^+ < \infty \qquad (2.2\text{-}4)$$

for some K and σ^+. Thus a function is of exponential type if its magnitude can be fitted under an exponential function. To see that the $\mathscr{L}$-transform exists for an exponential type function consider from (2.2-1)

$$|F(s)| = \left| \int_0^\infty f(t) e^{-st}\, dt \right| \qquad (2.2\text{-}5)$$

We have then

$$|F(s)| \leq \int_0^\infty |f(t)||e^{-st}| dt \qquad (2.2\text{-}6)$$

or

$$|F(s)| \leq \int_0^\infty K e^{\sigma^+ t} |e^{-st}| dt = \int_0^\infty K e^{(\sigma^+ - \sigma)t} dt \qquad (2.2\text{-}7)$$

The integral on the RHS (right-hand side) of (2.2-7) converges for

$$(\sigma^+ - \sigma) < 0$$

or

$$\sigma > \sigma^+ \qquad (2.2\text{-}8)$$

Thus for an exponential type function, $|F(s)|$ is finite and hence $F(s)$ exists for all s where (2.2-8) holds. This means $F(s)$ exists in the s-

plane for all s to the right of a vertical line through σ^+ on the real axis. For a given function $f(t)$, the lower limit of the values of σ for which (2.2-4) holds is called the abscissa of convergence σ^+ of $f(t)$.

It should be noted that almost all the functions that are used and are useful in the analysis of linear time-invariant systems are of exponential type. Hence, generally, a finite σ^+ exists for the functions we shall

Table 2.2-1   Properties of Laplace Transform

Given: $\mathscr{L}\{f(t)\} = F(s)$

t-domain	s-domain
1. $1(t-a)f(t-a)$	$e^{-as}F(s)$
2. $e^{-\alpha t}f(t)$	$F(s+\alpha)$
3. $t^n f(t)$	$(-1)^n \dfrac{d^n}{ds^n}\{F(s)\} \begin{matrix} n \geqslant 0 \\ n \text{ integer} \end{matrix}$
4. $\dfrac{f(t)}{t}$	$\int_s^\infty F(s)\, ds$
5. $f(at)$	$\dfrac{1}{a}F\left(\dfrac{s}{a}\right)$
6. (a) $\dfrac{df(t)}{dt}$	$sF(s) - f(0^+)$
(b) $\dfrac{d^n f(t)}{dt^n}$	$s^n F(s) - s^{n-1}f(0^+)\cdots - f^{(n-1)}(0^+)$
7. $\int_{-\infty}^t f(\tau)\, d\tau$	$\dfrac{F(s)}{s} + \dfrac{f^{(-1)}(0^+)}{s}$
8. $\int_0^t f_1(\tau)f_2(t-\tau)\, d\tau$	$F_1(s)F_2(s)$
9. $f(t) = f(t+T), \quad T = \text{period}$	$\dfrac{\int_0^T f(t)e^{-st}\, dt}{1 - e^{-sT}}$
10. $\lim\limits_{t \to 0} f(t)$	$\lim\limits_{s \to \infty} sF(s)$
11. $\lim\limits_{t \to \infty} f(t)$	$\lim\limits_{s \to 0} sF(s)$, if $sF(s)$ is analytic in the right half s-plane
12. $f_1(t)f_2(t)$	$\dfrac{1}{2\pi j}\int_{\sigma_1 - j\infty}^{\sigma_1 + j\infty} F_1(p)F_2(s-p)\, dp$ $\sigma > \sigma_1 + \sigma_{II}$ $\sigma_I < \sigma_1 < \sigma - \sigma_{II}$

be dealing with; in the applications of the $\mathscr{L}$-transform, this is the tacit assumption. Table 2.2-1 gives some useful properties of the $\mathscr{L}$-transform,[1] and Table 2.2-2 gives some common functions and their corresponding $\mathscr{L}$-transforms.

Table 2.2-2   Useful Laplace Transforms

$f(t)$	$F(s)$
1. $\delta(t)$	1
2. 1	$\dfrac{1}{s}$
3. t^n	$\dfrac{\Gamma(n+1)}{s^{n+1}}$, $\Gamma =$ Gamma function. $n > -1$ $\Gamma(n+1) = n!$ for n an integer
4. $e^{-\alpha t}$	$\dfrac{1}{s+\alpha}$
5. $\sin \beta t$	$\dfrac{\beta}{s^2+\beta^2}$
6. $\cos \beta t$	$\dfrac{s}{s^2+\beta^2}$
7. $\sinh \beta \tau$	$\dfrac{\beta}{s^2-\beta^2}$
8. $\cosh \beta t$	$\dfrac{s}{s^2-\beta^2}$
·9. $e^{-\alpha t}\sin \beta t$	$\dfrac{\beta}{(s+\alpha)^2+\beta^2}$
10. $e^{-\alpha t}\cos \beta t$	$\dfrac{s+\alpha}{(s+\alpha)^2+\beta^2}$
11. $\dfrac{t^n e^{-\alpha t}}{n!}$	$\dfrac{1}{(s+\alpha)^{n+1}}$, $n \geqslant 0$, integer
12. $t \sin \beta t$	$\dfrac{2\beta s}{(s^2+\beta^2)^2}$
13. $t \cos \beta t$	$\dfrac{s^2-\beta^2}{(s^2+\beta^2)^2}$
14. $\displaystyle\int_0^t \dfrac{\sin a\tau}{\tau}\, d\tau$	$\dfrac{1}{s}\tan^{-1}\dfrac{a}{s}$
15. $te^{-\alpha t}\sin \beta t$	$\dfrac{2\beta(s+\alpha)}{[(s+\alpha)^2+\beta^2]^2}$
16. $te^{-\alpha t}\cos \beta t$	$\dfrac{(s+\alpha)^2-\beta^2}{[(s+\alpha)^2+\beta^2]^2}$

[1]For the derivation of these properties, see S. C. Gupta, *Transform and State Variable Methods in Linear Systems*, John Wiley and Sons, New York, 1966.

Inverting the $\mathscr{L}$-transform

Equation (2.2-1) shows how to find the $\mathscr{L}$-transform, $F(s)$, of a given function $f(t)$. Also important in applications, is the inverse problem, finding $f(t)$ given its $\mathscr{L}$-transform $F(s)$. In the case where $F(s)$ is a rational function of s, a very important case in practice, partial fraction expansion allows us to express a complicated $F(s)$ as a sum of terms, each of which can be inverted easily using a table as simple as Table 2.2-2. To see how this is done consider $F(s)$ of the form

$$F(s) = \frac{a_m s^m + a_{m-1}s^{m-1} + \cdots + a_1 s + a_0}{s^n + b_{n-1}s^{n-1} + \cdots + b_0} = \frac{A(s)}{B(s)}$$

$$= \frac{A(s)}{\displaystyle\prod_{i=1}^{n}(s - p_i)} \tag{2.2-9}$$

where the p_i, $i = 1, 2,..., n$ are the roots of the denominator polynomial $B(s)$. If $B(s)$ has N distinct roots p_i, $i = 1, 2,..., N$ and each p_i is of multiplicity m_i, partial fraction expansion of $F(s)$ in (2.2-9) is

$$F(s) = \begin{cases} \dfrac{K_{11}}{s - p_1} + \dfrac{K_{12}}{(s - p_1)^2} + \cdots + \dfrac{K_{1m_1}}{(s - p_1)^{m_1}} \\[2ex] + \dfrac{K_{21}}{s - p_2} + \dfrac{K_{22}}{(s - p_2)^2} + \cdots + \dfrac{K_{2m_2}}{(s - p_2)^{m_2}} \\[1ex] \quad\vdots \\[1ex] + \dfrac{K_{N1}}{s - p_N} + \dfrac{K_{N2}}{(s - p_N)^2} + \cdots + \dfrac{K_{Nm_N}}{(s - p_N)^{m_N}} \end{cases} \tag{2.2-10}$$

where

$$K_{ij} = \frac{1}{(m_i - j)!} \frac{d^{m_i-j}}{ds^{m_i-j}}(s - p_i)^{m_i}F(s)\bigg|_{s=p_i} \qquad \begin{matrix} i = 1, 2, \ldots, N \\ j = 1, 2, \ldots, m_i \end{matrix}$$

Applying function-transform pair (11) from Table 2.2-2 to each term of (2.2-10) gives

$$f(t) = \begin{cases} K_{11}e^{p_1 t} + K_{12}te^{p_1 t} + K_{13}\dfrac{t^2}{2!}e^{p_1 t} + \cdots + K_{1m_1}\dfrac{t^{m_1-1}}{(m_1-1)!}e^{p_1 t} \\[2ex] + K_{21}e^{p_2 t} + K_{22}te^{p_2 t} + K_{23}\dfrac{t^2}{2!}e^{p_2 t} + \cdots + K_{2m_2}\dfrac{t^{m_2-1}}{(m_2-1)!}e^{p_2 t} \\[1ex] \quad\vdots \\[1ex] + K_{N1}e^{p_N t} + K_{N2}te^{p_N t} + K_{N3}\dfrac{t^2}{2!}e^{p_N t} + \cdots + K_{Nm_N}\dfrac{t^{m_N-1}}{(m_N-1)!}e^{p_N t} \end{cases} \tag{2.2-11}$$

which gives us the function of time we are seeking.

A more general method for evaluating an inverse $\mathscr{L}$-transform is by use of the inversion integral given next.

Inversion Integral

The inversion integral for the $\mathscr{L}$-transform is given by

$$f(t) = \frac{1}{2\pi j} \int_{\sigma_1-j\infty}^{\sigma_1+j\infty} F(s) e^{st} ds \qquad (2.2\text{-}12)$$

where $F(s) = \mathscr{L}\{f(t)\}$ and the integral is a complex integral along the vertical straight line with abscissa of integration σ_1; all the singularities lie to the left of this vertical line. The inversion integral must be evaluated along the contour where $F(s)$ exists for the given $f(t)$. In general a satisfactory choice of σ_1 is such that $\sigma_1 > \sigma^+$.

II. THE IMPULSE FUNCTION

The impulse function, usually called the δ-function, has been tremendously useful in analyzing linear time-invariant systems of the type considered here. The δ-function is intuitively defined as a function for which

$$\delta(t) = 0, \qquad t \neq 0 \qquad \text{and} \qquad \int_b^c \delta(t)\, dt = 1 \qquad (2.2\text{-}13)$$

where $[b, c]$ is any interval that includes the origin. Such a function is difficult to visualize; however, it can be conceived as the limit of a set of real functions. Such a set of functions might be as follows:

$$h(a, t) = \frac{1}{a}[1(t) - 1(t - a)], \qquad a > 0$$

where $1(t)$ is the unit step function. The function $h(a, t)$ is shown in Fig. 2.2-1. From this figure, we see that the area under $h(a, t)$ is

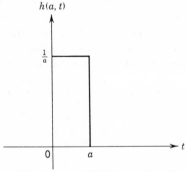

Figure 2.2-1 An approximation to the δ-function.

unity for all values of $a \neq 0$. And clearly as $a \to 0$, (2.2-13) is satisfied
by $h(a, t)$ and thus

$$\delta(t) = \lim_{a \to 0} h(a, t)$$

Properties of the Impulse Function

  1.   THE SIFTING PROPERTY. To obtain the sifting property, we
consider the integral

$$\int_b^c \delta(t) f(t)\, dt \triangleq \int_b^c \lim_{a \to 0} h(a, t) f(t)\, dt \tag{2.2-14}$$

where $b < 0 < c$, i.e., we integrate the product of the δ-function with
some arbitrary function $f(t)$. We assume that $f(t)$ is continuous.
Using $h(a, t)$, as defined above, by reversing the order of integration
and going to the limit, we have

$$\int_b^c \delta(t) f(t)\, dt = \lim_{a \to 0} \int_b^c \frac{1}{a} [1(t) - 1(t - a)] f(t)\, dt$$

$$= \lim_{a \to 0} \frac{1}{a} \int_0^a f(t)\, dt \tag{2.2-15}$$

where we use the fact that $[1(t) - 1(t - a)]$ has value unity in the
interval $0 < t < a$ and is zero everywhere else. Now, since $f(\cdot)$ is
assumed continuous we apply the mean-value theorem to the RHS
of (2.2-15):

$$\int_b^c \delta(t) f(t)\, dt = \lim_{a \to 0} \frac{1}{a} [f(\zeta) \cdot a], \qquad 0 \leqslant \zeta \leqslant a$$

Going to the limit on the RHS gives

$$\int_b^c \delta(t) f(t)\, dt = f(0) \tag{2.2-16}$$

the *sifting property* of the δ-function which we were seeking. In
other words, the product of a continuous function with a δ-function
integrated over any interval that includes the origin sifts out the
value of the continuous function at the origin.

  2.   THE $\mathscr{L}$-TRANSFORM OF THE IMPULSE FUNCTION. We can write

$$\mathscr{L}[\delta(t)] = \int_0^\infty \lim_{a \to 0} h(a, t) e^{-st} dt$$

$$= \lim_{a \to 0} \frac{1}{a} \int_0^a e^{-st} dt$$

$$= \lim_{a \to 0} \frac{1 - e^{-sa}}{sa} \tag{2.2-17}$$

Using L'Hospital's rule on the RHS of (2.2-17) gives a simple result,

$$\mathscr{L}[\delta(t)] = 1 \qquad (2.2\text{-}18)$$

The Impulse Response of a Linear Time-Invariant System

Next consider a single-input, single-output, linear time-invariant system (Fig. 2.2-2). Let the input be a unit impulse (a δ-function) and the corresponding output response be $g(t)$. We write

$$\delta(t) \rightarrow g(t) \qquad (2.2\text{-}19)$$

The output $g(t)$ is then said to be the *impulse response* of the system. Since the system is assumed time invariant, (2.2-19) gives

$$\delta(t-\tau) \rightarrow g(t-\tau) \qquad (2.2\text{-}20)$$

Given an input function $u(t)$, $-\infty < t < \infty$, then $u(\tau)$ for a given τ will be a constant, and by linearity (homogeneity)

$$u(\tau)\delta(t-\tau) \rightarrow u(\tau)g(t-\tau) \qquad (2.2\text{-}21)$$

Again by linearity (superposition) we have

$$\int_{-\infty}^{\infty} u(\tau)\delta(t-\tau)\,d\tau \rightarrow \int_{-\infty}^{\infty} u(\tau)g(t-\tau)\,d\tau \qquad (2.2\text{-}22)$$

But now applying the sifting property of the δ-function to the LHS (left-hand side) of (2.2-22) we have

$$u(t) \rightarrow \int_{-\infty}^{\infty} u(\tau)g(t-\tau)\,d\tau \qquad (2.2\text{-}23)$$

The integral on the RHS of (2.2-23) is called the *convolution integral* or simply the convolution of $u(t)$ with $g(t)$. In words, we have:

The output of a linear time-invariant system is the convolution of the input with the impulse response of that system.

For a system of input $u(t)$ and output $c(t)$

$$c(t) = \int_{-\infty}^{\infty} u(\tau)g(t-\tau)\,d\tau \qquad (2.2\text{-}24)$$

or, equivalently, by simple change of variable,

$$c(t) = \int_{-\infty}^{\infty} u(t-\tau)g(\tau)\,d\tau \qquad (2.2\text{-}25)$$

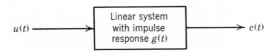

Figure 2.2-2   A linear time-invariant system.

Considering (2.2-24) and (2.2-25), we see that for the input-output characteristic of the system, *the impulse response* g(t) *completely characterizes the system.* For if we know the response of the system to a unit impulse, the response of the system to any input may be found by use of (2.2-24) or (2.2-25).

For a real system where all inputs are assumed to start at $t = 0$,

$$g(t) = u(t) = 0, \qquad t < 0 \tag{2.2-26}$$

and hence

$$c(t) = \int_0^t u(\tau)g(t-\tau)\,d\tau = \int_0^t g(\tau)u(t-\tau)\,d\tau \tag{2.2-27}$$

Let us emphasize again that the result in (2.2-27) applies only if the system is completely undisturbed for all $t < 0$.

The Transfer Function

The result of (2.2-27), which gives the output of a linear time-invariant system as the convolution of the input and the impulse response of the system, may be used to work with this type of system. In general this is not convenient; it is easier to use the $\mathscr{L}$-transform and to work with the transforms of the input and output. To find out why, consider the $\mathscr{L}$-transform of (2.2-27):

$$\mathscr{L}\{c(t)\} = \int_0^\infty e^{-st}dt \int_0^t u(\tau)g(t-\tau)\,d\tau$$

$$= \int_0^\infty e^{-st}dt \int_0^\infty u(\tau)g(t-\tau)\,d\tau$$

since $g(t) = 0$ for $t < 0$. Changing the order of integration gives

$$\mathscr{L}\{c(t)\} = \int_0^\infty u(\tau)\,d\tau \int_0^\infty e^{-st}g(t-\tau)\,dt$$

$$= \int_0^\infty u(\tau)\,d\tau \int_{-\tau}^\infty e^{-s(\theta+\tau)}g(\theta)\,d\theta, \qquad \text{where } \theta = t - \tau$$

$$= \int_0^\infty u(\tau)e^{-s\tau}d\tau \int_0^\infty e^{-s\theta}g(\theta)\,d\theta$$

which gives

$$\mathscr{L}\{c(t)\} = \mathscr{L}\{u(t)\} \cdot \mathscr{L}\{g(t)\} \tag{2.2-28}$$

This is generally written

$$C(s) = U(s)G(s) \tag{2.2-29}$$

Equation (2.2-29) is a simple result, which is the basis of the popularity of the $\mathscr{L}$-transform in considering linear time-invariant systems and at the same time a reason for all the attention that is given to these systems. In words, (2.2-29) says that the $\mathscr{L}$-transform of the output of a given linear time-invariant system is found by multiplying the $\mathscr{L}$-transform of the input excitation with the $\mathscr{L}$-transform of the impulse response. As the impulse response $g(t)$ may be said to characterize a given system, so also $G(s) = \mathscr{L}\{g(t)\}$ may be said to characterize the system.

The $\mathscr{L}$-transform of the impulse response is the *transfer function* of the system. The transfer function is such a handy device for working with linear time-invariant systems that frequently the control engineer, when working with this type of system, will think of the system or its subsystems simply as blocks with given transfer functions.

Determining the Transfer Function of a Given System

From (2.2-29) we see that if we can observe the input $u(t)$ and output $c(t)$ of a system (or plant; see Fig. 2.2-2), which is quiescent for $t < 0$ [$t = 0$ when $u(t)$ becomes nonzero], then

$$G(s) = \frac{\mathscr{L}\{c(t)\}}{\mathscr{L}\{u(t)\}}\bigg|_{u(t)\,=\,c(t)\,=\,0 \text{ for } t<0} \qquad (2.2\text{-}30)$$

In fact, (2.2-30) may be used as a definition of the transfer function for linear time-invariant systems. It also may be used to determine the impulse response, for once we have $G(s)$, we can find

$$g(t) = \mathscr{L}^{-1}\{G(s)\} \qquad (2.2\text{-}31)$$

Thus the impulse response of a system may be found by applying an arbitrary input to the system. An impulse is usually difficult to produce practically.

An important case of linear time-invariant systems that occurs frequently in practice is where the input and output variables satisfy a linear, constant-coefficient differential equation. These are often called lumped-parameter systems, and linear lumped-parameter circuits fall within this class. In this case, the differential equation may be written in the following form:

$$\frac{d^n}{dt^n}c(t) + b_{n-1}\frac{d^{n-1}}{dt^{n-1}}c(t) + \cdots + b_0 c(t) = a_m \frac{d^m}{dt^m}u(t) + \cdots + a_0 u(t),$$

$$m \leq n \qquad (2.2\text{-}32)$$

where the coefficients b_i and a_i are constants. Since the coefficients are constant and the differential equation is linear, the corresponding system is linear and time invariant. To determine the transfer function here, we take the $\mathscr{L}$-transform of (2.2-32), assuming that the system is initially quiescent, i.e., all initial conditions are zero. Under this assumption it is a simple matter to apply property 6 in Table 2.2-1 to get

$$(s^n + b_{n-1}s^{n-1} + \cdots + b_0)C(s) = (a_m s^m + a_{m-1}s^{m-1} + \cdots + a_0)U(s)$$

whence by (2.2-30)

$$G(s) = \frac{a_m s^m + a_{m-1}s^{m-1} + \cdots + a_0}{s^n + b_{n-1}s^{n-1} + \cdots + b_0}, \qquad m \leqslant n \qquad (2.2\text{-}33)$$

It is apparent that for this type of system the transfer function takes the form of a rational function. This is the usual practical case and, for the most part, the only one to be considered.

For a $G(s)$ of rational function form as in (2.2-33), the numerator and denominator polynomials may be factored to give

$$G(s) = \frac{a_m \prod\limits_{i=1}^{m}(s - z_i)}{\prod\limits_{i=1}^{n}(s - p_i)}$$

We see immediately that $G(z_i) = 0$, $i = 1, 2, \ldots, m$, and so z_i, $i = 1, 2, \ldots, m$ are called the zeros of $G(s)$. Similarly, $|G(s)| \rightarrow \infty$ as $s \rightarrow p_i$, $i = 1, 2, \ldots, n$ and so p_i, $i = 1, \ldots, n$ are called the poles of $G(s)$. Thus the poles and zeros determine the transfer function to within a multiplicative constant. A convenient way of displaying a transfer function graphically is to give an s-plane plot showing the poles and zeros of the transfer function. We make frequent use of such plots in this book.

If the differential equation of the system is not available and the system is known to be linear and time-invariant, the transfer function may be approximated by a function of the form of (2.2-33) by frequency-response methods. These will be discussed in a later chapter.

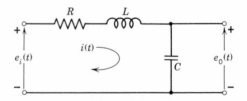

Figure 2.2-3 *RLC*-network.

Example 2.2-1

Determine the transfer function of the RLC network shown in Fig. 2.2-3. The input here is given by the voltage generator, $e_i(t)$. Writing the loop equation for the single loop with the loop current $i(t)$ as shown, we have

$$e_i(t) = Ri(t) + L\frac{di(t)}{dt} + \frac{1}{C}\int_0^t i(\tau)\,d\tau + v_{c0}$$

and

$$e_o(t) = \frac{1}{C}\int_0^t i(\tau)\,d\tau + v_{c0}$$

where v_{c0} is the initial voltage on the capacitor, i.e., $e_o(0)$. To find the transfer function we take the $\mathscr{L}$-transform of the preceding two equations term by term under the assumption of zero initial conditions [i.e., $v_{c0} = 0$, $i(0) = 0$]. We thus have

$$E_i(s) = \left(R + sL + \frac{1}{sC}\right)I(s)$$

$$E_o(s) = \frac{1}{sC}I(s)$$

Now the transfer function by (2.2-30) is $E_o(s)/E_i(s)$. The last two equations give

$$\frac{E_o(s)}{E_i(s)} = \frac{1}{s^2LC + sRC + 1}$$

which is the transfer function desired.

Example 2.2-2

Consider the mechanical system diagrammed in Fig. 2.2-4. It is simply a mass (M) attached to a spring (constant K) and damper (dynamic damping B) which the force $f(t)$ operates on. Displacement is x, positive in the direction shown. The zero position is taken to be

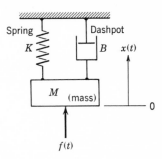

Figure 2.2-4   A damped spring-bob.

at the point where the spring and mass are in static equilibrium. Using D'Alembert's principle, we sum to zero the forces acting upon the mass, thus getting

$$f(t) - M\frac{d^2x(t)}{dt^2} - B\frac{dx(t)}{dt} - Kx(t) = 0$$

as the differential equation describing the dynamics of the system.

For this system the input may be taken as the force $f(t)$ and the output the displacement x of the mass. For this choice of input and output the transfer function is $X(s)/F(s)$. We proceed by taking the $\mathscr{L}$-transform of the preceding equation and solving for $X(s)/F(s)$. Assuming zero initial conditions, we get

$$F(s) - s^2MX(s) - sBX(s) - KX(s) = 0$$

or

$$\frac{X(s)}{F(s)} = \frac{1}{s^2M + sB + K}$$

which is the transfer function desired.

Example 2.2-3

Now consider the rotational electromechanical system of Fig. 2.2-5. It is simply a current generator driving a d-c motor which has an armature (moment of inertia J) with its shaft attached to a spring (constant K) and with some damping (dynamic damping B) due to the bearings, etc. It is a simple positional control system. The input, the current $i(t)$, produces a torque on the armature of the d-c motor that may be used to control the output, the shaft position θ. For a fixed-field d-c motor the torque on the armature shaft, over a considerable range, is directly proportional to the armature current:

$$T = K_T i$$

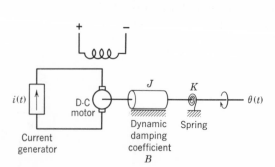

Figure 2.2-5   A rotational electromechanical system.

where T is armature torque, i the armature current, and K_T the constant relating the two. We may now sum the individual torques on the shaft to zero to get

$$K_T i(t) - J\frac{d^2\theta(t)}{dt^2} - B\frac{d\theta(t)}{dt} - K\theta(t) = 0$$

which is the differential equation relating the input and the output for this particular system. We assume that the zero for the angle $\theta(t)$ is taken where the armature is in static equilibrium.

The transfer function is now easily found:

$$\frac{\theta(s)}{I(s)} = \frac{K_T}{s^2 J + sB + K}$$

which is the desired transfer function.

It is worthwhile to note that all the three systems considered have the same form of transfer function.

Manipulation of Block Diagrams Involving Linear Time-Invariant Systems

Most systems of any consequence are made up by the interconnection of much simpler subsystems. Let us now consider the problem of determining the overall transfer function of a system made up of noninteracting interconnected subsystems for which the individual transfer functions are known. A few very basic and very simple cases will be discussed.

Consider first the case of the series connection of n subsystems as shown in Fig. 2.2-6.
Repeated application of (2.2-29) gives

$$C(s) = G_1(s)G_2(s)\cdots G_n(s)U(s)$$

whence by (2.2-30)

$$G(s) = \frac{C(s)}{U(s)} = G_1(s)G_2(s)\cdots G_n(s) = \prod_{i=1}^{n} G_i(s) \qquad (2.2\text{-}34)$$

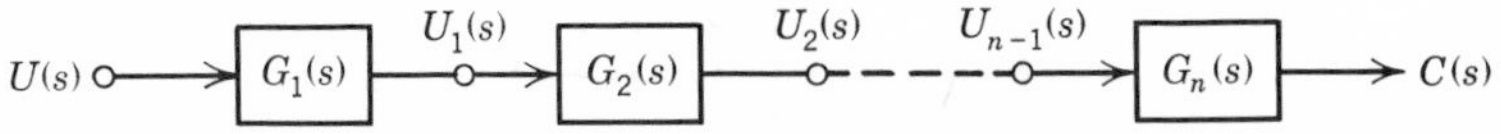

Figure 2.2-6 Series connection of n linear time-invariant subsystems.

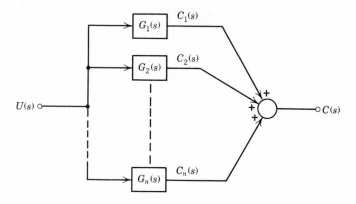

Figure 2.2-7   Parallel connection of n linear time-invariant subsystems.

Now consider the parallel interconnection of n subsystems, as in Fig. 2.2-7. In general,

$$C_i(s) = U(s)G_i(s)$$

$$C(s) = \sum_{i=1}^{n} C_i(s) = U(s) \sum_{i=1}^{n} G_i(s)$$

or

$$G(s) = \frac{C(s)}{U(s)} = \sum_{i=1}^{n} G_i(s) \qquad (2.2\text{-}35)$$

The two systems just considered are both open-loop systems. Let us now consider a simple closed-loop (or feedback) system as in Fig. 2.2-8 with zero initial conditions. The overall transfer function is found as follows:

$$\begin{aligned} E(s) &= R(s) - C(s)H(s) \\ &= R(s) - E(s)G(s)H(s) \\ &= \frac{R(s)}{1 + G(s)H(s)} \end{aligned} \qquad (2.2\text{-}36)$$

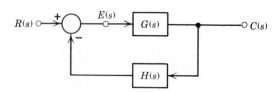

Figure 2.2-8   Basic feedback interconnection of subsystems.

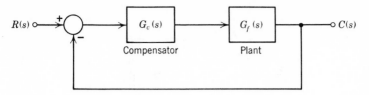

Figure 2.2-9   Linear plant with cascade compensator and unity feedback.

Then

$$C(s) = E(s)G(s) = \frac{R(s)G(s)}{1+G(s)H(s)}$$

and

$$\frac{C(s)}{R(s)} = \frac{G(s)}{1+G(s)H(s)} \tag{2.2-37}$$

Equation (2.2-37) then gives the transfer function of the system of Fig. 2.2-8.

An interesting special case is the system of Fig. 2.2-9. For this case, applying (2.2-37) we have

$$\frac{C(s)}{R(s)} = \frac{G_c(s)G_f(s)}{1+G_c(s)G_f(s)} \tag{2.2-38}$$

2.3   TRANSIENT ANALYSIS BY USE OF THE TRANSFER FUNCTION

Transient responses of linear dynamic systems are usually judged on the basis of their response to a unit step input. The response of a higher-order system is, in turn, judged on a comparative basis with the response of first-order and second-order systems. Here we consider the step responses of first-, second-, and higher-order systems and introduce the methods by which these systems are characterized and the terms by which these characteristics are specified.

The Response of a First-Order System

Consider a simple system with a transfer function

$$T(s) = \frac{a}{s+a}, \qquad a > 0 \tag{2.3-1}$$

Note that we will use $T(s)$ to denote the overall transfer function of the system. For a step input, we have

$$U(s) = \frac{1}{s}$$

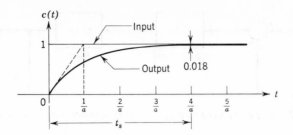

Figure 2.3-1   The step response of a first-order system.

Hence the output is given by

$$C(s) = T(s)U(s) = \frac{a}{s(s+a)} = \frac{1}{s} - \frac{1}{s+a} \tag{2.3-2}$$

Using Table 2.2-2, we have

$$c(t) = (1 - e^{-at})1(t) \tag{2.3-3}$$

which is plotted in Fig. 2.3-1. This is simply a waveform which rises monotonically (exponentially) from zero at the initial instant to the final value of 1. It may be seen that if the initial slope of the output waveform is extended until it intercepts the final value of the output (1 in this case), the intersection occurs at time $t = 1/a$. The value $1/a$ is the time constant τ of the system, i.e.,

$$\tau = \frac{1}{a} \tag{2.3-4}$$

We see then that for small a, which corresponds to a pole on the real axis in the LHP near the origin, the time constant is large and the system is a slow-responding system. On the other hand, for large a, which corresponds to the system pole on the real axis far into the LHP, the time constant is small and it is a quick-responding system. As far as the transient response of the system is concerned, the time constant completely characterizes the first-order system. We note only that the farther from the origin in the LHP the system pole occurs, the faster the exponential term in the response decays, i.e., the system has a "faster response."

Another way of characterizing the first-order system is by what is termed the *settling time*. The settling time of a system is the time required by a system, given a disturbance, to settle to within some fixed percentage of its final value. For the waveform in Fig. 2.3-1 it is seen that after a time duration of four time constants $(4/a)$ the waveform is within 1.8% of its final value. For our purposes the settling time for this system is defined as follows:

$$t_s \triangleq \text{settling time} \triangleq 4\tau = \frac{4}{a} \qquad (2.3\text{-}5)$$

Again in characterizing the speed of response of a system by its settling time, from (2.3-5) we see that for the system pole far into the LHP (large a) the settling time is small, indicating a quick-responding system. We may also consider the steady-state response of this type of system, i.e., the response of the system as $t \to \infty$. We see from Fig. 2.3-1 that as $t \to \infty$, the output $c(t)$ approaches the input $u(t)$:

$$c(\infty) = u(\infty)$$

We thus say that this system tracks a step input with zero steady-state error. It should be appreciated that this desirable state of affairs came about because the multiplying constant was adjusted properly. For any other value of the multiplying constant there will be steady-state error.

Now let us consider whether or not this system, (2.3-1), will track a unit ramp. For then

$$u(t) = t1(t)$$
$$U(s) = \frac{1}{s^2} \qquad (2.3\text{-}6)$$

and the output satisfies

$$C(s) = T(s)U(s) = \frac{a}{s^2(s+a)} \qquad (2.3\text{-}7)$$

We can define the error $e(t)$ as the difference between the input and output, i.e., $e(t) = u(t) - c(t)$. Using (2.3-6) and (2.3-7), we have

$$E(s) = \mathscr{L}\{e(t)\} = U(s) - C(s) \qquad (2.3\text{-}8)$$
$$= \frac{1}{s(s+a)}$$

Using this in the final-value theorem, we now get

$$e(t) = \lim_{s \to 0} sE(s) = \frac{1}{a}, \qquad a > 0 \qquad (2.3\text{-}9)$$

Thus the first-order system of (2.3-1) will track a unit ramp input with steady-state error $1/a$, which is coincidentally the time constant of the system. In fact, this is one way of measuring the time constant of a first-order system. Therefore putting the system pole far into the

LHP (large a) results in not only a faster responding system, but one that will track a unit ramp input with smaller error. Again we note that the steady-state error will be as small as the value given in (2.3-9) only if the gain of the system is adjusted properly. It is a simple exercise to show that the steady-state error to a ramp input is infinite if the constant a in the numerator on the RHS of (2.3-1) is replaced by K, $K \neq a$.

The Response of a Second-Order System

The first-order system is simple and hence easy to work with. Unfortunately, in practice one is forced to deal with higher-order systems. Basic to understanding the response of higher-order systems and to the problem of designing compensators for these systems is an understanding of the response of a second-order system. Thus we consider a transfer function of the form

$$T(s) = \frac{\omega_n^2}{s^2 + 2\zeta\omega_n s + \omega_n^2} \tag{2.3-10}$$

In practice, we will have

$$0 \leq \zeta \leq 1, \qquad \omega_n > 0$$

Zeta (ζ) is termed the *damping factor* and ω_n is termed the *undamped natural frequency*.

$T(s)$ of (2.3-10) has poles at

$$s = -\zeta\omega_n \pm j\omega_n\sqrt{1-\zeta^2} \tag{2.3-11}$$

The location of these poles is shown in the s-plane plot of Fig. 2.3-2. From the figure we see that the distance of the poles from the origin

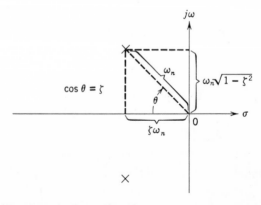

Figure 2.3-2   Pole location for a second-order system.

is ω_n and the angle of the pole with the negative real axis is given by $\cos \theta = \zeta$. For $\zeta = 1$, we have a double pole on the real axis at $\sigma = -\omega_n$. When $\zeta = 1$ we have what is termed *critical damping*. For the case $\zeta = 0$, we see that the two poles are on the $j\omega$-axis at $\pm j\omega_n$. This is the *undamped* case. For the case $0 < \zeta < 1$, there are two complex poles, one each in the second and third quadrants. This is the *underdamped* case. For $\zeta > 1$, the system is termed *overdamped*.

Let us now consider the effect of pole position on the response of the system. First we look at the transient response. The step response is calculated from

$$C_{\text{step}}(s) = U(s)T(s)$$

$$= \frac{1}{s}\left(\frac{\omega_n^2}{s^2 + 2\zeta\omega_n s + \omega_n^2}\right) \tag{2.3-12}$$

where $U(s) = 1/s$ and $T(s)$ is given by (2.3-10). This may now be expanded by partial fraction expansion to

$$C_{\text{step}}(s) = \frac{1}{s} - \frac{s + 2\zeta\omega_n}{s^2 + 2\zeta\omega_n s + \omega_n^2}$$

$$= \frac{1}{s} - \frac{s + \zeta\omega_n}{(s + \zeta\omega_n)^2 + \omega_n^2(1 - \zeta^2)} - \frac{\zeta}{\sqrt{1 - \zeta^2}} \frac{\omega_n \sqrt{1 - \zeta^2}}{(s + \zeta\omega_n)^2 + \omega_n^2(1 - \zeta^2)} \tag{2.3-13}$$

The transforms on the RHS may now be inverted using Table 2.2-2 to obtain

$$C_{\text{step}}(t) = \left[1 - e^{-\zeta\omega_n t}\left(\cos \omega_n \sqrt{1 - \zeta^2}t + \frac{\zeta}{\sqrt{1 - \zeta^2}} \sin \omega_n \sqrt{1 - \zeta^2}t\right)\right]1(t)$$

$$= \left[1 - e^{-\zeta\omega_n t} \frac{\sin(\omega_n \sqrt{1 - \zeta^2}t + \cos^{-1}\zeta)}{\sqrt{1 - \zeta^2}}\right]1(t) \tag{2.3-14}$$

The step response, (2.3-14), is plotted for various values of ζ, in the range $0 \leq \zeta \leq 1$, in Fig. 2.3-3. The character of the step response differs markedly from the step response of the first-order system in that the output, for some values of ζ, overshoots the final value and even oscillates about this final value.

Let us now consider the characteristics of the step response. First we note from (2.3-14) that the oscillations have an angular frequency ω_d given by

$$\omega_d = \omega_n \sqrt{1 - \zeta^2} \tag{2.3-15}$$

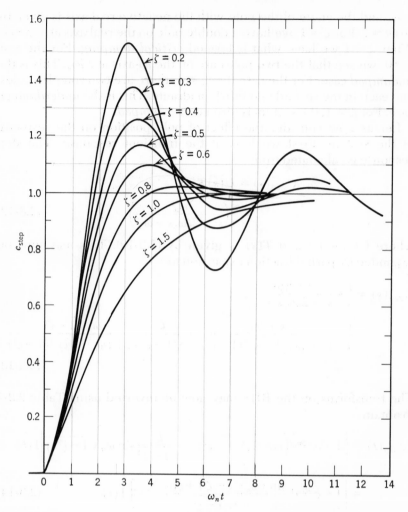

Figure 2.3-3 The step response of a second-order system for various values of damping factor ζ.

ω_d is called the damped natural frequency of the system. We note that $\omega_d \neq \omega_n$ except in the case of $\zeta = 0$, i.e., where there is no damping.

We see also from (2.3-14) that the magnitude of the oscillations decreases as $e^{-\zeta\omega_n t}$. In the same manner as the time constant ($\tau = 1/a$) was defined for first-order systems, we now define the *time constant* for the second-order system by

$$\tau = \frac{1}{\zeta\omega_n} \qquad (2.3\text{-}16)$$

In a time duration of four time constants the output will settle to within 1.8% of its final value, i.e., the amplitude of the oscillations will be $0.018 \times$ (final value). Thus the *settling time* for the second-order system may be defined by

$$t_s = 4\tau = \frac{4}{\zeta \omega_n} \tag{2.3-17}$$

Another characteristic of this step response is the time and magnitude of the first peak. As seen in Fig. 2.3-3, this is the highest peak. It is a matter of simple calculation to determine that

$$\frac{d}{dt} c_{\text{step}}(t) = 0$$

for

$$t = \frac{k\pi}{\omega_n \sqrt{1-\zeta^2}}, \quad k = 0, 1, 2$$

where $c_{\text{step}}(t)$ is as given in (2.3-14). The first peak occurs for $k = 1$; thus the time of the first peak (t_p) is given by

$$t_p = \frac{\pi}{\omega_n \sqrt{1-\zeta^2}} \tag{2.3-18}$$

Using this value in (2.3-14) gives

$$c_{\text{step}}(t_p) = 1 + \exp\left[-\left(\frac{\zeta\pi}{\sqrt{1-\zeta^2}}\right)\right] \tag{2.3-19}$$

We now define

$$\text{overshoot} \overset{\Delta}{=} M_o \overset{\Delta}{=} \text{peak value} - \text{final value} = \exp\left[-\left(\frac{\zeta\pi}{\sqrt{1-\zeta^2}}\right)\right] \tag{2.3-20}$$

$$\% \text{ overshoot} \overset{\Delta}{=} \frac{M_0 \times 100}{\text{final value}} \tag{2.3-21}$$

Time to first peak t_p and overshoot M_o are plotted in Fig. 2.3-4. Note that M_o, the overshoot, is a function of the damping factor ζ only, i.e., it does not involve ω_n. We note from Fig. 2.3-4 that a damping factor $\zeta = 0.7$ corresponds to an overshoot of approximately 7%, whereas a damping factor of $\zeta = 0.3$ corresponds to an overshoot of approximately 35%. For most systems in practice, an overshoot in this range is found tolerable or even desirable. Thus for practical purposes the range of damping factors of interest is

$$0.3 \leqslant \zeta \leqslant 0.7$$

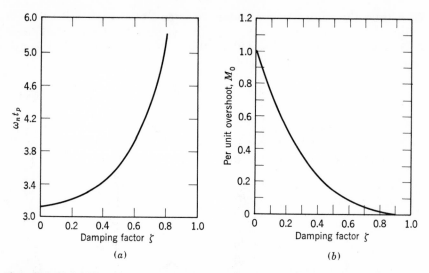

Figure 2.3-4 (a) Time to the first peak and (b) overshoot versus damping for a second-order system.

The range of desirable pole locations in the s-plane for second-order systems is thus restricted to the shaded areas of Fig. 2.3-5. Now consider (2.3-18), where we see that for a fixed value of ζ, t_p varies inversely as ω_n. Thus a high or large value for ω_n means a small t_p, and we have a quick-responding system. On the other hand, a small value of ω_n means a large t_p and consequently a slower responding system. From Fig. 2.3-2 we see that ω_n gives the distance of the poles from the origin. Thus in general the farther away from the origin the system pole is located, the faster the system responds.

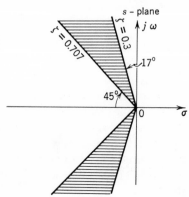

Figure 2.3-5 Regions of desirable pole locations in the left half s-plane.

Effect of Adding a Zero to the System

Let a zero be added to $T(s)$ of (2.3-10). Then we have

$$T(s) = \frac{(s+a)(\omega_n^2/a)}{s^2 + 2\zeta\omega_n s + \omega_n^2} \qquad (2.3\text{-}22)$$

Note that multiplication term in the numerator of $T(s)$ has been adjusted so that the *steady-state gain* or the *d-c gain* $[T(0)]$ of the system is unity [in mathematical terms $T(0) = 1$]. This makes the final value of the unit step response also 1, and thus the transfer function has been adjusted so that the system will track a step with zero steady-state error. The step response of this one-zero, two-pole system may be determined from

$$C_{\text{step}}^0(s) = \frac{(s+a)(\omega_n^2/a)}{s(s + 2\zeta\omega_n s + \omega_n^2)}$$

$$= \frac{\omega_n^2}{s(s^2 + 2\zeta\omega_n s + \omega_n^2)} + \frac{s}{a}\frac{\omega_n^2}{s(s^2 + 2\zeta\omega_n s + \omega_n^2)} \qquad (2.3\text{-}23)$$

where now the superscript 0 on the LHS indicates the system with the added zero. Comparing (2.3-23) with (2.3-13), we have

$$C_{\text{step}}^0(s) = C_{\text{step}}(s) + \frac{s}{a}C_{\text{step}}(s)$$

where $C_{\text{step}}(s)$ is taken from (2.3-13). We may now use the property of the $\mathcal{L}$-transform on the derivative of a function to see that

$$c_{\text{step}}^0(t) = c_{\text{step}}(t) + \frac{1}{a}\cdot\frac{d}{dt}\{c_{\text{step}}(t)\} \qquad (2.3\text{-}24)$$

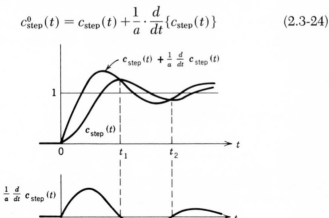

Figure 2.3-6   Effect of an additional zero on a second-order step response.

where $c_{step}(t)$ is given by (2.3-14). The effect of this added derivative term may be seen by examining Fig. 2.3-6, where a case for a fairly typical value of ζ is considered. We see from the figure that the effect of the zero is to add a pronounced early peak to the system's response. The overshoot may be increased appreciably. From (2.3-24) and Fig. 2.3-6 the smaller the value of a, and hence the closer the zero to the origin and the $j\omega$-axis, the more pronounced is this peaking phenomenon.

In general, this peaking phenomenon is considered a degradation of the system's response and thus zeros on the real axis near the origin are avoided. However, in a sluggish system the artful introduction of a zero at the proper position can improve the step response.

We can also see from (2.3-24) that as a increases, and the zero moves farther into the LHP, its effect becomes much less pronounced. For sufficiently large a, the effect of the zero is hardly noticeable. Figure 2.3-7 shows the effect of the zero in a quantitative manner.

Effect of a Real Pole on the Response of a System with a Pair of Complex Poles: A Third-Order System

Now let us consider the effect of an additional real pole on the response of a system with two complex poles. To this end we consider a transfer function of the form

$$T(s) = \frac{\omega_n^2 p}{(s+p)(s^2+2\zeta\omega_n s+\omega_n^2)} \tag{2.3-25}$$

The step response is found by taking the inverse of $(1/s)T(s)$ to be

$$c_{step}^p(t) = [1+K_p e^{-pt}+2|K_{\omega_n}|e^{-\zeta\omega_n t}\cos(\omega_n\sqrt{1-\zeta^2}\,t+\angle K_{\omega_n})]1(t) \tag{2.3-26}$$

where

$$K_{\omega_n} = \frac{\omega_n^2 p}{4\zeta\omega_n^3(1-\zeta^2)-2p\omega_n^2(1-\zeta^2)+j[2\omega_n^3\sqrt{(1-\zeta^2)^3}+2p\zeta\omega_n^2\sqrt{1-\zeta^2}-2\zeta^2\omega_n^3\sqrt{1-\zeta^2}]}$$

$$K_p = \frac{-\omega_n^2}{p^2-2\zeta\omega_n p+\omega_n^2} \tag{2.3-27}$$

Now the effect of the pole at $-p$ will depend upon the magnitude of p. From (2.3-27) it can be seen that

$$\lim_{p\to 0}|K_{\omega_n}| = 0$$

$$\lim_{p\to 0}|K_p| = 1$$

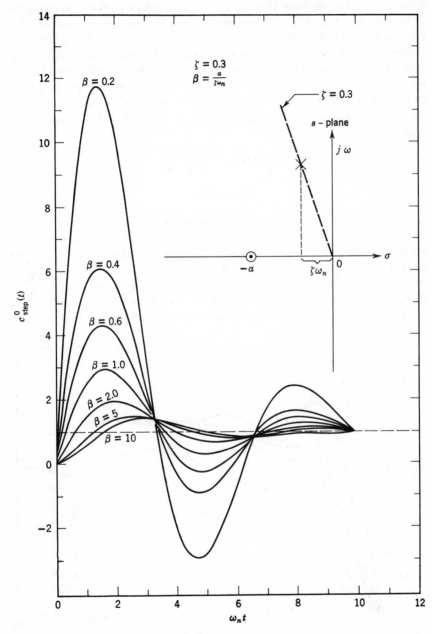

Figure 2.3-7 The effect of a zero on the negative real axis on the step response of a second-order system.

from which we deduce that as $p \to 0$, the effect of the two complex poles at $-\zeta\omega_n \pm j\omega_n\sqrt{1-\zeta^2}$ diminishes and the transient response becomes dominated by the pole at $-p$, i.e., the term e^{-pt} of (2.3-26). This means that a pole on the negative real axis near the origin will dominate the response of the system. This domination is enhanced by the fact that if $p \ll \zeta\omega_n$, the effect of the term e^{-pt} will last much longer than that of the term whose magnitude is governed by $e^{-\zeta\omega_n t}$. Now let us see what happens as p becomes large. A simple calculation shows that

$$\lim_{p \to \infty} |K_{\omega n}| = \frac{1/2}{\sqrt{1-\zeta^2}}$$

$$\lim_{p \to \infty} K_p = 0$$

from which we see that as p increases an opposite effect is observed. The complex poles at $-\zeta\omega_n \pm j\omega_n\sqrt{1-\zeta^2}$ now dominate the response, while the contribution of the pole at $-p$ diminishes in significance. Further, if $\zeta\omega_n \ll p$, the effect of the term e^{-pt} will be over much more quickly than the term whose magnitude decreases as $e^{-\zeta\omega_n t}$. All the preceding effects may be seen quite clearly in Fig. 2.3-8 where for $\zeta = 0.3$ the step responses of a system whose transfer function is given

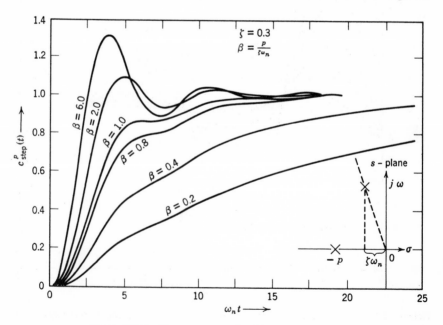

Figure 2.3-8   Step response of a third-order system.

by (2.3-25) are shown for $\beta = p/\zeta\omega_n$ with the values $0.2, 0.4, 0.8, 1, 2, 6$. It may be clearly seen that for $p/\zeta\omega_n = 6$, the response is for all practical purposes the response of the two-pole system of (2.3-10) as seen in Fig. 2.3-3. The pole on the real axis for this case contributes a negligible amount to the transient response of the system. For $p/\zeta\omega_n = 0.2$, however, the step response is effectively dominated by the real pole at $-p$.

Steady-State Error Response

As a final consideration in this examination of a second-order system, let us now look at the steady-state error response of these systems. From Fig. 2.3-3 it may be seen that the steady-state error to a unit step input is zero. Thus the second-order system, (2.3-10), will track a step input with zero steady-state error. Now, however, let the input be $U(s) = 1/s^2$, i.e., a unit ramp. Using (2.3-10) we have

$$C_{\text{ramp}}(s) = \frac{\omega_n^2}{s^2(s^2 + 2\zeta\omega_n s + \omega_n^2)} \qquad (2.3\text{-}28)$$

and

$$E(s) = U(s) - C_{\text{ramp}}(s) = \frac{s + 2\zeta\omega_n}{s(s^2 + 2\zeta\omega_n s + \omega_n^2)} \qquad (2.3\text{-}29)$$

and using the final value theorem we have

$$e_{ss}(\text{ramp}) = \lim_{s \to 0} sE(s) = \frac{2\zeta}{\omega_n} \qquad (2.3\text{-}30)$$

Now we can see from (2.3-16) that a large value of ζ means a faster responding system; however, this faster response is paid for, as seen in (2.3-30), by a larger value of steady-state error to a ramp. Thus in choosing a value of ζ one trades off quick response for increased steady-state error to a ramp.

The Response of Higher-Order Systems

As can be seen readily from the discussion of the second-order system, the higher-order systems will offer greater difficulty in estimating the peak time, overshoot, and settling time. Numerical calculations and plots by the digital computer, of course, remove this difficulty. All we need is a Laplace inversion and plot programs once the transfer function is known.

Higher-Order Systems with Two Dominant Poles

The easiest method of analyzing and designing higher-order systems is to assume that the higher-order system has a quadratic model i.e., to assume the response is dominated by the two complex poles near the origin. A pole-zero configuration of such a system is shown in Fig. 2.3-9.

Assuming that the dominant poles are given by

$$-\sigma_0 \pm j\omega_0 = -\zeta\omega_n \pm j\omega_n \sqrt{1-\zeta^2} \tag{2.3-31}$$

it can be shown[2] that the settling time t_s, peak time t_p, number of oscillations to settling time N, and the peak overshoot M of the higher-order system can be estimated from the following equations:

$$t_s = \frac{4}{\zeta\omega_n} \tag{2.3-32}$$

$$\omega_0 t_p = \frac{\pi}{2} - (\text{sum of angles from zeros to the predominant pole} - \sigma_0 + j\omega_0) + (\text{sum of angles from the other poles to the predominant pole} - \sigma_0 + j\omega_0) \tag{2.3-33}$$

$$N = \frac{2}{\pi}\frac{\sqrt{1-\zeta^2}}{\zeta} \tag{2.3-34}$$

[2]Y. Chu, "Synthesis of Feedback Control System by Phase-Angle Loci," *Trans. AIEE,* 1952, **71** (2), 330–339.

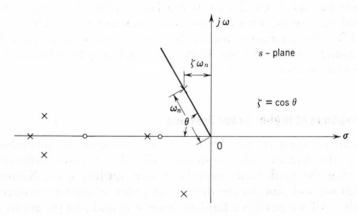

Figure 2.3-9   Higher-order system pole-zero configuration with two dominant poles.

and

$$M = \frac{\begin{pmatrix} \text{product of distances from all poles to} \\ \text{the origin excluding distances from} \\ \text{the two predominate poles to origin)} \end{pmatrix}}{\begin{pmatrix} \text{product of distances from all poles} \\ \text{to the predominate poles} - \sigma_0 + j\omega_0) \\ \text{excluding the distance between the} \\ \text{predominate poles)} \end{pmatrix}} \times \frac{\begin{pmatrix} \text{product of distance from} \\ \text{zeros to the predominate} \\ \text{pole} - \sigma_0 + j\omega_0) \end{pmatrix}}{\begin{pmatrix} \text{product of distance from} \\ \text{all zeros to origin)} \end{pmatrix}} \times \epsilon^{-\sigma_0 t_p}$$

$$(2.3\text{-}35)$$

PROBLEMS

2.1. Find the $\mathcal{L}$-transform and the abscissa of convergence for the following functions. Use the properties given in Table 2.2-1.
  (a) $\sin t$ (e) $\sqrt{t}\cos t$
  (b) $\cos t$ (f) $1(t) - 1(t-T)$
  (c) $t^3 e^{-4t}$
  (d) $\sqrt{t}$ (g) $\int_0^t \frac{\sin t}{t} \, dt$

2.2. Invert the following $\mathcal{L}$-transforms.
  (a) $\dfrac{1}{(s+a)^3}$ (e) $\dfrac{s}{(s^2+a^2)^2}$
  (b) $\dfrac{s}{(s+1)^3(s+2)^2(s+3)}$ (f) $\dfrac{e^{-sT}}{\sqrt{s+a}}$
  (c) $\dfrac{\omega}{s^2+2as+a^2+\omega^2}$ (g) $\dfrac{1}{1-e^{-sT}}$
  (d) $\dfrac{e^{-2s}}{(s+3)(s^2+2s+5)}$ (h) $\dfrac{1-e^{-s}-e^{-2s}+e^{-3s}}{s^2(1-e^{-6s})}$

2.3. For a system as shown in Fig. P2.3, $T(s) = 1/(s+a)$. The system has been disturbed so that $c(0) = -1$. When an input $u(t) = t1(t)$ (unit ramp) is applied, what is $c(t)$ for $t > 0$?

2.4. Given the network as in Fig. P2.4.
  (a) What is the transfer function $E_0(s)/E_i(s)$?
  (b) What is $e_0(t)$ for $e_i(t) = 1(t)$?
  (c) What is $e_0(t)$ for $e_i(t) = \delta(t)$?

2.5. Determine the transfer function $\theta(s)/E_i(s)$ for the system of Fig. P2.5.

2.6. Given the servomotor-generator set as in Fig. P2.6, find the transfer function $\theta(s)/E_i(s)$.

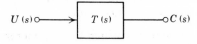

Figure P2.3

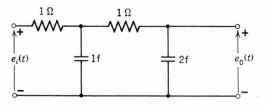

Figure P2.4

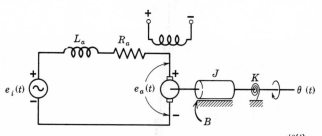

Figure P2.5   Armature torque $= T_a(t) = K_a i_a(t)$; $e_a(t) = K_m \dfrac{d\theta(t)}{dt}$.

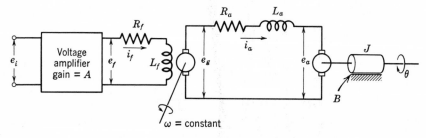

Figure P2.6 $e_g = K_g i_f$, $e_a = K_m \dot{\theta}$,   Torque $= K_a i_a$.

2.7. Given the shock absorber system as shown in Fig. P2.7, $v(t)$ is a velocity source applied to the massless bar whose position is given by x_3. For the bar $v(t) = (dx_3/dt)$, find the transfer function

$$\frac{X_1(s)}{V(s)}$$

2.8. A single-loop, unity-feedback, second-order system is described by the equations

$$J\ddot{\theta}_c(t) + B\dot{\theta}_c(t) = Ke(t)$$
$$e(t) = \theta_r(t) - \theta_c(t)$$

Consider three systems for which the parameter values are $K = 334$, $J = 1.5$, $B = 17.9$; $K = 166.7$, $J = 3.0$, $B = 17.9$; $K = 1334$, $J = 1.5$, $B = 35.8$. For each system assume:

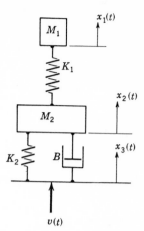

Figure P2.7

(a) A step-input displacement of 15°. Compute the maximum over-shoot, time to reach maximum overshoot, time for error to stay within 1.8%, settling time and steady-state error.

(b) A step-velocity input of 36°/sec. Compute the maximum overshoot, time to reach maximum overshoot, settling time, steady-state error.

2.9. For the electromechanical system shown in Fig. P2.9:

(a) What value should A be to give 10° of steady-state shaft deflection, θ, for an e of 1 volt?

(b) What value should B have to give 15% overshoot (of the shaft position θ) over its final steady-state value if the switch is closed?

(c) For $B = \frac{1}{2}$, how long will it take for the shaft position to settle to within 1.8% of its final value after the switch is closed?

(d) If $B = \frac{1}{2}$ and the voltage applied to the current amplifier is $e(t) = t1(t)$ (unit ramp) and the desired response to this is $\theta(t) = 10t1(t)$

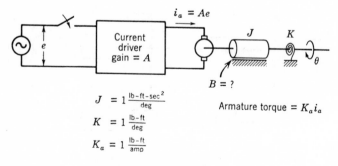

$$i_a = Ae$$

$$J = 1 \frac{\text{lb-ft-sec}^2}{\text{deg}}$$

$$K = 1 \frac{\text{lb-ft}}{\text{deg}}$$

$$K_a = 1 \frac{\text{lb-ft}}{\text{amp}}$$

Armature torque $= K_a i_a$

$B = ?$

Figure P2.9

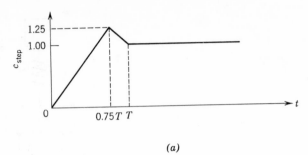

(a)

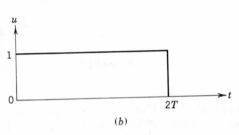

(b)

Figure P2.10

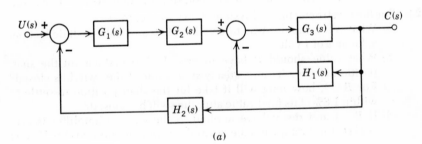

(a)

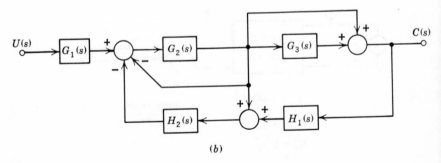

(b)

Figure P2.11

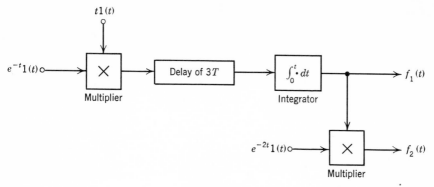

Figure P2.12

(ramp of slope 10), what should A be adjusted to and what will the steady-state difference between the actual and desired output be for this value of A?

2.10. A system has the unit-step response as shown in Fig. P2.10a.
(a) What is its impulse response?
(b) What is its response to the input function shown in Fig. P2.10b.

2.11. What is $C(s)/U(s)$ for the system of:
(a) Fig. P2.11a.
(b) Fig. P2.11b.

2.12. Find the following for the system with inputs as shown in Fig. P2.12:
(a) $F_2(s) = \mathscr{L}[f_2(t)] = ?$
(b) $\lim_{t \to \infty} f_1(t) = ?$

2.13. Find final settling time t_s, peak time t_p, overshoot M_0, and number of oscillations N for the systems with following transfer functions:

(a) $\dfrac{1}{(s^2+s+1)}$

(b) $\dfrac{(s+0.75)}{(s^2+s+1)}$

(c) $\dfrac{1}{(s+0.75)(s^2+s+1)}$

(d) $\dfrac{(s+0.6)(s+2)}{(s+0.63)(s^2+s+1)(s+2.5)(s+4)}$

3

ANALYSIS OF CONTINUOUS-TIME SYSTEMS

3.1 INTRODUCTION

In Chapter 2 we defined the transfer function for linear time-invariant systems and showed how to use it to find the time response of the output for any given input. The approach so far has been an essentially open-loop one. Now we begin to refine the analysis to consider the closed-loop case and the very important problem of predicting closed-loop behavior from open-loop characteristics. The first consideration in closing the loop around a system must be whether or not the resulting system will be stable; thus we define stability more specifically in this chapter. Working from this definition we find that any linear time-invariant system (open or closed loop) is stable if and only if *all* its poles are in the LHP.

We discuss two methods for finding whether or not a given transfer function has all its poles in the LHP (i.e., if it is stable): the Routh criterion and the Nyquist criterion. The Routh criterion is used relative to the denominator of the overall transfer function (it works only for the case where this denominator is a polynomial in the Laplace variable s) and is a purely algebraic test. The Nyquist criterion is based on use of the frequency response (i.e., the response of the system to sinusoidally varying inputs) and can be used to predict closed-loop stability from only open-loop measurements upon the system.

From stability we go to methods of analysis which are basically means of relating the closed-loop response of systems to a parameter (or parameters) which it is assumed the system designer is free to choose. These are the frequency-response, root locus, and Mitrovic methods. These form the basic classical tools by which linear time-invariant systems are analyzed and compensated. The frequency-response method is discussed at the end of this chapter; the root locus and Mitrovic's methods are discussed in Chapter 4.

3.2 STABILITY

Definition of Stability

We define a system as *stable* if

$$\text{Bounded Input} \Rightarrow \text{Bounded Output} \qquad (3.2\text{-}1)$$

In words, *the system will be considered stable if a bounded input always produces a bounded output*. By bounded, we mean bounded in magnitude. Intuitively this is a satisfying definition of stability; it also has the advantage that in the case of the linear time-invariant system that we are interested in, it is easily applicable mathematically.

Now considering the implications of boundedness in the s-domain, take the function $x(t)$, with $\mathscr{L}$-transform $X(s)$, which we shall assume to be bounded in magnitude for all t. We make the reasonable and practical assumption that $x(t) = 0$ for $t < 0$ and that $X(s)$ is a rational function of s. Thus $X(s)$ is of the form

$$X(s) = \frac{a_m s^m + a_{m-1} s^{m-1} + \cdots + a_0}{s^n + b_{n-1} s^{n-1} + \cdots + b_0}, \quad m \leq n \qquad (3.2\text{-}2)$$

Expanding $X(s)$ by partial fractions gives

$$\begin{aligned}
X(s) = K_\infty &+ \frac{K_{11}}{(s-p_1)} + \frac{K_{12}}{(s-p_1)^2} + \cdots + \frac{K_{1m_1}}{(s-p_1)^{m_1}} \\
&+ \frac{K_{21}}{(s-p_2)} + \frac{K_{22}}{(s-p_2)^2} + \cdots + \frac{K_{2m_2}}{(s-p_2)^{m_2}} \qquad (3.2\text{-}3) \\
&+ \cdots \\
&+ \frac{K_{l1}}{(s-p_l)} + \frac{K_{l2}}{(s-p_l)^2} + \cdots + \frac{K_{lm_l}}{(s-p_l)^{m_l}}
\end{aligned}$$

where the K_{ij} are the partial fraction expansion constants.

We have assumed that $X(s)$ has l distinct poles $p_1, p_2, \ldots, p_l$ of multiplicity $m_1, m_2, \ldots, m_l$, respectively. Inverting (3.2-3) using Table 2.2-2 gives

$$
x(t) = K_\infty \delta(t) + K_{11}e^{p_1 t} + \frac{K_{12}te^{p_1 t}}{1!} + \cdots + \frac{K_{1m_1}}{(m_1-1)!}t^{m_1-1}e^{p_1 t}
$$

$$
+ K_{21}e^{p_2 t} + \frac{K_{22}te^{p_2 t}}{1!} + \cdots + \frac{K_{2m_2}}{(m_2-1)!}t^{m_2-1}e^{p_2 t}
$$

$$
+ \cdots
$$

$$
+ K_{l_1}e^{p_l t} + \cdots + \frac{K_{lm_l}}{(m_l-1)!}t^{m_l-1}e^{p_l t}, \qquad t > 0 \qquad (3.2\text{-}4)
$$

Now consider (3.2-4) keeping in mind that $x(t)$ is bounded for all t. The first conclusion we draw from the RHS of (3.2-4) is that K_∞ must be zero if $x(t)$ is to be bounded since the δ-function is of undefined magnitude at $t = 0$. We note from (3.2-3) that

$$
K_\infty = \lim_{s \to \infty} X(s) \qquad (3.2\text{-}5)
$$

Thus $X(s)$ must have a zero at infinity. The implication of this on $X(s)$ is that $m < n$. Thus if x(t) *is to be bounded, the degree of the numerator of* X(s) *must be less than the degree of the denominator.*

Next consider a typical term on the far RHS of (3.2-4):

$$
\frac{K_{im_i}t^{m_i-1}}{(m_i-1)!}e^{p_i t}
$$

If this term is to be bounded for all t, Re $p_i \leq 0$. If Re $p_i < 0$, the term is bounded for $t > 0$ without any further consideration. If Re $p_i = 0$, the term is bounded as $t \to \infty$ only if

$$
m_i - 1 = 0 \qquad \text{or} \qquad m_i = 1
$$

That is, the multiplicity of the pole with Re $p_i = 0$ must be one if $x(t)$ is to be bounded.

The implications of the boundedness of $x(t)$ on $X(s)$ may then be summarized as follows:

1. $\displaystyle\lim_{s \to \infty} X(s) = 0$, which means $m < n$.

2. The poles of $X(s)$ are either in the LHP or on the $j\omega$-axis.

3. The poles on the $j\omega$-axis are of multiplicity one only.

$\qquad\qquad\qquad\qquad\qquad\qquad\qquad\qquad\qquad\qquad\qquad\quad$ (3.2-6)

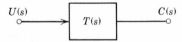

Figure 3.2-1   System considered for stability.

What are the implications of stability on the transfer function of a system? Consider the system of Fig. 3.2-1. We have

$$C(s) = T(s)\, U(s) \tag{3.2-7}$$

If the system is to be stable by the definition of (3.2-1), then for $U(s)$ satisfying the conditions of (3.2-6), $C(s)$ must satisfy them also. The immediate implications on $T(s)$ are

$$\lim_{s \to \infty} |T(s)| = k < \infty \tag{3.2-8}$$

and $T(s)$ must have no poles on the $j\omega$-axis or in the RHP of the s-plane. It may be seen that $T(s)$ can have no poles on the $j\omega$-axis since if it did, $U(s)$ could have a pole at the same point on the $j\omega$-axis and $C(s)$ would then have a double pole at that common point.[1] Hence $C(t)$ would be unbounded as $t \to \infty$.

Since the violation of (3.2-8) would imply a pole at $s = +\infty$, we can summarize the preceding discussion by saying: *A system is stable only when all its poles are in the LHP.* Or,

$$\operatorname{Re} p_i < 0, \qquad i = 1, 2, ..., n \tag{3.2-9}$$

where p_i is the ith pole of $T(s)$.

We see another practical advantage in using the $\mathscr{L}$-transform in the study of linear time-invariant systems. The question of stability reduces itself to determining whether the corresponding transfer function has all its poles in the LHP. Besides being easy to remember, this has the advantage of well-developed techniques that tell when a transfer function has all its poles in the LHP. The two most useful for our purposes are the Routh criterion and the Nyquist criterion. We consider both in the following sections.

The Routh Stability Criterion

With the Routh criterion we consider rational transfer functions i.e., functions of the form

$$T(s) = \frac{N(s)}{F(s)} = \frac{A(s)}{a_n s^n + a_{n-1} s^{n-1} + \cdots + a_0} \tag{3.2-10}$$

where $N(s)$ and $F(s)$ are polynomials in s. To determine whether $T(s)$ is stable we need only determine whether the degree of $N(s)$ is less than or equal to the degree of $F(s)$ and whether $F(s)$ has all

---

[1]An example of this is $T(s) = U(s) = 1/s$. Then $C(s) = U(s)\, T(s) = 1/s^2$, which implies $c(t) = t1(t)$, which is unbounded as $t \to \infty$.

its roots in the LHP. The Routh criterion is a method of determining whether a given polynomial has any roots in the RHP or on the $j\omega$-axis. In our case this polynomial is

$$F(s) = a_n s^n + a_{n-1} s^{n-1} + \cdots + a_1 s + a_0 = 0 \qquad (3.2\text{-}11)$$

The Routh test involves constructing a table. However, before we construct the table, we make $a_n > 0$ in (3.2-11) and note the following:

1. If any coefficient of any power of s is either negative or zero then there is at least one root in the RHS or on the imaginary axis of the s-plane. The system is then *unstable*.

2. In case there is neither a negative nor a zero coefficient, the system might still be unstable and we proceed to apply the *Routh* criterion.

Starting with the first coefficient a_n, the alternate coefficients are arranged in the first row and all the remaining coefficients starting with coefficient a_{n-1} are arranged in the second row as shown in (3.2-12). On the left-hand side, we designate various powers of s starting with s^n in the first row.

$$
\begin{array}{c|cccc}
s^n & a_n & a_{n-2} & a_{n-4} & \cdots \\
s^{n-1} & a_{n-1} & a_{n-3} & a_{n-5} & \cdots \\
s^{n-2} & b_{n-1} & b_{n-3} & b_{n-5} & \cdots \\
s^{n-3} & c_{n-1} & c_{n-3} & c_{n-5} & \cdots \\
\cdot & \cdot & \cdot & \cdot & \\
\cdot & \cdot & \cdot & \cdot & \\
\cdot & \cdot & \cdot & \cdot & \\
s^0 & r_{n-1} & \cdot & \cdots
\end{array}
\qquad (3.2\text{-}12)
$$

Then we proceed to form a row of b's from the a's as follows:

$$b_{n-1} = \frac{a_{n-1} a_{n-2} - a_n a_{n-3}}{a_{n-1}}$$

$$b_{n-3} = \frac{a_{n-1} a_{n-4} - a_n a_{n-5}}{a_{n-1}} \cdots \qquad (3.2\text{-}13)$$

We continue to make up rows, any row being dependent upon the preceding two rows as given by (3.2-13), until we get a row of all zeros. Then we stop. The number of rows obtained must be $n+1$ for the nth-order characteristic equation. *The condition of stability is given by the requirement that all terms of the first column are positive. The number of changes of sign in this column is equal to the number of roots of* $F(s) = 0$ *in the RHS of the s-plane.* We can lessen the numerical work if we recognize that any positive factor can be taken out of any row at any time.

Example 3.2-1
Given
$$F(s) = s^4 + 6s^3 + 2s^2 + s + 3 = 0$$

Find if there are any roots in the right half of the s-plane by using the Routh criterion.

We make the table as explained above; the number of rows is 5:

$$
\begin{array}{c|ccc}
s^4 & 1 & 2 & 3 \\
s^3 & 6 & 1 & \\
s^2 & \dfrac{11}{6} & 3 & \\
s^1 & -\dfrac{97}{11} & & \\
s^0 & 3 & &
\end{array}
$$

There are two roots in the RHS of s-plane. Therefore the system is unstable.

TWO EXCEPTIONS TO THE ROUTH TABLE

From the rules for constructing the Routh table in the previous section we know that if a zero appears in the first column, we cannot complete the table. There are two different cases when this occurs:

1. First column entry zero, nonzero entry elsewhere in the same row.
2. All-zero row.

In the first case we replace the zero in the first column by ϵ where ϵ is taken as a small positive number. The table is then completed and the sign of the first column entries is then taken in the limit as $\epsilon \to 0+$. Let us consider an example.

Example 3.2-2
Consider $s^4 + s^3 + s^2 + s + 2$. The Routh table is then

$$
\begin{array}{c|ccc}
s^4 & 1 & 1 & 2 \\
s^3 & 1 & 1 & \\
s^2 & 0 \to \epsilon & 2 & \\
s^1 & \dfrac{\epsilon - 2}{\epsilon} & & \\
s^0 & 2 & &
\end{array}
$$

In the limit as $\epsilon \to 0, \epsilon > 0$ the first column becomes

$$
\begin{array}{c}
1 \\
1 \\
0 \\
-\infty \\
2
\end{array}
$$

There are two sign changes in going down the column, i.e., from 0 to $-\infty$ and from $-\infty$ to 2. Thus $s^4 + s^3 + s^2 + s + 2$ has two RHP roots. The corresponding system would be unstable.

The second case is of more practical interest since the all-zero row indicates the fact that the polynomial under test has an even factor.[2] We note that if an even factor has a root at s_0, then it also has a root at $-s_0$. Since the polynomials we are considering have real coefficients, the even factors of such a polynomial will also have real coefficients, so the even factor will have roots symmetrically placed w.r.t. (with respect to) the σ-axis. Thus an even factor will have roots symmetrically placed w.r.t. both the σ- and the $j\omega$-axes. The presence of an all-zero row hence tells us that the polynomial under test either has roots on the $j\omega$-axis or roots symetrically spaced w.r.t. the $j\omega$-axis in the RHP and LHP. In either case, if the polynomial of interest is the denominator of a transfer function, the corresponding system is *unstable*.

The Routh table, however, gives more information than this, for if the table has an all-zero row, the row just above the all-zero row gives the coefficients of the even factor. Thus one has additional knowledge concerning the roots of the polynomial which are symmetrically placed w.r.t. the σ- and $j\omega$-axes. Let us consider such an example.

Example 3.2-3

Consider $s^6 + 2s^5 + 5s^4 + 8s^3 + 8s^2 + 8s + 4$ and the corresponding table:

$$
\begin{array}{c|cccc}
s^6 & 1 & 5 & 8 & 4 \\
s^5 & 2 & 8 & 8 & \\
s^4 & 1 & 4 & 4 & \\
s^3 & 0 & 0 & &
\end{array}
$$

The table has an all-zero row, the s^3 row. The s^4 row has entries 1, 4, 4. Hence the even factor is

$$
s^4 + 4s^2 + 4 = (s^2 + 2)^2
$$

---

[2] An even factor is one that has only even-power terms.

And we see that the polynomial under test has double roots at $s = \pm j\sqrt{2}$, i.e., it has four roots on the $j\omega$-axis.

If we wish to continue the table to determine how many RHP roots the polynomial has, the zero row is replaced by the coefficients of the derivative of the even factor. In this case the derivative of the even factor is $4s^3 + 8s$. Thus the table is continued as

s^6	1	5	8	4
s^5	2	8	8	
s^4	1	4	4	
s^3	4	8		
s^2	2	4		
s^1	0			

Again we have an all-zero row which we replace by the derivative of the even factor given by the row just above to get

s^6	1	5	8	4
s^5	2	8	8	
s^4	1	4	4	
s^3	4	8		
s^2	2	4		
s^1	4	←— coefficient of the derivative of $2s^2 + 4$		
s^0	4			

The table is now complete and we see that the first column does not have any sign changes so the polynomial under test does not have any RHP roots. However, there was an even factor of the fourth order. We conclude that all four roots of the even factor are on the $j\omega$-axis. Thus given a transfer function which is a rational function, we can use a Routh table to determine the number of RHP roots. The question of stability for the corresponding system is then answered.

DETERMINING STABLE RANGES FOR THE PARAMETERS OF A CLOSED-LOOP SYSTEM

In designing a control system the first and most basic requirement is the stability of the system. For any parameter of the closed-loop system, the minimum requirement is that it fall into the range of values for which the system is stable. Since we are primarily interested in designing feedback systems, the requirement is that the system be closed-loop stable. If the open-loop transfer function—the transfer function of the plant in the forward and feedback paths—is given and it is in the form of a rational function of s, then there is no difficulty in applying (2.2-37) to obtain the corresponding closed-loop transfer function. Let us consider the case where the denominator of the open-loop (hence also the closed-loop) transfer function is a function of some parameter. This means that the coefficients of the denominator polynomial are functions of this parameter. Let us now consider determining the range of this parameter over which the closed-loop system is stable. We set these coefficients of the denominator polynomial of the closed-loop transfer function into a Routh table. The entries in the first column are now functions of the parameter under test. Imposing the condition that these first-column entries be positive gives the conditions that the parameter must satisfy in order for the system to be stable. From these, the stable range of the parameter may be determined, as we see in the following example.

Example 3.2-4

Assume the system given in Fig. 3.2-2 and let us find the range of the parameter K over which the closed-loop system is stable. Applying (2.2-37),

$$T(s) = \frac{C(s)}{R(s)} = \frac{G(s)}{1+G(s)} = \frac{K/(s^3+3s^2+3s+1)}{1+K/(s^3+3s^2+3s+1)}$$

and therefore

$$\frac{C(s)}{R(s)} = \frac{K}{s^3+3s^2+3s+K+1}$$

Now we construct the Routh table for the denominator of this transfer function:

$$
\begin{array}{c|cc}
s^3 & 1 & 3 \\
s^2 & 3 & K+1 \\
s^1 & \dfrac{9-(K+1)}{3} & \\
s^0 & K+1 &
\end{array}
$$

The conditions that determine the stable range of the parameter K

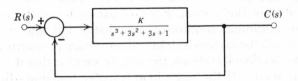

Figure 3.2-2

are found by setting the entries in the first column (which involve K) positive:

$$\frac{9-(K+1)}{3} > 0$$

$$K+1 > 0$$

which give

$$-1 < K < 8$$

Thus we note that for K between -1 and $+8$, the system of Fig. 3.2-2 is stable. It is interesting to consider K at the extreme ends of the range of stability. Consider $K = 8$ first. For $K = 8$ the s^1 row of the table is all-zero and hence the polynomial, the denominator of the transfer function, has an even factor. From the s^2 row of the table we determine that

$$3s^2 + K + 1 = 3s^2 + 9$$

is the even factor. Thus for $K = 8$ the system has poles on the $j\omega$-axis at $\pm j\sqrt{3}$. Now let us look at the other end of the stable range, $K = -1$. We note that for this value of K the denominator of the transfer function has a root at $s = 0$, again on the $j\omega$-axis.

In general, for a parameter at the extreme ends of the range of stability, the system will have a pole or poles on the $j\omega$-axis. This is because the $j\omega$-axis is the boundary between the stable and unstable regions of the s-plane.

The Nyquist Plots and the Stability Criterion

THE NYQUIST PLOT PRESENTATION OF THE FREQUENCY RESPONSE OF SYSTEMS

The Nyquist criterion is probably the most useful stability criterion in practice. As compared to the analytic method evolved by Routh, the Nyquist criterion is graphic in nature. Basic to the use of this criterion is an understanding of the frequency response of a system $[G(s)]$. Given the transfer function of a system, the frequency response is simply the transfer function evaluated along the $j\omega$-axis:

$$\text{Frequency Response of } G(s) \triangleq G(j\omega), \qquad 0 \leqslant \omega < \infty \qquad (3.2\text{-}14)$$

The frequency response[3] is considered only over the positive

---

[3] The name frequency response comes from the fact that if a system with transfer function $G(s)$ is excited with a unit sinusoid of frequency ω, the output (after all transients have disappeared) will be another sinusoid with magnitude $|G(j\omega)|$ and phase, relative to the input, of $\angle G(j\omega)$. We show this later.

frequency (ω) range since for a real system

$$G(-j\omega) = \overline{G}(j\omega) \qquad (3.2\text{-}15)$$

where the bar denotes the conjugate. Thus given the positive frequency (ω) range response, the frequency response over the whole (positive and negative) frequency range is easily constructed.

A convenient graphic method for displaying the frequency response is a Nyquist plot. This is a plot in the complex plane of $\operatorname{Re} G(j\omega)$ versus $\operatorname{Im} G(j\omega)$ for the parameter ω in the range $0 \le \omega < \infty$. This will result in a curve in the complex plane (the G-plane) which is usually tagged with the parameter ω. It will be seen later that the closed-loop stability of a system may be readily determined from the Nyquist plot of the open-loop frequency response. However, before considering stability, let us consider a few simple examples of Nyquist plots to illustrate what is involved here.

In Fig. 3.2-3 are shown the Nyquist plots for simple lag and lead networks with the corresponding frequency-response function, $G(j\omega)$. These are simple enough so that it can be seen directly that the Nyquist plot corresponds to the given frequency-response function.

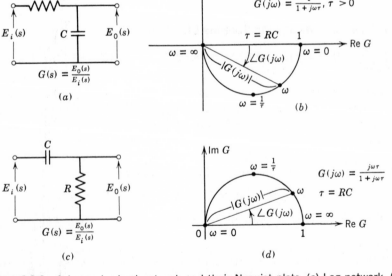

Figure 3.2-3 A lag and a lead network and their Nyquist plots. (a) Lag network. (b) Nyquist plot for lag network. (c) Lead network. (d) Nyquist plot for lead network.

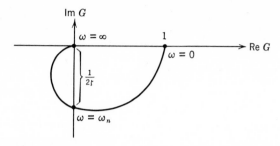

Figure 3.2-4 The Nyquist plot of a second-order system.

As a slightly more complicated example let us consider the Nyquist plot of the typical second-order system considered in Section 2.3:

$$G(s) = \frac{\omega_n^2}{s^2 + 2\zeta\omega_n s + \omega_n^2} \tag{3.2-16}$$

From (3.2-16) we see directly that $G(0) = 1$, $\underset{\omega \to \infty}{G(j\omega)} = 0$. Also for $\omega = \omega_n$,

$$G(j\omega_n) = -\frac{j}{2\zeta}$$

and

$$\underset{\omega \to \infty}{G(j\omega)} \to -\frac{\omega_n^2}{\omega^2}$$

Thus as $\omega \to \infty$, $G(j\omega)$ approaches the origin from the direction of the negative real axis. With these hints, the Nyquist plot for (3.2-16) may be sketched as in Fig. 3.2-4.

Of special interest in considering Nyquist plots is the behavior of $G(j\omega)$ as $\omega \to 0$ and as $\omega \to \infty$. Consider first the case as $\omega \to 0$. If a transfer function has n_p poles at the origin, then

$$\underset{\omega \to 0}{G(j\omega)} \to \frac{K_0}{(j\omega)^{n_p}}$$

where

$$K_0 = \lim_{s \to 0} s^{n_p} \cdot G(s)$$

Thus as $\omega \to 0$ the Nyquist plot of $G(j\omega)$ goes off to infinity at an angle of $-90°.n_p$. This is as shown in Fig. 3.2-5 for the cases $n_p = 1$ and $n_p = 3$.

For a system with n_z zeros at the origin we have

$$\underset{\omega \to 0}{G(j\omega)} = K_0(j\omega)^{n_z}$$

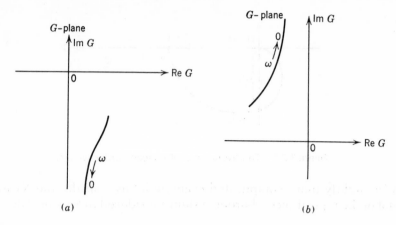

Figure 3.2-5 Nyquist plot for a G(s) with (a) one pole and (b) three poles at the origin.

where now

$$K_0 = \lim_{s \to 0} \left[\frac{G(s)}{s^{n_z}} \right]$$

For this case as $\omega \to 0$ the Nyquist plot approaches the origin at an angle of $+90° \cdot n_z$. An example of this is seen in Fig. 3.2-3d for the case $n_z = 1$.

Now consider the practical case where $G(s)$ approaches a finite constant value or zero as $s \to \infty$. Thus we have a $G(s)$ of the form

$$G(s) = \frac{a_m s^m + a_{m-1} s^{m-1} + \cdots + a_0}{s^n + b_{n-1} s^{n-1} + \cdots + b_0}, \qquad m < n$$

and

$$G(j\omega) \underset{\omega \to \infty}{\longrightarrow} \frac{a_m}{(j\omega)^{n-m}}$$

Thus as $\omega \to \infty$ the Nyquist plot approaches the origin from an angle of $-90° \cdot (n-m)$.[4] An example of this for the case $n - m = 1$ has been given in Fig. 3.2-3b.

Thus the gross characteristics of the Nyquist plot may be determined from the behavior of $G(j\omega)$ at the extreme ends of the frequency range of interest, i.e., at $\omega = 0$ and $\omega \to \infty$. Some illustrations of this are shown in Fig. 3.2-6.

In constructing Nyquist plots a problem that requires special atten-

---

[4] $n - m$ where n and m are the number of system poles and zeros, respectively, is termed the number of excess poles if $n - m > 0$.

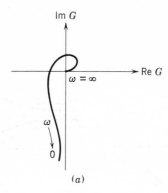

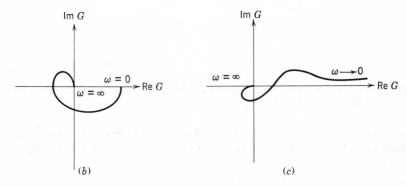

Figure 3.2-6 Nyquist plots of various $G(s)$ poles at origin and excess poles as indicated. (a) Single pole at the origin; four excess poles. (b) No poles at origin; three excess poles. (c) Four poles at origin; six excess poles.

tion arises if the $G(s)$ for which the Nyquist plot is being constructed has a pole on the $j\omega$-axis. We have briefly considered the case of poles and zeros at the origin, but now let us assume that the $G(s)$ of interest has a pole at $s = j\omega_0$, of order N, say, and consider $G(s)$ in the neighborhood of $j\omega_0$. Then by partial fraction expansion we have

$$G(s) = \frac{K_{\omega_0}}{(s - j\omega_0)^N} + \{\text{other terms involving poles not at } \omega_0\} \qquad (3.2\text{-}17)$$

where

$$K_{\omega_0} = \lim_{s \to j\omega_0} \left[(s - j\omega_0)^N \cdot G(s) \right]$$

For ω near ω_0

$$G(j\omega) \cong \frac{K_{\omega_0}}{(j\omega - j\omega_0)^N} = \frac{K_{\omega_0}}{(j)^N(\omega - \omega_0)^N} \qquad (3.2\text{-}18)$$

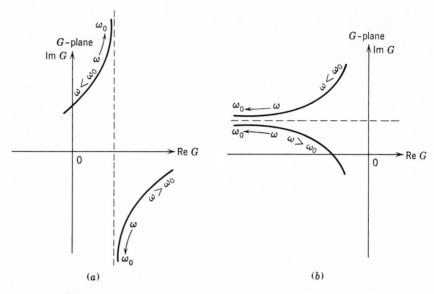

Figure 3.2-7 Nyquist plots for a $G(s)$ with a pole on the $j\omega$-axis. (a) Odd-order pole at $s = j\omega_0$. (b) Even-order pole at $s = j\omega_0$.

since the other terms in the expansion of $G(s)$ will have negligible magnitude for ω in this neighborhood. From (3.2-18) we see that as $\omega \rightarrow \omega_0$

$$|G(j\omega)| \rightarrow \infty$$

for N even,

$$\angle G(j\omega) \rightarrow -(90°) \cdot N$$

and for N odd,

$$\angle G(j\omega) \rightarrow -(N+2) \cdot (90°) \qquad \text{for } \omega < \omega_0$$

$$\rightarrow -(N) \cdot (90°) \qquad \text{for } \omega > \omega_0$$

This behavior of the Nyquist plot is illustrated in Fig. 3.2-7 for both cases.

NYQUIST PLOT WITH POLES ON THE $j\omega$-AXIS

The Nyquist criterion to be discussed in the next section is a method for determining the number of RHP zeros $1 + G(s)$ has from the Nyquist plot of $G(s)$, i.e., from $G(j\omega)$, $0 \leq \omega < \infty$.

To do this, however, one must have a continuous curve, the Nyquist plot must be a connected curve which covers the whole frequency range of interest. It may be seen from Fig. 3.2-7 that with a pole on the $j\omega$-axis, the Nyquist plot is not a connected curve for all values of ω.

In fact at $\omega = \omega_0$ there is a break. To get around this difficulty the path over which $G(s)$ is evaluated, which is otherwise just the $j\omega$-axis, is altered slightly to avoid poles on the $j\omega$-axis. This avoidance of the poles is done by choosing the path of evaluation as a small semicircle, wherever $j\omega$-axis poles occur, into the RHP. This is shown in Fig. 3.2-8a.

The path into the RHP could as well have been into the LHP; however, conventionally it is chosen into the RHP. This RHP semicircular path is shown in expanded form in Fig. 3.2-8b. Also shown is a general point s on this path, and the vector $(s-j\omega_0)$. We take the radius of the semicircle to be ϵ and the angle of $(s-j\omega_0)$ to be θ; then we write $s-j\omega_0$ in the form

$$(s-j\omega_0) = \epsilon\, e^{j\theta} \tag{3.2-19}$$

As s traverses the semicircular path in the direction of increasing ω, indicated by the arrowheads on the path in Fig. 3.2-8b, we see that θ traverses the range $-90° \leqslant \theta \leqslant 90°$. Thus substituting (3.2-19) into (3.2-17) gives

$$G(s) = K_{\omega_0}\epsilon^{-N}e^{-jN\theta} + \{\text{terms negligible near } s = j\omega_0\}$$

for s along the semicircle and ϵ small. We recall that N is the order of the pole at $j\omega_0$. Thus along the semicircle we have

$$|G(s)| \cong |K_{\omega_0}\epsilon^{-N}| = \text{constant}$$
$$\angle G(s) \cong -N\theta \tag{3.2-20}$$

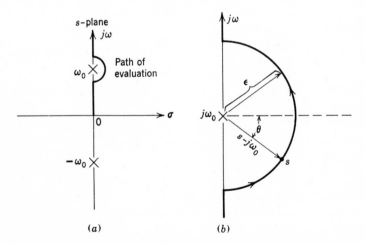

Figure 3.2-8 (a) The path of evaluation of G(s) for the case of a pole on the $j\omega$-axis. (b) Expanded view of the path of evaluation about a pole at $s = j\omega_0$.

From (3.2-20) we see that as the semicircle is traversed from $\theta = -\pi/2$ rad. to $\theta = +\pi/2$ rad., $G(s)$ moves along a circular arc with radius $= |K_{\omega_0}\epsilon^{-N}|$ and through the angle $N \cdot 90°$ to $-N \cdot 90°$. $\angle G(s)$ thus is decreasing as θ increases and the radius of this circular path is large for ϵ small. The Nyquist plots for the case of a first- and second-order pole at $j\omega_0$ are shown in Fig. 3.2-9. The arrows along the circular path indicate the direction of increasing θ, which corresponds to increasing ω. We note that the radius of the circular paths is simply assumed to be very large.

With this background we should now be able to see the general shape of the Nyquist plots for the usually encountered $G(s)$. To construct a precise Nyquist plot for a given $G(s)$, a useful and valid

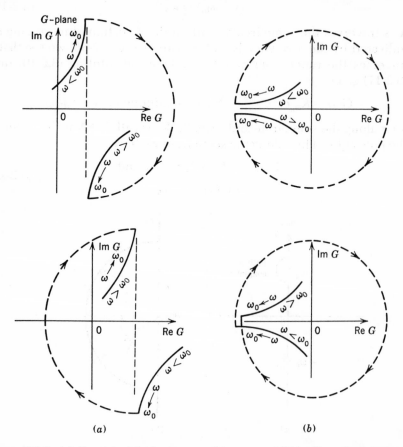

Figure 3.2-9   (a) Nyquist plots for two possible cases of $G(s)$ with a single pole at $s = j\omega_0$. (b) Nyquist plots for two possible cases of $G(s)$ with a double pole at $s = j\omega_0$.

method is to select a grid of points which covers the finite $j\omega$-axis — ω_1, ω_2, ω_3, ... ω_r may be selected — and $G(j\omega_1)$, $G(j\omega_2)$, ..., $G(j\omega_r)$ may then be calculated and the results plotted on the G-plane. By selecting the grid fine enough, the Nyquist plot may be constructed to any desired degree of accuracy. This will work even if $G(s)$ is not the usual rational function of s. It is, of course, laborious to do this by hand, but the construction of Nyquist plots is now so common a problem that computer programs, which are usually based upon this grid method, are available in most computer libraries.

Based on this preliminary discussion, we may consider the determination of closed-loop stability by Nyquist's criterion.

THE NYQUIST CRITERION

The basic unity-feedback system (Fig. 3.2-10) is used often in any consideration of the Nyquist criterion. In general we ask whether, given $G(s)$, the closed-loop system is stable. This we shall see can be answered right from the Nyquist plot constructed in the previous section.

From Fig. 3.2-10, the closed-loop transfer function is given by

$$T(s) = \frac{G(s)}{1 + G(s)} \tag{3.2-21}$$

Thus the poles of the closed-loop system will be the zeros of $1 + G(s)$. With the Nyquist criterion we determine how many zeros $1 + G(s)$ has in the RHP. We also determine if $1 + G(s)$ has any zeros on the $j\omega$-axis. Thus the question of closed-loop stability is answered.

First consider

$$F(s) = K\frac{(s - z_1)}{(s - p_1)}$$

which is a simple, single-pole, single-zero rational function with a pole at p_1 and a zero at z_1.

Now let us see what happens in the case of a contour, Γ_1, which encloses both z_1 and p_1. This is shown in Fig. 3.2-11a. Now we see as

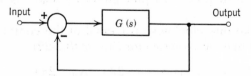

Figure 3.2-10 The basic system considered with the Nyquist criterion.

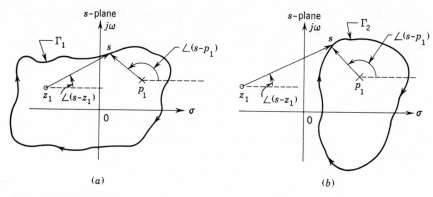

Figure 3.2-11   (a) A contour which encloses both the pole and the zero. (b) A contour which encloses only the pole.

Γ_1 is traversed in the clockwise direction, both $\angle(s-z_1)$ and $\angle(s-p_1)$ decrease continuously. In one revolution about Γ_1, then

$$\delta_{\Gamma_1}\angle(s-z_1) = -2\pi$$
$$\delta_{\Gamma_1}\angle(s-p_1) = -2\pi$$

where $\delta_{\Gamma_1}\angle$ indicates the change in angle as Γ_1 is traversed,

$$\delta_{\Gamma_1}\angle F(s) = \delta_{\Gamma_1}\angle(s-z_1) - \delta_{\Gamma_1}\angle(s-p_1) = (-2\pi) - (-2\pi) = 0$$

Thus the change in $\angle F(s)$ in traversing Γ_1, which is a completely general contour except that it encloses both z_1 and p_1, is zero. Using these same arguments on the contour Γ_2, which encloses only p_1 and not z_1 (Fig. 3.2-11b), we see that

$$\delta_{\Gamma_2}\angle F(s) = (0) - (-2\pi) = 2\pi$$

We thus see that the change in $\angle F(s)$ that one gets in traversing a contour in the s-plane for one clockwise revolution is strictly a function of how many poles and zeros are enclosed. Let us consider a general $F(s)$ with m zeros and n poles such as

$$F(s) = K\frac{(s-z_1)(s-z_2)\cdots(s-z_m)}{(s-p_1)(s-p_2)\cdots(s-p_n)}, \qquad m < n \qquad (3.2\text{-}22)$$

for which a typical pole-zero plot might be made as shown in Fig. 3.2-12.

Now if we make one clockwise traversal of the contour Γ as shown in Fig. 3.2-12 it may be seen that for $F(s)$ of (3.2-22)

$$\delta_\Gamma \angle F(s) = N_z(-2\pi) - N_p(-2\pi)$$
$$= (N_p - N_z)\, 2\pi \qquad (3.2\text{-}23)$$

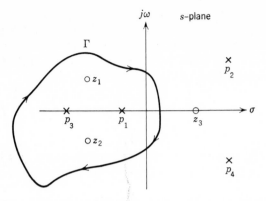

Figure 3.2-12 A general pole-zero plot with contour Γ which encloses N_z zeros and N_p poles.

where now N_z and N_p are the number of zeros and poles, respectively, enclosed by Γ. Thus evaluating $F(s)$ around any enclosed contour in the s-plane and considering $\angle F(s)$ around the contour and by using (3.2-23), we just get the difference in the number of poles and zeros enclosed by the contour multiplied by 2π.

Another way of looking at the result of (3.2-23) is to consider $F(s)$, of (3.2-22) say, evaluated on the contour Γ of Fig. 3.2-12 and plotted in the F-plane, Re F versus Im F. If we consider $F(s)$ evaluated for every point of the contour Γ in the s-plane, we get another contour in the F-plane. This might be as shown in Fig. 3.2-13. The arrowheads on the contour indicate the direction that $F(s)$ moves as s moves about Γ in the s-plane in a clockwise direction. Now as we traverse the

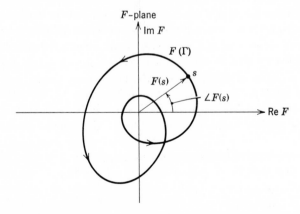

Figure 3.2-13 $F(s)$ evaluated on the contour Γ of Fig. 3.2-12.

contour in the F-plane [let us call it $F(\Gamma)$] once in the direction indicated by the arrows, contour Γ in the s-plane is traversed once in the clockwise direction. For this example, traversing $F(\Gamma)$ once gives a change in $\angle F(s)$ of 4π since the origin is encircled twice in the positive direction. Thus using (3.2-23) gives

$$\delta_\Gamma \angle F(s) = 2\pi (N_p - N_z) = 4\pi$$

or

$$N_p - N_z = 2 = N$$

where N is the number of positive encirclements of the origin that $F(\Gamma)$ makes in the F-plane. In general, we conclude from (3.2-23) that

$$N = N_p - N_z \tag{3.2-24}$$

where we consider a simple, closed contour Γ in the s-plane, and N_p and N_z are then the number of poles and zeros, respectively, enclosed by Γ and N is the number of encirclements of the origin that $F(\Gamma)$ makes in the F-plane.

DETERMINATION OF CLOSED LOOP STABILITY FROM NYQUIST PLOT ENCIRCLEMENTS OF THE $-1 + j0$ POINT

We now consider the determination of the number of RHP poles of the closed-loop system of Fig. 3.2-10, i.e., the stability of the closed-loop system. We have seen from (3.2-21) that the closed-loop poles are given by the zeros of $1 + G(s)$. Hence our problem may be reduced to the problem of determining the RHP zeros of $1 + G(s)$. The approach to this will be to determine $\angle[1 + G(s)]$ around a contour which encloses the whole RHP. We apply (3.2-24), then, by counting the number of encirclements of the origin made by $1 + G(s)$ in going around this contour, which encircles the RHP, we determine $N = N_p - N_z$, where N_p and N_z are the number of RHP poles and zeros, respectively, of $1 + G(s)$. We note that the poles of $G(s)$ and the poles of $1 + G(s)$ occur at the same points. Hence we determine N_p as the number of RHP poles of $G(s)$ and then $N_z = N - N_p$. Thus the question of closed-loop stability is answered, for any $N_z \geqslant 1$ indicates a RHP closed-loop pole or poles, which means the corresponding system is unstable.

Let us now consider the contour that encircles the whole RHP (Fig. 3.2-14). This contour is called the Nyquist contour and it is just the $j\omega$-axis and a semicircle of infinite radius which goes off into the RHP. Now let us consider $1 + G(s)$ along this contour. We see first that for a real physical system, the corresponding $G(s)$ will have no more

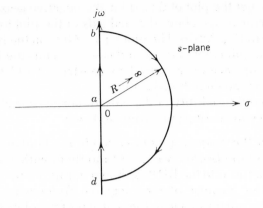

Figure 3.2-14 The Nyquist contour.

zeros than poles; therefore $1 + G(s) \rightarrow$ constant as $s \rightarrow \infty$ and hence $\angle[1 + G(s)] = 0$ or $180°$ along the infinite semicircle. Along the $j\omega$-axis, $1 + G(s)$ is just $1 + G(j\omega)$. To construct $1 + G(j\omega)$ on the $j\omega$-axis we make a Nyquist plot of $G(j\omega)$, $0 \leqslant \omega < \infty$. In the G-plane we see that the vector from the $-1 + j0$ point gives $|1 + G(j\omega)|$ and $\angle[1 + G(j\omega)]$ (this may be seen from Fig. 3.2-15). Thus given the Nyquist plot for a given $G(j\omega)$, $1 + G(j\omega)$ is easily determined. Furthermore, given $G(j\omega)$, $0 \leqslant \omega < \infty$, $1 + G(j\omega)$ for $0 > \omega > -\infty$ is constructed by realizing

$$G(-j\omega) = \overline{G}(j\omega)$$

where the bar denotes conjugate. Thus given the Nyquist plot for

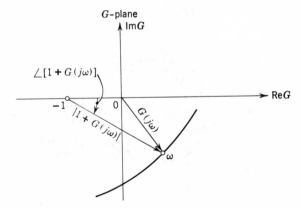

Figure 3.2-15 Evaluating $1 + G(j\omega)$ from the Nyquist plot of $G(s)$.

$0 \leqslant \omega < \infty$, we get the plot of $G(j\omega)$ for the negative ω-axis by simply taking the mirror image about the real axis of the plot for positive ω. This is shown in Fig. 3.2-16. The plot for $1 + G(s)$ on the infinite semi-circle joins the points for $\omega = -\infty$ and $\omega = +\infty$. Thus the Nyquist plot gives all the information required to construct $1 + G(s)$ about the Nyquist contour shown in Fig. 3.2-14.

To determine the stability of the closed-loop system shown in Fig. 3.2-10 by the Nyquist criterion, given $G(s)$, we:

1. Construct the Nyquist plot of $G(j\omega)$ in the G-plane. If $G(s)$ has any poles along the $j\omega$-axis, we go around them with a semicircle of infinitesimal radius into the RHP as shown in Fig. 3.2-9.
2. Extend the Nyquist plot to negative frequencies ($\omega < 0$) by constructing the mirror image in the real axis of the plot of Step 1.
3. Count the encirclements (N) of the $-1 + j0$ point while going in the direction of increasing ω around the extended Nyquist plot.
4. Determine the number (N_p) of RHP poles of $G(s)$. Then

$$N_z = N_p - N \qquad\qquad (3.2\text{-}25)$$

and

$$N_z = 0 \Rightarrow \text{stable closed-loop system}$$
$$N_z \geqslant 1 \Rightarrow \text{unstable closed-loop system}$$

If $N_z < 0$, an error has been made in the procedure.

It should be noted, in going through Step 4, that if $G(s)$ corresponds to a stable system, then $N_p = 0$. This is the usual case in practice. This important case will be discussed further in the next section.

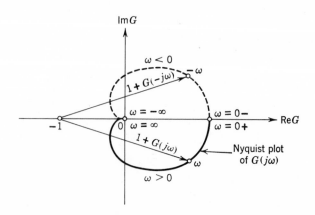

Figure 3.2-16   The extended Nyquist plot.

With the Nyquist criterion as given in Steps 1-4, we determine the number of RHP closed-loop poles. To completely answer the question of stability, we must also know whether the closed-loop system has poles on the $j\omega$-axis. This is also easily answered from the Nyquist plot, for if the closed-loop system has a pole on the $j\omega$-axis (at ω_1, say), then

$$1 + G(j\omega_1) = 0$$

or

$$G(j\omega_1) = -1$$

Thus the Nyquist plot passes through the $-1 + j0$ point in the G-plane for $\omega = \omega_1$. Given the Nyquist plot, this is readily seen.

To sum up the results of this section: *Given a system which is open-loop stable and which hence has no open-loop RHP poles, the system will be unity-feedback, closed-loop stable if its Nyquist plot does not pass through or encircle the $-1 + j0$ point in the G-plane.*

Example 3.2-5

The open-loop transfer function of the system of Fig. 3.2-17 is given by

$$G(s) = \frac{K_p}{1 + \tau s}, \qquad \tau > 0$$

(a) Assuming $K_p > 0$, we plot the Nyquist plot as shown in Fig. 3.2-18. $[1/(1 + j\omega\tau)$ has been plotted in Fig. 3.2-3b.] It is a circle centered at $K_p/2$ and has an intersection with the real axis at K_p. We conclude that for any $K_p > 0$ the $-1 + j0$ point will not be encircled and hence $N = 0$, since $G(s)$ has no RHP poles, $N_p = 0$ and hence $N_z = N_p - N = 0$ and the closed-loop system will have no RHP poles for any positive K_p. This means that the gain of the amplifier can be turned up as high as we wish it (as long as it is positive) and the resulting closed-loop system will still be stable.

(b) Assuming $K_p < 0$, we plot the Nyquist plot as shown in Fig. 3.2-19. Here the circle, which is the Nyquist plot, is over in the LHP and

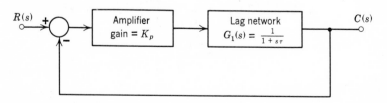

Figure 3.2-17 Closed-loop system with amplifier and lag network in forward path.

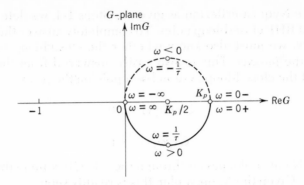

Figure 3.2-18   Extended Nyquist plot of $K_p/(1+s\tau)$.

there is a possibility of encircling the $-1+j0$ point. In fact, for $K_p < -1$ the $-1+j0$ is encircled. For this case $N=-1$ [the minus sign since it is encircled in the clockwise (negative) direction going in the direction of increasing ω] and since $G(s)$ still has no poles in the RHP, we have from (3.2-25) $N_z = 1$. The closed-loop system as shown in Fig. 3.2-17 has one RHP pole for $K_p < -1$, and it is hence unstable. Furthermore, for $K_p = -1$ the closed-loop system has a $j\omega$-axis pole at $\omega = 0$, i.e., at the origin. For $-1 < K_p < 0$, the $-1+j0$ point is not encircled and the system is hence closed-loop stable.

We have just seen that the system is closed-loop stable for all $K_p > 0$; therefore we conclude that the system of Fig. 3.2-17 is stable for all $K_p > -1$.

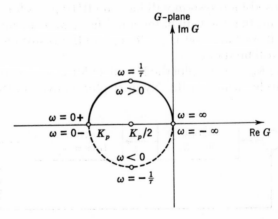

Figure 3.2-19   Extended Nyquist plot for $G(s) = K_p/(1+\tau s)$, $K_p < 0, \tau > 0$.

Example 3.2-6

As a little more complex example, one which involves poles on the $j\omega$-axis, let us consider now

$$G(s) = \frac{K_a(1+sT_4)}{s^2(1+sT_1)(1+sT_2)(1+sT_3)}$$

where we assume that

$$0 < \frac{1}{T_4} < \frac{1}{T_1} < \frac{1}{T_2} < \frac{1}{T_3}$$

We thus see that this system has a double pole at the origin, a zero at $-1/T_4$, and three poles more at $-1/T_1$, $-1/T_2$, $-1/T_3$ in order of increasing magnitude. We now assume that

$$\frac{1}{T_4} \ll \frac{1}{T_1} \quad \text{and} \quad \frac{1}{T_2} \ll \frac{1}{T_3}$$

By direct substitution we find that

$$G(j\omega)\Big|_{\omega = \frac{1}{\sqrt{T_1 T_2}}} \cong \frac{-K_a T_1 T_2 T_4}{T_1 + T_2}$$

Now, $G(s)$ has two poles at the origin and four excess poles, hence

$$\angle G(j\omega) \underset{\omega \to 0}{\to} -180°$$

$$\angle G(j\omega) \underset{\omega \to \infty}{\to} -360°$$

With this information we may conclude that the Nyquist plot is of the general configuration shown in Fig. 3.2-20b. The Nyquist contour is shown in Fig. 3.2-20a. The solid-line portion of the Nyquist plot shows the Nyquist plot for $\omega > 0$, the dashed curve shows the extension for $\omega < 0$. Since $G(s)$ has a pole on the $j\omega$-axis, extending the Nyquist plot to $\omega < 0$ does not result in a closed curve in the G-plane. We have a break at the point the pole occurs, the origin in this case. We get around this by going around the $j\omega$-axis pole along an infinitesimal semicircle into the RHP as shown on the Nyquist contour of Fig. 3.2-20a. The result is that in the G-plane the Nyquist plot for s along this semicircle goes along an arc of a circle of infinite radius. The arc moves in the clockwise direction as we go in the direction of increasing ω. This is indicated in Fig. 3.2-9a. The result is the circular arc shown in Fig. 3.2-20b, which goes as indicated by the arrowheads, as ω goes from $0-$ to $0+$.

The Nyquist plot is hence complete and we are now in a position to

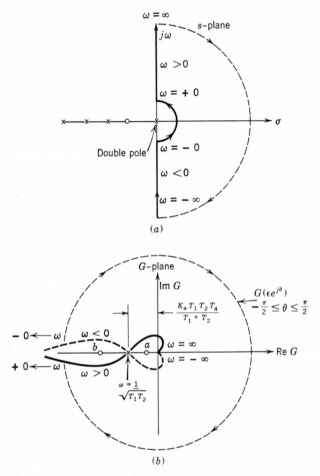

Figure 3.2-20 (a) Nyquist contour. (b) Extended Nyquist plot for
$G(j\omega) = K_a(1 + j\omega T_4)/(j\omega)^2(1 + j\omega T_1)(1 + j\omega T_2)(1 + j\omega T_3)$.

determine closed-loop stability by the Nyquist criterion. Two
different cases now appear:

$$1. \quad \frac{K_a T_1 T_2 T_4}{T_1 + T_2} < 1$$

$$2. \quad \frac{K_a T_1 T_2 T_4}{T_1 + T_2} > 1$$

Looking at case 1, let us assume that the scale of the Nyquist plot of
Fig. 3.2-20b is such that the $-1 + j0$ point appears at point b on the

negative real axis. Immediately it is seen that this point is not encircled, hence we have $N = 0$. We know

$$N = N_p - N_z$$

and that $G(s)$ has no RHP poles. Hence $1 + G(s)$ has no RHP poles and we conclude $N_z = 0$, $1 + G(s)$ has no RHP zeros, the closed-loop system hence has no RHP poles, and it is therefore stable. However, now consider case 2, $(K_a T_1 T_2 T_4)/(T_1 + T_2) > 1$. Here we can assume for the sake of argument that the scale in Fig. 3.2-20b is such that the $-1 + j0$ point appears at point a. This time as we go from $\omega = -\infty$ to $\omega = +\infty$ we see that point a is encircled twice in the clockwise direction, thus $N = 2$. As before, $G(s)$ has no poles in the RHP, so $N_p = 0$. Hence $N_z = 2$, i.e., $1 + G(s)$ has two RHP zeros and the closed-loop system will have two RHP poles, i.e., it is unstable.

The only remaining possibility for this example is

$$\frac{K_a T_1 T_2 T_4}{T_1 + T_2} = 1$$

Here the Nyquist plot goes through the $-1 + j0$ point twice. The result is that the system has two closed-loop poles on the $j\omega$-axis at $\omega = \pm 1/\sqrt{T_1 T_2}$. Thus the closed-loop system will oscillate at this frequency if it is disturbed.

We see from these two examples exactly what type of information can be extracted from the Nyquist plot of a system using the Nyquist criterion. This plot provides solid conditions which the parameters of the system must satisfy if the system is to be closed-loop stable. The Nyquist criterion can also be used in case of multiple-loop control systems.

Gain Margin and Phase Margin in the G-Plane

Let us now consider the practical, important case of $G(s)$ which has no RHP poles, allowing, however, the existence of $j\omega$-axis poles. In this case $1 + G(s)$ also has no RHP poles, so that $N_p = 0$ in (3.2-25). We have seen that if the system is to be closed-loop stable, then $G(j\omega)$ must not encircle or pass through the $-1 + j0$ point on the G-plane. This means $N = 0$ in (3.2-21). Again in the practical case

$$\lim_{s \to \infty} G(s) = 0$$

or $G(s)$ in general has more poles than zeros. Thus if the system is to be closed-loop stable, as $\omega \to \infty$, the Nyquist plot approaches the origin in such a way that the $-1+j0$ point is not encircled. Gain margin and phase margin are terms which have been defined to measure how close the Nyquist plot comes to encircling the $-1+j0$ point as it enters the unit circle and crosses the negative real axis.

Figure 3.2-21 shows what might be a typical Nyquist plot. Note that gain margin and phase margin are not related to each other but are independent performance measures. From the figure we see that the $-1+j0$ point is not encircled and hence this system will be closed-loop stable if $G(s)$ has no RHP poles. The gain and phase margins for this system are defined by

$$\text{Gain Margin} \triangleq \frac{1}{a}$$

$$\text{Phase Margin} \triangleq \phi$$

(3.2-22)

Phi (ϕ) is measured positive down from the negative real axis, i.e., it is positive for stable, closed-loop systems. These definitions may become clear if we consider the Nyquist plot of $(1/a)G(j\omega)$; we see it is of the same shape as the plot of $G(j\omega)$ except that magnitudes will be multiplied by the factor $1/a$. In this figure, a is less than 1, so $1/a$ is greater than 1. The effect will be to expand the Nyquist plot in such a way that $(1/a)G(j\omega)$ will just pass through the $-1+j0$ point. This is the dashed-line case shown in Fig. 3.2-21. The corresponding system will be closed-loop unstable. Thus adding an amplifier of gain $1/a$ in

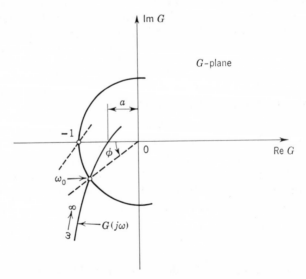

Figure 3.2-21   Gain and phase margins from the Nyquist plot.

series with $G(s)$ (we say we add the gain $1/a$ to the loop gain) will result in an unstable closed-loop system. Any added positive gain less than $1/a$ gives a stable closed-loop system. Hence $1/a$ is our *gain margin*.

Just as adding a gain may affect the Nyquist plot to make it pass through the $-1 + j0$ point, something that does not affect $|G(j\omega)|$ but adds phase shift of an amount ϕ will also result in a Nyquist plot that passes through the $-1 + j0$ point and hence an unstable closed-loop system. Any phase shift of lesser magnitude will result in a Nyquist plot that does not encircle the $-1 + j0$ point, the system will be closed-loop stable, and hence ϕ is the *phase margin*.

The gain and phase margins have long been used to measure the relative stability of a closed-loop system. A large gain and large phase margin indicate a very stable closed-loop system (usually also a very sluggish one). In turn, a gain margin slightly larger than 1 and a small positive phase margin will mean a closed-loop system which is relatively near to being an unstable system. A long settling time and a highly oscillatory step response may be expected. As a point of reference, a gain margin of approximately 3 and a phase margin of 30–35° will result in a reasonable degree of closed-loop stability. Any gain and phase margin much greater than this will in general indicate a sluggish system. For gain and phase margins much less than this the system may be expected to be nearing instability. It should also be mentioned that both gain and phase margins can be unreliable under some conditions.

The Nyquist Criterion as Applied to Nonunity Feedback Systems

The Nyquist criterion as it has been developed is for the unity-feedback system as shown in Fig. 3.2-10. Let us now consider the case where the feedback is not unity but some function $H(s)$ as shown in Fig. 3.2-22. For this system, we have

$$\frac{C(s)}{R(s)} = \frac{G(s)}{1 + G(s)H(s)} \qquad (3.2\text{-}23)$$

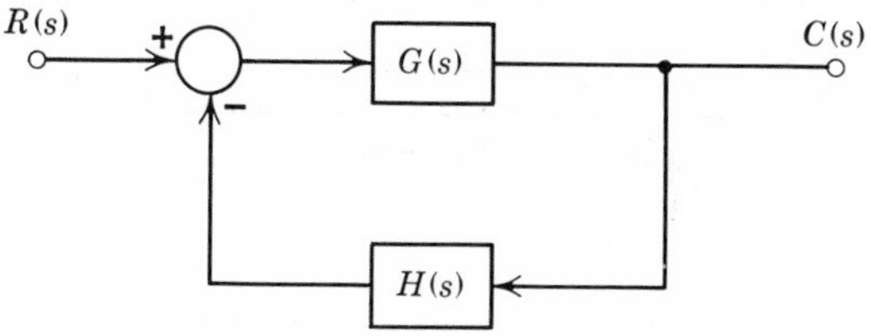

Figure 3.2-22 Nonunity-feedback system.

Thus now the closed-loop poles of the system will be given by the zeros of $1 + G(s)H(s)$. To apply the Nyquist criterion to this system (Fig. 3.2-22), we must make a Nyquist plot of $G(j\omega)H(j\omega)$ and count the encirclements of the $-1 + j0$ point in the GH-plane to determine the number of RHP closed-loop poles and hence closed-loop stability. In general, the nonunity feedback case is handled exactly in the same manner as the unity feedback system, only $G(s)H(s)$ [correspondingly $G(j\omega)H(j\omega)$] replaces $G(s)$ [correspondingly $G(j\omega)$].

3.3 THE STEADY-STATE ERROR CONSTANTS: K_p, K_v, AND K_a

The Error Constants as Functions of the Open-Loop Transfer Function

We saw in Chapter 2 how the overall response of a control system may be judged by its transient response and its steady-state error response. Recall that the steady-state error is defined as the difference between input and output as $t \to \infty$. In Chapter 2, the transient response was discussed at length, and a discussion of what constitutes a good transient response was initiated. Now we consider steady-state error characteristics.

The system we consider here is the unity-feedback, linear time-invariant system as shown in Fig. 3.3-1. The input is $R(s)$, the output $C(s)$, and the difference between input and output is $E(s)$. Here we discuss steady-state error characteristics for various inputs as a function of $G(s)$. From Fig. 3.3-1 we see that

$$\frac{C(s)}{R(s)} = \frac{G(s)}{1 + G(s)} \tag{3.3-1}$$

and

$$\frac{E(s)}{R(s)} = \frac{1}{1 + G(s)} \tag{3.3-2}$$

since $C(s) = E(s)G(s)$. Note from (3.3-2) that if the steady-state error is to approach a fixed value, then the closed-loop system of Fig. 3.3-1 must be stable. This implies that $1 + G(s)$ must have its zeros in the

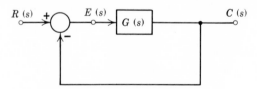

Figure 3.3-1   Unity-feedback system.

LHP. This is an overriding consideration for the entire discussion of steady-state error and hence closed-loop stability will be assumed in this discussion.

From (3.3-2):

$$E(s) = \frac{R(s)}{1+G(s)}$$

The steady-state error (e_{ss}) may now be found by use of the final-value theorem as follows:

$$e_{ss} = \lim_{t \to \infty} [e(t)] = \lim_{s \to 0} sE(s) = \lim_{s \to 0} \frac{sR(s)}{1+G(s)} \qquad (3.3\text{-}3)$$

We recall that for the final-value theorem to give a finite constant value e_{ss}, $sE(s)$ must have no poles on the $j\omega$-axis or in the RHP. Equation (3.3-3) may now be used to determine steady-state error. Let us consider first the steady-state error to a step input.

THE POSITION ERROR CONSTANT K_p

For a unit step input,

$$R(s) = \frac{1}{s}$$

Applying this in (3.3-3),

$$e_{ss}(\text{step}) = \lim_{s \to 0} \frac{s(1/s)}{1+G(s)} = \frac{1}{1 + \lim_{s \to 0} G(s)} \qquad (3.3\text{-}4)$$

The limit quantity in the denominator of (3.3-4) is called the *position error constant K_p*. It is defined as

$$K_p \triangleq \lim_{s \to 0} G(s) \qquad (3.3\text{-}5)$$

Using (3.3-5) gives the simple formula

$$e_{ss}(\text{step}) = \frac{1}{1+K_p} \qquad (3.3\text{-}6)$$

We thus see that the steady-state error response to a step input depends upon the value of $G(s)$ evaluated at the origin. We note two cases.

1. No pole at the origin, i.e., a type 0 system.[5] Here $K_p = G(0) =$ finite constant and hence $e_{ss}(\text{step}) \neq 0$.

---

[5] A system with N open-loop poles at the origin ($s = 0$) is termed a "type N" system.

2. A pole at the origin, i.e., a type 1 system hence $\lim\limits_{s \to 0} G(s) \to \infty$. In this case we have $K_p \to \infty$ and using this in (3.3-6) we see that $e_{ss}(\text{step}) = 0$.

Thus a system which has a forward-path pole at the origin will track a unit step with zero steady-state error if it is closed-loop stable.

THE VELOCITY ERROR CONSTANT K_v

Now let us look at the steady-state error to a ramp input, i.e., e_{ss} for $R(s) = 1/s^2$. Applying this in (3.3-3),

$$e_{ss}(\text{ramp}) = \lim_{s \to 0} \frac{s(1/s^2)}{1 + G(s)} = \frac{1}{\lim\limits_{s \to 0} sG(s)} \qquad (3.3\text{-}7)$$

The limit quantity in the denominator is called the *velocity error constant K_v* and is defined by

$$K_v \overset{\Delta}{=} \lim_{s \to 0} sG(s) \qquad (3.3\text{-}8)$$

Using (3.3-8) in (3.3-7) gives a simple formula:

$$e_{ss}(\text{ramp}) = \frac{1}{K_v} \qquad (3.3\text{-}9)$$

We note three cases:

1. $G(s)$ has no pole at the origin, i.e., a type 0 system. From (3.3-8) we have $K_v = 0$ and from (3.3-9) it is seen that $e_{ss}(\text{ramp}) \to \infty$. The system will not track a ramp with finite error.

2. $G(s)$ has a single pole at the origin, i.e., a type 1 system. In this case from (3.3-8) we see that K_v is finite and thus so is $e_{ss}(\text{ramp})$. We note from (3.3-9) that the larger K_v, the smaller $e_{ss}(\text{ramp})$.

3. $G(s)$ has a multiple pole at the origin, i.e., a type 2 or higher system. In this case we see from (3.3-8) that $K_v \to \infty$. Hence $e_{ss}(\text{ramp}) = 0$. Thus in the steady-state a system with a multiple pole at the origin will, in closed-loop, track a ramp with zero error.

Again we see the importance of the character of $G(s)$ at the origin in determining steady-state error performance. Finally let us consider the acceleration error constant.

THE ACCELERATION ERROR CONSTANT K_a

Consider the steady-state error to a unit parabolic input (termed a unit acceleration input). A unit parabolic input is the integral w.r.t.

time of the unit ramp function and is given by $r(t) = (t^2/2)1(t)$; transformed, which becomes

$$R(s) = \frac{1}{s^3}$$

Using this in (3.3-3) gives

$$e_{ss}(\text{accel}) = \lim_{s \to 0} \frac{s(1/s^3)}{1 + G(s)} = \frac{1}{\lim_{s \to 0} s^2 G(s)} \qquad (3.3\text{-}10)$$

Now we define the *acceleration error constant* K_a by

$$K_a \overset{\Delta}{=} \lim_{s \to 0} s^2 G(s) \qquad (3.3\text{-}11)$$

Using this in (3.3-10) gives

$$e_{ss}(\text{accel}) = \frac{1}{K_a} \qquad (3.3\text{-}12)$$

which is a simple expression for steady-state error. Again we distinguish three cases:

1. $G(s)$ has at most one pole at the origin, i.e., a type 0 or type 1 system. In this case $K_a = 0$, and from (3.3-12) we get that steady-state error to a unit parabolic input is infinite.

2. $G(s)$ has a double pole at the origin, i.e., a type 2 system. From (3.3-11) we have that $K_a =$ finite, nonzero constant. Then $e_{ss}(\text{accel})$ is at the same time a finite constant. We see from (3.3-12) that the larger K_a, the smaller $e_{ss}(\text{accel})$.

3. $G(s)$ has a pole of order three or higher at the origin, i.e., a type 3 or higher system. From (3.3-11) we see that $K_a \to \infty$. From (3.3-12), $e_{ss}(\text{accel}) = 0$, thus in the steady-state this type of system will track a parabolic input with zero error, again provided that the system is closed-loop stable. However, stability may be a large order for an open-loop transfer function that has a triple pole on the $j\omega$-axis.

These three types of response to a unit parabolic input are illustrated in Fig. 3.3-2.

Table 3.3-1 summarizes the results of this section. The table may be extended to higher-order inputs and higher numbers of open-loop poles at the origin by defining higher-order error constants. Systems with more than two poles at the origin do not occur often in practice, so this would not be too useful. Let us emphasize again that the table applies only when the system is closed-loop stable.

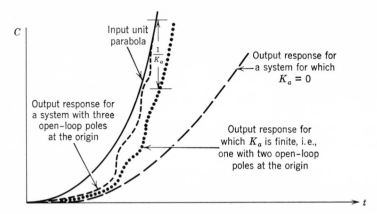

Figure 3.3-2   Three types of response to a unit parabola input.

Table 3.3-1   Steady-State Error for Various Inputs and Systems

	Steady-State Error			
Type of input	Type 0 System (no open-loop poles at $s = 0$)	Type 1 System (one open-loop pole at $s = 0$)	Type 2 System (two open-loop poles at $s = 0$)	More than Two Open-Loop Poles at $s = 0$
Unit step	$\dfrac{1}{1+K_p}$	0	0	0
Unit ramp	∞	$\dfrac{1}{K_v}$	0	0
Unit parabola	∞	∞	$\dfrac{1}{K_a}$	0
	$K_p = G(0)$	$K_v = \lim_{s \to 0} sG(s)$	$K_a = \lim_{s \to 0} s^2 G(s)$	

The Error Constants as Functions of the Closed-Loop Transfer Function

We consider a closed-loop system as in Fig. 2.2-8 in which the input is $R(s)$, the output is $C(s)$, and the transfer function relating the two is of the form

$$T(s) = \frac{C(s)}{R(s)} = \frac{K_c \prod_{i=1}^{m} (s - z_i)}{\prod_{i=1}^{n} (s - p_i)}, \qquad m < n \qquad (3.3\text{-}13)$$

or

$$T(s) = \frac{C(s)}{R(s)} = \frac{a_m s^m + a_{m-1} s^{m-1} + \cdots a_1 s + a_0}{b_n s^n + b_{n-1} s^{n-1} + \cdots + b_1 s + b_0} \qquad (3.3\text{-}14)$$

We define the error for this closed-loop system as the difference between the input and the output, i.e., if $E(s)$ is the error, then

$$E(s) \overset{\Delta}{=} R(s) - C(s)$$

$$= R(s)\left[1 - \frac{C(s)}{R(s)}\right]$$

Since we are interested here in the steady-state error (e_{ss}), we use the final-value theorem, if the system is closed-loop stable, to get

$$e_{ss} = \lim_{s \to 0}\left\{sR(s)\left[1 - \frac{C(s)}{R(s)}\right]\right\} \tag{3.3-15}$$

which may now be used in conjunction with (3.3-13) and (3.3-14) to find the steady-state error to a unit step [e_{ss}(step)], ramp [e_{ss}(ramp)], or parabolic input [e_{ss}(accel)]. Having these values, we can readily determine the steady-state error constants K_p, K_v, and K_a in terms of the z_i and p_i or the a_i and b_i, i.e., the closed-loop transfer function parameters.

THE POSITION ERROR CONSTANT K_p

To determine K_p we first consider steady-state error for a step input [e_{ss}(step)], i.e., for $R(s) = 1/s$. Using $R(s) = 1/s$ in (3.3-15),

$$e_{ss}(\text{step}) = 1 - \frac{C(0)}{R(0)} \tag{3.3-16}$$

Now using $C(0)/R(0)$ from (3.3-13),

$$e_{ss}(\text{step}) = 1 - \frac{K_c \prod_{i=1}^{m} (-z_i)}{\prod_{i=1}^{n} (-p_i)}.$$

Using this value of e_{ss}(step) in (3.3-6),

$$K_p = \frac{K_c \prod_{i=1}^{m} (-z_i)}{\prod_{i=1}^{n} (-p_i) - K_c \prod_{i=1}^{m} (-z_i)} \tag{3.3-17}$$

Similarly using $C(0)/R(0)$ from (3.3-14) in (3.3-16) and the result in (3.3-6) gives

$$K_p = \frac{a_0}{b_0 - a_0} \tag{3.3-18}$$

Equations (3.3-17) and (3.3-18) are the desired result, the position error constant in terms of closed-loop parameters.

THE VELOCITY ERROR CONSTANT K_v

To determine K_v we consider the steady-state error for a unit ramp input, i.e., for $R(s) = 1/s^2$. Using this $R(s)$ in (3.3-15), we have

$$e_{ss}(\text{ramp}) = \lim_{s \to 0} \left[\frac{1 - C(s)/R(s)}{s} \right] \tag{3.3-19}$$

$$= \frac{1}{K_v}$$

using (3.3-9). From (3.3-19) we see that for a finite $e_{ss}(\text{ramp})$ and a nonzero K_v we must have

$$\frac{C(0)}{R(0)} = 1 \tag{3.3-20}$$

This is the case of interest here, so let us assume (3.3-20), which implies that $K_p \to \infty$. Equation (3.3-20) and L'Hospital's rule used on (3.3-19) give

$$\frac{1}{K_v} = -\lim_{s \to 0} \frac{d}{ds} \left[\frac{C(s)}{R(s)} \right] \tag{3.3-21}$$

This may be written

$$\frac{1}{K_v} = -\lim_{s \to 0} \frac{(d/ds)\,[C(s)/R(s)]}{C(s)/R(s)} = -\lim_{s \to 0} \frac{d}{ds} \log_e \frac{C(s)}{R(s)} \tag{3.3-22}$$

Using $C(s)/R(s)$ as given in (3.3-13), we have

$$\log_e \frac{C(s)}{R(s)} = \log_e K_c + \sum_{i=1}^{m} \log_e (s - z_i) - \sum_{i=1}^{n} \log_e (s - p_i) \tag{3.3-23}$$

Putting (3.3-23) in (3.3-22) and evaluating gives

$$\frac{1}{K_v} = \sum_{i=1}^{m} \frac{1}{z_i} - \sum_{i=1}^{n} \frac{1}{p_i} \tag{3.3-24}$$

Now using $C(s)/R(s)$ as given in (3.3-14) in (3.3-21) and evaluating yields

$$\frac{1}{K_v} = \frac{a_1 b_0 - a_0 b_1}{-b_0^2}$$

Since we already have assumed (3.3-20), we have $a_0 = b_0$ and hence the preceding result simplifies to

$$\frac{1}{K_v} = \frac{a_1 - b_1}{-b_0}$$

or

$$K_v = \frac{-b_0}{a_1 - b_1} \tag{3.3-25}$$

Equations (3.3-24) and (3.3-25) are the desired results.

THE ACCELERATION ERROR CONSTANT K_a

To evaluate the acceleration error constant K_a we consider steady-state error with a unit parabolic input, i.e., for $R(s) = 1/s^3$. Using this $R(s)$ in (3.3-15) gives

$$e_{ss}(\text{accel}) = \lim_{s \to 0} \frac{[1 - C(s)/R(s)]}{s^2} \tag{3.3-26}$$

$$= \frac{1}{K_a}$$

using (3.3-12). We now see that if $e_{ss}(\text{accel})$ is to be finite, then $[1 - C(s)/R(s)]$ must have at least a double zero at $s = 0$. Lacking this, $e_{ss}(\text{accel}) \to \infty$ and $K_a = 0$. We are hence interested in the case where $[1 - C(s)/R(s)]$ does have a double zero at $s = 0$. This implies that

$$\frac{C(0)}{R(0)} = 1 \tag{3.3-27}$$

and

$$\left. \frac{d}{ds} \frac{C(s)}{R(s)} \right|_{s=0} = 0 \tag{3.3-28}$$

Now under the assumption[6] of a double zero of $[1 - C(s)/R(s)]$ at $s = 0$, we may apply L'Hospital's rule to (3.3-26) to get

$$\frac{1}{K_a} = -\frac{1}{2} \lim_{s \to 0} \left[\frac{d^2}{ds^2} \frac{C(s)}{R(s)} \right] \tag{3.3-29}$$

Using (3.3-27) in (3.3-29),

$$\frac{1}{K_a} = -\frac{1}{2} \lim_{s \to 0} \left\{ \frac{(d^2/ds^2)[C(s)/R(s)]}{C(s)/R(s)} \right\} \tag{3.3-30}$$

We note that

$$\frac{d^2}{ds^2} \left[\log_e \frac{C(s)}{R(s)} \right] = \frac{(d^2/ds^2)[C(s)/R(s)]}{C(s)/R(s)} - \left\{ \frac{(d/ds)[C(s)/R(s)]}{C(s)/R(s)} \right\}^2 \tag{3.3-31}$$

[6]This assumption also means $K_p \to \infty$, $K_v \to \infty$ and hence $e_{ss}(\text{step}) = e_{ss}(\text{ramp}) = 0$.

Substituting (3.3-31) in (3.3-30) gives

$$\frac{1}{K_a} = -\frac{1}{2}\lim_{s\to 0}\left\{\frac{d^2}{ds^2}\left[\log_e\frac{C(s)}{R(s)}\right] + \left[\frac{(d/ds)[C(s)/R(s)]}{C(s)/R(s)}\right]^2\right\} \qquad (3.3\text{-}32)$$

We now may use (3.3-13) and (3.3-28) in (3.3-32) to get

$$\frac{1}{K_a} = \frac{1}{2}\left(\sum_{i=1}^{m}\frac{1}{z_i^2} - \sum_{i=1}^{n}\frac{1}{p_i^2}\right) \qquad (3.3\text{-}33)$$

Similarly, using $C(s)/R(s)$ of (3.3-14) in (3.3-29) gives

$$\frac{1}{K_a} = \frac{a_1 b_1 b_0 + a_0 b_2 b_0 - a_2 b_0^2 - a_0 b_1^2}{b_0^3} \qquad (3.3\text{-}34)$$

From (3.3-27) and (3.3-28) we may determine that

$$a_0 = b_0$$
$$a_1 = b_1$$

This result in (3.3-34) gives

$$K_a = \frac{b_0}{b_2 - a_2} \qquad (3.3\text{-}35)$$

Equations (3.3-33) and (3.3-34) are the desired results.
  The results of this section are summarized in Tables 3.3-2 and 3.3-3.

Table 3.3-2   The Steady-State Error Constants as Functions of the Closed-Loop Pole-Zero Configuration

For a closed-loop system with transfer function $\quad T(s) = \dfrac{C(s)}{R(s)} = \dfrac{K_c \prod\limits_{i=1}^{m}(s - z_i)}{\prod\limits_{i=1}^{n}(s - p_i)}$

System Type	Restrictions		K_p	K_v	K_a
0	$\mathrm{Re}\,p_i < 0,\ i = 1, 2, \ldots, n$	None	$\dfrac{K_c \prod\limits_{i=1}^{m}(-z_i)}{\prod\limits_{i=1}^{n}(-p_i) - K_c \prod\limits_{i=1}^{m}(-z_i)}$	0	0
1		$T(0) = \dfrac{C(0)}{R(0)} = 1$	∞	$\dfrac{1}{\sum\limits_{i=1}^{m}\dfrac{1}{z_i} - \sum\limits_{i=1}^{n}\dfrac{1}{p_i}}$	0
2		$T(0) = \dfrac{C(0)}{R(0)} = 1$ $T'(0) = \dfrac{d}{ds}\dfrac{C(s)}{R(s)}\bigg|_{s=0} = 0$	∞	∞	$\dfrac{2}{\sum\limits_{i=1}^{m}\dfrac{1}{z_i^2} - \sum\limits_{i=1}^{n}\dfrac{1}{p_i^2}}$

Table 3.3-3 The Steady-State Error Constants as Functions of the Coefficients of the Closed-Loop Transfer Function

For a closed-loop system with transfer function

$$T(s) = \frac{C(s)}{R(s)} = \frac{a_m s^m + a_{m-1} s^{m-1} + \cdots + a_1 s + a_0}{b_n s^n + b_{n-1} s^{n-1} + \cdots + b_1 s + b_0}$$

System Type	Restrictions	K_p	K_v	K_a
0	All poles of $T(s)$ in LHP — None	$\dfrac{a_0}{b_0 - a_0}$	0	0
1	$a_0 = b_0$	∞	$\dfrac{b_0}{b_1 - a_1}$	0
2	$a_0 = b_0$ $a_1 = b_1$	∞	∞	$\dfrac{b_0}{b_2 - a_2}$

Example 3.3-1

The open-loop transfer function of a unity-feedback system is given by

$$G(s) = \frac{4(s+1)}{s^2(s+3)}$$

Determine the error constants.

We first note that the system is of type 2. Therefore $K_p = K_v = \infty$. Also,

$$K_a = \lim_{s \to 0} s^2 G(s) = \frac{4}{3}$$

These results imply that there will be zero steady-state error for step and ramp inputs but a steady-state error of 0.75 for a unit acceleration input.

Example 3.3-2

A second-order system has overall transfer function

$$T(s) = \frac{\omega_n^2}{s^2 + 2\zeta\omega_n s + \omega_n^2}$$

Find the velocity error constant K_v.

For this case $T(s)$ is known as ratio of polynomials with $b_0 = a_0 = \omega_n^2$, $a_1 = 0$, $b_1 = 2\zeta\omega_n$, and $b_2 = 1$. Using Table 3.3-3, we immediately get $K_v = \zeta/2\omega_n$.

3.4 FREQUENCY-RESPONSE METHODS

In Section 3.2 we defined the frequency response of a system as the transfer function of the system evaluated on the $j\omega$-axis for positive ω. In that section the frequency response of the open-loop system was used to determine closed-loop stability by the Nyquist criterion. Though the determination of closed-loop stability is an important application of the frequency response, its real usefulness goes far beyond this. It is also a very useful tool in designing compensators for systems. Moreover, given an actual physical system, its frequency response can be measured quite easily.

To see how this is done consider a simple system with transfer function $G(s)$, input $U(s)$, and output $C(s)$. We know that

$$C(s) = G(s)U(s) \tag{3.4-1}$$

Let us assume first that our system is a stable one, so that all the poles of $G(s)$ are in the LHP. Second, let us assume that the input is a sine wave of unit amplitude and frequency ω. For this case

$$U(s) = \frac{\omega}{s^2 + \omega^2}$$

and the output is

$$C(s) = \frac{\omega}{s^2 + \omega^2}G(s)$$

We expand $C(s)$ into a partial fraction expansion as

$$C(s) = \frac{G(j\omega)}{2j(s - j\omega)} - \frac{G(-j\omega)}{2j(s + j\omega)} + \left[\text{terms of the form } \frac{K_{p_i}}{(s - p_i)^m} \right] \tag{3.4-2}$$

where p_i is a typical pole of $G(s)$.
Using the inverse $\mathcal{L}$-transform on (3.4-2) gives

$$c(t) = \frac{G(j\omega)}{2j}e^{j\omega t} - \frac{G(-j\omega)}{2j}e^{-j\omega t} + \left[\begin{array}{c} \text{terms of the form} \\ K_{p_i}\dfrac{t^{m-1}}{(m-1)!}e^{p_i t} \end{array} \right], t > 0$$

Now consider $c(t)$ as $t \to \infty$. We see because of the stability of $G(s)$, $\operatorname{Re} p_i < 0$ and the term in brackets will go to zero. Therefore

$$c(t) \underset{t\to\infty}{\to} \frac{G(j\omega)}{2j}e^{j\omega t} - \frac{G(-j\omega)}{2j}e^{-j\omega t}$$

Using

$$G(j\omega) = |G(j\omega)|e^{j\angle G(j\omega)}$$

$$G(-j\omega) = |G(j\omega)|e^{-j\angle G(j\omega)}$$

gives immediately

$$c(t) \underset{t\to\infty}{\to} |G(j\omega)| \sin[\omega t + \angle G(j\omega)] \qquad (3.4\text{-}3)$$

From (3.4-3) we see that if a sine wave of unit amplitude is impressed upon the input of a stable, linear time-invariant system, after the transient component due to the system's poles dies out the output again will be a sine wave whose amplitude is now $|G(j\omega)|$ and whose phase (relative to the input sine wave) is $\angle G(j\omega)$. Thus to measure the frequency response of a system we apply a variable frequency oscillator at the input to the system and then sweep the variable frequency over the range of desired ω.

$|G(j\omega)|$ and $\angle G(j\omega)$ may then be recorded at as many points as required to give a plot to the accuracy desired. Equipment is commercially available for recording the case where the input and output can be obtained in the form of electrical signals. Recording $|G(j\omega)|$ and $\angle G(j\omega)$ directly onto a G-plane (polar coordinate paper facilitates this) gives the Nyquist plot.

Many other ways have been devised for displaying the frequency response; each has a particular characteristic making it advantageous for some particular purpose. We have already seen the Nyquist plot which gave a closed-loop stability criterion. We consider two other presentations, Bode plots and Nichols charts. Bode plots have the advantage of being easily constructed for a given rational $G(s)$. Nichols charts have the additional property that given the open-loop frequency response, the closed-loop, unity-feedback frequency response can be read off directly. This feature is helpful for designing a system compensator.

In general, frequency-response methods are important because a system can be identified from its frequency response. It is easy to determine whether or not it will be closed-loop stable. And, finally, we will see that a system compensator may be designed using only the frequency-response information. Thus with frequency response one can identify, stabilize, and compensate linear time-invariant systems. This covers the scope of the control engineer's interest in this type of system.

Bode Plots

Bode plots are companion plots of $20 \log_{10}|G(j\omega)|$ and $\angle G(j\omega)$ versus $\log_{10}\omega$. These two parts are called the magnitude and phase plots, respectively.

Let us first consider a typical rational transfer function $G(s)$ factored in the following form:

$$G(s) = \frac{K_N(1+sT_a)(1+sT_b)\cdots}{s^N(1+sT_1)(1+sT_2)\cdots[1+(2\zeta/\omega_n)s+(s^2/\omega_n^2)]}. \qquad (3.4\text{-}4)$$

This transfer function now has zeros at $s = -1/T_a, -1/T_b, \ldots$; N poles at the origin; real poles at $s = -1/T_1, -1/T_2, \ldots$; and complex poles at $s = -\zeta\omega_n \pm j\omega_n\sqrt{1-\zeta^2}, \ldots$. We note that the constant multiplier K_N is given by

$$K_N = \lim_{s\to 0} s^N G(s) \qquad (3.4\text{-}5)$$

where N is the numbers of poles of $G(s)$ at the origin, i.e. N is the type number for this system. Thus for a type 0 system, $K_p = K_0$. For a type 1 system, $K_v = K_1$, and for a type 2 system $K_a = K_2$. Any rational $G(s)$ can be factored in the form of (3.4-4). If $G(s)$ has complex zeros, then even though it is not shown in (3.4-4), a quadratic term of the form $[1+(2\zeta/\omega_n)s+s^2/\omega_n^2]$ will also appear in the numerator. As the first step in constructing the Bode plot for $G(s)$, we consider

$$20 \log_{10}|G(j\omega)| = 20 \left\{ \begin{aligned} &\log_{10} K_N + \log_{10}|1+j\omega T_a| + \log_{10}|1+j\omega T_b|\cdots \\ &-N \log_{10}|\omega| - \log_{10}|1+j\omega T_1| - \log_{10}|1+j\omega T_2|\cdots \\ &-\log_{10}\left|1+j\frac{2\zeta\omega}{\omega_n}-\left(\frac{\omega}{\omega_n}\right)^2\right| \end{aligned} \right\}$$

$$(3.4\text{-}6a)$$

and

$$\angle G(j\omega) = \tan^{-1}\omega T_a + \tan^{-1}\omega T_b + \cdots$$

$$- N(90°) - \tan^{-1}\omega T_1 - \tan^{-1}\omega T_2 \cdots \qquad (3.4\text{-}6b)$$

$$- \tan^{-1}\left(\frac{2\zeta\omega\omega_n}{\omega_n^2 - \omega^2}\right)\cdots$$

From (3.4-6) we see that both $20 \log_{10}|G(j\omega)|$ and $\angle G(j\omega)$ are the sum of four kinds of terms:

1. $20 \log_{10} K_N$ for which the corresponding angle contribution is zero.

2. $20N \log_{10} |\omega|$ for which the corresponding angle contribution is $N\,(90°)$.

3. $20 \log_{10} |1+j\omega T|$ for which the corresponding angle contribution is $\tan^{-1} \omega T$.

4. $20 \log_{10} \left|1+j\dfrac{2\zeta\omega}{\omega_n}-\left(\dfrac{\omega^2}{\omega_n{}^2}\right)\right|$ for which the corresponding angle contribution is $\tan^{-1}\left(\dfrac{2\zeta\omega\omega_n}{\omega_n{}^2-\omega^2}\right)$.

We next consider the plotting of each of these terms and then the construction of a Bode plot for a given $G(j\omega)$ as the sum of combinations of these terms.

TERMS OF THE FORM 20 LOG₁₀Kₙ AND 20N LOG₁₀ω AND THE CORRESPONDING ANGLE CONTRIBUTION

We consider these two terms together for we can see from (3.4-6a) that

$$20 \log_{10} |G(j\omega)| \underset{\omega\to 0}{\to} 20 \log_{10} K_N - 20 N \log_{10} \omega \qquad (3.4\text{-}7)$$

That is, as $\omega \to 0$, $20 \log_{10} |G(j\omega)|$ approaches an asymptote given by the function of ω on the RHS of (3.4-7). It is called the low-frequency asymptote. Let us look more closely at the function:

$$f(\log_{10} \omega) = 20 \log_{10} K_N - 20N \log_{10} \omega \qquad (3.4\text{-}8)$$

We see that as a function of $\log_{10} \omega$ it is just a straight line with slope $-20N$, which goes through $20 \log_{10} K_N$ where $\log_{10} \omega = 0$, or where $\omega = 1$. This is as shown in Fig. 3.4-1. Of special interest in the figure are the points where $f(\log_{10} \omega)$ of (3.4-8) has its roots, i.e., where

$$20 \log_{10} K_N - 20N \log_{10} \omega = 0$$

These are found directly to give

$$\omega^N = K_N \text{ or } \omega = (K_N)^{1/N} \qquad (3.4\text{-}9)$$

The horizontal axis is shown with ω plotted with a $\log_{10}$ scale, which is customary. The vertical scale is expressed in units of db (decibels). For instance,

$$20 \log_{10} K_N = 10 \text{ db}$$

implies

$$K_N = 10^{10/20} = 10^{1/2} \cong 3.16$$

It is common to speak of a gain for $|G(j\omega)|$ as being, say, x db. This means that $20 \log_{10} |G(j\omega)| = x$ db and hence $|G(j\omega)| = 10^{x/20}$ in

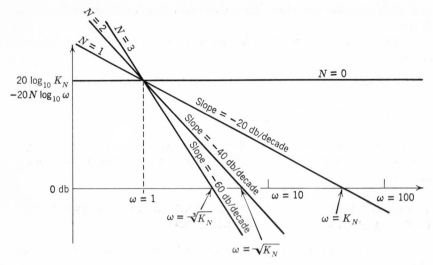

Figure 3.4-1 Low frequency asymptotes for $N = 0, 1, 2$, and 3.

whatever units are appropriate for $|G(j\omega)|$. The slopes indicated in Fig. 3.4-1 are in terms of db/decade. A decade is defined as a change in ω by a factor of 10. A factor of 10 change in ω implies a difference of 1 in $\log_{10} \omega$:

$$\omega_1 = 10\omega_2 \Rightarrow \log_{10} \omega_1 = 1 + \log_{10} \omega_2$$

We see from Fig. 3.4-1 that the contribution of a term of the form $20 N \log_{10} K_N - 20 \log_{10} \omega$ is just a straight line with a slope of $20 N$ db/decade in the plot of $20 \log_{10} |G(j\omega)|$ versus $\log_{10} \omega$. As $\omega \rightarrow 0$, $20 \log_{10} |G(j\omega)|$ approaches this line.

The angle contribution of a term involving constant K_N is obviously zero. Corresponding to $-20 N \log_{10} \omega$ it is $\angle (j\omega)^{-N}$ which is simply $-N(90°)$. This is easy to evaluate.

TERMS OF THE FORM $(1 + j\omega T)$

First we note the asymptotic behavior of $20 \log_{10} |1 + j\omega T|$. We have

$$20 \log_{10} (1 + j\omega T) \underset{\omega \rightarrow 0}{\rightarrow} 0 \qquad (3.4\text{-}10)$$

On the other hand, as ω increases to where $\omega T \gg 1$, we get

$$20 \log_{10} |1 + j\omega T| \rightarrow 20 \log_{10} \omega T = 20 \log_{10} \omega + 20 \log_{10} T \qquad (3.4\text{-}11)$$

$20 \log_{10} \omega + 20 \log_{10} T$ plotted as a function of $\log_{10} \omega$ is a straight line of slope 20 db/decade, which goes through the point $20 \log_{10} T$ where

$\omega = 1$. Looking for the root of this function, we have

$$20 \log_{10} \omega = - 20 \log_{10} T$$

or

$$\omega = \frac{1}{T}$$

Thus the high-frequency asymptote intercepts the low-frequency asymptote at $\omega = 1/T$. This frequency is termed the *corner frequency* (Fig. 3.4-2).

Also shown in Fig. 3.4-2 is the plot of $20 \log_{10} |1 + j\omega T|$, which is seen to be a smooth curve approaching the low-frequency and high-frequency asymptotes at the low and high frequencies, respectively. It differs from the asymptote by a maximum of 3 db at the corner frequency. Further points to note are that the curve differs from the asymptotes by 1 db at $\omega = 1/2T$ and $\omega = 2/T$ (an octave off the corner frequency in either direction). It should be noted that it is very easy to make a sketch of this curve. We simply draw in the asymptotes (straight lines of slope zero and 20 db/decade which intersect at the corner frequency $\omega = 1/T$). By putting in one point 3 db above the asymptotes at the corner frequency and two points 1 db above the asymptotes at frequencies 1 octave above and below the corner

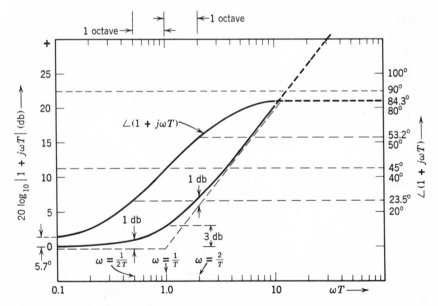

Figure 3.4-2   Magnitude and phase plot of the $1 + j\omega T$ term.

frequency and drawing a smooth curve through these three points which approaches the asymptotes at the low and high frequencies, we can draw a very accurate sketch.

Turning now to $\angle(1+j\omega T)$, we know that

$$\angle(1+j\omega T) = \tan^{-1}\omega T \qquad (3.4\text{-}12)$$

and so it is seen that

$$\angle(1+j\omega T) \underset{\omega \to 0}{\to} 0$$

and $\qquad\qquad\qquad\qquad\qquad\qquad\qquad\qquad\qquad\qquad (3.4\text{-}13)$

$$\angle(1+j\omega T) \underset{\omega \to \infty}{\to} 90°$$

This gives the low- and high-frequency asymptotes for the phase function. A third point of interest is the corner frequency. Putting $\omega = 1/T$ into the RHS of (3.4-12) gives

$$\angle(1+j\omega T) = 45° \qquad \text{for} \qquad \omega = 1/T \qquad (3.4\text{-}14)$$

The plot of $\angle(1+j\omega T)$ versus $\log_{10}\omega$ is seen in Fig. 3.4-2. We see that the phase curve is a smooth one which goes from 0 to 90° through 45° at the corner frequency. Plotting against a log scale has resulted in a curve that is symmetric about the corner frequency. This is helpful in the construction of such a curve. Also shown in the figure are the amounts that the curve differs from the asymptotes at frequencies an octave and a decade above and below the corner frequency. Again with these points and the asymptotes an accurate sketch can be quickly drawn.

The Bode plot of the term $(1+j\omega T)$ shown in Fig. 3.4-2 is the contribution of a simple zero at $-1/T$. In (3.4-6) each real zero will make a contribution of the form shown in Fig. 3.4-2. Each zero contribution (both in magnitude and in phase) will be the same, only the corner frequency will be different.

TERMS OF THE FORM $\dfrac{1}{1 + j\omega T}$

It is seen from (3.4-6) that the contribution of a real pole at $-1/T$ [a term of the form $1/(1+sT)$ in $G(s)$ of (3.4-4)] will make a contribution of exactly the same form as the contribution of a zero, but with the opposite sign. Thus to see the form of the plots due to a real pole term one simply reflects the magnitude and phase plot of Fig. 3.4-2 in the 0-db, 0-degree line (the horizontal axis) (Fig. 3.4-3). Here the magnitude plot goes down to $-\infty$ at a slope of -20 db/decade,

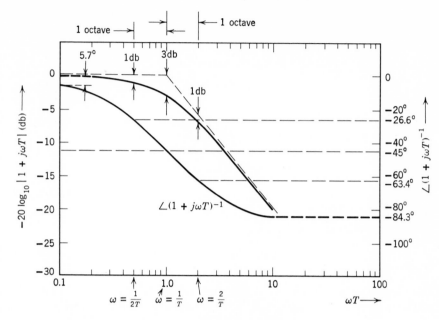

Figure 3.4-3   Magnitude and phase plots for the $(1 + j\omega T)^{-1}$ term.

being down 3 db at the corner frequency. The phase plot goes then from 0 to $-90°$ passing through $-45°$ at the corner frequency.

TERMS OF THE FORM $\dfrac{1}{1+j(2\zeta/\omega_n)\omega-\omega^2/\omega_n{}^2}$

This last term to be considered is slightly more complicated since besides being a function of ω as before it is also a function of the variable ζ (the damping factor). The range of ζ of interest is $0 \leqslant \zeta \leqslant 1$ and, as may be expected, the shape of the Bode plot both in magnitude and phase depends strongly upon what value of damping factor is being considered.

Let us first consider the magnitude part of the plot, i.e., the plot of

$$-20 \log_{10}\left|1+2\zeta\,\frac{j\omega}{\omega_n}-\frac{\omega^2}{\omega_n{}^2}\right| \qquad (3.4\text{-}15)$$

The asymptotes are given by

$$-20 \log_{10}\left|1+2\zeta\,\frac{j\omega}{\omega_n}-\frac{\omega^2}{\omega_n{}^2}\right| \underset{\omega\to0}{\longrightarrow} 0 \qquad (3.4\text{-}16)$$

and

$$-20\log_{10}\left|1+2\zeta\frac{j\omega}{\omega_n}-\frac{\omega^2}{\omega_n{}^2}\right| \underset{\omega\to\infty}{\to} -40\log_{10}\frac{\omega}{\omega_n}$$

$$\to -40\log_{10}\omega+40\log_{10}\omega_n$$

$$(3.4\text{-}17)$$

From this we see first that the asymptotic behavior of the magnitude plot does not depend upon ζ at either low or high frequency. From

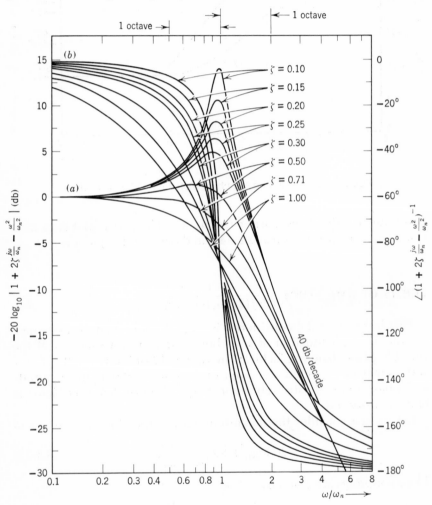

Figure 3.4-4 Magnitude and phase plots for a term of the form $[1+2\zeta(j\omega/\omega_n)-(\omega^2/\omega_n{}^2)]^{-1}$.

(3.4-17) we see that the low-frequency asymptote is just the 0 db line. Thus as $\omega \to 0$ the magnitude plot of this term approaches a horizontal line through 0 db. From the RHS of (3.4-17), the high-frequency asymptote is seen to be a straight line (recall that we are plotting $\log_{10}\omega$ on the horizontal axis) which goes through the 0 db line (the low-frequency asymptote) at $\omega = \omega_n$ and has a slope of -40 db/decade. Thus the corner frequency in this case is ω_n.

This asymptotic structure may be seen in Fig. 3.4-4 where plots for $-20 \log_{10}|1 + 2\zeta(j\omega/\omega_n) - \omega^2/\omega_n^2|$ are shown for various values of ζ in the range $0 < \zeta \leq 1$. The plots for these complex pole terms are essentially of the same form as the magnitude plots for real poles (see Fig. 3.4-3), except that for small values of ζ it is seen that the curves have a pronounced peak at a frequency of slightly less than ω_n. In fact, the smaller the value of ζ, the less the damping (see the figure), the higher the peak. We shall discuss this peak in some detail. Consider Fig. 3.4-5, which shows what might be a typical plot of $-20 \log_{10}|1 + 2\zeta(j\omega/\omega_n) - \omega^2/\omega_n^2|$. The peak here occurs at a frequency which is labeled ω_m and the height of the peak is M_m. This peak will occur at an ω where $|1 + 2\zeta(j\omega/\omega_n) - \omega^2/\omega_n^2|$ is at a minimum. The minimum can be found by simple differentiation and is given by

$$\omega_m = \omega_n\sqrt{1 - 2\zeta^2} \qquad (3.4\text{-}18)$$

The corresponding peak value M_m is then

$$M_m = -20 \log_{10}\left|1 + 2\zeta\frac{j\omega_m}{\omega_n} - \frac{\omega_m^2}{\omega_n^2}\right| = -20 \log_{10} 2\zeta\sqrt{1 - \zeta^2} \quad (3.4\text{-}19)$$

Note that as ζ goes to zero, the peak value goes off to infinity and the peak frequency approaches ω_n. This can be seen in Fig. 3.4-4. We see from (3.4-18) that for $\zeta = 1/\sqrt{2} \cong 0.707$, $\omega_m = 0$. Thus for $\zeta = 0.707$ the magnitude part of the Bode plot does not have a peak [$M_m = 0$ in (3.4-19) for this value of ζ]. In network parlance, this value of ζ gives

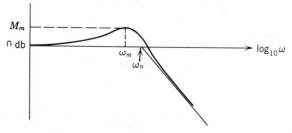

Figure 3.4-5 A typical magnitude frequency response of a term of the form $-20 \log_{10}|1 + 2\zeta(j\omega/\omega_n) - (\omega^2/\omega_n^2)|$.

a maximally flat frequency response. For $\zeta > 0.707$, the magnitude curves (see Fig. 3.4-4) can be seen to be increasingly rounded with increasing ζ.

Now let us look at $\angle \, [1 + 2\zeta \, (j\omega/\omega_n) - \omega^2/\omega_n^2]^{-1}$, the angle contribution corresponding to the magnitude term just considered. We have

$$ - \angle \left(1 + 2\zeta \frac{j\omega}{\omega_n} - \frac{\omega^2}{\omega_n^2} \right) = -\tan^{-1} \left(\frac{2\zeta\omega\omega_n}{\omega_n^2 - \omega^2} \right). $$

The curves for this for various values of ζ are also shown in Fig. 3.4-4. The low-frequency asymptote is the horizontal line through $0°$, while the high-frequency asymptote is the horizontal line through $-180°$. For all values of ζ, the curves are all seen to pass through $-90°$ at $\omega = \omega_n$, the corner frequency. The curves become sharper in going from the low-frequency range ($0°$) to the high-frequency range ($-180°$) as ζ decreases until for $\zeta = 0$ the curve jumps discontinuously from $0°$ down to $-180°$ at $\omega = \omega_n$.

Construction of a Bode Plot for a Given G(s)

To construct a Bode plot for a given $G(s)$, it should first be factored into the form of (3.4-4). We see that there are four types of factors. The

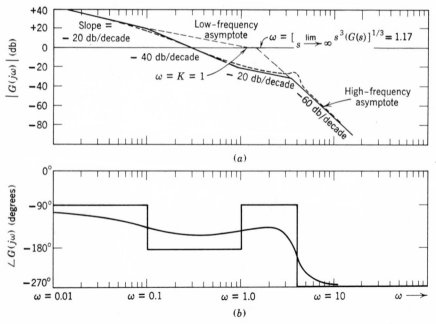

Figure 3.4-6 Bode plot for $G(j\omega) = (1 + j\omega)/j\omega(1 + 10j\omega)(1 + j\omega/8 - \omega^2/16)$.

contribution of each type has been discussed in the preceding section. The problem in constructing the Bode plot now is to algebraically add the contributions due to factors indicated by (3.4-6). The method of doing this is the same for both the magnitude and phase plots. We can summarize the steps as follows:

1. Construct a straight-line approximation using the asymptotes for each particular term.

2. At enough points (to permit the drawing of the curve to the desired accuracy), determine the corrections to the straight line approximation.

3. Draw a smooth curve through the points as determined in Step 2 and asymptotic to the straight-line approximation of Step 1.

4. Once all curves are drawn, the algebraic values at various ω's are added to give a composite curve both for magnitude and phase.

The following example will illustrate the technique.

Example 3.4-1

As a simple example let us consider

$$G(s) = \frac{1.6\,(s+1)}{s\,(s+0.1)\,(s^2+2s+16)}$$

$$= \frac{(1+s)}{s\,(1+10s)\,(1+s/8+s^2/16)}$$

We have a simple real pole at the origin and at $s = -0.1$, complex poles with $\zeta = 0.25$ and $\omega_n = 4$, and a zero at $s = -1$. The information required for the construction of the asymptotic approximation to both the magnitude and phase curves is given in Table 3.4-1. The resulting asymptotes are shown in Fig. 3.4-6a. It should be appreciated that the construction of the asymptotes is very routine once the corner frequencies are known. Remember that each pole contributes -20 db/decade of slope to the magnitude asymptote and each zero $+20$ db/decade. After establishing the low-frequency asymptote [slope $-20N$ db/decade through $\omega = (K_N)^{1/N}$, where N is the number of poles at the origin], continue construction to the right, increasing the slope by 20 db/decade if a zero is passed and -20 db/decade if a pole is passed.

In the same manner, the asymptotic approximation to the angle curve is constructed as shown in Fig. 3.4-6b. We start constructing from the left at the low-frequency asymptote ($-N90°$, where N is the number of poles at the origin) and jump $+90°$ as each zero is passed (proceeding to the right on the $\log_{10} \omega$-axis) and $-90°$ as each pole is passed. The corrected points are shown by the curve in Fig. 3.4-6 and

Table 3.4-1 Asymtotic Approximation Table for Construction of Bode Plot for

$$G(j\omega) = \frac{1 + j\omega}{j\omega(1 + 10j\omega)(1 + (j\omega/8) - (\omega^2/16))}$$

Term	Corner Frequency	Magnitude of Slope Contribution	Sum of Magnitude of slope	Angle Contribution	Sum of Angle Contribution
$K_N = 1$	none	0 db/dec	0 db/dec	0°	0°
$\dfrac{1}{j\omega}$	0	−20 db/dec	−20 db/dec	−90°	−90°
$\dfrac{1}{(1 + 10j\omega)}$	0.1	−20 db/dec	−40 db/dec	−90°	−180°
$(1 + j\omega)$	1.0	+20 db/dec	−20 db/dec	+90°	−90°
$\dfrac{1}{(1 + (j\omega/8) - (\omega^2/16))}$	$\omega_n = 4.0$ $\zeta = 0.25$	−40 db/dec	−60 db/dec	−180°	−270°

the smooth curves for both the magnitude and phase parts of the plot have been drawn through these points to give the curves shown in the figure. As a general comment, we note that the magnitude plot follows the asymptotic approximation much more closely than does the angle plot.

The cases we have considered heretofore have all been what are termed *minimum-phase transfer functions*, i.e., those with all poles and zeros in the LHP. This is the common case; however, let us briefly note the case of RHP poles and zeros.

RHP POLES AND ZEROS

We look first at the case of a zero on the positive real axis. This will contribute a factor of the form $(1-sT)$, $T > 0$, in the numerator of a $G(s)$ factored as in (3.4-4). For the Bode plot we consider the frequency response, the term $(1-j\omega T)$. We simply need note that

$$|1-j\omega T| = |1+j\omega T|$$
$$\angle (1-j\omega T) = -\angle (1+j\omega T) \tag{3.4-20}$$

to determine the contribution of a RHP zero. We see from (3.4-20) that the magnitude contribution is of exactly the same form as the magnitude contribution of a real LHP zero. This is shown in Fig. 3.4-2. The angle contribution of a real RHP zero, however, appears with the opposite sign of the contribution of the LHP zero. Thus the angle contribution of the RHP zero goes from 0 to $-90°$ for $0 \leq \omega < \infty$. The form of the angle contribution is hence the same as shown in Fig. 3.4-3 for the LHP real pole. So we see that for a magnitude plot, a real RHP zero behaves exactly like a real LHP zero. In the angle part of the plot, however, a RHP zero behaves like a LHP pole. Similarly, a RHP pole behaves like a LHP zero in this part. Moreover, similar statements may be made about RHP complex poles and zeros, i.e., the magnitude contribution is the same as LHP poles and zeros, and the angle contribution is opposite in sign.

Note that the plots can be made easily by using the digital computer. Most computing centers have a program for computing the frequency responses for a rational $G(s)$ and the presentation from the machine is usually in the form of a Bode plot. This is, of course, the preferred source of Bode plots if one intends any extensive work with them.

Closed-Loop Stability by Bode Plots

In the development of Nyquist criterion, we determined that the unity-feedback closed-loop system of Fig. 3.4-7 was closed-loop stable

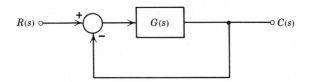

Figure 3.4-7   Unity-feedback closed-loop system.

if the frequency-response curve did not encircle the $-1+j0$ point in the G-plane for a $G(s)$ with no RHP poles. The two possible cases are as shown in Fig. 3.4-8. The Bode plot is also a presentation of the frequency response, so it is natural to consider next the Nyquist criterion in terms of Bode plots.

We note that in Fig. 3.4-8a the $-1+j0$ point is encircled if when $\angle\, G(j\omega) = -180°$, $|G(j\omega)| > 1$. In terms of the Bode plot curves, this means that the $-1+j0$ point is encircled if when $\angle\, G(j\omega) = -180°$, $|G(j\omega)| > 0$ db. This is illustrated in Fig. 3.4-9a. Encirclement is likewise implied by $\angle\, G(j\omega) < -180°$ when $|G(j\omega)| = 1$. This corresponds to $\angle\, G(j\omega) < -180°$ and $|G(j\omega)| = 0$ db on the Bode plot (Fig. 3.4-9a). The result will be an unstable system if the loop is closed about this particular system as in Fig. 3.4-7.

Now let us consider a case where the frequency response $G(j\omega)$ does not encircle the $-1+j0$ point in the G-plane. This happens when $\angle\, G(j\omega) = -180°$, $|G(j\omega)| < 1$, and also when $|G(j\omega)| = 1$, $\angle\, G(j\omega) > -180°$. This corresponds, on the Bode plot, to $\angle\, G(j\omega) = -180°$, $|G(j\omega)| < 0$ db, or $|G(j\omega)| = 0$ db, $\angle\, G(j\omega) > -180°$. For this situa-

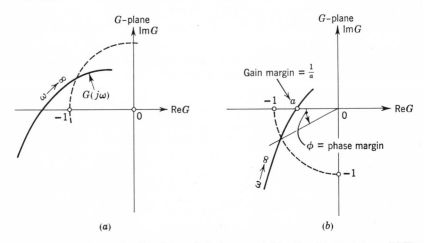

(a)   (b)

Figure 3.4-8   (a) The $-1+j0$ point encircled; an unstable closed-loop system. (b) The $-1+j0$ point not encircled; a stable closed-loop system.

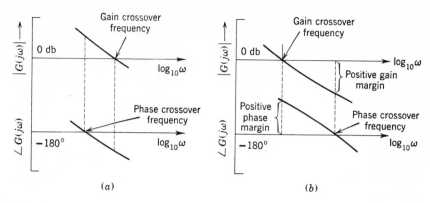

(a)

(b)

Figure 3.4-9 (a) $|G(j\omega)| > 0$ db when $\angle G(j\omega) = -180°$. $\angle G(j\omega) < -180°$ when $|G(j\omega)| = 0$ db. An unstable closed-loop system. (b) $|G(j\omega)| < 0$ db when $\angle G(j\omega) = -180°$. $\angle G(j\omega) > -180°$ when $|G(j\omega)| = 0$ db. A stable closed-loop system.

tion, which is indicated in Fig. 3.4-9b, the system will be closed-loop stable.

Gain margin and phase margin as they were defined in Section 3.2 are indicated in Fig. 3.4-8b. The corresponding values may be picked off the Bode plot as shown in Fig. 3.4-9b. It has been noted that a gain margin greater than 1 corresponds to a stable closed-loop system. In terms of Bode plot quantities, this corresponds to a gain margin greater than 0 db. Thus the gain margin as shown in Fig. 3.4-9b corresponds to a positive gain margin in db. Also positive phase margin was defined in Section 3.2 to mean a stable closed-loop system. Thus the phase margin shown in Fig. 3.4-9b is positive.

It should be noted that a positive gain margin coupled with a negative phase margin (and vice versa) is impossible. Zero gain (in db) and phase margins imply a frequency response which passes through the point $-1 + j0$. This also indicates an unstable system by our definition. Hence for this case the closed-loop system will have a pole on the $j\omega$-axis at the crossover frequency, i.e., where $\angle G(j\omega) = -180°$, $|G(j\omega)| = 0$ db.

Also shown in Fig. 3.4-9 are the gain and phase crossover frequencies. The gain crossover frequency is the frequency at which the magnitude plot crosses the 0 db line. The phase crossover frequency is the frequency at which the phase plot crosses the $-180°$ line. Another usable stability criterion may be given in terms of these frequencies by

(gain crossover freq.) < (phase crossover freq.) → stability

(phase crossover freq.) < (gain crossover freq.) → instability

As an example of how closed-loop stability is determined by this method, consider the Bode plot as presented in Fig. 3.4-6. We see that the magnitude plot crosses the 0-db line at $\omega \cong 0.32$. For this ω, $\angle G(j\omega) \cong -147°$. This is greater than $-180°$, hence the system will be closed-loop stable. The phase margin in this case is then $+33°$. The $\angle G(j\omega)$ curve crosses the $-180°$ line at approximately $\omega = 3.7$. At this ω, $|G(j\omega)| = -24$ db. We hence have $+24$ db of gain margin. With this amount of gain and phase margin, the closed-loop system should be solidly stable.

We also note that the gain crossover frequency is less than the phase crossover frequency, so closed-loop stability is also implied by this criterion.

Nichols Charts

LOG MAGNITUDE VERSUS ANGLE PLOTS

Log magnitude versus angle plots are another means of presenting response of linear time-invariant systems. The usual procedure is to plot $20 \log_{10}|G(j\omega)|$ in db on the vertical axis and $\angle G(j\omega)$ in degrees on the horizontal axis. With this plot, the frequency appears as a parameter along the curve in the same manner as in the G-plane presentation of the frequency response. A typical example of such a plot is shown in Fig. 3.4-10. Typically, if $G(s)$ has a pole at the origin, the frequency response begins at the top of the plot for ω near zero and goes down the chart as ω increases. The example shown in Fig. 3.4-10 begins at the top, at the low-frequency end of the curve. The curve

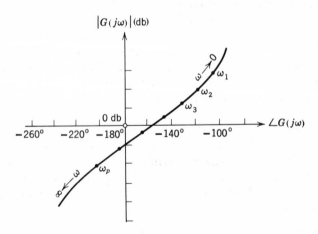

Figure 3.4-10 A typical log magnitude versus angle plot.

is approaching the $-90°$ line as $\omega \to 0$, hence we can conclude that the $G(s)$ corresponding to this frequency-response plot has a single pole at the origin. As ω increases, $|G(j\omega)|$ is seen to decrease. We hence see that the corresponding $G(s)$ has more poles than zeros.

THE NYQUIST CRITERIA IN TERMS OF LOG MAGNITUDE – ANGLE PLOTS

It will be recalled from the previous section that given a stable open-loop $G(s)$, the closed-loop system of Fig. 3.4-7 will be stable if the frequency response does not encircle the $-1+j0$ point in the G-plane. To see what this means in the log magnitude-angle plane, we note the $-1+j0$ point in the G-plane corresponds to the (0 db, $\pm k180°$), $k = 1, 3, 5, \ldots$, point in the log magnitude-angle plane. The point of interest for control system design work is the (0 db, $-180°$) point. From Fig. 3.4-8b we have noted that the $-1+j0$ point is not encircled if when $\angle G(j\omega) = -180°$, $|G(j\omega)| < 1$. Or, in terms of log magnitude, the $-1+j0$ point is not encircled when $\angle G(j\omega) = -180°$, $20 \log_{10} |G(j\omega)| < 0$ db. We also see (again from Fig. 3.4-8b) that the $-1+j0$ point is not encircled when $|G(j\omega)| = 1$ [or $20 \log_{10}|G(j\omega)| = 0$ db], $\angle G(j\omega) > -180°$. This means that the frequency response curve in the log magnitude-angle plane must pass to the right of the (0 db, $-180°$) point. This is as shown in Fig. 3.4-11a, which is a frequency-response plot for a system that would be stable when operated in a closed loop. The case of a closed-loop unstable frequency response is shown in Fig. 3.4-11b. Gain and phase margins as defined in Section 3.2 for the G-plane are also easily seen on the

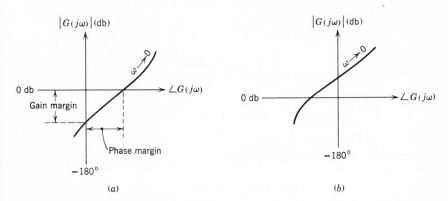

Figure 3.4-11 (*a*) Frequency response for a $G(s)$ which will be closed-loop stable. (*b*) Frequency response for a $G(s)$ which will be closed-loop unstable.

log magnitude-angle plot. These are also shown in Fig. 3.4-11*a*. Gain margin in db and phase margin in degrees are positive here, indicating a stable closed-loop system. The system of Fig. 3.4-11*b* would have negative gain and phase margins, which means the corresponding closed-loop system is unstable. We now look at the Nichols Chart.

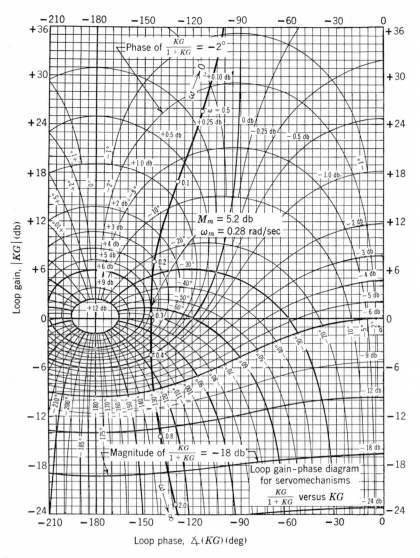

Figure 3.4-12   Nichols chart plot for $G(j\omega) = (1+j\omega)/j\omega(1+10j\omega)(1+j\omega/8-\omega^2/16)$.

NICHOLS CHARTS

The Nichols chart is a log magnitude-angle plane of the type just discussed; however, it has an additional feature which is very useful in analyzing and designing compensators for *unity-feedback* control systems. This feature is an additional set of axes (a nonrectangular set). If one now considers a point x in the log magnitude-angle plane, the nonrectangular set of coordinate axes gives the value of $20 \log_{10} |x/(1+x)|$ and $\angle [x/(1+x)]$. This nonrectangular set of axes is called the closed-loop axes. Such a chart is shown in Fig. 3.4-12. Thus we plot an open-loop frequency response $G(j\omega)$, say, on the Nichols chart, using the rectangular coordinates. For every point on the plotted frequency response we read off on the closed-loop axes the value of $20 \log_{10} |G(j\omega)/[1+G(j\omega)]|$ and $\angle \{G(j\omega)/[1+G(j\omega)]\}$ i.e., the corresponding unity-feedback, closed-loop frequency-response values for that particular ω. To illustrate, the frequency response of $G(s)$ of Example 3.4-1, plotted in Bode plot form in Fig. 3.4-6, has been transferred to the Nichols chart of Fig. 3.4-12. This is done by taking values of $20 \log_{10} |G(j\omega)|$ and $\angle G(j\omega)$ from the Bode plot of Fig. 3.4-6 for given values of ω. Then the corresponding point on the log magnitude angle plot of Fig. 3.4-12 is found by using the rectangular or open-loop set of coordinate axes. For each point transferred, the corresponding value of ω is recorded next to the point. Enough points are established on the Nichols chart to permit the drawing of a smooth curve, as shown in Fig. 3.4-12. The final step is to read off the corresponding unity-feedback, closed-loop frequency response from the closed-loop (nonrectangular) axes.

A Bode plot of this closed-loop frequency response may be constructed using these values (Fig. 3.4-13). An important point on the magnitude part of the Bode plot of Fig. 3.4-13 is the maximum height point of the curve. This has been labeled M_m on the figure with the corresponding value of frequency, ω_m.[7] This point is easily distinguishable on the Nichols chart (of Fig. 3.4-12) because it occurs where the frequency-response curve is just tangent to a $|G/(1+G)|$ curve. This is also labeled on the Nichols chart of Fig. 3.4-12. The corresponding ω_m is read from the curve itself. This point should be noted, because it is important in estimating the characteristics of the time response of the closed-loop system and hence is also important in designing closed-loop compensators for the system. We see from Fig. 3.4-13 that $M_m = 5.2$ db and $\omega_m = 0.28$ rad/sec.

---

[7] M_m and ω_m have been defined for a second-order frequency response in Fig. 3.4-5.

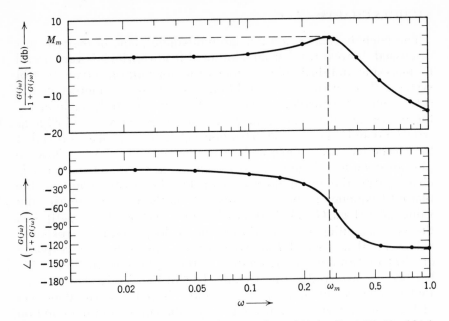

Figure 3.4-13 Closed-loop Bode plot for open-loop $G(j\omega) = (1+j\omega)/j\omega(1+10j\omega)$ $(1+j\omega/8-\omega^2/16)$.

PROBLEMS

3.1. For the following characteristic polynomials, determine the ranges of the parameter X for which the system is stable. Where are the $j\omega$-axis poles for X on the limits of the stable ranges?
(a) $s^3 + Xs^2 + 5s + 1$
(b) $s^3 + (3+X)s^2 + (3+25X)s + 1 + 150X$
(c) $s^4 + (10+X)s^3 + (17+13X)s^2 + (-28+46X)s + 48X$

3.2. Find the transfer function for the circuit of Fig. P3.2 and determine the ranges of values of G_1 for which there is stability.

3.3. How many roots do the following polynomials have to the right of the vertical line through $\sigma = -1$?
(a) $s^5 + 8s^4 + 17s^3 - 2s^2 - 24s$
(b) $s^4 + 6.1s^3 + 9.6s^2 + 4.9s + 0.4$

3.4. A system will oscillate with frequency ω if it has poles at $s = \pm j\omega$ and no RHP poles. Given the system of Fig. P3.4 and

$$G(s) = \frac{K(s+1)}{s^3 + as^2 + 2s + 1}$$

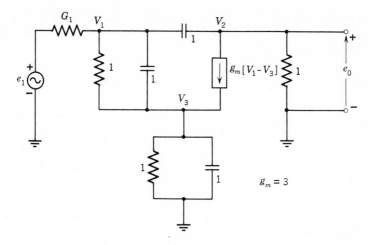

Figure P3.2

We want an oscillator that oscillates with frequency $\omega = 2$ rad/sec. What values of K and a should be chosen?

3.5. For the following open-loop transfer functions operated in a unity negative-feedback configuration, compute e_{ss}(step), e_{ss}(ramp), e_{ss}(accel), and the steady-state error constants. Assume the resulting closed-loop system is stable.

(a) $G(s) = \dfrac{a_m \prod\limits_{i=1}^{m} (s - z_i)}{\prod\limits_{i=1}^{n} (s - p_i)}$

(d) $G(s) = \dfrac{a_m s^m + \cdots + a_1 s}{s^n + b_{n-1} s^{n-1} + \cdots + b_2 s^2}$

(b) $G(s) = \dfrac{K(s+1)}{s^2(s+3)}$

(e) $G(s) = \dfrac{K e^{-as}}{s(s+b)}$

(c) $G(s) = \dfrac{A}{s+1}$

(f) $G(s) = \dfrac{\phi(s)}{\psi(s)}$, $\phi(0) \neq 0,$ $\psi(0) = 0,$
   $\phi'(0) \neq 0,$ $\psi'(0) \neq 0$

Figure P3.4

3.6. For the following closed-loop transfer functions determine e_{ss}(step) e_{ss}(ramp), and e_{ss}(accel) and the corresponding steady-state error constants.

(a) $T(s) = \dfrac{8(s+1)}{(s+2)(2s+1)(s+4)}$

(d) $T(s) = \dfrac{a}{s+a}$

(b) $T(s) = \dfrac{4s+1}{s^4+2s^3+2s^2+4s+1}$

(e) $T(s) = \dfrac{(s+a)(\omega_n^2/a)}{s^2+2\zeta\omega_n s+\omega_n^2}$
$\zeta, \omega_n > 0$

(c) $T(s) = \dfrac{\omega_n^2}{s^2+2\zeta\omega_n s+\omega_n^2}$, $\zeta, \omega_n > 0$

(f) $T(s) = \dfrac{K(s+1)}{s^2+s+1}$

3.7. Given the Nyquist plots (Fig. P3.7) of two minimum-phase open-loop transfer functions.
(a) At what values of ω does the root locus cross the $j\omega$-axis?
(b) If the loop is closed (negative feedback) through an amplifier of gain A, over what values of $A > 0$ is the closed-loop system stable? Over what values is it unstable?
(c) How many excess poles does the system have?

3.8. Given a system with Nyquist plot as shown in Fig. P3.8. The system is known to have more poles than zeros.
(a) How many poles does this system have at the origin?
(b) How many excess poles does this system have?
(c) How many zeros does this system have?

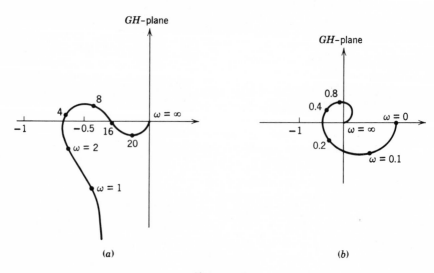

(a)

(b)

Figure P3.7

GH-plane

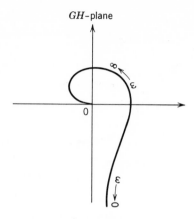

Figure P3.8

3.9. Draw the Bode plots for the following open-loop transfer functions. Indicate the gain and phase margins on each plot.

(a) $\dfrac{(s-1)}{s^2(s+1)}$

(e) $\dfrac{e^{-0.1s}}{s^2+2\,s+100}$

(b) $\dfrac{1}{s(s+1)(s+5)}$

(f) $\dfrac{(s+1)(1+0.1\,s)}{s(1+0.2\,s)^2(1+0.01\,s)}$

(c) $\dfrac{(1+0.25\,s+0.25\,s^2)}{s(1+0.25\,s)(1+0.1\,s)}$

(g) $\dfrac{10\,(s+1)^2}{s^2(s^2+2\,s+100)}$

(d) $\dfrac{s}{s^2+s+9}$

(h) $\dfrac{16000(s+1)}{s(s+10)\,(s+40)^2}$

3.10. Given the minimum phase magnitude part of the Bode plot in Fig. P3.10.

(a) This system operated in a unity negative-feedback loop will have how much steady-state error to a unit step? ramp? parabola?

(b) For the quadratic factor at $\omega = 16$, what is ζ?

(c) The system has how many excess poles?

(d) Draw roughly the corresponding minimum phase Nyquist plot.

3.11. Given the Nyquist plot of Fig. P3.11.

(a) Determine the type of this system.

(b) It has how many excess poles?

(c) As ω increases from zero, does a pole or a zero appear first?

(d) What is the minimum number of poles that this system can have?

3.12. In Fig. P3.12 are shown the magnitude asymptotes and phase for an open-loop frequency response. Assume that the asymptotes are an adequate approximation for the magnitude plot.

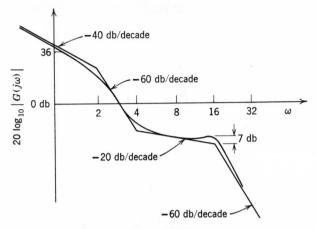

Figure P3.10

(a) What is K_a for the given system?
(b) What are the gain and phase margins for this system?
(c) If a gain margin of 20 db is desired, what is the corresponding K_a?
(d) What is the corresponding phase margin for 20 db of gain margin?

3.13. Draw the Nyquist plots for the open-loop transfer functions in Problem 3.9. Indicate the gain and phase margins on the plot.

3.14. In taking a frequency response of a system, it was found that in a certain frequency range the slope of the magnitude curve increased by 20 db/dec, while the phase decreased by 90°. What does this indicate about the $G(s)$ of the system being measured?

3.15. Given the system of Problem 2.9 with $B = 0.1$ lb-ft-sec/deg. If $e(t) = \sin \omega t$, for what ω will maximum deflection (θ) occur?

3.16. Given

$$G(j\omega) = \frac{10}{j\omega(0.05\,j\omega + 1)(0.01j\omega + 1)}$$

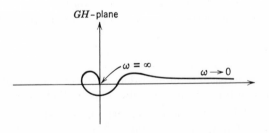

Figure P3.11

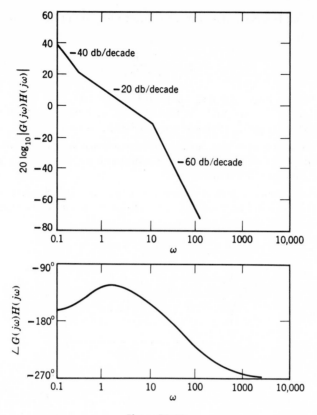

Figure P3.12

to be operated in a unity, negative-feedback, closed loop. $M_m \cong 1.25$ db is desired.

(a) How much loop again (in db) must be added to achieve the desired M_m?

(b) What will ω_m be for this desired M_m?

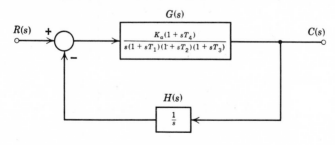

Figure P3.18

(c) What is steady-state error to a unit ramp input for this desired M_m?

3.17. Given

$$G(s) = \frac{K}{s(s+3)(s+9)}$$

(a) Make a Bode plot of $G(j\omega)$ for $0.1 \leq \omega \leq 10$.

(b) Plot $G(j\omega)$ on a Nichols chart with a K such that $M_m = 1$ db.

(c) What are gain and phase margins for this system with $M_m = 1$ db?

3.18. Draw a Nyquist plot for the system shown in Fig. P3.18. Use the assumptions of Example 3.2-6. Comment upon the stability and phase and gain margins of this system.

3.19. A closed-loop system is shown in Fig. P3.19a with the Nyquist plot of $G(s)$ shown in Fig. P3.19b. $G(s)$ in closed-loop without compensation is

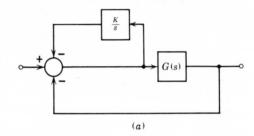

(a)

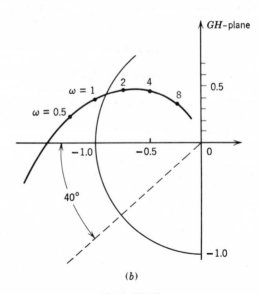

(b)

Figure P3.19

unstable. It is proposed to stabilize it by the addition of an integrator with variable gain K as shown in Fig. P3.19a.

(a) Find the value of K that will just stabilize the system.

(b) Find the value of K that will give 40° of phase margin.

(c) What value of gain margin does the system have when it has 40° of phase margin?

4

MORE ON THE ANALYSIS OF CONTINUOUS-TIME SYSTEMS

4.1 INTRODUCTION

This chapter is a continuation of the discussion of the analysis of continuous-time systems. The frequency-response techniques as developed in the last chapter are graphical methods which primarily consider the open-loop transfer function. There are two other graphical techniques, but these are in the s-domain: the root locus method and Mitrovic's method. The root locus method uses the open-loop transfer function, whereas Mitrovic's method is based on the use of the characteristic equation of the closed-loop transfer function. We develop these two basic techniques and their use in the analysis of continuous-time systems here.

4.2 ROOT LOCUS METHOD

Basic Technique

The basic idea in root locus is to determine the closed-loop pole (and incidentally closed-loop zero) configuration from the configuration of the open-loop poles and zeros. We have already seen that with the Nyquist criterion we can determine the number of closed-loop poles in the RHP. Similarly, with the Routh criterion we can find the number of closed-loop poles in the RHP. In fact, with the Routh criterion we can find the number of RHP poles given the denominator of the transfer function. The root locus is a much more powerful method in that it allows us to determine not just whether the closed-loop system has any poles in the RHP but exactly where the closed-loop poles are going to be. The great power of this method lies in the fact that it furnishes so much more than just information relative to stability. Knowing the closed-loop poles will allow us to quickly judge the closed-loop time response for a given system, a great aid in compensating the system.

To begin, consider Fig. 4.2-1, which shows the closed-loop system of interest here. For this system,

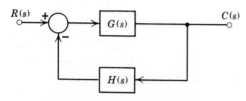

Figure 4.2-1 Closed-loop system.

$$\frac{C(s)}{R(s)} = \frac{G(s)}{1 + G(s)H(s)} \qquad (4.2\text{-}1)$$

From (4.2-1) we see that the closed-loop poles can occur only for those values of s where

$$G(s)H(s) = -1 \qquad (4.2\text{-}2)$$

The root locus is based upon the simple observation that closed-loop poles will occur only for those values of s where

$$\angle G(s)H(s) = \pm 180^\circ i, \quad i = 1,3,5,\ldots \qquad (4.2\text{-}3)$$

and

$$|G(s)H(s)| = 1$$

It should be appreciated that not all points in the s-plane satisfy (4.2-3). The root locus consists of those points where (4.2-3) is satisfied.

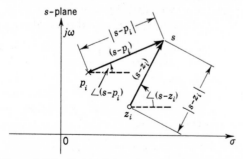

Figure 4.2-2 Determining $\angle(s - z_i)$ and $|s - z_i|$ graphically.

Before defining the root locus more exactly, let us consider the form of $G(s)H(s)$ [pole-zero plot of $G(s)H(s)$ is shown in Fig. 4.2-2]:

$$G(s)H(s) = K\frac{A(s)}{B(s)}$$

$$= K\frac{\displaystyle\prod_{i=1}^{m}(s - z_i)}{\displaystyle\prod_{i=1}^{n}(s - p_i)} \tag{4.2-4}$$

where z_i, $i = 1, 2, \ldots, m$ and p_i, $i = 1, 2, \ldots, n$ give the locations of the m open-loop zeros and n open-loop poles, respectively. Thus $A(s)$ is of degree m and $B(s)$ of degree n. It should further be noted from (4.2-4) that the coefficient of highest power of s for both $A(s)$ and $B(s)$ is unity. With (4.2-4) in mind, we arrive at the following definitions:

The root locus is the set of points in the s-plane where

$$\angle G(s)H(s) = \pm 180° \, i, \qquad i = 1,3,5,\ldots \tag{4.2-5a}$$

The parameter K is defined for each point of the root locus by

$$K = \frac{|B(s)|}{|A(s)|}, \qquad 0 \leqslant K < \infty \tag{4.2-5b}$$

To give some notion of what the root locus for a given $G(s)H(s)$ might look like, consider two examples of constructing root loci.

Example 4.2-1

Let

$$G(s) = \frac{K}{s(s + a)}, \qquad a > 0, \quad H(s) = 1 \tag{4.2-6}$$

To construct the root locus we find those values of s (as a function of K and a) for which (4.2-2) is satisfied, i.e., those values of s for which

$$\frac{K}{s(s + a)} = -1$$

This gives directly

$$s^2 + as + K = 0$$

The solution of the quadratic gives

$$s_1, s_2 = \frac{-a \pm \sqrt{a^2 - 4K}}{2} \tag{4.2-7}$$

Now s_1 and s_2 as given by (4.2-7) are on the root locus of the $G(s)$ of (4.2-6). Let us consider K in the range $0 \leq K < \infty$. For K in the range $0 \leq K \leq a^2/4$, s as given by (4.2-7) is real and lies in the range $-a \leq s \leq 0$. For $K > a^2/4$, s is complex. The real part is $-a/2$ and the imaginary part increases in magnitude as K increases (see Fig. 4.2-3). The root locus starts ($K = 0$) at the two open-loop poles at $s = 0$ and $s = -a$ and is in the LHP for all $K > 0$. Thus the $G(s)$ of (4.2-6) can be put into a unity-feedback, closed-loop system and it will be stable for $K \geq 0$.

As an example of the use of the root locus, the line in the LHP which corresponds to a damping factor $\zeta = 0.707$ has been drawn (Fig. 4.2-3). From Fig. 2.3-3 we see that for this damping factor the step response of a system with two poles will have approximately 6% overshoot and a quick settling time, i.e., this is a desirable damping factor. The K at the point where the $\zeta = 0.707$-line crosses the root locus is found by setting the imaginary part of (4.2-7) equal to the real part and solving for K:

$$K = \frac{a^2}{2}$$

This has been indicated on the root locus in Fig. 4.2-3. Thus if we use the $G(s)$ of (4.2-6) with $K = a^2/2$, the system will have a damping factor such that a desirable step response is obtained. If we were to operate such a system, we would in all likelihood choose a K near $a^2/2$ unless some other consideration prevented us from using that value.

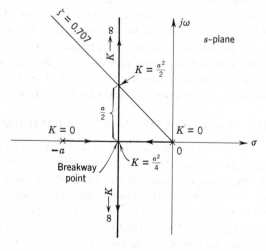

Figure 4.2-3   Root locus for $G(s) = K/s(s+a)$, $H(s) = 1$.

Example 4.2-2

Consider the root locus of

$$G(s) = \frac{K}{(s-\sigma_0)^N}, \quad N \geq 1, \quad H(s) = 1 \qquad (4.2\text{-}8)$$

This is simply an Nth-order pole on the real axis at σ_0. This case is simple but it will be seen to have important implications when the rules for constructing root loci are considered in the next section.

We look for those values of s where (4.2-5a) is satisfied. If

$$r = |s - \sigma_0|$$

and

$$\theta = \angle(s - \sigma_0)$$

we have from (4.2-8)

$$G(s) = Kr^{-N}e^{-jN\theta} \qquad (4.2\text{-}9)$$

thus

$$\angle G(s) = -N\theta$$

and hence the root locus occurs where

$$N\theta = k180°, k = \pm 1, \pm 3, \ldots \qquad (4.2\text{-}10)$$

The root locus is shown in Fig. 4.2-4.

We should have firmly in mind that *given a value of s on the root locus, with K adjusted as in (4.2-5b), (4.2-2) is satisfied and hence the closed-loop system of Fig. 4.2-1 has a pole at this value of s.*

The problem of finding closed-loop poles is thus reduced to finding those values of s for which (4.2-5a) is satisfied. Given such an s, K of (4.2-5b) can be computed. Our approach to doing both shall be graphical. We begin by showing how $\angle G(s)H(s)$ and K [as given by (4.2-5b)] can be determined for any point in the s-plane.

DETERMINING $|G(s)H(s)|$ AND $\angle G(s)H(s)$ GRAPHICALLY

We have a pole-zero plot of $G(s)H(s)$ in the s-plane, as in Fig. 4.2-2. Consider determining $|G(s)H(s)|$ and $\angle G(s)H(s)$ at a general point s. In Fig. 4.2-2, this point is shown in the first quadrant; it may, of course, be in any quadrant. Also shown are what can be considered to be a typical pole and zero, p_i, and z_i, respectively. The vector from z_i to the point s represents the factor $(s - z_i)$. The angle that this vector makes with the positive real axis is then $\angle(s - z_i)$. Its length is $|s - z_i|$. A similar situation occurs with the vector from p_i to s and the factor

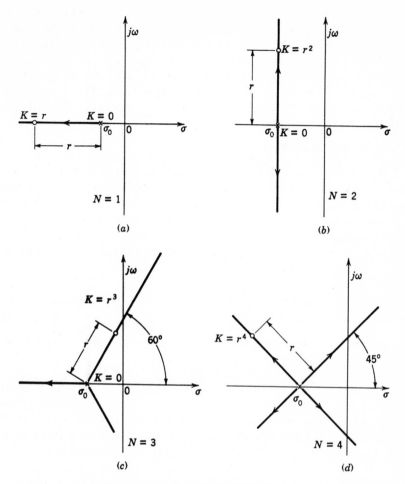

Figure 4.2-4   Root loci for $G(s) = K/(s - \sigma_0)^N$.

$(s - p_i)$. In general, then, with $G(s)H(s)$ in the form of (4.2-4), we have

$$\angle G(s)H(s) = \sum_{i=1}^{m} \angle (s - z_i) - \sum_{i=1}^{n} \angle (s - p_i) \qquad (4.2\text{-}11a)$$

and

$$|G(s)H(s)| = \frac{K \prod_{i=1}^{m} |s - z_i|}{\prod_{i=1}^{n} |s - p_i|} \qquad (4.2\text{-}11b)$$

Since $\angle(s-z_i)$, $\angle(s-p_i)$, $|s-z_i|$, and $|s-p_i|$ can be determined graphically, $\angle G(s)H(s)$ can be determined as can $|G(s)H(s)|$ up to the factor K. We assume K is adjustable to give closed-loop poles where desired. As will be seen later, the spirule is a device that measures the angles and lengths of the vectors of the form $(s-z_i)$ and $(s-p_i)$ and then assembles the sums and products on the RHS of (4.2-11) to permit one to determine $|G(s)H(s)|$ and $\angle G(s)H(s)$ for any point in the s-plane.

Rules for Constructing Root Loci

To help in the rough drafting of the general form of a root locus, construction rules have been devised. We shall discuss seven of them, first stating the rule, then giving a justification for its validity, and finally citing an example if appropriate. We assume that $G(s)H(s)$ for which the root locus is desired, is of the form of (4.2-4), i.e., with m open-loop zeros at $s = z_i$, $i = 1, 2, \ldots, m$ and n open-loop poles at $s = p_i$, $i = 1, 2, \ldots, n, m < n$.

Rule 1. *The root locus has* n *branches where* n *is the number of open-loop poles.*

To see this, consider $1 + G(s)H(s)$ for $G(s)H(s)$ as given by (4.2-4):

$$1 + G(s)H(s) = 1 + K\frac{A(s)}{B(s)} = \frac{B(s) + KA(s)}{B(s)} \qquad (4.2\text{-}12)$$

$B(s)$ is of degree n, hence so is $B(s) + KA(s)$ and we see that $1 + G(s)H(s)$ will have n zeros. Each will occur on a separate branch of the root locus. As examples of this see Figs. 4.2-3 and 4.2-4.

Rule 2. *The root locus branches begin* (K $= 0$) *on the open-loop poles and end* (K $= \infty$) *on the open-loop zeros.*

From (4.2-12) we see that $1 + G(s)H(s)$ will have zeros where

$$B(s) + KA(s) = 0 \qquad (4.2\text{-}13)$$

For $K = 0$, $1 + G(s)H(s) = 0$ where $B(s) = 0$. The roots of $B(s)$ are the open-loop poles. For $K \rightarrow \infty$, the term $KA(s)$ will dominate the LHS of (4.2-13). Thus $1 + G(s)H(s) = 0$ where $A(s) = 0$. The roots of $A(s)$ are the open-loop zeros. We note from (4.2-4) that if $m < n$, then $G(s)H(s)$ has $n - m$ zeros at infinity. Thus $n - m$ branches of the root locus go off to infinity as $K \rightarrow \infty$. For examples see Figs. 4.2-3 and 4.2-4.

Rule 3. *The root locus exists at a point on the real axis if there is an odd number of poles plus zeros on the real axis to the right of this point.*

To see this consider Fig. 4.2-5. We consider a general point s on the real axis. For each pole and zero to the right of the point s and on the real-axis,

$$\angle(s - z_i) = \angle(s - p_i) = 180° \tag{4.2-14}$$

For each pole and zero on the real-axis to the left of s,

$$\angle(s - z_i) = \angle(s - p_i) = 0 \tag{4.2-15}$$

Now

$$\angle G(s)H(s) = \sum_{i=1}^{m} \angle(s - z_i) - \sum_{i=1}^{n} \angle(s - p_i) \tag{4.2-16}$$

We note that complex poles and zeros appear symmetrically above and below the real axis, thus their net contribution to $\angle G(s)H(s)$ as given by (4.2-16) is zero for a point on the real axis. Thus for s on the real axis

$$\angle G(s)H(s) = (N_{zR} - N_{pR})(180°) \tag{4.2-17}$$

where N_{zR} and N_{pR} are the number of real zeros and poles, respectively, to the right of s. Thus (4.2-5a) is satisfied for $N_{zR} - N_{pR}$ odd. Our rule now follows immediately if it is noted that $N_{zR} + N_{pR}$ is odd if $N_{zR} - N_{pR}$ is odd.

As an example of the use of the rule, consider the root loci in Figs. 4.2-3 and 4.2-4.

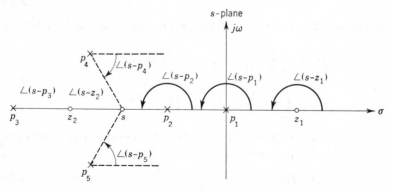

Figure 4.2-5 Angle contributions for a point on the real axis.

Rule 4. *The branches of the root locus which go off to the zeros at infinity approach asymptotically the straight lines with angles*

$$\theta = \frac{k(180°)}{n-m}, \qquad k = \pm 1, \pm 3, \pm 5, \ldots \qquad (4.2\text{-}18)$$

These asymptotes intersect the real axis at

$$\sigma_0 = \frac{\displaystyle\sum_{i=1}^{n} p_i - \sum_{i=1}^{m} z_i}{n-m} \qquad (4.2\text{-}19)$$

Equation (4.2-18) follows from the fact that if we consider $G(s)H(s)$ for $|s| \to \infty$, the finite poles and zeros all appear to be at the same point. First we note that this point will be on the real axis since all poles and zeros are distributed symmetrically with respect to the real axis. Second, since for s large all the poles and zeros appear to be at the same point, the zeros will cancel out the poles. One pole will be canceled for each finite zero. Hence for $|s|$ large, $G(s)H(s)$ appears to be $n-m$ poles on the real axis at some point σ_0. The root locus for this case has been considered in Example 4.2-2. The angle that the branches of the root locus of N poles on the real-axis makes with the positive real axis is given by (4.2-10). For $n-m$ poles on the real axis we simply let $N = n-m$ in (4.2-10) to obtain (4.2-18). Examples of asymptotes for the cases $n-m = 1, 2, 3, 4$, are given in Fig. 4.2-4.

The problem remains of finding where on the real axis the $n-m$ poles will appear, i.e., of finding σ_0. To do this observe first that

$$G(s)H(s) \underset{s\to\infty}{\longrightarrow} \frac{K}{(s-\sigma_0)^{n-m}} = \frac{K}{s^{n-m} - (n-m)\sigma_0 s^{n-m-1} + \cdots + (-\sigma_0)^{n-m}}$$

$$(4.2\text{-}20)$$

Considering $G(s)H(s)$ as in (4.2-4) and multiplying out the RHS gives

$$G(s)H(s) = \frac{K\left[s^m - \left(\displaystyle\sum_{i=1}^{m} z_i\right)s^{m-1} + \cdots + \prod_{i=1}^{m}(-z_i) \right]}{s^n - \left(\displaystyle\sum_{i=1}^{m} p_i\right)s^{n-1} + \cdots + \prod_{i=1}^{n}(-p_i)} \qquad (4.2\text{-}21)$$

If we divide by synthetic division both numerator and denominator in (4.2-21) by the numerator, the result is

$$G(s)H(s) = \frac{K}{s^{n-m} - \left(\displaystyle\sum_{i=1}^{n} p_i - \sum_{i=1}^{m} z_i\right)s^{n-m-1} + \cdots} \qquad (4.2\text{-}22)$$

Now comparing (4.2-20) and (4.2-22) and noting that as $s \to \infty$, the higher powers in the denominator will dominate, from the coefficients of s^{n-m-1} we get

$$(n-m)\sigma_0 = \sum_{i=1}^{n} p_i - \sum_{i=1}^{m} z_i$$

which immediately leads to (4.2-19). The use of this rule is illustrated by Example 4.2-3.

Example 4.2-3

Here $G(s)$ is given by (4.2-6) and $n = 2$, $m = 0$ so that $n - m = 2$. Using this in (4.2-18) gives

$$\theta = \frac{180°}{2} = 90° \qquad \text{for} \quad k = 1$$

$$\theta = \frac{3 \cdot 180°}{2} = 270° \qquad \text{for} \quad k = 3$$

For higher odd values of k the angles obtained differ from 90 and 270° by multiples of 360°. Thus the two branches of the root locus that go off to infinity as $s \to \infty$ approach asymptotes which make angles 90 and 270° with the positive real axis. We now calculate where the asymptotes intersect the real axis (σ_0) by (4.2-19). From (4.2-6) we see that

$$p_1 = 0$$

$$p_2 = -a$$

and there are no finite zeros. Therefore

$$\sigma_0 = \frac{\sum_{i=1}^{n} p_i - \sum_{i=1}^{m} z_i}{n-m} = \frac{0-a+0}{2} = -\frac{a}{2}$$

i.e., the asymptotes intersect the real axis at $-a/2$.

From Fig. 4.2-3, which shows the root locus for this example, we see that the root locus and the asymptotes actually coincide and they do indeed intersect the real axis at $s = -a/2$. The angles of intersection are 90 and 270°.

Rule 5. *Root locus branches intersect the real axis at points where* K *is at an extremum for real* s.

Let us consider the root locus along the real axis and examine, especially, those segments where root locus exists between two poles or two zeros (Fig. 4.2-6). First consider the root locus between the two poles. By Rule 2 we know that a branch starts on each pole. Thus

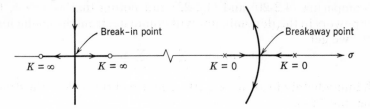

Figure 4.2-6   Breakaway and break-in points.

the segment between the two poles must be part of two different branches. It cannot be one complete branch since branches can end only on zeros. Thus somewhere between the two poles, the two branches must break away from the real axis, i.e., we have a break-away point. K increases as we proceed on the branches away from the poles and it keeps on increasing until the breakaway point is reached. It is thus seen that K is a maximum for real values of s at the breakaway point. It would be nice if the breakaway point always occurred midway between the two poles, which is the case shown in Fig. 4.2-3; but this is unfortunately not the general case.

Now let us look at the case where the root locus on the real axis lies between two zeros. Again consider Fig. 4.2-6. Since the part of the locus between the two zeros cannot be one complete branch, it must be part of two different branches that break into the real axis coming from the complex part of the plane. There is thus a break-in point somewhere between the two zeros. Since K increases along the root locus as a zero is approached, we see that at the break-in point, K is a minimum for real s. Again it would be nice if the break-in point occurred halfway between the zeros, but in general this is not the case. The fact that K is at an extremum where the breakaway or break-in point occurs does give us a method for finding the breakaway and break-in points. The method is basically to find K using (4.2-2) and to then find the points where for real s, $dK/ds = 0$.

Example 4.2-4

As an example of the use of Rule 5 consider

$$G(s) = \frac{K(s+2)}{s(s+1)}, \qquad H(s) = 1 \qquad (4.2\text{-}23)$$

Using Rule 3, we note that the root locus exists on the real axis between the two poles at $s = 0$ and $s = -1$. There is hence a break-away point somewhere between $s = -1$ and $s = 0$. Similarly, by Rule 3, root locus exists on the real axis between the zero at $-\infty$ and the zero at $s = -2$. There is then a break-in point somewhere within this range. To find this we use the fact that the root locus exists for those values of s which satisfy (4.2-2). For this case, using (4.2-23) in (4.2-2) gives

$$\frac{K(s+2)}{s(s+1)} = -1.$$

which in turn gives

$$K = -\frac{s(s+1)}{s+2} \qquad (4.2\text{-}24)$$

and

$$\frac{dK}{ds} = -\frac{s^2+4s+2}{(s+2)^2}$$

It is now routine to find that $dK/ds = 0$ at

$$s = -0.586 \qquad \text{and} \quad s = -3.414$$

The point $s = -0.586$ lies between the two poles at $s = 0$ and $s = -1$, hence this is the breakaway point. The point $s = -3.414$ lies between the zero at $-\infty$ and the zero at $s = -2$. This is the break-in point. The root locus for this case is drawn in Fig. 4.2-7.

The values of K at the breakaway and break-in points are found by using $s = -0.586$ and $s = -3.414$, respectively, in (4.2-24).

Rule 6. *The angle of departure from a complex pole (or entry to a complex zero) can be found by subtracting from 180° the angle contribution at this pole (or zero) of all other finite poles and zeros.*

To see this consider Fig. 4.2-8 and the complex pole at p. If the circle about p has a radius which is small relative to the distance to the other poles and zeros, it can be seen that the angle contribution of those other poles and zeros essentially will be constant for every point on this circle. Let us assume this is the case for the circle about

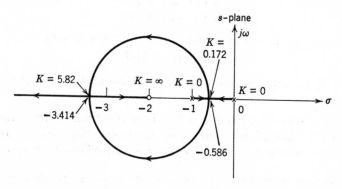

Figure 4.2-7   Root locus for $K(s+2)/s(s+1)$.

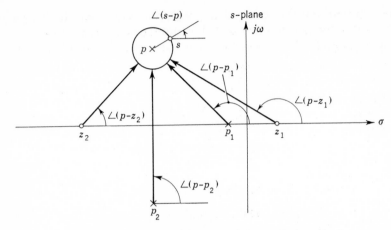

Figure 4.2-8   Angle of departure from a complex pole.

p. The root locus intersects this circle at a point where the angle contribution of the pole at p plus the angle contribution of all other poles and zeros adds up to $180°$, or some odd multiple thereof.

Now considering s as the point where the root locus crosses the circle, we have[1]

$$\angle(p-z_1) + \angle(p-z_2) - \angle(p-p_1) - \angle(p-p_2) - \angle(s-p) = \pm 180°$$

or

$$-\angle(s-p) = \pm 180° - \angle(p-z_1) - \angle(p-z_2) + \angle(p-p_1) + \angle(p-p_2)$$

$\angle(s-p)$ gives the angle of departure from the complex pole.

For the example of Fig. 4.2-8,

$$\angle(p-z_1) \cong 150°$$
$$\angle(p-z_2) \cong 45°$$
$$\angle(p-p_1) \cong 135°$$
$$\angle(p-p_2) = 90°$$

thus
$$-\angle(s-p) \cong \pm 180° - 150° - 45° + 135° + 90°$$
$$\angle(s-p) \cong 150°$$

Rule 7. $j\omega$-*axis crossings may be determined by use of the Routh criterion.*

It was shown in Example 3.2-4. how the Routh criterion may be used for a closed-loop system, to determine the value of K which gives a closed-loop pole on the $j\omega$-axis and where this pole occurs.

---

[1]The root locus exists where $\angle G(s)H(s) = +180°$ or where $\angle G(s)H(s) = -180°$. We write here $\pm 180°$ and use the sign which is convenient.

The point at which the closed-loop pole, for a given $G(s)H(s)$, occurs on the $j\omega$-axis gives the point where the root locus crosses the $j\omega$-axis. To find these crossover points, a Routh table is constructed for the numerator polynomial of $1 + G(s)H(s)$ leaving K as a free parameter. The values of K for which this Routh table has an all-zero row are the values for which $1 + G(s)H(s)$ has a zero on the $j\omega$-axis, and these are the $j\omega$-axis crossing values of K. The row above the all-zero row gives the coefficients of the even factor, which has roots where the root locus crossings occurs (see Example 3.2-4).

Let us now try to use the seven rules to determine the general configuration of the root locus for a given $G(s)$

Example 4.2-5

Consider

$$G(s) = \frac{K}{s(s+1)(s+2)}, \qquad H(s) = 1$$

There are open-loop poles at $s = 0$, $s = -1$, and $s = -2$, and no finite open-loop zeros.

Rule 1 tells us that the root locus has three branches since $G(s)$ has three poles.

Rule 2 tells us that the branches start on the three open-loop poles.

Rule 3 tells us the root locus exists on the real axis for $-1 \leqslant s \leqslant 0$ and for $-\infty < s \leqslant -2$.

Rule 4 tells us how to locate the three asymptotes for the three branches which go off to infinity; $n - m = 3$ in this case. Using (4.2-18) gives the angles of the asymptotes as

$$\theta = \frac{180°}{3} = 60° \qquad \text{for } k = 1$$

$$\theta = \frac{3 \cdot 180°}{3} = 180° \qquad \text{for } k = 3$$

$$\theta = \frac{5 \cdot 180°}{3} = 300° \qquad \text{for } k = 5$$

By (4.2-19), we find that these asymptotes intersect the real axis at

$$\sigma_0 = \frac{0 - 1 - 2}{3} = -1$$

Rule 5 tells us that the breakaway point, which must occur somewhere between the pole at $s = -1$ and $s = 0$, may be found by using (4.2-2) to get

$$K = -s(s+1)(s+2) = -(s^3 + 3s^2 + 2s)$$

and thus $dK/ds = 0$ at $s = -0.423$ and $s = -1.577$. The point $s = -0.423$ occurs between the two poles at $s = -1$ and $s = 0$, hence this must be the breakaway point of the root locus. The point $s = -1.577$ lies between the poles at $s = -2$ and $s = -1$. No root locus exists here so there cannot be any breakaway or break-in point at this point.

Rule 6 is not applicable since there are no complex poles or zeros.

Rule 7 tells us to construct the Routh table for the numerator of $1 + G(s)H(s)$ to find the $j\omega$-axis crossings. Thus

$$1 + G(s)H(s) = \frac{s^3 + 3s^2 + 2s + K}{s(s+1)(s+2)}$$

and the Routh table for the numerator polynomial is

s^3	1	2
s^2	3	K
s^1	$\dfrac{6-K}{3}$	
s^0	K	

The Routh table s^1 row is all zero for $K = 6$.[2] For this value of K, the unity-feedback, closed-loop system with the $G(s)H(s)$ under consideration has a pole on the $j\omega$-axis, i.e., the root locus has a $j\omega$-axis crossing for this value of K and the row above the all-zero row gives the coefficients of the even factor whose roots give the location of the $j\omega$-axis crossing. From the s^2 row we have the even factor as $3s^2 + K$. For $K = 6$, we calculate $3s^2 + 6 = 0$ or $s = \pm j\sqrt{2}$.

The root locus thus crosses the $j\omega$-axis at $\omega = \pm\sqrt{2}$. With the information obtained through the use of these rules, the root locus can be roughly sketched as in Fig. 4.2-9, from which we see directly that for $K > 6$ the closed-loop system will have two RHP poles and is thus unstable.

Example 4.2-6

As another example consider

$$G(s)H(s) = \frac{K}{(s+3)(s+5)(s^2+2s+2)}$$

A pole-zero map for this is shown in Fig. 4.2-10.

Rule 1 tells us the root locus has four branches since there are four open-loop poles.

---

[2]The s^0 row is also all zero for $K = 0$. We see directly that for this value of K there is a zero of the characteristic polynomial at $s = 0$.

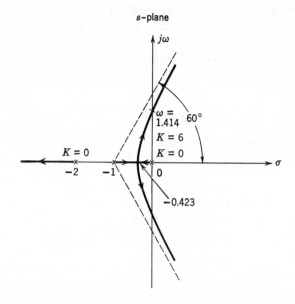

Figure 4.2-9 Root locus of $K/s(s+1)(s+2)$.

Rule 2 tells us the branches start on the open-loop poles and end on the open-loop zeros. In this case all the open-loop zeros are at infinity, so none of the four branches ends in the finite plane.

Rule 3 tells us the root locus exists on the real axis between the poles at $s = -5$ and $s = -3$.

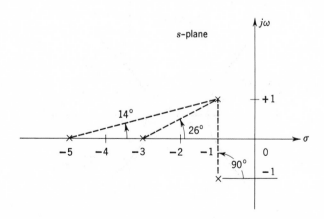

Figure 4.2-10 Pole-zero map of $K/(s+3)(s+5)(s^2+2s+2)$.

Rule 4 tells us there are four asymptotes that make angles with the positive real axis of $\theta = 45, 135, 225,$ and $315°$ and these intersect the real axis at $\sigma_0 = -2.5$.

Rule 5 tells us there is a breakaway point at $s = -4.29$.

Rule 6 tells us that the branch departs the pole at $s = -1 + j$ at an angle given by

$$-\theta - 90° - 26° - 14° = -180° \quad \text{or} \quad \theta = 50°$$

Rule 7 tells us to construct the Routh table for the numerator of $1 + G(s)H(s)$: $s^4 + 10s^3 + 33s^2 + 46s + 30 + K$

$$
\begin{array}{c|ccc}
s^4 & 1 & 33 & 30 + K \\
s^3 & 10 & 46 & \\
s^2 & 28.4 & 30 + K & \\
s^1 & \dfrac{1006.4 - 10K}{28.4} & & \\
s^0 & 30 + K & &
\end{array}
$$

The s^1 row is all-zero for $K = 100.64$. From the s^2 row we find that for this value of K, the root locus crosses the $j\omega$-axis at the roots of

$$28.4s^2 + 130.64 = 0$$

or at

$$s = \pm j2.15$$

With this information, the root locus may be sketched (Fig. 4.2-11).

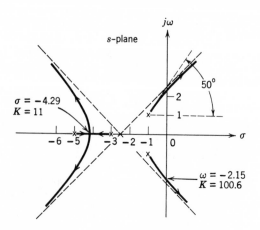

Figure 4.2-11   Root locus of $K/(s + 3)(s + 5)(s^2 + 2s + 2)$.

Here again we see that the root locus is in the RHP for $K > 100.64$, thus for K above this value the closed-loop system will be unstable.

SUMMARY: RULES FOR CONSTRUCTING THE ROOT LOCUS

Rule 1. The root locus has n branches where n is the number of open-loop poles.

Rule 2. The branches begin ($K = 0$) on the open-loop poles and end ($K = \infty$) on the open-loop zeros.

Rule 3. The root locus exists on the real axis if there is an odd number of poles and zeros to the right on the real axis.

Rule 4. The branches of the root locus which go off to the open-loop zeros at infinity approach asymptotically the straight lines with angles

$$\theta = \frac{\pm k(180°)}{n - m}, \qquad k = 1, 3, 5, \ldots$$

where m is the number of finite, open-loop zeros. These asymptotes intersect the real axis at

$$\sigma_0 = \frac{\sum_{i=1}^{n} p_i - \sum_{i=1}^{m} z_i}{n - m}$$

where p_i and z_i are the open-loop poles and zeros, respectively.

Rule 5. Branches intersect the real axis at points where K is at an extremum for real values of s.

Rule 6. The root locus angle of departure from a complex pole (or zero) can be found by subtracting from 180° the angle contribution at this pole (or zero) of all the other finite poles (−) and zeros (+).

Rule 7. $j\omega$-axis crossings may be determined by use of the Routh criterion. Values of $K > 0$ which give all-zero rows in the Routh table indicate the root locus crossings.

The Use of the Spirule

The spirule (Fig. 4.2-12) is a device invented by W. R. Evans to graphically determine $\angle\, G(s)H(s)$ and

$$K = \frac{\prod_{i=1}^{n} |s - p_i|}{\prod_{i=1}^{m} |s - z_i|} \qquad (4.2\text{-}25)$$

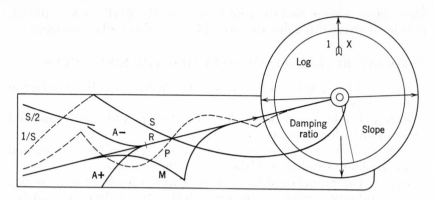

Figure 4.2-12   The spirule.

for any point in the s-plane given a pole-zero plot of $G(s)H(s)$ where $p_i, i=1, 2, \ldots, n$ and $z_i, i=1, \ldots, m$ are the poles and zeros, respectively, of $G(s)H(s)$. The spirule is capable of doing more than determining angle and magnitude contributions graphically, however, it is only these two functions we make use of here. We assume that the reader has a spirule and its instruction booklet at hand. A list of spirule manipulations is included to indicate specifically how one may use the spirule to determine $\angle G(s)H(s)$ (and incidentally if the point s is on the root locus) and K for any point s in the s-plane.

SPIRULE MANIPULATIONS

To find $\angle G(s)H(s)$ at point s:

1. Place the X-1 arrow on the R-line.
2. Set the eyelet on the point s.
3. Align the R-line along the horizontal with the arm pointing off to the left.
4. Fix the disk.
5. Swing the arm of the spirule until the R-line passes through the first pole.
6. Release the disk.
7. Repeat steps 3–6 for each pole.
8. Align the R-line on the first zero.
9. Fix the disk.
10. Swing the arm of the spirule until the R-line is aligned with the horizontal.

11. Repeat steps 8–10 for each zero.
12. Read $-\angle\,G(s)\,H(s)$ on the protractor disk at the R-line.

If the test point lies on the root locus, the 180° arrow should pass through the R-line. (If there is a variation of a degree or so either way, say that the point lies on the root locus.)

To determine K at a point s:

1. Set the X-1 arrow on the R-line, set the eyelet on the point s.
2. Align the R-line with the first pole.
3. Fix the disk.
4. Swing the arm of the spirule until the S-curve passes through the first pole.
5. Release the disk.
6. Repeat steps 2–5 for every pole.
7. Align the S-curve with the first zero.
8. Fix the disk.
9. Swing the arm of the spirule until the R-line passes through the first zero.
10. Release the disk.
11. Repeat steps 7–10 for every zero.
12. Write down the reading on the scale arm (blue numbers) to which an arrow points.
13. Write down the size of the arrow pointing to the blue number. (This will be 0.1, 1, or 10, the index factor.)
14. Compute the scale factor: S.F. $= (5'' = X)^N$

where $N = n - m =$ no. of poles $-$ no. of zeros.

$$K = (\text{Reading}) \cdot (\text{Index Factor}) \cdot (\text{Scale Factor})$$

USE OF THE SPIRULE TO DETERMINE A ROOT LOCUS

In general, to draw a root locus we use the rules outlined earlier and draw a rough sketch of the root locus. Once we have this rough sketch, it is a simple matter to correct it by trial and error using the spirule. It will be seen in Chapter 8, where root locus systems techniques are developed, that only small portions of the entire root locus are required to perform compensation. Hence little time is required for this. Once the location of the root locus is known, it is routine to determine the parameter K for any point on the locus using the spirule.

As an exercise, try using the spirule to correct some of the sketches of root loci that are drawn as examples earlier in this section.

THE USE OF THE SPIRULE TO DETERMINE FREQUENCY RESPONSE

To see how this is done consider a $G(s)$ of the form

$$G(s)H(s) = \frac{K_G \prod_{i=1}^{m} (s - z_i)}{\prod_{i=1}^{n} (s - p_i)} = K_G \frac{A(s)}{B(s)} \qquad (4.2\text{-}26)$$

where we have added the subscript G (K_G) to the multiplying constant K of (4.2-25) to distinguish it from K of (4.2-5b). To determine the frequency response we simply need realize that $G(j\omega)H(j\omega)$ is just $G(s)H(s)$ evaluated along the $j\omega$-axis. Thus if we have a pole-zero map drawn for a given $G(s)H(s)$ and consider a point on the $j\omega$-axis for this ω, we can determine $\angle G(j\omega)H(j\omega)$ by the rules for determining the angle given in the list of spirule manipulations above. For this same point, $|G(j\omega)H(j\omega)|$ is found by determining K, again as is shown in the list of spirule manipulations, for this point. Using (4.2-5b), we have

$$K = \frac{\prod_{i=1}^{n} |j\omega - p_i|}{\prod_{i=1}^{m} |j\omega - z_i|} = \frac{K_G}{|G(j\omega)H(j\omega)|}$$

hence

$$|G(j\omega)H(j\omega)| = \frac{K_G}{K} \qquad (4.2\text{-}27)$$

This process may be repeated for as many points along the $j\omega$-axis as desired to give the range and accuracy of the frequency response desired. If one has polar plot paper, $\angle G(j\omega)H(j\omega)$ and $|G(j\omega).H(j\omega)|$ may be used directly to make a G-plane plot of the frequency response. If a Bode plot is desired, $20 \log_{10} |G(j\omega)H(j\omega)|$ must be taken for the magnitude portion of the plots. This is a fairly practical way of getting a frequency-response plot for a limited range of ω; however, it has the disadvantage of giving very poor accuracy when poles or zeros are very near to or on the $j\omega$-axis. This is a general failing of the spirule for any point in the s-plane, i.e., it gives poor accuracy near poles and zeros.

THE DETERMINATION OF PARTIAL FRACTION EXPANSION COEFFICIENTS WITH THE SPIRULE

Let us now consider the problem of expanding a rational function $\mathscr{L}$-transform [say $X(s)$] into a partial fraction expansion.

$$X(s) = \frac{K_x A(s)}{B(s)} = K_x \frac{\prod_{j=1}^{m}(s-z_j)}{\prod_{j=1}^{n}(s-p_j)}, \qquad m < n \qquad (4.2\text{-}28)$$

Let us make the practical assumption that the poles are all distinct, i.e., $p_i \neq p_j$ if $i \neq j$. $X(s)$ may then be expanded into a partial fraction expansion

$$X(s) = \frac{K_{11}}{s-p_1} + \frac{K_{21}}{s-p_2} + \cdots + \frac{K_{n1}}{s-p_n}$$

$$= \sum_{i=1}^{n} \frac{K_{i1}}{s-p_i} \qquad (4.2\text{-}29)$$

We now must determine the constants K_{i1}, $i = 1, 2, \ldots, n$. We know that

$$K_{i1} = (s-p_i)X(s)\big|_{s=p_i}$$

Using (4.2-28) this gives

$$K_{i1} = K_x \frac{\prod_{j=1}^{m}(p_i-z_j)}{\prod_{\substack{j=1 \\ j\neq i}}^{n}(p_i-p_j)} \qquad (4.2\text{-}30)$$

In general, if p_i is complex, K_{i1} will also be complex, and hence to determine K_{i1} we must know both $|K_{i1}|$ and $\angle K_{i1}$. From (4.2-30), we have

$$|K_{i1}| = |K_x| \frac{\prod_{j=1}^{m}|p_i-z_j|}{\prod_{\substack{j=1 \\ j\neq i}}^{n}|p_i-p_j|} \qquad (4.2\text{-}31)$$

Thus we may make a pole-zero map of $X(s)$ and determine K at the pole p_i, $|K_{i1}|$ is then determined by

$$K = \frac{\prod_{\substack{j=1 \\ j\neq i}}^{n}|p_i-p_j|}{\prod_{j=1}^{m}|p_i-z_j|}$$

and hence, using (4.2-31),

$$|K_{i1}| = \frac{|K_x|}{K} \tag{4.2-32}$$

the K here is determined straightforwardly with the spirule as shown in the list of spirule manipulations with $s = p_i$, ignoring the pole at p_i itself. The $\angle K_{i1}$ is seen, from (4.2-30) to be given by

$$\angle K_{i1} = \sum_{j=1}^{m} \angle (p_i - z_j) - \sum_{\substack{j=1 \\ j \neq i}}^{n} \angle (p_i - p_j) \tag{4.2-33}$$

Thus again using the spirule and the pole-zero map for $X(s)$, $\angle K_{i1}$ is easily determined by the rule for determining $\angle G(s)H(s)$ at $s = p_i$ in the list of spirule manipulations. Again we ignore the pole at $s = p_i$ itself.

In this manner we may determine $\angle K_{i1}$ and $|K_{i1}|$ at each pole, i.e., for $i = 1, 2, \ldots, n$. In general, then,[3]

$$K_{i1} = |K_{i1}| \exp (j \angle K_{i1}) \tag{4.2-34}$$

and using this in (4.2-29) gives

$$X(s) = \sum_{i=1}^{n} \frac{|K_{i1}| \exp (j \angle K_{i1})}{s - p_i} \tag{4.2-35}$$

which may now be easily inverted term by term to give

$$x(t) = \left(\sum_{i=1}^{n} |K_{i1}| e^{p_i t} \exp (j \angle K_{i1}) \right) 1(t) \tag{4.2-36}$$

We note two cases: p_i is either (1) real or (2) complex.

1. If p_i is real, then $\angle K_{i1}$ will either be 0 or 180°. If $\angle K_{i1} = 0°$, then the sign of K_{i1} is plus, i.e., $\exp (j \angle K_{i1}) = 1$ in (4.2-35). If $\angle K_{i1} = 180°$, then the sign of K_{i1} is minus, i.e., $\exp (j \angle K_{i1}) = -1$ in (4.2-35).

2. If p_i is complex, then there is also a pole at $\bar{p}_i$ (the bar indicates conjugate); i.e., if

$$p_i = \mathrm{Re}\, p_i + j\, \mathrm{Im}\, p_i = |p_i| \exp (j \angle p_i) \tag{4.2-37}$$

Then there is also a pole at

$$p_i = \mathrm{Re}\, p_i - j\, \mathrm{Im}\, p_i = |p_i| \exp (-j \angle p_i) \tag{4.2-38}$$

Similarly, the constant K_{i1} corresponding to the pole at $\bar{p}_i$ will be the conjugate of the K_{i1} corresponding to the pole at p_i. Thus $X(s)$ as

---

[3] The j in (4.2-34 and 4.2-35) is the imaginary operator.

given in (4.2-35) will have the form

$$X(s) = \frac{|K_{i1}| \exp (j \angle K_{i1})}{s - p_i} + \frac{|K_{i1}| \exp (-j \angle K_{i1})}{s - p_i} \qquad (4.2\text{-}39)$$

$$+ \text{(partial fraction terms due to other poles)}$$

Now inverting $X(s)$ as given here and using (4.2-39) gives

$$x(t) = |K_{i1}| \exp (\text{Re } p_i t) \{ \exp [j(\text{Im } p_i t + \angle K_{i1})]$$
$$+ \exp [-j(\text{Im } p_i t + \angle K_{i1})] \} 1(t)$$
$$+ \text{(terms due to other poles)} \ 1(t)$$

which may be written as

$$x(t) = [2|K_{i1}| \exp (\text{Re } p_i t) \cos (\text{Im } p_i t + \angle K_{i1})] 1(t)$$
$$+ \text{(terms due to other poles)} \ 1(t) \qquad (4.2\text{-}40)$$

If we have M pairs of complex poles and N real poles, so that in general $n = 2M + N$, using (4.2-40) we may write (4.2-36) as

$$x(t) = \left[\sum_{i=1}^{M} 2|K_{i1}| \exp (\text{Re } p_i t) \cos (\text{Im } p_i t + \angle K_{i1}) \right] 1(t)$$

$$+ \left[\sum_{i=1}^{N} K_{i1} \exp (p_i t) \right] 1(t) \qquad (4.2\text{-}41)$$

We note that to determine $x(t)$ as in (4.2-41) we need only determine $|K_{i1}|$ and $\angle K_{i1}$ at one of the poles of each complex pair.

Example 4.2-7

As an example of the use of (4.2-41) to determine a time response consider the step response of a system with transfer function

$$G(s) = \frac{\omega_n{}^2}{s^2 + 2\zeta \omega_n s + \omega_n{}^2}$$

For a step-input the output $X(s)$ will be given by

$$X(s) = \frac{\omega_n{}^2}{s(s^2 + 2\xi \omega_n s + \omega_n{}^2)} \qquad (4.2\text{-}42)$$

This $X(s)$ has a real pole at $s = 0$ and a complex pair at $s = -\zeta \omega_n \pm j\omega_n \sqrt{1 - \zeta^2}$. The pole-zero plot for the $X(s)$ of (4.2-42) is shown in Fig 4.2-13, as are all distances and angles of interest. For this example the magnitudes and angles may be determined directly from the figure. The three poles have been labeled p_1, p_2, p_3, with p_1 at the origin. Let us first determine K_{11}, the coefficient corresponding to p_1, the pole at the origin.

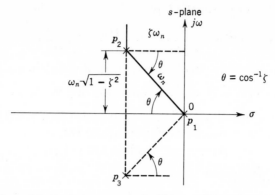

Figure 4.2-13   Pole-zero map for $X(s) = \omega_n^2/s(s^2 + 2\zeta\omega_n s + \omega_n^2)$.

Using (4.2-32),

$$|K_{11}| = \frac{|K_x|}{K} = \frac{\omega_n^2}{|p_1 - p_2||p_1 - p_3|} = \frac{\omega_n^2}{\omega_n^2} = 1$$

Now using (4.2-33),

$$\angle K_{11} = -\angle(p_1 - p_2) - \angle(p_1 - p_3) = -(-\theta) - \theta = 0$$

Similarly,

$$|K_{21}| = \frac{|K_x|}{K} = \frac{\omega_n^2}{|p_2 - p_1||p_2 - p_3|} = \frac{1}{2\sqrt{1 - \zeta^2}}$$

and

$$\angle K_{21} = -\angle(p_2 - p_1) - \angle(p_2 - p_3) = -(180° - \theta) - 90°$$

$$= -270° + \theta = 90° + \theta$$

where $\theta = \cos^{-1}\zeta$

Using these values in (4.2-41) gives

$$x(t) = \left[1 + \frac{e^{-\zeta\omega_n t}}{\sqrt{1 - \zeta^2}}\cos(\omega_n\sqrt{1 - \zeta^2}t + 90° + \theta)\right]1(t)$$

which becomes

$$x(t) = \left[1 + \frac{e^{-\zeta\omega_n t}}{\sqrt{1 - \zeta^2}}\sin(\omega_n\sqrt{1 - \zeta^2}t + \theta)\right]1(t)$$

This result agrees with that obtained in (2.3-14).

Example 4.2-8

A second example of determining these coefficients is finding the closed-loop step response of the system with open-loop transfer function

$$G(s) = \frac{K}{s(s + 1)(s + 2)}, \qquad H(s) = 1 \qquad (4.2\text{-}43)$$

The root locus will now be used to find the closed-loop pole configuration with a desirable damping factor for the complex poles. When we have the closed-loop pole configuration, the partial fraction expansion coefficients of the output step-response transform will be determined. With these coefficients the step response will be written out using (4.2-41). The root locus for the $G(s)H(s)$ of (4.2-43) has been sketched in Fig. 4.2-14. Let us now adjust K so that two closed-loop poles will be achieved near the $j\omega$-axis with a damping factor $\zeta = 0.5$. It will be seen shortly that these poles will dominate the step response of the closed-loop system. We first draw a line in the second quadrant of the s-plane which gives the loci of all points with damping factor $\zeta = 0.5$, as shown in the figure. From a rough sketch of the root locus the approximate point where the root locus crosses the $\zeta = 0.5$ line is estimated. Using this as a first guess, the spirule is used in a trial and error manner and it is found that the root locus, $\zeta = 0.5$-line crossing occurrs at $s = -0.33 + j0.58$. At this point, again using the spirule, it is found that $K \cong 1.1$. Thus with $G(s)H(s)$ as given in (4.2-43) and $K \cong 1.1$, a closed-loop pole will occur at $s = -0.33 + j0.58$ and also at the conjugate point $s = -0.33 - j0.58$. A third closed-loop pole will occur with this value of K on the branch along the negative

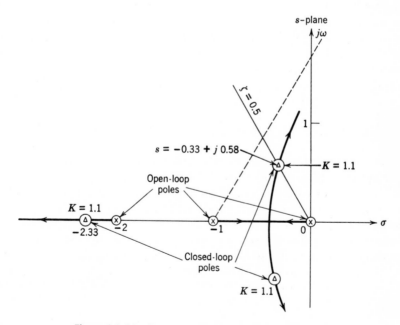

Figure 4.2-14   Root locus for $G(s) = K/s(s+1)(s+2)$.

real axis out beyond the pole at $s = -2$. By determining K at several points along this branch it is found that at $s \cong -2.33$, $K = 1.1$. Thus the third closed-loop pole occurs at this point for $K = 1.1$.

Now using the $G(s)H(s)$ of (4.2-43) in the closed-loop system of Fig. 4.2-1,

$$\frac{C(s)}{R(s)} = \frac{G(s)}{1 + G(s)H(s)} = \frac{K}{s(s+1)(s+2)+K} \qquad (4.2\text{-}44)$$

We have just determined from the root locus of Fig. 4.2-14 that for $K \cong 1.1$ the closed-loop system has poles at $s = -0.33 \pm j0.58$ and at $s = -2.33$. Using this and (4.2-44) we determine $C(s)/R(s)$ for $K = 1.1$ as

$$\frac{C(s)}{R(s)} = \frac{1.1}{(s+0.33-j0.58)(s+0.33+j0.58)(s+2.33)}$$

For a step input, $R(s) = 1/s$. Thus the output step response is given by

$$C(s) = \frac{1.1}{s(s+0.33-j0.58)(s+0.33+j0.58)(s+2.33)} \qquad (4.2\text{-}45)$$

In order to write out the time response, we now make a pole-zero map of $C(s)$ and determine the partial fraction expansion coefficients using the spirule. The pole-zero map for this $C(s)$ is shown in Fig. 4.2-15 where the poles have been labeled p_1 through p_4 starting with the driving pole at the origin. There are no finite zeros. Now using (4.2-32) and (4.2-33), $|K_{i1}|$ and $\angle K_{i1}$ can be determined by using the spirule at each of the poles. This has been done and the results are given next to each pole. Note from (4.2-45) that $K_x = 1.1$ is used here (4.2-32). Using these values in (4.2-41) gives the output [$c(t)$ is the time function here instead of $x(t)$ as in (4.2-41)] as

$$c(t) = [1 + 1.34e^{-0.33t} \cos(0.58t + 134°) - 0.11e^{-2.33t}] \, 1(t) \qquad (4.2\text{-}46)$$

And from this the time response may be computed and plotted directly. We may note that the term in (4.2-46) due to pole p_4 (at $s = -2.33$) has a coefficient which is much smaller in magnitude than the coefficients of the other terms. This pole will then have a relatively minor effect on the time response of this system and the step response will be dominated by the two complex poles. It will hence look very much like the response of the two-pole system with damping factor $\zeta = 0.5$ considered in Section 2.3. From Fig. 2.3-3 we can see the step response will have approximately 15% overshoot and the settling time t_s will be given by

$$t_s = 4\tau = \frac{4}{\zeta \omega_n} = \frac{4}{0.33} = 12 \text{ sec.}$$

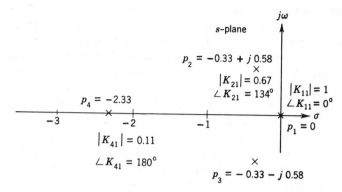

Figure 4.2-15   Pole-zero map for $C(s)$ of (4.2-45).

In general the relative magnitude of a partial fraction expansion co-efficient gives the relative importance of the effect of the corresponding pole on the time response. The poles for which the magnitudes of the coefficients are relatively small generally can be ignored in (4.2-41) if only an approximate time response is required. The spirule hence can be used to tell quickly which poles have an important effect on the time response of a system. We should, however, also note that the nearness of a pole to the $j\omega$-axis also has a large effect upon its importance in the time response. In general the time constant τ_i of a pole p_i is given by

$$\tau_i = \frac{1}{|\text{Re } p_i|}$$

if the pole is in the LHP[4]. Thus a pole near the $j\omega$-axis has a long time constant, affecting the time response over a long period of time. On the other hand, a pole far into the LHP has a short time constant and thus has an effect over only a short period of time. Hence in constructing a time response by use of (4.2-41), those poles having relatively small partial fraction expansion coefficients and those relatively far into the LHP are ignored.

4.3 MITROVIC'S METHOD

Mitrovic's method is more of an approach to the problem of linear system analysis and design than a method. It is basically a two-

---

[4]If a pole is in the RHP it is, of course, always important in the time response.

parameter approach, whereas the frequency-response and root locus methods considered earlier are basically one-parameter methods, that parameter being the frequency-independent loop gain. The frequency-response and root locus methods both indicate how system response varies as this parameter (frequency-independent gain) is varied, and since this parameter is usually a very important one with a direct effect upon the steady-state error performance, these methods are very useful. With Mitrovic's method, however, we are not as limited in the choice of the parameters. Almost any parameter that will control system pole location will work with Mitrovic's method, and this is sometimes an advantage.Another advantage is working with two parameters rather than just one parameter.

Basic Technique

In Mitrovic's method, rather than considering the open-loop transfer function and trying to control the closed-loop response thereby, we consider the closed-loop, or the overall transfer function directly. It really does not matter if the system under consideration is a closed-loop one or not. More specifically, the denominator of the overall transfer function is what we work with. If $T(s)$[5] is the overall transfer function and it is of the usual rational function form, then

$$T(s) = \frac{N(s)}{F(s)} \qquad (4.3\text{-}1)$$

We here consider $F(s)$. Thus the emphasis is all placed upon controlling the poles of the system. This works because, as we have already seen, the stability and the general character of the response are determined by the position of these poles. For a system for which the transfer function is as in (4.3-1), $F(s)$ is called the *characteristic polynomial of the system*; and

$$F(s) = 0 \qquad (4.3\text{-}2)$$

is then called the characteristic equation. It is obvious that the roots of the characteristic polynomial [those values of s that satisfy the characteristic equation (4.3-2)] are the poles of the system. With Mitrovic's method what we do is control the coefficients of the characteistic polynomial and thereby the poles of the system transfer function. The success of the method depends upon the fact that it

[5]If the system of interest should happen to be a closed-loop system (as in Fig. 4.2-1), then $T(s) = G(s)/[1 + G(s)H(s)]$.

does show us how to control these coefficients so as to achieve a desired system pole location.

Our approach is to consider the characteristic equation along lines of constant damping factor (ζ) in the second quadrant of the s-plane. We use a change of variable of the form

$$s = -\zeta\omega_n + j\omega_n \sqrt{1-\zeta^2} \qquad (4.3\text{-}3)$$

which we have seen before (in Chapter 2). Recall that ζ is the damping factor and ω_n is the undamped natural frequency. Using (4.3-3) in the characteristic equation gives a complex equation in the two variables ζ and ω_n and the coefficients of the characteristic polynomial. Setting the real and the imaginary parts equal to zero results in two equations which are linear in the coefficients of the characteristic polynomial. With these two equations we may control system pole location as a function of the coefficients. The remainder of the section shows how this may be done and an example of the use of the method on a system transfer function is included.

The Two Basic Equations

Let us begin by assuming a definite form for the characteristic equation. Thus

$$F(s) = a_n s^n + a_{n-1} s^{n-1} + \cdots + a_1 s + a_0 = 0 \qquad (4.3\text{-}4)$$

We assume without loss of generality that $a_n > 0$. We want to consider roots of the characteristic equation along lines of constant damping factor ζ (Fig. 4.3-1). The significance of ζ and ω_n is seen from the figure; it has also been seen earlier, in Section 2.3. We may convert

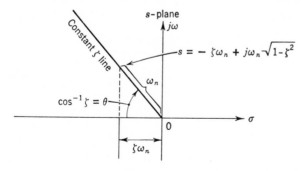

Figure 4.3-1 Line of constant damping factor ζ in the second quadrant.

the characteristic equation from an equation in the variable s to one in the variables ζ and ω_n by substituting (4.3-3) into (4.3-4):

$$a_n(-\zeta\omega_n + j\omega_n\sqrt{1-\zeta^2})^n + a_{n-1}(-\zeta\omega_n + j\omega_n\sqrt{1-\zeta^2})^{n-1} + \cdots$$

$$+ a_1(-\zeta\omega_n + j\omega_n\sqrt{1-\zeta^2}) + a_0 = 0 \tag{4.3-5}$$

Now by multiplying out the factors as indicated in (4.3-5) and then setting the real and imaginary parts separately to zero and after eliminating common factors we get

$$a_0 + a_2\omega_n^2\phi_1(\zeta) + a_3\omega_n^3\phi_2(\zeta) + \cdots + a_n\omega_n^n\phi_{n-1}(\zeta) = 0$$

$$-a_1 + a_2\omega_n\phi_2(\zeta) + a_3\omega_n^2\phi_3(\zeta) + \cdots + a_n\omega_n^{n-1}\phi_n(\zeta) = 0 \tag{4.3-6}$$

where now

$$\phi_0(\zeta) = 0$$
$$\phi_1(\zeta) = -1 \tag{4.3-7}$$

and

$$\phi_k(\zeta) = -[2\zeta\phi_{k-1}(\zeta) + \phi_{k-2}(\zeta)] \tag{4.3-8}$$

for $k > 2$. Values of the most often used ϕ-functions for the most popular values of ζ are given in Table 4.3-1. If higher-order ϕ-functions are required, the table may be easily extended by using (4.3-8).

The two equations as shown in (4.3-6) are the basis of Mitrovic's method. Let us consider how they might be used. First, let us say that we want a system pole at a given point in the second quadrant of the s-plane. Let us say this gives a desirable time and/or frequency response. We can determine the corresponding ζ and ω_n and plug these values into (4.3-6). The result will be two equations in the $n+1$ unknowns $a_0, a_1, \cdots, a_n$ which must be satisfied if the system is to have a pole at the desired location. From (4.3-6) we know that these equations are linear in the coefficients. So if any two of these coefficients are free and the remaining $n-1$ are fixed, we may use these

Table 4.3-1 $\phi_k(\zeta)$ **for Selected Values of ζ and k**

	$\zeta=0$	$\zeta=0.1$	$\zeta=0.2$	$\zeta=0.3$	$\zeta=0.4$	$\zeta=0.5$	$\zeta=0.6$	$\zeta=0.7$	$\zeta=0.8$	$\zeta=1.0$
ϕ_1	-1	-1	-1	-1	-1	-1	-1	-1	-1	-1
ϕ_2	0	0.2	0.4	0.6	0.8	1.0	1.2	1.4	$+1.6$	2.0
ϕ_3	1.0	0.96	0.84	0.64	0.36	0	-0.44	-0.96	-1.56	-3.0
ϕ_4	0	-0.392	-0.736	-0.984	-1.088	-1.000	-0.672	-0.056	$+0.896$	$+4.000$
ϕ_5	-1.000	-0.882	-0.546	-0.050	$+0.510$	$+1.000$	$+1.246$	$+1.038$	$+0.1264$	-5.000
ϕ_6	0	$+0.568$	$+0.954$	$+1.014$	$+0.680$	0	-0.824	-1.398	-1.098	$+6.000$
ϕ_7	$+1.000$	$+0.768$	$+0.164$	-0.164	-1.054	-1.000	-0.258	$+0.918$	$+1.630$	-7.000
ϕ_8	0	-0.722	-1.020	-0.679	$+0.164$	$+1.000$	$+1.133$	$+0.112$	-1.510	$+8.000$
ϕ_9	-1.000	-0.624	$+0.244$	$+0.966$	$+0.923$	0	-1.102	-1.075	$+0.7860$	-9.000
ϕ_{10}	0	$+0.847$	$+0.922$	$+0.099$	-0.902	-1.000	$+0.189$	$+1.393$	$+0.2524$	$+10.000$

two equations to establish the values of the free coefficients (these we assume to be functions of the free parameters) so as to realize a pole at the desired location. Since the equations are linear in the two free coefficients, the solution is straightforward. To see how this might work consider example 4.3-1.

Example 4.3-1

Let us say that for a given system the characteristic polynomial is

$$F(s) = s^3 + a_2 s^2 + a_1 s + 15$$

We then have the two coefficients a_1 and a_2, which may be adjusted to any desired value. Let us say then that through some consideration we may want to have

$$\zeta = 0.5 \quad \text{and} \quad \omega_n = 2$$

to give a settling time and transient response that satisfy system transient specifications. Setting these values of ζ and ω_n into (4.3-6) with $n = 3$ and the remaining coefficients as given in $F(s)$ above and the values of the ϕ-function from Table 4.3-1, we get

$$15 - a_2 \cdot 4 + 8 = 0$$

and

$$-a_1 + a_2 \cdot 2 + 0 = 0$$

whence

$$a_1 = 23/2$$

$$a_2 = 23/4$$

For these values of a_1 and a_2 our system has complex poles at $-\zeta\omega_n \pm j\omega_n\sqrt{1-\zeta^2} = -1 \pm j\sqrt{3}$. The third pole may be found easily by dividing out the factor corresponding to the two complex poles from the characteristic polynomial. The third pole is found to be at $s = -15/4$. This third pole is now far enough into the LHP relative to the complex poles so that the response will be dominated essentially by the complex poles. The values of the coefficients a_1 and a_2 as determined above should be satisfactory. If, on the other hand, we find that the two equations in the two coefficients are singular, we can conclude that the desired pole position is unrealizable with only the two coefficients free to be adjusted.

The convenience of this method is that we have complete freedom in the choice of the two coefficients. In addition, higher-order characteristic equations may be handled in the same way with little additional difficulty. The disadvantages to this approach can be seen to be that we have no control or knowledge of what the transfer function zeros are doing when we adjust free parameters in this way. Nor do we have any direct indication where the system poles other than the two complex poles with desired ζ and ω_n are falling. These points, it should be noted, require separate investigation after the parameters which give the desired location for the one complex pair have been found.

The system designer, generally does not control the values of the coefficients of the characteristic equation. More often the designer controls some system parameters. Let us say that there are two of these free parameters, x_1 and x_2. The usual case is then

$$a_i = f_i(x_1, x_2), \qquad i = 0, 1, \ldots, n \tag{4.3-9}$$

i.e., the coefficients are functions of the free parameters. Substituting (4.3-9) into (4.3-6), we have

$$f_0(x_1, x_2) + f_2(x_1, x_2)\omega_n^2\phi_1(\zeta) + \cdots + f_n(x_1, x_2)\omega_n^n\phi_{n-1}(\zeta) = 0 \tag{4.3-10}$$

$$-f_1(x_1,x_2) + f_2(x_1,x_2)\omega_n\phi_2(\zeta) + \cdots + f_n(x_1,x_2)\omega_n^{n-1}\phi_n(\zeta) = 0$$

which are now for a given ζ and ω_n functions of the two parameters x_1 and x_2. Determining x_1 and x_2 so that (4.3-10) is satisfied results in a pole location with the desired ζ and ω_n. The difficulty depends then upon how complicated the functions $f_i(x_1,x_2)$, $i = 0, 1, \ldots, n$ are. If they are linear functions, (4.3-10) is in turn linear in x_1 and x_2 and the solution gives no difficulty. If they are not linear, the difficulty is greater. Usually, by resorting to numerical methods and the digital computer, a solution can be found. An important special case is when the $f_i(x_1,x_2)$ are bilinear:

$$f_i(x_1,x_2) = c_{i0} + c_{i1}x_1 + c_{i2}x_2 + c_{i12}x_1x_2, \qquad i = 0,1,2,\ldots,n \tag{4.3-11}$$

This occurs in the case of linear networks where the parameters x_1 and x_2 are element values such as an R or C. Substitution of (4.3-11) into (4.3-10) results, for a given ζ and ω_n, in a pair of simultaneous, bilinear equations. As it turns out, a solution[6] is also possible in this case.

---

[6]For this solution and an application of the general method see D. McDermott, S. C. Gupta, and L. Hasdorff, "Realizable Regions for Feedback Circuits," *IEEE Region Six Conference Proceedings*, April 1966, Vol. II.

Determination of Characteristic Polynomial Coefficients to Achieve a Desirable System Response

As an example of the type of thing that can be done using the two equations of (4.3-6), consider what is called the *parameter plane* in which the coefficients of the characteristic polynomial a_0 and a_1 are taken as the free variables. The plane to be considered will have a_0 on the vertical axis and a_1 on the horizontal, i.e., it is the (a_1, a_0)-plane. Again, any other pair of coefficients could be used.

We begin by solving (4.3-6) for a_1 and a_0, which gives

$$a_0 = -a_2\omega_n^2\phi_1(\zeta) - a_3\omega_n^3\phi_2(\zeta) - \cdots - a_n\omega_n^n\phi_{n-1}(\zeta)$$

$$a_1 = a_2\omega_n\phi_2(\zeta) + a_3\omega_n^2\phi_3(\zeta) + \cdots + a_n\omega_n^{n-1}\phi_n(\zeta)$$

$$(4.3\text{-}12)$$

We have assumed that a_0 and a_1 are the free coefficients (parameters) so the coefficients on the RHS of (4.3-12) $a_2, a_3, \ldots, a_n$ must hence be fixed for the system under consideration. Thus (4.3-12) may be used to plot lines of constant ζ(damping factor) in the (a_1, a_0)-plane. We get a line here for a given value of ζ as the variable ω_n on the RHS of (4.3-12) is allowed to vary over the range of interest $0 \leq \omega_n < \infty$. This is a very useful idea since it allows us to map lines of constant ζ in the s-plane (see Fig. 4.3-1) into the (a_1, a_0)-plane.

To see this consider the $\zeta = 0$ line in the s-plane. This is the $j\omega$-axis, which is the demarcation line between the values of a_1 and a_0 for which our system will be stable and those for which it will be unstable. What one may typically expect if one maps in the (a_1, a_0)-plane a line for $\zeta = 0$ using (4.3-12) is shown in Fig. 4.3-2. It may be

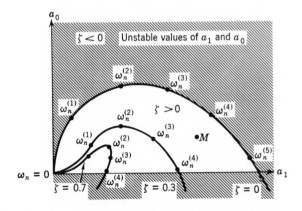

Figure 4.3-2   Unstable and stable regions in the (a_1, a_0)-plane.

seen from (4.3-12) that for $\omega_n = 0$ all the curves go through the origin. The figure shows that the $\zeta = 0$ line divides the plane into two regions. The region where $\zeta > 0$ may be found by choosing a value for $\zeta > 0$ and mapping the corresponding line into the plane. In Fig. 4.3-2 the $\zeta = 0.3$ line is shown; this serves to show the region where $\zeta > 0$. Thus above the $\zeta = 0$ line $\zeta < 0$ and a pair of values chosen from this region will result in a system pole in the RHP of the s-plane. Hence the system will be unstable.

Our region of stable values of a_1 and a_0 may be further limited since we know that in order for the characteristic polynomial to have all LHP roots all coefficients must be positive. Hence the stable values of (a_1, a_0) can only occur in the first quadrant of the (a_1, a_0)-plane. To understand this, recall the Routh criterion (Section 3.2), which says that a polynomial must have all positive coefficients to have all LHP roots. This means that in general we need consider only the first quadrant of any (a_i, a_j) parameter plane we are concerned with. From Fig. 4.3-2 we see that the region of possible values of a_1 and a_0 has been greatly restricted by the use of (4.3-12) and a simple plot in the (a_1, a_0)-plane. This is very useful design information. However, we can do even more than this. If we wish to restrict our poles to fall below the $\zeta = x$ line in the s-plane, we need simply plot the corresponding $\zeta = x$ line in the (a_1, a_0)-plane and then restrict the values of possible (a_1, a_0) combinations to that region where $\zeta \geqslant x$. If for the system of Fig. 4.3-2 we wish to have all poles with damping factor $\zeta \geqslant 0.3$, we see immediately that (a_1, a_0) would have to come from the region enclosed by the $\zeta = 0.3$ line and the positive real axis.

If now we choose a pair of values for a_1 and a_0, this defines a point in the (a_1, a_0)-plane, which we shall call the M-point (Fig. 4.3-2). It should not be difficult now to see that where the M-point is located determines how the system under consideration is going to respond. If a_0 and a_1 are chosen in the stable region the M-point is up near the $\zeta = 0$ line. We know (from our considerations of two-pole systems in Section 2.3) that two poles will result, contributing terms to the output response which will be highly oscillatory in nature. The response will have high overshoots and in general exhibit an undesirable type of behavior. These poles near the $j\omega$-axis will dominate the response and so the behavior of the system may be expected to be unsatisfactory.

For a more satisfactory response we must bring the M-point farther down into the stable region and in general below the $\zeta = 0.3$ line. On the other hand, from the consideration of the two-pole system in Section 2.3, we know that for a damping factor $\zeta > 0.8$ a sluggish type

of response may be expected. The most desirable location for the
M-point is then somewhere between the $\zeta = 0.3$ and $\zeta = 0.8$ curves.
With the (a_1, a_0)-plane another consideration is necessary, however.
From the characteristic equation in (4.3-4) we know that when
$a_0 = 0$, a root will occur at the origin, i.e., we will have a pole at $s = 0$.
Consequently, if the M-point is chosen near the a_1-axis, a real pole
near the origin may be expected to occur. This pole will be additional
to the complex pair, which will be indicated by the constant ζ-lines.
It is also seen in Section 2.3 that such a pole near the origin will
dominate the response and a very sluggish type of behavior will
result. Thus the most desirable region for the location of the M-
point is between the $\zeta = 0.3$ and $\zeta = 0.7$ lines and up away from the
a_1-axis.

In general, a_0 and a_1 will not be available for direct choice. The
usual case is to have two (or even more) free parameters, say x_1
and x_2. Then a_0 and a_1 are functions of these two parameters:

$$a_0 = a_0(x_1, x_2), \qquad a_1 = a_1(x_1, x_2)$$

The problem is to choose x_1 and x_2 so that an M-point in the desirable
region results. To see how these considerations may be used in working with a system, let us consider the following example.

Example 4.3-2

Consider operating a unity-feedback system with

$$G(s) = \frac{K(s + z_0)}{s(s + 1)(s^2 + 2s + 2)}$$

in the forward path. We assume that K and z_0 are the free system
parameters which may be adjusted to control the response. The
overall transfer function for this case is

$$T(s) = \frac{G(s)}{1 + G(s)} = \frac{N(s)}{F(s)}$$

The characteristic polynomial becomes

$$F(s) = s^4 + 3s^3 + 4s^2 + (K + 2)s + Kz_0$$

so that now by adjusting K and z_0 we may control a_1 and a_0. All other
coefficients of the characteristic equation are unaffected by variations in K and z_0. Thus it is logical to consider the (a_1, a_0)-plane. Using
(4.3-12), we may calculate lines of constant ζ in the (a_1, a_0)-plane.
These lines for $\zeta = 0.0, 0.3$, and 0.7 are shown in Fig. 4.3-3. We see
immediately the region of values of (a_1, a_0) for which the system is

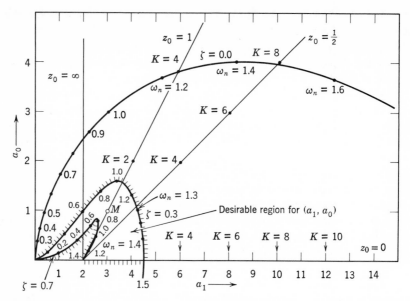

Figure 4.3-3 (a_1, a_0)-plane for the characteristic equation $F(s) = s^4 + 3s^3 + 4s^2 + a_1 s + a_0$

closed-loop stable. It is the region below the $\zeta = 0$ curve in the first quadrant. The desirable values of (a_1, a_0) are between the $\zeta = 0.3$ and $\zeta = 0.7$ curves, as indicated in the figure. It should be noted that the possible values for (a_1, a_0) have been greatly restricted by limiting ourselves to values of ζ in the range $0.3 \leq \zeta \leq 0.7$.

From $F(s)$, we note

$$a_1 = K + 2$$
$$a_0 = K z_0$$

These relations permit us to plot lines in the (a_1, a_0)-plane along which z_0 is a constant. We may use K as a parameter along these curves. These curves for $z_0 = 0, \frac{1}{2}, 1,$ and ∞ are drawn on top of the constant-ζ lines in the (a_1, a_0)-plane of Fig. 4.3-3. We see that for $0 \leq K \leq 2$ and $z_0 > 0$ practically the entire region of desirable (a_0, a_1) is covered and hence desirable response characteristics may be expected for K and z_0 in this range. This type of information is of course very helpful for designing a system specifying desirable ranges for adjustable system parameters.

For the case at hand, if we wished to adjust the free parameters of the system to achieve a desirable response, putting the M-point at $a_0 = 1$, $a_1 = 3$ (Fig. 4.3-3) should result in a satisfactory response. The point is far enough above the a_1-axis that a real pole too near the origin should not be encountered. For this choice of M-point, $\zeta = 0.5$, $\omega_n = 1$ so the time constant of the system $\tau = 1/\zeta\omega_n = 2$ sec. A settling time of approximately 8 sec. may be expected. We can also see from the figure that $K = 1$, $z_0 = 1$ realizes this M-point. We see that for this choice of K and z_0 the overall transfer function is

$$T(s) = \frac{(s+1)}{s^4 + 3s^3 + 4s^2 + 3s + 1}$$

$$= \frac{1}{(s+1)(s^2+s+1)}$$

(4.3-13)

Thus we see by factoring the characteristic polynomial obtained, that the zero at $s = -1$ has been cancelled by a pole at the same point. Hence besides the complex poles with $\zeta = 0.5$, $\omega_n = 1$, we see from (4.3-13) that there is a real pole at $s = -1$. This real pole is fairly close to the $j\omega$-axis relative to the two dominant complex poles which

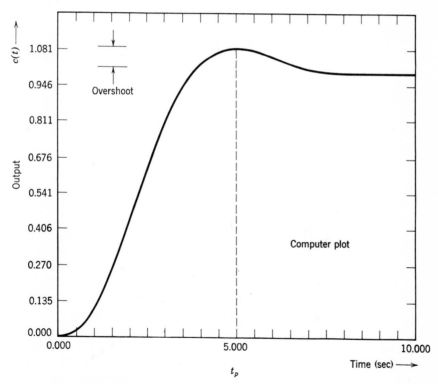

Figure 4.3-4   Step response for $T(s) = 1/(s+1)(s^2+s+1)$.

have real part $\zeta \omega_n = \frac{1}{2}$. As mentioned in Section 2.3, this should result in step response which is somewhat more sluggish than would be the response of a simple two-complex-poles system. Overshoot should not be as high and the time to the peak (t_p) should be longer than if the real pole were not there.

The step response is shown in Fig. 4.3-4. We see a fairly good step response has been achieved for which $t_p = 5$ sec. and overshoot $= 8\%$. This compares with $t_p = 3.6$ sec. and overshoot $= 15\%$ which a two-pole system with $\zeta = 0.5$, $\omega_n = 1$ would have. Thus having the pole on the real axis gives a slightly slower-responding system, but lower overshoot. The relative importance of these two character-istics determine what other values of K and z_0 should be tried in the neighborhood of the values chosen here. However, the step response achieved is a good one, so the values for K and z_0 that have been found may be satisfactory.

PROBLEMS

4.1. Draw the root loci for the following open-loop transfer functions:

(a) $\dfrac{K(s-1)}{s(s+1)}$ (c) $\dfrac{Ks}{s^2+4s+8}$ (e) $\dfrac{K(s+10)(s+15)}{s^3+3s^2+3s+1}$

(b) $\dfrac{K(s+1)}{s^2}$ (d) $\dfrac{Ks}{(s^2+4s+8)(s^2-2s+8)}$ (f) $\dfrac{K(s^2+s+4)}{s(s+4)(s+10)}$

4.2. Draw the zero-degree loci for the open-loop transfer functions of Problem 4.1.

4.3. Given $G(s) = K/(s+1)^3$, $H(s) = 1$ in a standard negative-feedback configuration.

(a) What value of K gives $\zeta = 0.5$ for the two dominant poles?

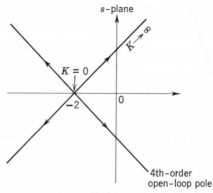

Figure P4.4

(b) What is the approximate step-response overshoot for this value of K?

(c) What is the minimum value of steady-state error to a unit step input that can be achieved with this closed-loop system?

4.4. The root locus for a $G(s)H(s)$ is shown in Fig. P4.4.

(a) What value of K gives a damping factor $\zeta = 0.7$ for the two poles nearest the $j\omega$-axis when the system is operated closed-loop?

(b) What is the settling time for the closed-loop system with the dominant two poles adjusted to have $\zeta = 0.7$?

(c) What K makes the closed-loop system go unstable?

4.5. For $G(s) = K/s(s+1)(s+2)$ operated in a unity-feedback system:

(a) What value of K should be chosen to have steady-state error of 0.5 to a unit ramp input?

(b) What is the minimum value of steady-state error to a ramp which can be achieved with this system?

(c) For what value of K will the closed-loop system have approximately 15% of overshoot to a step input?

(d) For what value of K do we have an essentially critically damped closed-loop response, i.e., a dominant double-pole on the negative real axis?

4.6. In an (x_0,x_1)-plane draw the lines for the following characteristic polynomials which:

(I) Separate the stable from the unstable values of x_0 and x_1.

(II) Give a pair of poles with $\zeta = 0.7$.

  (a) $s^3 + 3s^2 + x_1 s + x_0$

  (b) $s^4 + 22s^3 + 172s^2 + x_1 s + x_0$

  (c) $s^4 + 2.6s^3 + 3.35s^2 + (2.6 + x_1)s + x_0$

  (d) $s^4 + 24s^3 + (10x_1 + x_0)s^2 + 560s + (20x_0 + 55x_1)$

  (e) $s^4 + x_1{}^2 s^3 + 264s + 520s + x_0$

4.7. Given the system as shown in Fig. P4.7. Find the values of K_f and the corresponding ω_n which give $\zeta = 0.4$, 0.5, 0.6 for the dominant closed-loop poles.

4.8. Determine the parameters x_0 and x_1 such that the given characteristic polynomials will have roots with ζ and ω_n as indicated:

(a) $s^4 + x_0 s^3 + 264s^2 + 520s + x_1; \zeta = 0.6, \omega_n = 2$

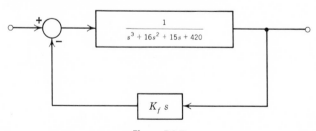

Figure P4.7

(b) $s^4 + 24s^3 + (10x_0 + x_1)s^2 + 560s + (20x_0 + 55x_1)$; $\zeta = 0.707$, $\omega_n = 2.828$

(c) $s^5 + 29s^4 + 308s^3 + (x_0 + x_1)s^2 + 3600s + (3x_0 + 2x_1)$; $\zeta = 0.707$, $\omega_n = 2.828$

4.9. For the following $G(s)$ and $H(s)$ operated in the closed-loop system of Fig. 4.2-1, determine, by use of the spirule, the response to a unit-step input.

(a) $G(s) = \dfrac{1}{s(s+1)(s+2)}$, $H(s) = 1$

(b) $G(s) = \dfrac{2}{s(s+2)}$, $H(s) = \dfrac{1}{s+1}$

(c) $G(s) = \dfrac{15}{(s+5)(s+3)(s^2+2s+2)}$, $H(s) = 1$

5

DISCRETE-TIME SYSTEMS AND THEIR RESPONSE

5.1 INTRODUCTION

We have examined the basic control problem when the data are available continuously within and outside the system. The analysis of such systems has been considered in Chapters 2, 3, and 4.

In Chapter 1 we noted that the control signals in a given system sometimes can be made to change only at discrete instances of time, as for instance when a digital computer is in the signal path of a control system. These requirements give rise to discrete-time or sampled-data control systems.

In this chapter we first develop the basic techniques required for analyzing such systems and then determine how these techniques can be used to find the response of the systems. We start by evolving a mathematical representation of a sampler which appears in all types of discrete-time systems. Later the concepts of Z-transform and modified Z-transform are introduced; these lead to the evaluation of system response analogous to that of continuous systems.

5.2 DEVELOPMENT OF BASIC Z-TRANSFORM AND CONCEPT OF TRANSFER FUNCTIONS IN THE z-DOMAIN

The Sampler and its Mathematical Representation

The sampler is simply a switch which closes every T seconds for one instant of time. To define the function of the switch in more specific mathematical terms, consider the sampler shown in Fig. 5.2-1. When $u(t)$ is sampled, the switch obviously cannot be closed and opened again in zero time. In practice it will take h seconds for this operation. This will give us pulses of width h at the output of the sampler. In order to make a mathematical analysis easier, we assume that these finite-width pulses can be replaced by impulses such that the strength of the impulses is equal to the product of the magnitude of the function at the initial instant of sampling time with the pulse width h. But for convenience, since h is a constant, it can be considered a separate

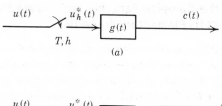

(a)

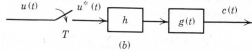

(b)

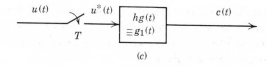

(c)

(d)

Figure 5.2-1 (a) Original finite-width sampling. (b) and (c) Equivalent approximate system. (d) Approximate sampler representation (ideal sampler).

multiplying element and combined with the following plant, as it is in Figs. 5.2-1a, b, and c. Thus with this equivalence we need consider only the "ideal sampler" of Fig. 5.2-1d. Thus we write

$$u^*(t) \overset{\Delta}{=} \sum_{n=0}^{\infty} u(nT)\, \delta(t-nT) \tag{5.2-1}$$

This equation implies that the output of the sampler is a string of impulses (δ-functions) starting at $t = 0$ spaced at intervals T seconds

165

and of amplitude (or power) $u(nT)$. The nature of $u(t)$ for $t < 0$ is of no consequence; we are interested only in $t \geqslant 0+$.

A typical input $u(t)$, the actually sampled $u(t)$, and the ideally sampled $u(t)$ which is used in mathematical analysis are shown in Fig. 5.2-2. The ideally sampled $u(t)$ is called the impulse modulation of $u(t)$. This mathematical representation paves the way for analyzing discrete-time systems.

Transforms of the Sampled Function and Introduction to the Z-Transform and Some of Its Elementary Properties

Once we accept the mathematical representation of the output of the ideal sampler, we proceed to develop its Laplace transform and try to develop a block diagram algebra as was done for continuous systems. We test whether the basic tools of analysis developed in Chapters 2, 3, and 4 can be used. We have

$$\mathscr{L}\left\{u^*(t)\right\} = \mathscr{L}\left\{\sum_{n=0}^{\infty} u(nT)\,\delta(t-nT)\right\}$$

$$= \sum_{n=0}^{\infty} u(nT)\,e^{-nTs} \tag{5.2-2}$$

We note that the Laplace transform of a sampled function is in the form of an infinite series and involves factors of the form e^{sT} and its various powers. This fact makes the analysis of sampled-data systems in the s-domain more involved.

In order to avoid this difficulty, a new variable, z, is introduced such that

$$z = e^{sT} \tag{5.2-3}$$

where this relationship provides a simple conformal mapping from s to the z-plane. With this transformation, we define a new transform

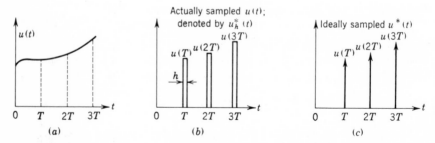

Figure 5.2-2 A typical $u(t)$, actually sampled $u^*(t)$, and ideally sampled $u^*(t)$.

called the Z-transform such that

$$Z\{u(t)\} = U(z) = \text{Z-transform of } u(t)$$

$$\overset{\Delta}{=} \mathscr{L}\{u^*(t)\}\Big|_{e^{sT}=z} = \sum_{n=0}^{\infty} u(nT)z^{-n}$$

(5.2-4)

This is called the one-sided Z-transform [because it enfolds $u(t)$ only over the range $t \geqslant 0+$, hence it entails integral values of $n \geqslant 0$]. We note from (5.2-4) that the Z-transform of a function depends only upon the values of the function at sampling instants. All functions which have the same values at sampling instants have the same Z-transforms. As an example, let us find the Z-transform of the unit-step function:

$$Z\{1(t)\} = \sum_{n=0}^{\infty} 1(nT) z^{-n} = \sum_{n=0}^{\infty} z^{-n} = \frac{1}{1-z^{-1}} = \frac{z}{z-1} \quad (5.2\text{-}5)$$

We can obtain the Z-transforms of the time functions in the closed form if we can sum the series as in (5.2-5).

We can now develop some *more commonly used Z-transform properties* which enable us to write the Z-transform of various commonly encountered time functions easily. Let

$$Z\{f(t)\} \overset{\Delta}{=} F(z) \overset{\Delta}{=} \sum_{n=0}^{\infty} f(nT) z^{-n} \quad (5.2\text{-}6)$$

1. COMPLEX TRANSLATION

$$Z\{e^{-at}f(t)\} = \sum_{n=0}^{\infty} e^{-anT}f(nT) z^{-n}$$

$$= \sum_{n=0}^{\infty} f(nT)(ze^{aT})^{-n}$$

Using (5.2-6), we note that

$$Z\{e^{-at}f(t)\} = F(e^{aT}z) \quad (5.2\text{-}7)$$

As an application, using (5.2-5), we write

$$Z\{e^{-at}\} = Z\{1(t)\}\Big|_{z \to ze^{aT}} = \frac{ze^{aT}}{ze^{aT}-1} = \frac{z}{z-e^{-aT}} \quad (5.2\text{-}8)$$

From (5.2-8),

$$Z\{\cosh at\} = Z\left\{\frac{e^{at}+e^{-at}}{2}\right\} = \frac{z(z-\cosh \alpha T)}{z^2 - 2z\cosh \alpha T + 1} \quad (5.2\text{-}9)$$

and

$$Z\{\sinh at\} = \frac{z\sinh \alpha T}{z^2 - 2z\cosh \alpha T + 1} \quad (5.2\text{-}10)$$

Also

$$Z\{\cos \beta t\} = Z\left\{\frac{e^{j\beta t} + e^{-j\beta t}}{2}\right\} = \frac{1}{2}\left[\frac{z}{z - e^{-j\beta T}} + \frac{z}{z - e^{j\beta T}}\right]$$

$$= \frac{z(z - \cos \beta T)}{z^2 - 2z \cos \beta T + 1} \tag{5.2-11}$$

and similarly,

$$Z\{\sin \beta t\} = \frac{z \sin \beta T}{z^2 - 2z \cos \beta T + 1} \tag{5.2-12}$$

2. COMPLEX DIFFERENTIATION

$$Z\{tf(t)\} = \sum_{n=0}^{\infty} nT f(nT) z^{-n}$$

$$= -T \sum_{n=0}^{\infty} f(nT) z \left\{\frac{d}{dz}(z^{-n})\right\}$$

$$= -T z \frac{d}{dz}\left[\sum_{n=0}^{\infty} f(nT) z^{-n}\right]$$

Again using (5.2-6),

$$Z\{tf(t)\} = -Tz \frac{d}{dz}[F(z)] \tag{5.2-13}$$

As an application consider

$$Z\{t\} = -Tz \frac{d}{dz}[Z\{1(t)\}]$$

$$= -Tz \frac{d}{dz}\left(\frac{z}{z-1}\right) = \frac{Tz}{(z-1)^2} \tag{5.2-14}$$

The preceding two properties are very basic. They enable us to develop Z transforms of most commonly used time functions. We will develop further properties as we need them.

Basic Block Diagram Relationship

Let us now consider the system of Fig. 5.2-3, a linear time-invariant plant with impulse response $g(t)$. The output is shown sampled through an assumed sampler, the output of which is $c^*(t)$. For this

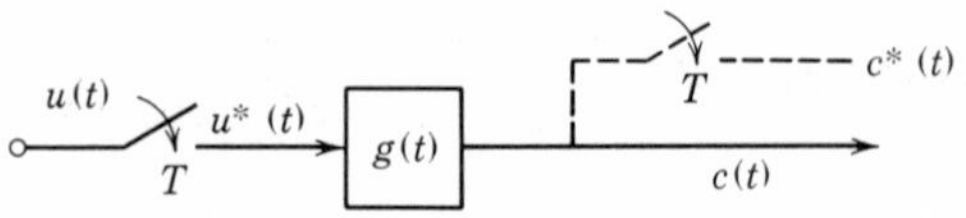

Figure 5.2-3   Linear time-invariant system with sampler at input.

system, the convolution integral may be used to determine the output $c(t)$. This gives

$$c(t) = \int_{-\infty}^{\infty} u^*(\tau) g(t-\tau)\, d\tau \qquad (5.2\text{-}15)$$

Now using (5.2-2),

$$c(t) = \int_0^{\infty} \sum_{n=0}^{\infty} u(nT)\delta(\tau-nT)g(t-\tau)\,d\tau \qquad (5.2\text{-}16)$$

since $u^*(t) = 0$ for $t < 0$. Applying the sifting property of the δ-function to (5.2-16), we get

$$c(t) = \sum_{n=0}^{\infty} u(nT)g(t-nT) \qquad (5.2\text{-}17)$$

which is the sampled-data equivalent to the convolution integral. Equation (5.2-17) is one way to calculate the time response of this type of system, i.e., the one of Fig. 5.2-1. However, it is more convenient to use the Z-transform approach.

Consider the Z-transform of the output of the system, $c(t)$. We have

$$C(z) = Z\{c(t)\} = \sum_{n=0}^{\infty} c(nT)z^{-n} \qquad (5.2\text{-}18)$$

Equation (5.2-17) substituted into (5.2-18) gives

$$C(z) = \sum_{n=0}^{\infty} \left[\sum_{m=0}^{\infty} u(mT)g(\overline{n-m}T) \right] z^{-n}$$

$$= \sum_{m=0}^{\infty} u(mT) \sum_{n=0}^{\infty} g(\overline{n-m}T)z^{-n} \qquad (5.2\text{-}19)$$

In the preceding summation, let

$$k = n - m$$

We have

$$C(z) = \sum_{m=0}^{\infty} \left[u(mT) \sum_{k=-m}^{\infty} g(kT)z^{-k-m} \right]$$

$$= \sum_{m=0}^{\infty} u(mT)\, z^{-m} \sum_{k=0}^{\infty} g(kT)z^{-k} \qquad (5.2\text{-}20)$$

since $g(kT) = 0$ for $k < 0$.

Using the definition of Z-transform, from (5.2-20) we get

$$C(z) = U(z)G(z) \qquad (5.2\text{-}21)$$

where $U(z)$ and $G(z)$ are the Z-transforms of $u(t)$ and $g(t)$, respectively. Equation (5.2-21) is now the sampled-data equivalent to $C(s) = U(s)G(s)$. It is again a very simple result which, in words, says that the Z-transform of the output of a sampled-data system (SDS) is obtained by multiplying the Z-transform of the input by the Z-transform of the impulse response of the linear plant portion of the system (that which appears after the sampler). In this case $G(z)$ is again called the transfer function of the system as shown in Fig. 5.2-3. We shall call it the z-transfer function if it is necessary to distinguish it from $G(s) = \mathscr{L}\{g(t)\}$.

We emphasized the fact that (5.2-21) holds only for the case of zero initial conditions at $t = 0^-$.

Example 5.2-1

The impulse response of a linear system is e^{-t}. If a sampled ramp function is used as an input to this system, find the Z-transform of the output.

We have here $u(t) = t$ and $g(t) = e^{-t}$, $t > 0$. Then using (5.2-14) and (5.2-8), respectively, we get

$$U(z) = \frac{Tz}{(z-1)^2} \quad \text{and} \quad G(z) = \frac{z}{z-e^{-T}}$$

Therefore

$$C(z) = U(z)G(z) = \frac{Tz^2}{(z-1)^2(z-e^{-T})}$$

$$= \frac{Tz^2}{z^3 - (2+e^{-T})z^2 + (2+e^{-T})z - e^{-T}}$$

$$= Tz^{-1} + T(2+e^{-T})z^{-2} + T(2+e^{-T})(1+e^{-T})z^{-3} + \cdots$$

Therefore using (5.2-4) we have

$$c(0) = 0, \quad c(T) = T, \quad c(2T) = T(2+e^{-T}),$$
$$c(3T) = T(2+e^{-T})(1+e^{-T}), \ldots$$

We note that the expansion of the rational function of z into the infinite series in z^{-1} can be accomplished by simple synthetic division. We may find as many terms in the infinite series in z^{-1} [and, consequently as many values of $c(nT)$] as desired by continuing the synthetic division. However, such expansion may be tedious except for simple cases.

The Determination of the Transfer Function

In the SDS case, as in the continuous-time case, the transfer function may be determined if the input and output of the system can be observed, starting with the system in a quiescent state. From (5.2-21) we have

$$G(z) = \frac{C(z)}{U(z)} \bigg|_{\text{zero initial conditions}} \tag{5.2-22}$$

This is, of course, simple and direct. Note that in order to find the z-transfer function for this type of system, $c(t)$ and $u(t)$ need be known only at sampling instances.

A case of practical interest in the use of (5.2-22) in determining transfer functions occurs for those linear time-invariant SDS, in which the input and output of the system (at the sampling instances) are related by a linear, constant-coefficient difference equation. This now may be written in the form

$$c(\overline{k+nT}) + b_{n-1} c(\overline{k+n-1T}) + b_{n-2} c(\overline{k+n-2T}) + \cdots + b_0 c(kT)$$
$$= a_m u(\overline{k+mT}) + a_{m-1} u(\overline{k+m-1T}) + \cdots + a_0 u(kT), \qquad (5.2\text{-}23)$$

where

$$m \leqslant n, k = 0^+, 1^+, 2^+, 3^+, \ldots$$

A system in which the inputs and outputs, at sampling instances, satisfy such a difference equation may be seen by inspection to be a linear time-invariant one. To find the z-transfer function of such a system simply take the Z-transform of both sides. However, before we do this we must establish how to find the Z-transform of shifted time functions of the form $f(t+nT)$. Using (5.2-4), we can write

$$Z\{f(t+nT)1(t)\} = \sum_{k=0}^{\infty} f(kT+nT)1(kT)z^{-k}$$

Let $k + n = p$; then we get

$$Z\{f(t+nT)1(t)\} = \sum_{p=n}^{\infty} f(pT)1(pT-nT)z^{-p+n}$$

$$= z^n \left[\sum_{p=0}^{\infty} f(pT)1(pT)z^{-p} - \sum_{p=0}^{n-1} f(pT)1(pT)z^{-p} \right]$$

$$= z^n F(z), \text{ with zero initial conditions.} \qquad (5.2\text{-}24)$$

Applying relation (5.2-24) to (5.2-23), we get

$$z^n C(z) + b_{n-1} z^{n-1} C(z) + \cdots + b_0 C(z) = a_m z^m U(z) + a_{m-1} z^{m-1} U(z)$$
$$+ \cdots + a_0 U(z) \qquad (5.2\text{-}25)$$

which gives the z-transfer function as

$$\frac{C(z)}{U(z)} = G(z) = \frac{a_m z^m + a_{m-1} z^{m-1} + \cdots + a_0}{z^n + b_{n-1} z^{n-1} + \cdots + b_0}, \ m \leqslant n \qquad (5.2\text{-}26)$$

We see that in this important case, the z-transfer function is the ratio of two polynomials in z, a rational function. This case, the most important one in practice, is dealt with exclusively here. Note that the z-transfer function of a system corresponding to a system whose transfer function in the s-domain is a rational function is in turn a rational function in the z-domain. That is, if

then
$$\mathscr{L}\{g(t)\} = (\text{rational function in } s)$$

$$Z\{g(t)\} = (\text{rational function in } z)$$

This may be seen in the tables of transform pairs given later.

Note also that if the transfer function of a given SDS is a rational function in z, then a constant-coefficient difference equation may be written relating the input and output (at sampling instances) of the system. We put the transfer function in the form of (5.2-26) and then write the corresponding difference equation as in (5.2-23).

Handling Block Diagrams in the z-Domain

Utilizing the basic relationship of (5.2-21), we can very easily determine the z-transfer function for various configurations of block diagrams. Let us consider a few specific cases.

1. Consider the system of Fig. 5.2-4a. We note that

$$C(s) = G_1(s)G_2(s)U(z) \tag{5.2-27}$$

$$\therefore C(z) = Z\{G_1(s)G_2(s)\}U(z)$$

$$= G_1G_2(z)U(z) \tag{5.2-28}$$

This implies that whenever there is more than one system function between two switches, one has to take the Z-transform of the product in the s-domain.

2. Consider the system in Fig. 5.2-4b. We note that

$$B(s) = U(z)G_1(s)$$

$$\therefore B(z) = U(z)G_1(z)$$

Also

$$C(s) = U(z)G_1(z)G_2(s)$$

$$\therefore C(z) = U(z)G_1(z)G_2(z) \tag{5.2-29}$$

In this case we have the product of the Z-transforms of $G_1(s)$ and $G_2(s)$ because of the sampler between them.

3. Consider the system of Fig. 5.2-4c. Here

$$C(s) = E(z)G(s) \tag{5.2-30}$$

and

$$E(s) = R(s) - H(s)C(s) \tag{5.2-31}$$

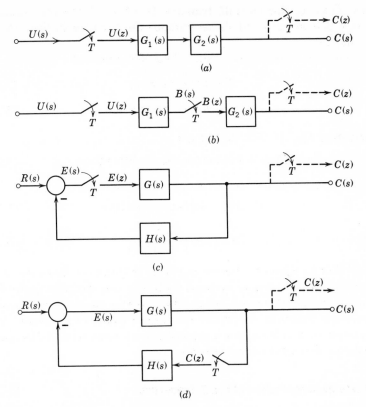

Figure 5.2-4 Some typical block diagrams of sampled-data or discrete-time systems.

Using (5.2-30), we get

$$E(s) = R(s) - E(z)G(s)H(s). \qquad (5.2\text{-}32)$$

Taking Z-transforms of both sides,

$$E(z) = R(z) - E(z)GH(z)$$
$$\therefore E(z) = \frac{1}{1 + GH(z)} R(z) \qquad (5.2\text{-}33)$$

Also from (5.2-30) by Z-transform

$$C(z) = E(z)G(z)$$

Substituting $E(z)$ from (5.2-33) gives

$$\frac{C(z)}{R(z)} = T(z) = \frac{G(z)}{1 + GH(z)} \qquad (5.2\text{-}34)$$

where $T(z)$ is the overall transfer function. Note the similarity between this system and those without the sampler in the s-domain (see Chapter 2).

4. Consider the system of Fig. 5.2-4d. We note that

$$C(s) = G(s)E(s) \tag{5.2-35}$$

and

$$E(s) = R(s) - H(s)\,C(z) \tag{5.2-36}$$

Substituting $E(s)$ in (5.2-35), we get

$$C(s) = R(s)\,G(s) - G(s)\,H(s)\,C(z) \tag{5.2-37}$$

Taking Z-transforms of both sides and manipulating,

$$C(z)[1 + GH(z)] = RG(z) \tag{5.2-38}$$

or

$$C(z) = \frac{RG(z)}{1 + GH(z)} \tag{5.2-39}$$

We see that we cannot separate out $R(z)$ in this case. Because of this, the analysis and design of systems with only one sampler in the feedback path are slightly more difficult because we cannot write the overall transfer function in terms of system parameters alone. More experience in handling block diagrams with samplers will be gained from the problems at the end of chapter.

Alternate Representation of the Z-Transform

We note from the foregoing manipulations of the block diagrams that in order to obtain the Z-transforms of the output, error, etc., we will need to obtain $G(z)$ given $G(s)$.

Let $G(s) = 1/(s+1)$. We can obtain $G(z)$ by first finding $g(t)$ as e^{-t} and then using (5.2-8) to get $G(z) = z/(z - e^{-T})$. This way we can even construct a table that will enable us to write the Z-transforms directly from the s-domain. This is usually quite satisfactory; however, we shall need alternate ways of obtaining the Z-transform from the s-domain in order to develop the modified Z-transform as well as the basic sampling theorem and some frequency-domain concepts.

The δ-function has the property that

$$u(t)\,\delta(t-\tau) = u(\tau)\,\delta(t-\tau)$$

We can rewrite (5.2-1) for the sampled signal as

$$u^*(t) \overset{\Delta}{=} \sum_{n=0}^{\infty} u(t)\,\delta(t - nT) \tag{5.2-40}$$

or

$$u^*(t) = u(t) \sum_{n=0}^{\infty} \delta(t - nT) \qquad (5.2\text{-}41)$$

We know from (5.2-4) that

$$Z\{u(t)\} = U(z) = \mathscr{L}\{u^*(t)\}\Big|_{z=e^{Ts}} = \mathscr{L}\left\{u(t) \sum_{n=0}^{\infty} \delta(t - nT)\right\}\Big|_{z=e^{Ts}}$$

Let us now examine $\qquad\qquad\qquad\qquad\qquad\qquad\qquad$ (5.2-42)

$$U(z) = \mathscr{L}\left\{u(t) \sum_{n=0}^{\infty} \delta(t - nT)\right\} \qquad (5.2\text{-}43)$$

We note that (5.2-43) is the Laplace transform of the product of two time functions $u(t)$ and $\sum_{n=0}^{\infty} \delta(t - nT)$. We can therefore use the complex convolution theorem of the Laplace transform as given in Table 2.2-1 to evaluate (5.2-43).

We now have

$$\mathscr{L}\{u(t)\} = U(s)$$

$$\mathscr{L}\left\{\sum_{n=0}^{\infty} \delta(t - nT)\right\} = \sum_{n=0}^{\infty} e^{-nTs} = \frac{1}{1 - e^{-Ts}}$$

Therefore

$$\mathscr{L}\{u^*(t)\} = U^*(s) = \mathscr{L}\left\{u(t) \sum_{n=0}^{\infty} \delta(t - nT)\right\}$$

$$= \frac{1}{2\pi j} \int_{\sigma_1 - j\infty}^{\sigma_1 + j\infty} \frac{U(p)}{1 - e^{-T(s-p)}} \, dp \qquad (5.2\text{-}44)$$

where

$$\sigma > (\sigma^{\mathrm{I}} + \sigma^{\mathrm{II}}) \qquad \text{and} \qquad \sigma^{\mathrm{I}} < \sigma_1 < \sigma - \sigma^{\mathrm{II}}$$

and σ^{I} and σ^{II} are abscissas of convergence of $U(p)$ and $1/[1 - e^{-T(s-p)}]$ respectively, in the p-plane.

Let us now evaluate the integral of (5.2-44). Normally, since $u(t)$ is well behaved (stable), the poles of $U(p)$ will be in the left half of the p-plane as shown in Fig. 5.2-5. The poles of $1/[1 - e^{-T(s-p)}]$ can be obtained by setting

$$1 - e^{-T(s-p)} = 0$$

which implies that the poles are at $-T(s - p) = j2\pi k$, k all integral values, or

$$p = s + j\frac{2\pi k}{T}$$

Since k can have an infinite number of integral values, the number of poles represented by $p = s + j(2\pi k/T)$ is infinite. These poles are shown in Fig. 5.2-5. We know from the complex convolution theorem that the σ_1-line has to be to the right of the poles of $U(p)$ and to the left

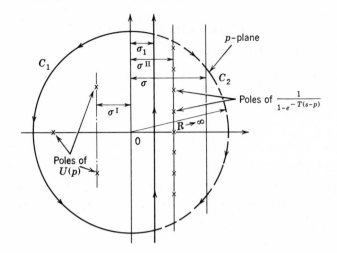

Figure 5.2-5   Location of σ_1-line for integration to determine $U^*(s)$.

of the real part of s. This condition insures the convergence of the convolution integral. The σ_1-line satisfying these requirements is shown in Fig. 5.2-5, along with the σ^{I}-, σ^{II}-, and σ-lines. To evaluate the integral of (5.2-44) we can close the σ_1-line either to the left or to the right and then evaluate the residues for enclosed poles.

1. Let us first close the line to the left as shown in Fig. 5.2-5. The value of the integral is zero along the circular arc C_1 provided

$$\left| \frac{U(p)}{1 - e^{-T(s-p)}} \right| \to 0$$

as $|p| \to \infty$ on C_1. We then write

$$I_1 = \frac{1}{2\pi j} \int_{\sigma_1 - j\infty}^{\sigma_1 + j\infty} \frac{U(p)}{1 - e^{-T(s-p)}} \, dp = \frac{1}{2\pi j} \oint \frac{U(p)}{1 - e^{-T(s-p)}} \, dp \qquad (5.2\text{-}45)$$

The integral on the right can be evaluated by Cauchy's theorem since the contour just encloses the poles of $U(p)$. Thus the value of the integral is the sum of residues of the poles of $U(p)$. Let the poles of $U(p)$ be $p_1, p_2, \ldots, p_k$ with corresponding multiplicity $n_1, n_2, \ldots, n_k$. Then using Cauchy's theorem we have

$$\frac{1}{2\pi j} \int_{\sigma_1 - j\infty}^{\sigma_1 + j\infty} \left\{ \frac{U(p)}{1 - e^{-T(s-p)}} \right\} dp = \sum_{i=1}^{k} \frac{1}{(n_i - 1)!} \frac{d^{n_i - 1}}{dp^{n_i - 1}}$$

$$\left[(p - p_i)^{n_i} U(p) \frac{1}{1 - e^{-T(s-p)}} \right]\Bigg|_{p = p_i}$$

where n_i is the order of pole p_i.

We now let $z = e^{Ts}$; we obtain

$$U(z) = \sum_{i=1}^{k} \frac{1}{(n_i - 1)!} \frac{d^{n_i-1}}{dp^{n_i-1}} \left[(p - p_i)^{n_i} U(p) \left\{ \frac{z}{z - e^{Tp}} \right\} \right] \Bigg|_{p=p_i} \quad (5.2\text{-}46)$$

Equation (5.2-46) obviously gives us a closed form of $U(z)$ for a finite number of poles of $U(p)$. The result is only valid if $|U(p)| \to 0$ as $|p| \to \infty$ on C_1 as noted.

2. Let us now close the contour to the right. This closed contour is also shown in Fig. 5.2-5. Note first that the singularities enclosed are those of $1/[1 - e^{-T(s-p)}]$ and, second, the contour is in a clockwise direction, which implies, by Cauchy's theorem, that the integral around the contour is minus the sum of the residues.[1] The enclosed poles are at $p = s + j(2\pi k/T)$. We can show that

$$U^*(s) = - \sum_{k=-\infty}^{\infty} \frac{U(p)}{(d/dp)[(1 - e^{-T(s-p)})]} \Bigg|_{p=s+j(2\pi k/T)} + \frac{u(0+)}{2}$$

$$= \frac{1}{T} \sum_{k=-\infty}^{\infty} U\left(s + j\frac{2\pi}{T}k\right) + \frac{u(0+)}{2} \quad (5.2\text{-}47)$$

where $u(0+)$ is the value of $u(t)$ at $t = 0+$.

In order to obtain $U(z)$, we can set $z = e^{Ts}$ in (5.2-47). Equations (5.2-46) and (5.2-47) give alternate ways of evaluating Z-transforms; however, only (5.2-46) will give results in closed form and directly from the s-domain. Table 5.2-1 is a short table of commonly used Z-transforms.

Example 5.2-2

Find the Z-transform of $G(s) = 1/(s+1)^2$, we get

$$G(z) = Z\left\{ \frac{1}{(s+1)^2} \right\} = \frac{1}{2\pi j} \int_{\sigma_1 - j\infty}^{\sigma_1 + j\infty} \frac{1}{(p+1)^2} \frac{1}{1 - e^{-T(s-p)}} dp \Bigg|_{z=e^{Ts}}$$

Applying (5.2-46),

$$G(z) = \frac{1}{2-1!} \frac{d}{dp}\left(\frac{z}{z - e^{Tp}} \right) \Bigg|_{p=-1} = \frac{Tze^{-T}}{(z - e^{-T})^2}$$

---

[1] For more details consult S. N. Carrol and W. L. McDaniel, "Use of Convolution Integral in the Sampled-Data Theory," *IEEE Trans. on Automatic Control*, 1966, **AC-11**, 328-329. More details are also available in S. C. Gupta, *Transform and State Variable Methods in Linear Systems*, John Wiley, New York, 1966, p. 143.

TABLE 5.2-1 Short Table of Z-Transforms

$F(s)$	$F(z)$	$f(t)$
$\dfrac{1-e^{-sT}}{s}F(s)$	$\dfrac{z-1}{z}Z\left\{\dfrac{F(s)}{s}\right\}$	
K	K	$K\delta(t)$
$\dfrac{1}{s}$	$\dfrac{z}{z-1}$	$1(t) \;\; =u(t)$
$\dfrac{1}{s^2}$	$\dfrac{Tz}{(z-1)^2}$	t
$\dfrac{1}{s^3}$	$\dfrac{T^2z(z+1)}{2(z-1)^3}$	$\dfrac{t^2}{2}$
$\dfrac{1}{s+\alpha}$	$\dfrac{z}{z-e^{-\alpha T}}$	$e^{-\alpha t}$
$\dfrac{1}{(s+\alpha)^2}$	$\dfrac{zTe^{-\alpha T}}{(z-e^{-\alpha T})^2}$	$te^{-\alpha t}$
$\dfrac{\alpha}{s^2+\alpha^2}$	$\dfrac{z\,\sin\alpha T}{z^2-2z\cos\alpha T+1}$	$\sin\alpha t$
$\dfrac{s}{s^2+\alpha^2}$	$\dfrac{z(z-\cos\alpha T)}{z^2-2z\cos\alpha T+1}$	$\cos\alpha t$
$\dfrac{\alpha}{(s+\beta)^2+\alpha^2}$	$\dfrac{ze^{-\beta T}\sin\alpha T}{z^2-2ze^{-\beta T}\cos\alpha T+e^{-2\beta T}}$	$e^{-\beta t}\sin\alpha t$
$\dfrac{s+\beta}{(s+\beta)^2+\alpha^2}$	$\dfrac{z(z-e^{-\beta T}\cos\alpha T)}{z^2-2ze^{-\beta T}\cos\alpha T+e^{-2\beta T}}$	$e^{-\beta t}\cos\alpha t$
$\dfrac{\alpha}{s^2-\alpha^2}$	$\dfrac{z\sinh\alpha T}{z^2-2z\cosh\alpha T+1}$	$\sinh\alpha t$
$\dfrac{s}{s^2-\alpha^2}$	$\dfrac{z(z-\cosh\alpha T)}{z^2-2z\cosh\alpha T+1}$	$\cosh\beta t$
$\dfrac{\alpha}{(s+\beta)^2-\alpha^2}$	$\dfrac{z\,e^{-\beta T}\sinh\alpha T}{z^2-2ze^{-\beta T}\cosh\alpha T+e^{-2\beta T}}$	$e^{-\beta t}\sinh\alpha t$
$\dfrac{s+\beta}{(s+\beta)^2-\alpha^2}$	$\dfrac{z(z-e^{-\beta T}\cosh\alpha T)}{z^2-2ze^{-\beta T}\cosh\alpha T+e^{-2\beta T}}$	$e^{-\beta t}\cosh\alpha t$
$\dfrac{1}{s^n}$	$\lim\limits_{a\to0}\dfrac{(-1)^{n-1}}{n-1!}\dfrac{\partial^{n-1}}{\partial a^{n-1}}\left(\dfrac{z}{z-e^{-aT}}\right)$	$\dfrac{t^{n-1}}{(n-1)!}$

5.3 THE BASIC PROCESS OF SAMPLING AND THE USE OF HOLD CIRCUITS

Simple Form of the Sampling Theorem

We have defined the effect of the sampler on a continuous input as given. How do we arrive at a suitable frequency of sampling? Should we apply the sampled signal directly to the controlled plant or should we try to reconstruct the signal? To answer these basic questions, we examine the frequency content of the presampled and postsampled signals.

The spectrum of the signal is given by $U(j\omega)$. The spectrum of the sampled signal is given by $U^*(j\omega)$. This can be obtained in terms of the input spectrum from (5.2-47). Assuming $u(0+) = 0$, we have

$$U^*(j\omega) = \frac{1}{T} \sum_{k=-\infty}^{\infty} U\left(j\omega + j\frac{2\pi}{T}k\right) \qquad (5.3\text{-}1)$$

Let the sampling frequency be $\omega_r = 2\pi/T$; then

$$U^*(j\omega) = \frac{1}{T} \sum_{k=-\infty}^{\infty} U(j\overline{\omega + k\omega_r}) = \frac{1}{T}[\cdots + U(j\overline{\omega - \omega_r}) + U(j\omega)$$
$$+ U(j\overline{\omega + \omega_r}) + \cdots] \qquad (5.3\text{-}2)$$

For practical purposes we can assume that $U(j\omega)$ is band limited, i.e., $|U(j\omega)| \simeq 0$ for $|\omega| > \omega_c$, since for the usual control signal the amplitude of high-frequency components decreases very rapidly with increasing frequency. A typical case is shown in Fig. 5.3-1a. From (5.3-2) we take the absolute values:

$$\left|U^*(j\omega)\right| \leqslant \cdots + \frac{1}{T}\left|U(j\overline{\omega - \omega_r})\right| + \frac{1}{T}\left|U(j\omega)\right| + \frac{1}{T}\left|U(j\overline{\omega + \omega_r})\right| + \cdots$$
$$(5.3\text{-}3)$$

The equality sign will hold if $\omega_c \leqslant \omega_r/2$. In that case we note that the magnitude spectrum of the sampled signal $U^*(j\omega)$ as plotted in Fig. 5.3-1b using (5.3-3) has a primary component centered at $\omega = 0$ and an infinite number of side components centered at steps of ω_r. These components are all of the same shape as $U(j\omega)$ but multiplied by a constant factor $1/T$. As shown, we note that if $\omega_c \leqslant \omega_r/2$, then there is no overlap of the sideband components of the sampled-signal spectrum. Thus in order to reconstruct $u(t)$ from $u^*(t)$, all we do is pass $u^*(t)$ through a lowpass filter which rejects the sideband components and amplifies by the factor T.

Unless we somehow get rid of the higher-order sidebands, they will

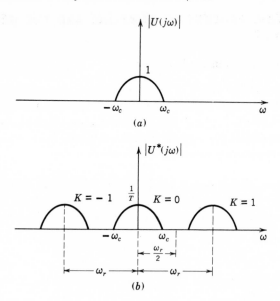

Figure 5.3-1   (*a*) A typical control signal spectrum. (*b*) Control signal spectrum after sampling.

go into the controlled system, where they act undesirably as high-frequency "noise," so normally the sampled signal has to be passed through some kind of bandpass filter (usually after it is processed through a discrete compensator). This removes the sidebands and provides a continuous input to the controlled elements.

The sampling frequency ω_r is usually taken much higher than twice the highest significant frequency ω_c of the reference signal in practice. This insures very little overlapping of spectra and makes the filtering job easier. (Choice of sampling frequency is also influenced by stability factors, as will be seen later.) A continuous signal from the ideally sampled signal can be approximately reconstructed by what is called a *hold circuit*, which we consider now.

The Use of the Hold Circuit

In practice, an ideal sampler (or its approximation) is synthesized in combination with a hold circuit. Although this is called a circuit, it need not necessarily be so. Whatever its nature, it is simply a device whose function is to reconstruct from the ideally sampled signal a good approximation to some useful continuous signal. Intuitively it should be obvious that we do not want to apply a sampled signal (a string of impulses) to the element being controlled (the plant). We want to reconstruct from the sampled signal a continuous signal before we apply it to the plant.

The simplest (and most frequently occurring) such device is the *zero-order hold* (ZOH), which accepts an impulse of strength K and puts out a fixed constant pulse of value K between samplings. Such a device may be realized as a linear time-invariant system with an impulse response as shown in Fig. 5.3-2. More precisely,

$$g_{oh}(t) = 1 \quad \text{for} \quad 0 \leqslant t \leqslant T$$

$$= 0 \quad \text{otherwise} \tag{5.3-4}$$

Therefore

$$G_{oh}(s) = \frac{1 - e^{-sT}}{s} \tag{5.3-5}$$

The effect of a sampler and zero-order hold may be seen in Fig. 5.3-3. We note that the Z-transform of the zero-order hold impulse response function is unity:

$$Z\{1(t) - 1(t - T)\} = Z\left\{\frac{1 - e^{-sT}}{s}\right\} = (1 - e^{-sT}) Z\left\{\frac{1}{s}\right\} = (1 - z^{-1})\frac{z}{z - 1} = 1 \tag{5.3-6}$$

Note that any factor of the form e^{sT} can be replaced by z before we take the Z-transform of any function in the s-domain.

The next higher order of sophistication in hold circuits is the *first-order hold* (FOH). The first-order hold utilizes strengths of impulses at the last two sampling instances and puts out a ramp which linearly extrapolates these strengths.

This ramp holds as the output until the next sampling occurs. At that time a new straight line (the updated one) is given as the output; this again goes through the last two sampling points, and so on (Fig. (5.3-4). The impulse response of a FOH can be obtained and the corresponding transfer function is

$$G_{1h}(s) = \frac{sT + 1}{s^2 T}(1 - e^{-sT})^2 \tag{5.3-7}$$

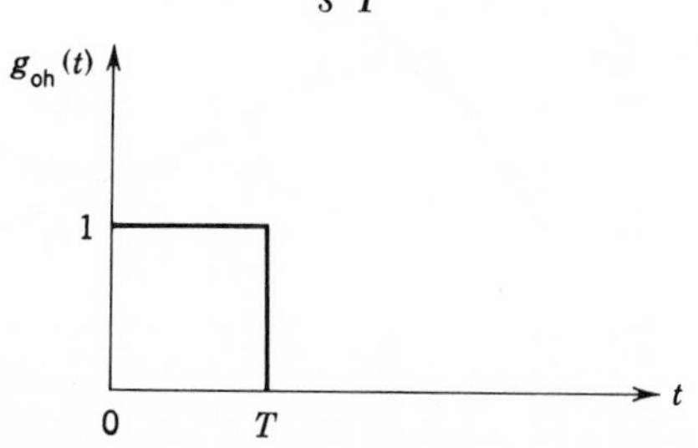

Figure 5.3-2   Impulse response of a zero-order hold.

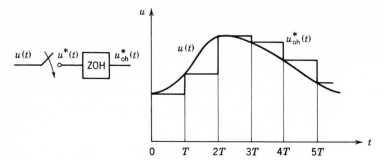

Figure 5.3-3   The effect of a sampler and a zero-order hold.

A second-order hold is one whose output is a second-degree curve that passes through the pertinent values at the last three sampling instances. This process is repeated for third-order and higher-order holds. The complexity of the corresponding transfer functions (and hence the difficulty of realization) increases as the order of the hold increases. Of course, the higher the order of the hold, the better should the output of the hold approximate the function being sampled, as is usually the case.

The amplitude frequency spectrum of both zero-order and first-order holds is shown in Fig. 5.3-5. Note that both act as lowpass filters. However, because of the lesser complexity, the zero-order hold is the most common in practice. As a rough general rule, if the frequency spectrum of the sampled signal contains only frequencies that are low (about 1/10 or less) relative to the sampling frequency, then increasing the order of the holds will result in a better approximation of the sampled signal. Moreover, other types of hold circuits such as fractional-order hold circuits and exponential hold circuits have also been proposed in the literature. A fractional hold circuit has an ampli-

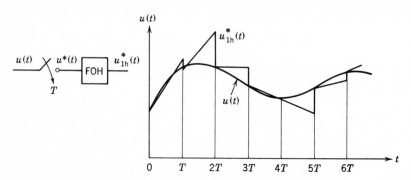

Figure 5.3-4   The effect of a sampler and first-order hold.

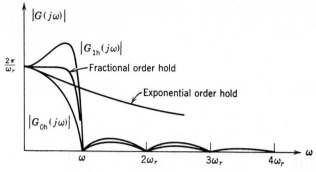

Figure 5.3-5   Magnitude frequency spectra of various order holds.

tude frequency characteristic that lies between $|G_{0h}(j\omega)|$ and $|G_{1h}(j\omega)|$ and is much closer to the ideal bandpass filter characteristic (see Problem 5.10). An exponential hold circuit exhibits an exponential decay in the frequency domain. Both these amplitude characteristics are shown in Fig. 5.3-5.

The Realization of the Zero-Order Hold Circuit

Hold circuits are not normally encountered in continuous-time systems. In order to develop a feel for such circuits we now consider the realization of the most commonly used hold circuit, the zero-order hold. The transfer function of a zero-order hold circuit is given from (5.3-5):

$$G_{0h}(s) = \frac{1 - e^{-sT}}{s} = \frac{1}{s}\left(1 - \frac{1}{e^{sT}}\right)$$

TWO-TERM APPROXIMATION

This transfer function can be approximated as

$$G_{0h}(s) \simeq \frac{1}{s}\left\{1 - \frac{1}{1+sT}\right\} = \frac{T}{1+sT} \qquad \text{if} \quad T^2 s^2 \ll 1 \qquad (5.3\text{-}8)$$

The realization of this $G_{0h}(s)$ is a simple RC network as shown in Fig. 5.3-6a.

THREE-TERM APPROXIMATION

A better approximation, if $T^3 s^3 \ll 1$, is

$$G_{0h}(s) \simeq \frac{1}{s}\left(1 - \frac{1}{1 + sT + s^2 T^2/2}\right) = \frac{T(1 + sT/2)}{1 + sT + s^2 T^2/2} \qquad (5.3\text{-}9)$$

This $G_{oh}(s)$ can be realized as an RLC network as shown in Fig. 5.3-6b. Still better approximations can be made by taking more and more terms of the series into account.

5.4 TRANSIENT ANALYSIS BY USE OF TRANSFER FUNCTIONS

In Section 5.2 we developed the basic Z-transform relationships. We are now in a position to obtain the Z-transform of any variable within a system (e.g., the output, the error) if we know the transfer functions of the various blocks and the input.

In order to develop the transient response, let us first obtain the inverse Z-transform, which enables us to get to the time domain from the z-domain. We know from (5.2-4) that

$$U(z) = \sum_{n=0}^{\infty} u(nT)\, z^{-n} \qquad (5.4\text{-}1)$$

From this expansion we note that the coefficients of various powers of z^{-1} in the expansion of $U(z)$ as a polynomial in z^{-1} give the value of the function $u(t)$ at the sampling instants. Hence we can see that all we need do is expand the function in z in powers of z^{-1} and pick off the coefficients of the powers of z^{-1}. This is workable, but we would have to expand into an infinite series, which usually involves considerable labor. The result obtained is not in a closed form. We cannot in general find $u(nT)$ for all n except in special cases. However, this technique of obtaining time response is very good where only a few terms are required. In order to obtain closed form expressions for $u(nT)$ we can either expand $U(z)$ in partial fractions so that the Z-transform tables can be used or we can use the *inversion integral* for the Z-transform.

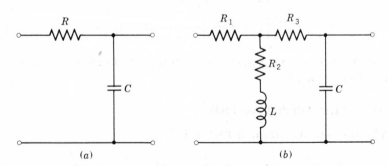

(a)   (b)

Figure 5.3-6   *RC* and *RLC* realization of a zero-order hold circuit.

The Inversion Integral for the Z-Transform

Consider (5.4-1) as

$$U(z) = u(0) + u(T)z^{-1} + \cdots + u(nT)z^{-n} + u(\overline{n+1}T)z^{-n-1} + \cdots \quad (5.4\text{-}2)$$

Let us multiply both sides of (5.4-2) by z^{n-1} to obtain

$$U(z)z^{n-1} = u(0)z^{n-1} + u(T)z^{n-2} + \cdots + u(nT)z^{-1} + u(\overline{n+1}T)z^{-2} + \cdots \quad (5.4\text{-}3)$$

We now integrate both sides in the complex z-plane so that the contour of integration is a circle with center at the origin which encloses all the singularities (poles) of $U(z)z^{n-1}$ (Fig. 5.4-1). We therefore write (5.4-3) as

$$\frac{1}{2\pi j} \oint U(z)z^{n-1}\, dz = \frac{1}{2\pi j} \left\{ \oint u(0)z^{n-1}\, dz + \oint u(T)z^{n-2}\, dz \right.$$

$$\left. + \cdots + \oint u(nT)z^{-1}\, dz + \oint u(\overline{n+1}T)z^{-2}\, dz + \cdots \right\} \quad (5.4\text{-}4)$$

By Cauchy's residue theorem, all the integrals on the RHS are zero except

$$\frac{1}{2\pi j} \oint u(nT)z^{-1}\, dz \quad (5.4\text{-}5)$$

which evaluates to $u(nT)$. Therefore

$$u(nT) = \frac{1}{2\pi j} \oint U(z)z^{n-1}\, dz \quad (5.4\text{-}6)$$

This integral is called the *inversion integral* for the Z-transform. It enables us to obtain the value of $u(t)$ at sampling instants for all n. We note that evaluation of the inversion integral involves the evaluation of residues at the singularities of $U(z)z^{n-1}$.

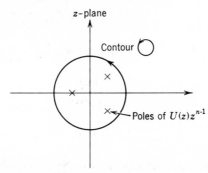

Figure 5.4-1   Contour of integration for the inverse Z-transform.

Assuming that the poles of $U(z)$ are simple and are at $z_1, z_2, \ldots, z_k$, then from (5.4-6)

$$u(nT) = \sum_{m=1}^{k} \alpha_m (z_m)^{n-1} \qquad (5.4\text{-}7)$$

where

$$\alpha_m = (z - z_m) \, U(z)|_{z \,=\, z_m}$$

Note that $u(nT)$ will be finite for all values of n only if

$$|z_m| \leqslant 1, \qquad m = 1,2, \ldots, k \qquad (5.4\text{-}8)$$

We can therefore say that the time response will be bounded only if the poles or singularities of the particular $U(z)$ lie within or on the unit circle if the poles are simple. If some of the poles are multiple, then boundedness is insured only if the multiple poles lie within the unit circle. We consider the problem of stability and boundedness in much greater detail in Chapter 6.

Example 5.4-1

Let

$$U(z) = \frac{z}{z+a}$$

Then

$$u(nT) = \frac{1}{2\pi j} \oint \frac{z}{z+a} z^{n-1} \, dz = z^n \Big|_{z=-a}$$

$$= (-a)^n$$

Example 5.4-2

A step input is applied to a system shown in Fig. 5.4-2. Find the output at sampling instants.

In this case we first find $G(z)$, the forward-path transfer function,

$$G(z) = Z\{G(s)\} = Z\left\{\frac{1-e^{-s}}{s(s+1)}\right\}$$

$$= [1 - z^{-1}]\left[Z\left\{\frac{1}{s(s+1)}\right\}\right]$$

$$= (1 - z^{-1})\left(\frac{z}{z-1} - \frac{z}{z-0.368}\right)$$

$$= \frac{0.632}{z-0.368}$$

Now

$$r(t) = 1$$

Therefore

$$R(z) = \sum_{n=0}^{\infty} r(nT)z^{-n} = \sum_{n=0}^{\infty} z^{-n} = \frac{z}{z-1}$$

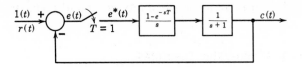

Figure 5.4-2   Example of a sampled-data system with hold circuit.

Using (5.2-34),

$$C(z) = \frac{G(z)}{1+G(z)} R(z)$$

$$= \frac{0.632z}{(z-1)(z+0.264)}$$

We can obtain $c(nT)$ from $C(z)$ by three techniques.

Expansion Method

In order to expand $C(z)$ in powers of z^{-1} we may perform synthetic division to obtain

$$C(z) = 0.632z^{-1}+0.465z^{-2}+0.511z^{-3}+\cdots$$

We find from the coefficients of the expansion in z^{-1} that $c(0) = 0$, $c(T) = 0.632$, $c(2T) = 0.465$, $\cdots$. However, we do not get $c(nT)$ for all n.

Partial Fraction Expansion

We can write

$$\frac{C(z)}{z} = \frac{0.632}{(z-1)(z+0.264)} = \frac{0.5}{z-1} - \frac{0.5}{z+0.264}$$

or

$$C(z) = \frac{0.5z}{z-1} - \frac{0.5z}{z+0.264}$$

Using Table 5.2-1 and Example 5.4-1, we get

$$c(nT) = 0.5-0.5(-0.264)^n, \qquad n = 0, 1, 2, \ldots$$

Inversion Integral Method

We get

$$c(nT) = \frac{1}{2\pi j} \oint \frac{0.632z}{(z-1)(z+0.264)} z^{n-1}dz = 0.5-0.5(-0.264)^n$$

We note that the partial fraction expansion and the inversion integral give the output at the nth sampling instant in closed form.

In this case we note the initial output $c(0)$ is zero (for $n = 0$) and the final steady-state output $c(\infty) = 0.5$ (for $n = \infty$). This means that the steady-state error is also 0.5 at the sampling instants. Considering the response only at the sampling instants, we would say the system is stable.

Initial-Value and Final-Value Theorems

As in the case of continuous systems, we can develop the initial-value and final-value theorems in the z-domain to determine initial and final values for z-transformed functions.

INITIAL-VALUE THEOREM

We have

$$U(z) = \sum_{n=0}^{\infty} u(nT)z^{-n}$$

$$= u(0) + u(T)z^{-1} + \cdots + u(nT)z^{-n} + \cdots \qquad (5.4\text{-}9)$$

Let $z \to \infty$ on both sides of the equation; we immediately obtain the initial value as

$$u(0) = \lim_{z \to \infty} U(z) \qquad (5.4\text{-}10)$$

Equation (5.4-10) is the initial-value theorem.

FINAL-VALUE THEOREM

Consider

$$Z\{u(\overline{n+1}T) - u(nT)\}$$

We know that

$$Z\{u(n+1T)\} = zU(z) - zu(0)$$

Therefore

$$Z\{u(\overline{n+1}T) - u(nT)\} = zU(z) - zu(0) - U(z)$$

$$= (z-1)U(z) - zu(0) \qquad (5.4\text{-}11)$$

Also

$$Z\{u(n+1)T - u(nT)\} = \sum_{n=0}^{\infty} [u(\overline{n+1}T) - u(nT)]z^{-n}$$

$$= \lim_{N \to \infty} \sum_{n=0}^{N} [u(\overline{n+1}T) - u(nT)]z^{-n} \qquad (5.4\text{-}12)$$

From (5.4-11) and (5.4-12), we get

$$\lim_{N \to \infty} \sum_{n=0}^{N} [u(\overline{n+1}T) - u(nT)]z^{-n} = (z-1)U(z) - zu(0) \qquad (5.4\text{-}13)$$

Let $z \to 1$ in (5.4-13). Then

$$\lim_{N \to \infty} u(NT) - u(0) = \lim_{z \to 1} (z-1)U(z) - u(0) \qquad (5.4\text{-}14)$$

Hence

$$u(\infty) = \lim_{z \to 1} (z-1)U(z) \tag{5.4-15}$$

which is the final-value theorem.

This relationship is valid only if the limit exists. The limit has finite value only if $(z-1)U(z)$ has all of its poles inside the unit circle.

Example 5.4-3

For the system of Example 5.4-2, find the initial and final values of the output directly from $C(z)$.

We evaluated

$$C(z) = \frac{0.632z}{(z-1)(z+0.264)}$$

Using (5.4-10), we have

$$c(0) = \lim_{z \to \infty} C(z) = 0$$

Using (5.4-15),

$$c(\infty) = \lim_{z \to 1} (z-1)C(z) = 0.5$$

These results check with values obtained in Example 5.4-2.

Introduction to the Modified Z-transform (Z_m-transform)

We have seen in Example 5.4-2 that the Z-transform method when applied to a sampled-data system gives the output values only at the sampling instants. However, if we examine the system at the output of the fixed plant, we see that the output appears continuously, as in the case of continuous systems. We must be able to find the output or the error between sampling instants in order to completely determine the behavior of the system.

In order to overcome this basic shortcoming of the Z-transform approach, let us consider an open-loop sampled-data system as shown in Fig. 5.4-3. $G(s)$ is the transfer function of the plant. In series with this we put a fictitious delay element, represented by $e^{-\Delta sT}$, and sampler. If $\Delta = 0$, then the output is available at the sampling instant

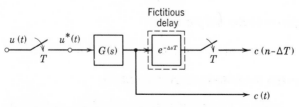

Figure 5.4-3   Sampled-data system with fictitious delay.

nT. If $\Delta = 1$, then the output is delayed by a full period and is available at $(n+1)T$. However, if Δ is varied between 0 and 1, we get the output between the sampling instants.

Let us define

$$G(s, \Delta) \overset{\Delta}{=} G(s)e^{-\Delta sT}, \qquad 0 < \Delta < 1 \qquad (5.4\text{-}16)$$

We need $G(z, \Delta)$ in order to get the output. However, as we substitute $G(s, \Delta)$ into the integral of (5.2-44) to obtain $G(z, \Delta)$, we find that the contour of integration cannot be closed because of nonconvergence of $G(s, \Delta)$ on the infinite circular arc to the left. We can avoid this problem by a simple substitution:

$$\Delta = 1 - m, \qquad 0 < m < 1 \qquad (5.4\text{-}17)$$

In this case we have

$$G(s, m) \overset{\Delta}{=} G(s, \Delta)\big|_{\Delta = 1 - m} = G(s)e^{-sT}e^{msT} \qquad (5.4\text{-}18a)$$

or

$$G(s, m)e^{sT} = G(s)e^{msT} \qquad (5.4\text{-}18b)$$

Taking the Z-transform of (5.4-18b) leads to

$$zG(z, m) = \frac{1}{2\pi j} \int_{\sigma_1 - j\infty}^{\sigma_1 + j\infty} \frac{G(p)e^{mpT}}{1 - e^{-T(s-p)}}\, dp, \qquad 0 < m < 1 \qquad (5.4\text{-}19)$$

Note that in taking the Z-transform of (5.4-18b), we substitute z for the e^{Ts} factor because of the shifting theorem. Convergence on the left infinite circular arc is insured if $G(p)$ has a denominator at least one degree higher than the numerator.

We define the integral of (5.4-19) as the modified Z-transform of $G(z)$ and write

$$Z_m\{G(s)\} \overset{\Delta}{=} G(z, m) \overset{\Delta}{=} \frac{z^{-1}}{2\pi j} \int_{\sigma_1 - j\infty}^{\sigma_1 + j\infty} \frac{G(p)e^{mpT}}{1 - e^{-T(s-p)}}\, dp, \qquad 0 < m < 1$$

$$(5.4\text{-}20)$$

Note that

$$G(z) = \lim_{m \to 0} zG(z, m) \qquad (5.4\text{-}21)$$

Now we can write the output as

$$C(z, m) = G(z, m)U(z), \qquad 0 < m < 1 \qquad (5.4\text{-}22)$$

We can use the inversion integral to obtain

$$c(nT, m) = c(\overline{n + m - 1}T) = \frac{1}{2\pi j} \oint C(z, m)z^{n-1}\, dz \qquad (5.4\text{-}23)$$

By varying m between 0 and 1, we get the output response between any two sampling instants and hence for all time, and so the modified Z-transform can be effectively utilized to determine the response between sampling instants. The response between sampling instants can also be obtained by state variable methods; however, this discussion is deferred to Chapter 7.

The properties of Z_m-transform can be developed as for the Z-transform. Some common modified Z-transform pairs are listed in Table 5.4-1.

Example 5.4-4

Consider an open-loop, sampled-data system as shown in Fig. 5.4-3 with $G(s) = 1/(s+1)$. Find the response to a step input for all instants of time. Assume $T = 1$ sec.

Since the response is required between sampling instants, we use the modified Z-transform:

$$G(z, m) = Z_m\{G(s)\} = Z_m\left\{\frac{1}{s+1}\right\} = \frac{z^{-1}}{2\pi j}\int_{\sigma_1-j\infty}^{\sigma_1+j\infty} \frac{e^{mpT}}{(p+1)(1-e^{-T(s-p)})}\,dp$$

$$= \frac{e^{-m}}{z-0.368}$$

Hence

$$C(z, m) = G(z, m)U(z)$$

$$= \left(\frac{e^{-m}}{z-0.368}\right)\left(\frac{z}{z-1}\right)$$

Then

$$c(\overline{n+m-1}T) = \frac{1}{2\pi j}\oint_\Gamma \frac{e^{-m}\cdot z^n}{(z-1)(z-0.368)}\,dz$$

$$= 1.58\,e^{-m}(1-0.368^n)$$

We know that between any two sampling instants, we can write

$$t = (n-\Delta)T = (n+m-1)T = (n+m-1)$$

valid for $n = 1, 2\ldots$. First we let $n = 1$ and vary m between 0 and 1; thus we get the output between $t = 0$ and 1 sec. Then we let $n = 2$ and vary m between 0 and 1. This gives the output between $t = 1$ and 2 sec. We can obtain output between all sampling instants the same way. The output is shown in Fig. 5.4-4.

Table 5.4-1 Short Table of Modified Z-Transforms

$F(s)$	$F(z, m)$	$f(t)$
$\dfrac{1-e^{-sT}}{s}F(s)$	$\dfrac{z-1}{z}Z_m\left\{\dfrac{F(s)}{s}\right\}$	
K	0	$K\delta(t)$
$\dfrac{1}{s}$	$\dfrac{1}{z-1}$	$1(t)$
$\dfrac{1}{s^2}$	$\dfrac{T}{(z-1)^2}\left[m(z-1)+1\right]$	t
$\dfrac{1}{s^3}$	$\dfrac{T^2}{2}\left[\dfrac{m^2}{z-1}+\dfrac{2m+1}{(z-1)^2}+\dfrac{2}{(z-1)^3}\right]$	$\dfrac{t^2}{2}$
$\dfrac{1}{s+\alpha}$	$\dfrac{e^{-\alpha mT}}{z-e^{-\alpha T}}$	$e^{-\alpha t}$
$\dfrac{1}{(s+\alpha)^2}$	$\dfrac{Te^{-\alpha mT}\left[e^{-\alpha T}+m(z-e^{-\alpha T})\right]}{(z-e^{-\alpha T})^2}$	$te^{-\alpha t}$
$\dfrac{\alpha}{s^2+\alpha^2}$	$\dfrac{z\sin\alpha mT+\sin(1-m)\alpha T}{z^2-2z\cos\alpha T+1}$	$\sin\alpha t$
$\dfrac{s}{s^2+\alpha^2}$	$\dfrac{z\cos\alpha mT-\cos(1-m)\alpha T}{z^2-2z\cos\alpha T+1}$	$\cos\alpha t$
$\dfrac{\alpha}{(s+\beta)^2+\alpha^2}$	$e^{-\beta mT}\dfrac{\left[z\sin\alpha mT+e^{-\beta T}\sin(1-m)\alpha T\right]}{z^2-2ze^{-\beta T}\cos\alpha T+e^{-2\beta T}}$	$e^{-\beta t}\sin\alpha t$
$\dfrac{s+\beta}{(s+\beta)^2+\alpha^2}$	$e^{-\beta mT}\dfrac{\left[z\cos\alpha mT-e^{-\beta T}\cos(1-m)\alpha T\right]}{z^2-2ze^{-\beta T}\cos\alpha T+e^{-2\beta T}}$	$e^{-\beta t}\cos\alpha t$
$\dfrac{\alpha}{s^2-\alpha^2}$	$\dfrac{z\sinh\alpha mT+\sinh(1-m)\alpha T}{z^2-2z\cosh\alpha T+1}$	$\sinh\alpha t$
$\dfrac{s}{s^2-\alpha^2}$	$\dfrac{z\cosh\alpha mT-\cosh(1-m)\alpha T}{z^2-2z\cosh\alpha T+1}$	$\cosh\beta t$
$\dfrac{\alpha}{(s+\beta)^2-\alpha^2}$	$e^{-\beta mT}\dfrac{\left[z\sinh\alpha mT+e^{-\beta T}\sinh(1-m)\,\alpha T\right]}{z^2-2ze^{-\beta T}\cosh\alpha T+e^{-2\beta T}}$	$e^{-\beta t}\sinh\alpha t$
$\dfrac{s+\beta}{(s+\beta)^2-\alpha^2}$	$e^{-\beta mT}\dfrac{\left[z\cosh\alpha mT-e^{-\beta T}\cosh(1-m)\alpha T\right]}{z^2-2ze^{-\beta T}\cosh\alpha T+e^{-2\beta T}}$	$e^{-\beta t}\cosh\alpha t$
$\dfrac{1}{s^n}$	$\displaystyle\lim_{a\to 0}\dfrac{(-1)^{n-1}}{(n-1)!}\dfrac{\partial^{n-1}}{\partial a^{n-1}}\left(\dfrac{e^{-amT}}{z-e^{-aT}}\right)$	$\dfrac{t^{n-1}}{(n-1)!}$

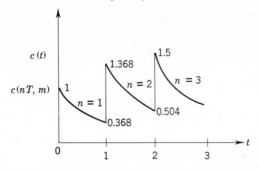

Figure 5.4-4 Response for all instants of time for Example 5.4-4.

Modified Z-transform for Feedback Sampled-Data Systems

We have seen how the Z_m-transform can be utilized to obtain the response between sampling instants for open-loop sampled-data systems. To obtain this response for feedback sampled-data systems, consider the system shown in Fig. 5.4-5. We have

$$C(s) = E(z) \, G(s) \qquad (5.4\text{-}24)$$

From (5.2-33)

$$E(z) = \frac{1}{1 + GH(z)} R(z) \qquad (5.4\text{-}25)$$

Therefore

$$C(s) = \frac{G(s)}{1 + GH(z)} R(z) \qquad (5.4\text{-}26)$$

Now we apply the modified Z-transform to (5.4-24):

$$C(z, m) = E(z) \, G(z, m)$$

and therefore

$$\frac{C(z, m)}{R(z)} = \frac{G(z, m)}{1 + GH(z)} \qquad (5.4\text{-}27)$$

Note that the Z_m-transform applies only to that portion of the system which appears between the last sampler and the point where the output is sought between samplings.

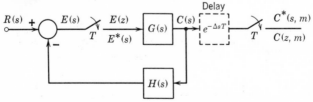

Figure 5.4-5 A fictitious delay element applied to a feedback sampled-data system to obtain the time response between sampling instants.

Response of a Second-Order System

To illustrate the effects of sampling, the hold circuit, and the use of Z_m-transform, let us consider a specific second-order system.

Example 5.4-5

Consider the second-order system shown in Fig. 5.4-6. We shall find the output for all instants of time for the following three cases:

1. If the sampler and the hold circuit are replaced by a "short circuit," thus making the system continuous.
2. If the hold circuit is replaced by a "short circuit" and the sampler remains.
3. If the hold circuit and sampler are both in place.

1. No Sampler — No Hold

We can write the output as

$$C(s) = \frac{1}{s^2+s+1} R(s) = \frac{1}{s(s^2+s+1)}$$

Inversion gives

$$c(t) = 1 - 1.16\, e^{-(1/2)t} \sin\left(\frac{\sqrt{3}t}{2} + \frac{\pi}{3}\right)$$

The plot of this is shown in Fig. 5.4-7.

2. Sampler Only — No Hold

To obtain $c(t)$ for all instants of time, we use the modified Z-transform. We have

$$G(s) = \frac{1}{s(s+1)}$$

Therefore

$$G(z, m) = Z_m\left\{\frac{1}{s} - \frac{1}{s+1}\right\} = \left(\frac{1}{z-1} - \frac{e^{-m}}{z-0.368}\right)$$

$$G(z) = \lim_{m \to 0} z\, G(z, m) = \frac{0.632z}{(z-1)(z-0.368)}$$

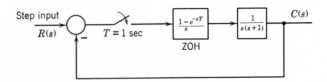

Figure 5.4-6   Feedback sampled-data system.

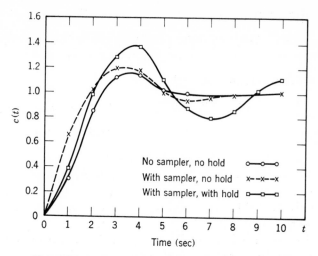

Figure 5.4-7 Step response of the system of Fig. 5.4-6.

Using (5.4-27), we now write $C(z, m)$ as

$$C(z, m) = \frac{(z-0.368) - e^{-m}(z-1)}{z^2 - 0.736z + 0.368} \frac{z}{z-1}$$

Using the inversion integral of (5.4-23), we obtain

$$c\,(\overline{n+m-1}T) = 1 - e^{-(1/2)n}\,[1.004\cos n\theta + (2.08\,e^{-m} - 0.766)\sin n\theta]$$

where $\theta = 52.6°$.

Now, by letting $n = 1, 2, \ldots$, and varying m between 0 and 1, we can plot $c(t)$. This is also shown in Fig. 5.4-7.

3. *With Hold Circuit and Sampler*

In this case

$$G(s) = \frac{(1 - e^{-sT})}{s^2(s+1)}$$

and

$$G(z, m) = Z_m\left\{\frac{1 - e^{-sT}}{s^2(s+1)}\right\}$$

$$= \frac{(m-1)\,(z^2 - 1.368z + 0.368) + (z - 0.368) + e^{-m}(z-1)^2}{z(z-1)\,(z-0.368)}$$

Further,

$$G(z) = \lim_{m\to 0} zG(z, m) = \frac{0.368z + 0.264}{(z-0.368)\,(z-1)}$$

Also, using (5.4-27), we get

$$C(z, m) = \frac{(m-1)(z^2 - 1.368z + 0.368) + (z - 0.368) + e^{-m}(z-1)^2}{(z^2 - z + 0.632)(z-1)}$$

Using the inversion integral on this gives

$$c(\overline{n+m-1}T) = 1 + (e^{-0.23})^{n-1} [(0.983\ e^{-m} + 0.980m - 1.97) \cos (n-1)\ \theta$$

$$+ (-0.796\ e^{-m} + 0.210m + 0.585) \sin (n-1)\theta]$$

where $\theta = 51°$. This is also plotted in Fig. 5.4-7.

From the figures we note that the sampled response has higher overshoot and takes longer to settle down than the unsampled response. With the hold circuit, the response is even worse–it takes still longer to settle and the overshoot is even higher.

We could have used the ordinary Z-transform and plotted the response at sampling instants and then joined these points with a smooth curve to get a plot sufficient for most practical purposes. This method is reasonably accurate if T is small relative to the dominant time constant and if the transfer function between the sampler and the output has a denominator which is of degree at least two higher than the numerator. The peak overshoot and peak time may be calculated from dominant pole and zero locations (this will be shown in Chapter 9). Once we know this, it is usually easy to draw a smooth curve through the sampled points, which is sufficient for most practical applications.

Use of Z_m-Transform for Analyzing Systems with a Pure-Delay Element

Laplace transform methods do not yield analytical results very easily for continuous-control systems with a pure-delay element in the loop. The reason for this is that the resulting transcendental transfer functions cannot be treated easily through the use of $\mathscr{L}$-transforms. The Z_m-transform offers a systematic way of handling such delay elements for sampled-data feedback systems. We demonstrate the technique by a specific example.

Example 5.4-6

Consider a feedback sampled-data system as shown in Fig. 5.4-8. Use the modified Z-transform to obtain the output for all instants of time.

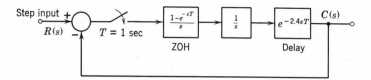

Figure 5.4-8   Feedback *SDS* with a delay element in the loop.

In this case, we can write

$$G(s) = \frac{(1-e^{-sT})\,e^{-2.4sT}}{s^2} = G_1(s)\,e^{-2.4sT}$$

and

$$\frac{C(z,m)}{R(z)} = \frac{1}{1+G(z)}\,G(z,m)$$

Now we have to find $G(z,m)$ and $G(z)$ when there is a delay factor in $G(s)$. Let us first obtain $G(z,m)$:

$$G(z,m) = Z_m\{G_1(s)\,e^{-2.4sT}\}$$

To obtain the Z_m-transform as above, we write the delay part as

$$e^{-\Delta sT} = e^{-(n+1-m')sT}$$

where n is an integer and $1-m'$ represents the nonintegral part of the delay. This gives the proper convergence conditions on the infinite semicircle when we close the path of integration on the integral which gives the Z_m-transform. In our case $n=2$, $m'=0.6$. Hence we get

$$G(z,m) = Z_m\{G_1(s)\,e^{-(n+1-m')sT}\}$$

We will substitute for n and m' later.

Before taking the Z_m-transform by using (5.4-19), we note:

1. As long as $0 < m \leq 1-m'$, then $m+m' \leq 1$ and the integral will converge on the infinite semicircle to the left in the substitution

$$e^{-(n+1-m')sT} = z^{-(n+1)}\,e^{m'sT}$$

is used in (5.4-19).

2. When $1-m' < m < 1$, then $m+m' > 1$. In this case the integral will converge if the substitution is

$$e^{-(n+1-m')sT} = z^{-n}\,e^{-(1-m')sT}$$

in (5.4-19).

We therefore have

$$G(z,m) = \frac{z^{-1}}{2\pi j}\int_{\sigma_1-j\infty}^{\sigma_1+j\infty} \frac{G_1(p)}{1-e^{-T(s-p)}}\,z^{-(n+1)}\,e^{(m+m')pT}\,dp, \qquad 0 < m \leq 1-m'$$

and

$$G(z, m) = \frac{z^{-1}}{2\pi j} \int_{\sigma_1 - j\infty}^{\sigma_1 + j\infty} \frac{G_1(p)}{1 - e^{-T(s-p)}} z^{-n} e^{(m+m'-1)pT} \, dp, \qquad 1 - m' < m < 1.$$

In each case the exponent of e in the numerator (either $m + m'$ or $m + m' - 1$) must be positive and less than unity to achieve convergence.

Substituting our values of $G_1(s)$, m', n, and T, we get

$$G(z, m) = \frac{z-1}{z} \frac{z^{-4}}{2\pi j} \int_{\sigma_1 - j\infty}^{\sigma_1 + j\infty} \frac{e^{(m+0.6)p}}{p^2[1 - e^{-(s-p)}]} \, dp, \qquad 0 < m \le 0.4$$

and

$$G(z, m) = \frac{z-1}{z} \frac{z^{-3}}{2\pi j} \int_{\sigma_1 - j\infty}^{\sigma_1 + j\infty} \frac{e^{(m+0.6-1)p}}{p^2[1 - e^{-(s-p)}]} \, dp, \qquad 0.4 < m < 1$$

Evaluation gives

$$G(z, m) = z^{-4} \left(m + 0.6 + \frac{1}{z-1} \right), \qquad 0 < m \le 0.4$$

and

$$G(z, m) = z^{-3} \left(m - 0.4 + \frac{1}{z-1} \right), \qquad 0.4 < m < 1$$

We get

$$G(z) = \lim_{m \to 0} z \, G(z, m) = z^{-3} \left(0.6 + \frac{1}{z-1} \right) = z^{-3} \frac{0.6z + 0.4}{z-1}$$

Hence

$$C(z, m) = \frac{G(z, m)}{1 + G(z)} R(z)$$

$$= \frac{\left[\begin{array}{l} z^{-4}[(m+0.6)(z-1)+1][1(m)-1(m-0.4)] \\ +z^{-3}[(m-0.4)(z-1)+1][1(m-0.4)-1(m-1)] \end{array} \right]}{(z-1) + z^{-3}(0.6z+0.4)} \times \frac{z}{z-1}$$

where $1(m)$ is the unit-step function with m as the argument. Trying to find $c(nT, m)$ by the inversion integral would obviously be very cumbersome since the denominator is fourth order. The power series expansion offers a possibility. In this case the coefficients will be functions of m. Expansion in powers of z^{-1} gives

$$\begin{aligned} C(z, m) = \; & z^{-3}\{(m-0.4)[1(m-0.4)-1(m-1)]\} + z^{-4}(m+0.6) \\ & + z^{-5}(m+1.6) + z^{-6}\{(m+2.6)[1(m)-1(m-0.4)] \\ & + (3.24-0.6m)[1(m-0.4)-1(m-1)]\} \\ & + z^{-7}(2.64-0.6m) + \cdots, \qquad 0 < m < 1 \end{aligned}$$

This expansion starts with z^{-3} due to the delay in the loop. Now we can plot the time response by evaluating the various coefficients as m is varied between 0 and 1, as shown in Fig. 5.4-9.

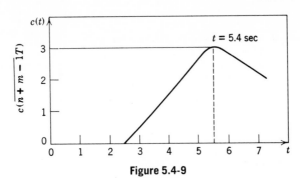

Figure 5.4-9

Thus a modified Z-transform can be effectively utilized to find transient response of sampled systems with pure-delay elements.

PROBLEMS

5.1. Find the Z-transform of following time functions:
(a) $t \sin t$ (d) $e^{-t} \cos 2t$
(b) $t^2 e^{-t}$ (e) $te^{-t} \cos t$
(c) t^3

5.2. The impulse response of a linear system is te^{-2t}. Find the output using the expansion method for the following sampled inputs with a sampling period $T = \log_e 2$ sec:
(a) t
(b) Step input
(c) e^{-2t}

5.3. The output-input for a sampled-data system is described by the following difference equation:
$$c(\overline{n+2T}) + 2c(\overline{n+1T}) + 3c(nT) = u(\overline{n+1T}) - u(nT)$$
Find the transfer function for this system in the z-domain.

5.4. The transfer function of a sampled-data system is given by
$$\frac{C(z)}{U(z)} = \frac{0.23(z - 0.712)}{z^2 - 0.372z + 0.637}$$
Determine the difference equation that describes this system.

5.5. Find the error function $E(z)$ as well as the output $C(z)$ for the systems shown in Fig. P5.5.

5.6. Find the closed-form representation of the Z-transform of the following functions:

(a) $\dfrac{1}{s(s+1)}$ (b) $\dfrac{1}{(s+2)(s+4)}$ (c) $\dfrac{sT+1}{s^2 T}(1 - e^{-sT})^2$

(d) $\dfrac{1}{s^2(s+1)}$ (e) $\dfrac{1}{s^2+9}$ (f) $\dfrac{1}{s^2(s+4)^2}$

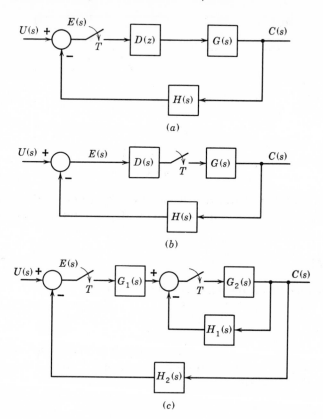

(a)

(b)

(c)

Figure P5.5

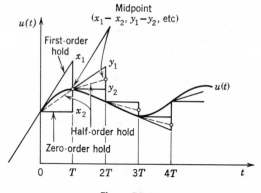

Figure P5.10

5.7. Show that

$$Z\left\{\frac{f(t)}{t+\lambda T}\right\} = \frac{z^\lambda}{T}\int_z^\infty \frac{F(z)}{z^{\lambda+1}}\,dz$$

5.8. Show that

$$Z\left\{f_1(t)f_2(t)\right\} = \frac{1}{2\pi j}\int_\Gamma p^{-1}F_1(p)F_2\left(\frac{z}{p}\right)dp$$

where Γ encloses the singularities of $p^{-1}F_1(p)$.

5.9. Show that a band-limited signal $u(t)$ $[-\omega_c, \omega_c]$ can be reconstructed by the knowledge of $u(t)$ and its first k derivatives known at sampling instants separated by $(k+1)\,\pi/\omega_c$ sec.

5.10. Develop the transfer function of a half-order hold circuit. The half-order hold circuit is defined as a hold circuit which holds the value as shown in Fig. P5.10. Find the Z-transform of this transfer function. Also plot $|G_{1/2h}(j\omega)|$.

5.11. Find the inverse Z-transform of following functions by expansion as well as by the inversion integral. Obtain the inverse also by use of tables wherever you can.

(a) $\dfrac{z-0.5}{z^2-z+2}$ (b) $\dfrac{T^2z(z+1)}{2(z-1)^3}$ (c) $\dfrac{z(z^2+2z+1)}{(z^2-z+1)(z^2+2z+3)}$

(d) $\dfrac{z(z+0.5)}{(z-0.3)^2(z-1)^2}$ (e) $\dfrac{2z}{(2z-1)^2}$ (f) $\dfrac{z^2}{(z-1)(z-0.2)}$

5.12. Find and plot the output at sampling instants for the systems shown in Fig. P5.12. Check the final value of the output using the final-value theorem.

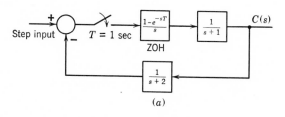

(a)

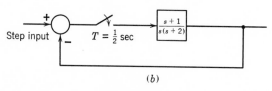

(b)

Figure P5.12

5.13. Find the modified Z-transform of the following functions:

(a) $\dfrac{\overline{1}}{s^2(s+1)}$ (c) $\dfrac{1}{s^2+s+2}$

(b) $\dfrac{1-e^{-Ts}}{s(s+4)}$ (d) $\dfrac{s+1}{(s+2)(s+3)}$

5.14. Use the modified Z-transform approach to find the output for systems shown in Fig. P5.12 for all instants of time. Plot this output.

5.15. Use the modified Z-transform to find and plot the output of a system with pure-delay element as shown in Fig. P5.15.

5.16. Realize a sampler and a zero-order hold if the sampling period is 0.001 sec. Check the accuracy of your realization in the laboratory.

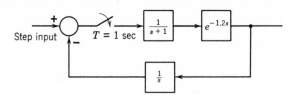

Figure P5.15

6

ANALYSIS OF DISCRETE-TIME SYSTEMS

6.1 INTRODUCTION

In Chapter 5 we considered the basic tools required for the analysis of sampled-data systems (SDS). Our main interest was determining the time response of these SDS. This led to the development of the Z-transform and the Z_m-transform. We note that the variable z plays the same role in the analysis of SDS as the variable s plays in the analysis of continuous-time systems.

Although SDS involve a sampler and hence their response is somewhat different from continuous-time systems, and although a different transform is used in their analysis, the basic considerations of the control system designer remain the same when one is working with an SDS. Thus considerations of stability, steady-state error, how the time response is related to pole-zero locations or to the frequency response all remain equally valid and important.

Here now we shall consider these topics and see what their implications are when we work in the z-domain (with SDS) rather than in the s-domain as in the case of continuous-time systems. The order here will be essentially the same as in Chapters 3 and 4 where these same topics were considered for continuous-time systems. We begin with considerations of stability.

6.2 STABILITY

In the development of the inversion integral we noted that the time response will be unbounded unless the poles of the corresponding Z-transform are all within the unit circle and/or are simple poles on the unit circle.

For a feedback SDS the poles of the transfer function are given by

$$1 + GH(z) = 0 \tag{6.2-1}$$

This is the *characteristic equation*. The roots of this equation must be within the unit circle if the system is to be stable. In that case, the output will be bounded for any bounded input.

In order to establish this basic requirement for stability in the z-domain, let us consider the basic transformation which gets us from the s- to the z-domain. Consider first a function $u(t)$. When this function is sampled assuming $u(0+) = 0$, we get $u^*(t)$ such that

$$U^*(s) = \frac{1}{T} \sum_{k=-\infty}^{\infty} U\left(s+j\frac{2\pi}{T}k\right) = \frac{1}{T} \sum_{k=-\infty}^{\infty} U(s+jk\omega_r), \quad (6.2\text{-}2)$$

Equation (6.2-2) implies that $U^*(s)$ is periodic with a period $j\omega_r$. Furthermore, if $U(s)$ has a pole at $s = p_1$, then $U^*(s)$ will have an infinite number of poles at $s = p_1+jk\omega_r$, $k = \pm 1, \pm 2, \ldots$.

Let us divide the s-plane as shown in Fig 6.2-1. We have a strip ω_r wide which we call the primary strip. If we know $U^*(s)$ in this strip, then $U^*(s)$ is known for all s because of the periodicity property.

From the discussion in Section 3.2 we know that for $u(t)$ to be

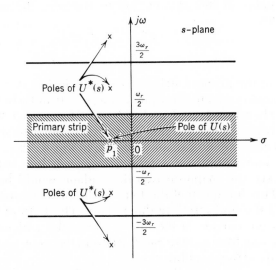

Figure 6.2-1 The primary strip in the s-plane.

205

bounded, $U(s)$ will have no RHP poles or multiple poles on the $j\omega$-axis. The same will apply now to a sampled-function transform $U^*(s)$. The poles of $U^*(s)$ will be infinite in number but they will all be periodic in the imaginary part to the poles in the primary strip. And we note that the poles in the primary strip are just the poles of the unsampled signal's transform. Thus for a given $u(t)$ that is bounded, the poles of $U^*(s)$ in the primary strip will all be in the LHP or be simple poles on the $j\omega$-axis.

We know that $U(z) = U^*(s)|_{z=e^{sT}}$, hence z-plane poles will just be poles of $U^*(s)$ under the mapping $z = e^{sT}$. Thus the question of boundedness of a $u(t)$ corresponding to a given $U(z)$ can be answered simply by considering where the LHP primary strip maps in to the z-plane.

Transformation $z = e^{Ts}$

First note that this transformation is also periodic in s with a period $j\omega_r$ since

$$e^{T(s + jk\omega_r)} = e^{Ts} \tag{6.2-3}$$

This implies that if we transform the primary strip from Fig. 6.2-1 from the s- into the z-plane, then all the other strips of this width shown in Fig. 6.2-1 will map right on top of the mapping of the primary strip. We are concerned with the mapping of the LHS of the primary strip because functions with bounded time response magnitude will have their poles in this region. Let us consider the mapping. Let $s = \sigma + j\omega$; then

$$z = e^{sT} = e^{(\sigma + j\omega)T} = e^{\sigma T}e^{j\omega T}$$

or

$$|z| = e^{\sigma T} \quad \text{and} \quad \angle z = \omega T \tag{6.2-4}$$

In the LHS of the primary strip we are considering, $\sigma \leqslant 0$. Consider first the imaginary axis in the s-plane, i.e., $\sigma = 0$ and ω between $-\omega_r/2$ and $+\omega_r/2$. Therefore $|z| = 1$ and $\angle z$ varies from $-\omega_r T/2$ to $\omega_r T/2$ or $-\pi$ to π. This means that the imaginary axis in the primary strip maps into the unit circle in the z-plane as shown in Fig. 6.2-2.

For each line where $\sigma = $ constant, i.e., each line parallel to the imaginary axis, the mapping into the z-plane would be a circle of radius $e^{\sigma T}$. This implies that the LHS of the primary strip maps into the inside of the unit circle in the z-plane. The RHS of the primary strip is then the outside of the unit circle in the z-plane. Any other strip, say the one from $\omega_r/2$ to $3\omega_r/2$, maps in the same way because of the periodicity of the mapping $z = e^{sT}$.

Hence we can say that all the poles of $U^*(s)$ from the LHS of the s-plane map into poles within the unit circle in the z-plane. All poles

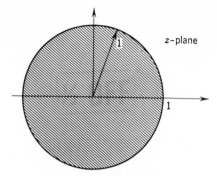

Figure 6.2-2 Mapping of the LHP primary strip in the s-plane into the unit circle in the z-plane.

on the $j\omega$-axis in the s-plane will map onto the unit circle in the z-plane. Thus if $U^*(s)$ corresponds to a function that is bounded in magnitude in the time domain, the $U(z)$ will have poles either inside or on the unit circle and the poles on the unit circle will have multiplicity one (they will be simple). Similarly, if $G(s)$ is a stable transfer function, it, along with $G^*(s)$, will have only LHP poles. Furthermore, $G(z)$ here will have all its poles inside the unit circle in the z-plane.

Now we know that the poles of the z-transfer function must lie inside the unit circle if a system is to be stable, how can we determine whether a given transfer function has its poles inside the unit circle without actually finding the location of these poles? Let the open-loop transfer function be $GH(z)$; then, using (5.2-34), we can write the denominator of the transfer function for the corresponding closed-loop transfer function:

$$F(z) = [1 + GH(z)] = a_n z^n + a_{n-1} z^{n-1} + \cdots + a_1 z + a_0 = 0 \quad (6.2\text{-}5)$$

This is called the *characteristic equation* for the system and its roots are the system poles. There are many analytical tests and graphical methods that can be used to determine whether the roots of the characteristic equation are within the unit circle or not. One graphical method, Mitrovic's method, will be developed in Section 6.6. We now consider two analytical tests.

Bilinear Transformation and the Routh Criterion

We have to see whether the roots of (6.2-5) are within the unit circle. If we can find a complex transformation which maps the inside of the unit circle in the z-plane into the left half of the new plane, then the roots within the unit circle will also go into this and the standard Routh criterion can be applied.

We know the transformation $z = e^{Ts} [s = (1/T) \log_e z]$ cannot be used because of the periodicity of e^{Ts} and multiple values of $(1/T) \log_e z$. We need a one to one transformation, so that one point inside the unit circle goes into one point in the left half of the new plane. A simple transformation that maps the inside of the unit circle in the z-plane to the LHS of the w-plane is given by the bilinear transformation:

$$w = \frac{z-1}{z+1}; \qquad z = \frac{1+w}{1-w} \qquad (6.2\text{-}6)$$

The characteristic equation of (6.2-5) in the z-plane changes to

$$a_n \left(\frac{1+w}{1-w}\right)^n + a_{n-1} \left(\frac{1+w}{1-w}\right)^{n-1} + \cdots + a_1 \left(\frac{1+w}{1-w}\right) + a_0 = 0 \quad (6.2\text{-}7)$$

or

$$\alpha_n w^n + \alpha_{n-1} w^{n-1} + \cdots + \alpha_1 w + \alpha_0 = 0 \qquad (6.2\text{-}8)$$

in the w-plane. Now the Routh criterion can be used to determine if the roots of (6.2-8) lie in the LHS of the w-plane. Note that the α_k's are functions of the a_k coefficients and they are found by expanding (6.2-7). This makes the algebraic work for the analysis very tedious and consequently this procedure is seldom used. A criterion that works directly in the z-plane is the Jury stability criterion.

Jury Stability Criterion

This criterion is based on the construction of a table of the coefficients of the characteristic equation

$$F(z) = a_n z^n + a_{n-1} z^{n-1} + \cdots + a_1 z + a_0 = 0, \qquad a_n > 0 \quad (6.2\text{-}9)$$

similar to the Routh table in the case of continuous-time systems. The table is given in (6.2-10) and the stability constraints are given in (6.2-11) and (6.2-12). If all these constraints are satisfied, then the characteristic equation has all its roots inside the unit circle and the corresponding SDS is stable.

Row	z^0	z^1	z^2	$\cdots$	z^{n-k}	$\cdots$	z^{n-1}	z^n
1	a_0	a_1	a_2	$\cdots$	a_{n-k}	$\cdots$	a_{n-1}	a_n
2	a_n	a_{n-1}	a_{n-2}	$\cdots$	a_k	$\cdots$	a_1	a_0
3	b_0	b_1	b_2	$\cdots$	$\cdots$	$\cdots$	b_{n-1}	
4	b_{n-1}	b_{n-2}	b_{n-3}	$\cdots$	$\cdots$	$\cdots$	b_0	
5	c_0	c_1	c_2	$\cdots$	$\cdots$	c_{n-2}		
6	c_{n-2}	c_{n-3}	c_{n-4}	$\cdots$	$\cdots$	c_0		
.	.	.	.	$\cdots$	.			
.	.	.	.	$\cdots$	.			
.	.	.	.	$\cdots$	.			
$2n-5$	s_0	s_1	s_2	s_3				
$2n-4$	s_3	s_2	s_1	s_0				
$2n-3$	r_0	r_1	r_2					

$$(6.2\text{-}10)$$

where

$$b_k = \begin{vmatrix} a_0 & a_{n-k} \\ a_n & a_k \end{vmatrix}, \qquad d_k = \begin{vmatrix} c_0 & c_{n-2-k} \\ c_{n-2} & c_k \end{vmatrix}$$

$$\vdots$$

$$c_k = \begin{vmatrix} b_0 & b_{n-1-k} \\ b_{n-1} & b_k \end{vmatrix}, \qquad r_0 = \begin{vmatrix} s_0 & s_3 \\ s_3 & s_0 \end{vmatrix}$$

$$\vdots$$

$$r_2 = \begin{vmatrix} s_0 & s_1 \\ s_3 & s_2 \end{vmatrix}$$

The stability constraints are

$$F(1) > 0, \qquad (-1)^n F(-1) > 0 \tag{6.2-11}$$

$$\left. \begin{aligned} &|a_0| < a_n, \\ &|b_0| > |b_{n-1}| \\ &|c_0| > |c_{n-2}| \\ &|d_0| > |d_{n-3}| \\ &\quad . \\ &\quad . \\ &\quad . \\ &|r_0| > |r_2| \end{aligned} \right\} \quad (n-1) \text{ constraints} \tag{6.2-12}$$

From (6.2-10) we note that an nth-order system will have $(2n-3)$ rows in the table. Furthermore, the $(2k+2)$th row has elements of the $(2k+1)$th row in reverse order $(k = 0,1,2,\ldots)$. We find that the number of constraint conditions is $n + 1$, which is not surprising since there are $n + 1$ coefficients.

In applying this criterion, we first apply constraints of (6.2-11). If these are satisfied, only then do we proceed to check the constraints needed from the table.

For commonly used low-order systems, the stability constraints can be deduced from the table in terms of the coefficients as follows.[1]

1. *First order*:

$$F(z) = a_1 z + a_0 = 0, \qquad a_1 > 0$$

Stability conditions:

$$\left| \frac{a_0}{a_1} \right| < 1 \qquad\qquad (6.2\text{-}13)$$

2. *Second order*:

$$F(z) = a_2 z^2 + a_1 z + a_0 = 0, \qquad a_2 > 0$$

Stability conditions:

$$a_2 + a_1 + a_0 > 0$$
$$a_2 - a_1 + a_0 > 0 \qquad\qquad (6.2\text{-}14)$$
$$a_0 - a_2 < 0$$

3. *Third order*:

$$F(z) = a_3 z^3 + a_2 z^2 + a_1 z + a_0 = 0, \qquad a_3 > 0$$

Stability conditions:

$$a_3 + a_2 + a_1 + a_0 > 0$$
$$a_3 - a_2 + a_1 - a_0 > 0 \qquad\qquad (6.2\text{-}15)$$
$$|a_0| < a_3$$
$$a_0{}^2 - a_3{}^2 < a_0 a_2 - a_1 a_3$$

In order to demonstrate the Jury stability criterion, let us consider some examples.

Example 6.2-1

The characteristic equation of a feedback SDS is given by

$$F(z) = 2z^4 + 7z^3 + 16z^2 + 4z + 1 = 0$$

[1]For complete derivations of these results and for higher-order systems, see E. I. Jury, *Theory and Application of the Z-Transform Method*, Wiley, New York, 1964.

In checking for system stability, since $n = 4$, we have five constraints and five rows in the table. Here

$$F(z) = 2z^4 + 7z^3 + 16z^2 + 4z + 1$$

Hence

$$F(1) = 30 > 0, \qquad (-1)^4 F(-1) = 8 > 0$$

The table is

Row	z^0	z^1	z^2	z^3	z^4
1	1	4	16	7	2
2	2	7	16	4	1
3	-3	-10	-16	-1	
4	-1	-16	-10	-3	
5	8	14	38		

For stability we must have

$$|1| < 2 \quad \leftarrow \text{satisfied}$$
$$|-3| > |-1| \leftarrow \text{satisfied}$$
$$|8| > |38| \leftarrow \text{not satisfied}$$

The system is therefore unstable, i.e. the characteristic equation has roots outside the unit circle.

Example 6.2-2

Consider a SDS as shown in Fig. 6.2-3. Use Jury's criterion to find the ranges of values of gain K and sampling period T that insure stability.

First, to find the characteristic equation:

$$GH(z) = Z\left\{\frac{K(1-e^{-sT})}{s^2(s+1)}\right\} = K\frac{z-1}{z}\left[\frac{z}{z-e^{-T}} + \frac{Tz}{(z-1)^2} + \frac{z}{z-1}\right]$$

$$= \frac{K[z(T+e^{-T}-1)+1-Te^{-T}-e^T]}{(z-e^{-T})(z-1)}$$

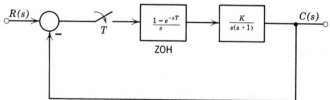

Figure 6.2-3 Sampled data system with unity feedback.

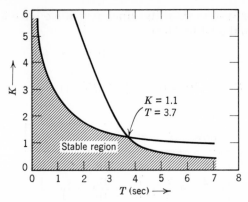

Figure 6.2-4   *K* versus *T* plot for Example 6.2-2.

The characteristic equation $1 + GH(z) = 0$ becomes

$$z^2 + z\,[K(e^{-T} + T - 1) - (1 + e^{-T})] + K(1 - Te^{-T} - e^{-T}) + e^{-T} = 0$$

Since this is a second-order equation, (6.2-14) gives the conditions

$$KT - KTe^{-T} > 0$$
$$2(1 + e^{-T}) + K(2 - 2e^{-T} - T - Te^{-T}) > 0$$

and

$$K(1 - Te^{-T} - e^{-T}) + e^{-T} - 1 < 0$$

These three inequalities imply that the system will be stable if

$$K > 0, \quad T > 0 \tag{I}$$

$$K < \frac{1 - e^{-T}}{1 - e^{-T} - Te^{-T}} \tag{II}$$

$$K < \frac{2(1 + e^{-T})}{2e^{-T} + Te^{-T} + T - 2} \tag{III}$$

Equations I, II, and III can be plotted on a *K* versus *T* plane. The region where all three are satisfied is the stable region (Fig. 6.2-4). We see directly from the figure what regions of *K* and *T* will give a stable closed-loop system.

Nyquist Criterion

Just as in the case of the continuous time systems, SDS stability can also be considered from the point of view of the open-loop transfer function $GH(z)$. The two methods usually used for this are the root locus method and the Nyquist Criterion. As seen in Section 4.2, the root locus technique essentially determines the location of the closed-loop poles from the open-loop poles. This point will be developed in Section 6.5. The Nyquist criterion for a SDS is as follows.

The Nyquist contour in the s-plane (as defined in Section 3.2) is shown in Fig. 6.2-5a. Note that if there are any poles of the open-loop system on the $j\omega$-axis, we avoid them by going around (but not enclosing) them on a small semicircle in the RHP. This Nyquist path can be transferred from the s- to the z-plane by the transformation $z = e^{Ts}$ which we discussed in the previous section. We have already seen that the RHP of the s-plane maps into the outside of the unit circle in the z-plane. This area can be encircled by a contour made up of two circles of radii unity and infinity (Fig. 6.2-5b). Note that for a pole on the $j\omega$-axis, there will be a pole on the unit circle which is avoided by going around it (not enclosing it) on a semicircle toward the outside of the unit circle.

With the path for Nyquist criterion as shown in Fig. 6.2-5b, the procedure and criterion itself can be easily stated as follows:

1. $GH(z)$ is plotted in the $GH(z)$-plane, [Re $GH(z)$ versus Im $GH(z)$] as z is varied on the Nyquist path shown in Fig. 6.2-5b.

2. For each zero and pole of $GH(z)$ in the z-plane enclosed by the contour, i.e., outside the unit circle, there is a net rotation of $GH(z)$ by an angle 2π (clockwise rotation) or an angle -2π (counterclockwise rotation) around the origin of the $GH(z)$-plane, respectively.

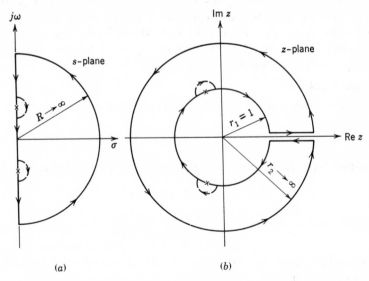

(a)   (b)

Figure 6.2-5 Nyquist contours in the s-plane and the z-plane.

3. If the origin is shifted to the $-1+j0$ point in the $GH(z)$-plane, the $GH(z)$ curve about the origin becomes the $1+GH(z)$ curve about the $-1+j0$ point.

4. The number of poles of $1+GH(z)$ outside the unit circle are the same as the number of poles of $GH(z)$ outside the unit circle. Let this number be N_p. In the $GH(z)$-plane the number of encirclements, N, of the $GH(z)$ curve around the $-1+j0$ point in the counterclockwise direction is the difference in the number of zeros and poles outside the unit circle of $1+GH(z)$. If N_z is the number of zeros (or the roots of the characteristic equation) of $1+GH(z)$ outside the unit circle, then

$$N = N_z - N_p \qquad (6.2\text{-}16)$$

5. For a system to be stable, $N_z = 0$. This implies that the number of counterclockwise encirclements of $GH(z)$ of the $-1+j0$ point in the $GH(z)$-plane must be equal to $-N_p$ if the corresponding closed-loop system is to be stable. The minus sign implies that the encirclements must be in the clockwise direction. We note that in order to apply the Nyquist criterion, we have to know the number of open-loop poles outside the unit circle, N_p, and the $GH(z)$-plot in the $GH(z)$-plane as z is varied on the Nyquist path.

We should note here that the Nyquist criterion in the s-plane and the Nyquist criterion in the z-plane are exactly analogous. This is seen by considering the Nyquist criterion in Section 3.2. The difference is that our independent variable is varied along the unit circle rather than along the $j\omega$-axis. In fact, if we make the substitution $z = e^{sT}$ and let $s = j\omega$, the Nyquist criterion of Section 3.2 may be applied directly as shown in the next example.

Example 6.2-3

The open-loop transfer function of a feedback SDS is given by

$$GH(s) = \frac{2(1-e^{-sT})}{s^2(s+1)}, \qquad T = 1 \sec$$

Using the Nyquist criterion to determine system stability, we have

$$GH(z) = Z\{G(s)H(s)\}$$

$$= \frac{0{\cdot}736(z+0{\cdot}716)}{(z-1)(z-0{\cdot}368)}$$

We note that there are no poles of $GH(z)$ outside the unit circle. Therefore $N_p = 0$.

We want to plot $GH(z)$ in the $GH(z)$-plane as z moves along the

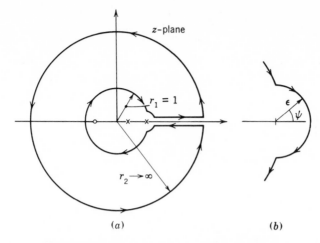

Figure 6.2-6 Nyquist contour for Example 6.2-3.

Nyquist contour. Since there is a pole of $GH(z)$ on the unit circle $(z = 1)$, we avoid it as shown in Fig. 6.2-6. Let us first consider the unit circle part of the Nyquist contour. Here we take

$$z = e^{j\varphi}$$

Where φ varies from 0 to -2π. On the small semicircle which avoids the $z = 1$ point, (See Fig. 6.2-6), we let

$$z = 1 + \epsilon e^{j\psi}$$

ψ varying from $\pi/2$ to $-\pi/2$ in the limit as $\epsilon \to 0$.

On the Unit Circle

Letting $z = e^{j\varphi}$ in $GH(z)$, we get

$$GH(z) = \frac{0.736(0.716 + e^{j\varphi})}{(e^{j\varphi} - 1)(e^{j\varphi} - 0.368)}$$

As φ is varied from 0 to -2π (actually 0 and -2π points are not taken because of the pole at $z = 1$) the plot appears as curve A in Fig. 6.2-7. This computation may be tedious since we must calculate $|GH(z)|$ and $\angle GH(z)$ for each φ. Here use of a digital computer is logical.

On the Small Semicircle

Here $z = 1 + \epsilon e^{j\psi}$, $\epsilon \to 0$. Therefore

$$GH(z) = \frac{0.736(1.716 + \epsilon e^{j\psi})}{\epsilon e^{j\psi}(\epsilon e^{j\psi} + 0.632)}$$

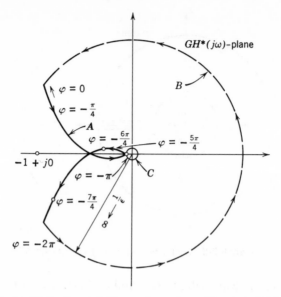

Figure 6.2-7 Nyquist diagram for system of Example 6.2-3.

Since $\epsilon \to 0$, we have

$$GH(z) \cong \frac{2}{\epsilon} e^{-j\psi}$$

Now as ψ varies from $+\pi/2$ to $-\pi/2$, we note that $GH(z)$ traces an infinite (or very large) semicircle from $-\pi/2$ to $\pi/2$. This is curve B in Fig. 6.2-7.

On the Infinite Circle

The third part of the Nyquist contour is the infinite circle. On this we let

$$z = r_2 e^{j\varphi}$$

$r_2 \to \infty$ and φ varies from 0 to 2π. We have

$$GH(z) = \frac{0.736(r_2 e^{j\varphi} + 0.716)}{(r_2 e^{j\varphi} - 1)(r_2 e^{j\varphi} - 0.368)} , \quad r_2 \to \infty$$

$$\cong \frac{0.736 \times e^{-j\varphi}}{r_2} , \quad r_2 \to \infty$$

This shows that $GH(z)$ traces a circle of very small radius from 0 to -2π. This is curve C in Fig. 6.2-7.

From Fig. 6.2-7 we note that there are no encirclements of the

$-1+j0$ point. Therefore $N = 0$. Since $N_p = 0$, $N_z = 0$ and the system is stable. From Fig. 6.2-7 we also see that the addition of frequency-independent gain $K \geq 1/0.45 = 2.22$ into the loop will cause the $GH(z)$ curve to enclose the $-1+j0$ point and hence the system will become unstable. The gain margin for this system is hence $2.22 = 6.9$ db. We could make a Bode plot or Nichols chart presentation of this frequency response and interpret stability from it in exactly the same manner as in the case of continuous systems. We will discuss this in a later section.

Hidden Oscillations in Sampled-Data Systems[2]

We have established the stability determination for feedback SDS entirely based on the Z-transform. We recall from (5.2-4) that the Z-transform gives information at the sampling instants only. Because of this limitation of the Z-transform, hidden oscillations may exist within the system which may not be detected by our analysis. In fact a system may show stability at the sampling instants but actually oscillate wildly between sampling instants. These oscillations will thus all be of some multiple of $\omega_r/2$ where ω_r is the sampling frequency.

To check for these hidden oscillations, the Z_m-transform may be used. The modified z-transfer function for a closed-loop system from (5.4-27) is

$$\frac{C(z, m)}{R(z)} = \frac{G(z,m)}{1 + GH(z)} \qquad (6.2\text{-}17)$$

We thus check the poles of $C(z,m)$ to determine true stability and these we note are the roots of $1 + GH(z) = 0$ and the poles of $G(z,m)$ which are not poles of $GH(z)$.

Normally hidden oscillations exist only if the open-loop transfer function is unstable. Thus hidden oscillations usually can be found by examining the roots of $1 + GH(z)$ in the normal way and then checking to see if $G(z, m)$ has any poles outside the unit circle.

6.3 ERROR CONSTANTS K_p, K_v, and K_a

In Chapter 5 we developed techniques for determinating the transient behavior of SDS. In the analysis and design of such systems, as in the case of continuous-time systems, the steady-state behavior is important. As in the case of continuous-time systems discussed in Section 3.3, steady-state error constants for SDS are defined from the

---

[2]E. I. Jury, "Hidden Oscillations in Sampled-Data Control Systems," *Trans. AIEE*, **75**, Part II, January 1956, pp. 391–395.

values of the steady-state error for unit step, ramp, and acceleration inputs.

Let us consider a unity-feedback SDS. From (5.2-33), letting $H(s) = 1$, we write

$$E(z) = \frac{1}{1+G(z)} R(z) \tag{6.3-1}$$

and

$$E(z,m) = R(z,m) - C(z,m) \tag{6.3-2}$$

where [from (5.4-27)]

$$C(z,m) = \frac{G(z,m)}{1+G(z)} R(z) \tag{6.3-3}$$

The steady-state error at sampling instants is given by the final-value theorem of Section 5.4.

$$\lim_{n\to\infty} e(nT) = \lim_{z\to 1} (z-1)E(z) \tag{6.3-4}$$

if $(z-1)E(z)$ has all its poles inside the unit circle. To obtain information on between-sampling instants or to determine the steady-state hidden oscillations, we can obtain

$$\lim_{n\to\infty} e(nT,m) = \lim_{z\to 1} (z-1) E(z,m), \qquad 0 < m < 1 \tag{6.3-5}$$

if $(z-1)E(z,m)$ has all its poles inside the unit circle.

Apart from hidden oscillation determination, the usual analysis and design are based on the sampled steady-state error as given by (6.3-4). This can be specified in terms of the error constants just as in the case of continuous-control systems.

In all that follows we assume the system is stable and hence steady-state error exists and the final-value theorem applies.

The Position Error Constant K_p

For a step input $R(z) = z/(z-1)$; by (6.3-1) and (6.3-4) we have

$$\lim_{n\to\infty} e(nT) = e_{ss}(nT) = \lim_{z\to 1} \frac{1}{1+G(z)} = \frac{1}{1+\lim_{z\to 1} G(z)} \tag{6.3-6}$$

$$= \frac{1}{1+K_p}$$

where we have defined

$$\lim_{z\to 1} G(z) \triangleq K_p \triangleq \text{position error constant} \tag{6.3-7}$$

The higher the value of K_p, the lower will be the steady-state error due to a step input. K_p will be infinite if $G(z)$ has a pole at $z=1$ and hence for this case the steady-state error due to a step input will be zero.

We can also express K_p in terms of poles and zeros of the closed-loop system. This representation is useful in z-plane synthesis as we shall see in Section 9.4. Thus consider an overall transfer function for a unity-feedback SDS given by

$$T(z) = \frac{C(z)}{R(z)} = \frac{G(z)}{1+G(z)} = \frac{K(z-z_1)\cdots(z-z_k)}{(z-p_1)\cdots(z-p_n)} \qquad (6.3\text{-}8)$$

Then

$$G(z) = \frac{T(z)}{1-T(z)} = \frac{K(z-z_1)\cdots(z-z_k)}{(z-p_1)\cdots(z-p_n)-K(z-z_1)\cdots(z-z_k)} \qquad (6.3\text{-}9)$$

Therefore using (6.3-7) we get

$$K_p = \lim_{z\to 1} G(z) = \frac{K(1-z_1)\cdots(1-z_k)}{(1-p_1)\cdots(1-p_n)-K(1-z_1)\cdots(1-z_k)} \qquad (6.3\text{-}10)$$

This expression gives the position error constant K_p in terms of the closed-loop poles and zeros and the frequency-independent gain K.

The Velocity Error Constant K_v

For the ramp input $R(z) = Tz/(z-1)^2$. Therefore by (6.3-1).

$$E(z) = \frac{1}{1+G(z)} \frac{Tz}{(z-1)^2}$$

We now use (6.3-4) to obtain

$$\lim_{n\to\infty} e(nT) = e_{ss}(nT) = \lim_{z\to 1} \frac{Tz}{(z-1)[1+G(z)]} \qquad (6.3\text{-}11)$$

$$= \lim_{z\to 1} \frac{1}{(1/T)(z-1)G(z)} = \frac{1}{K_v}$$

Where we define

$$\lim_{z\to 1} \frac{1}{T}(z-1)G(z) \triangleq K_v \triangleq \text{velocity error constant} \qquad (6.3\text{-}12)$$

The higher the value of K_v, the lower will be the steady-state error for a ramp input. Note that the units of K_v are sec^{-1}.

In order to find K_v in terms of closed-loop poles and zeros, we reconsider

$$E(z) = \frac{1}{1+G(z)} \frac{Tz}{(z-1)^2}$$

Using (6.3-9), we get

$$1 + G(z) = \frac{1}{1 - T(z)}$$

Therefore

$$E(z) = \frac{Tz}{(z-1)^2}[1 - T(z)] \qquad (6.3\text{-}13)$$

or, using (6.3-4),

$$\lim_{n \to \infty} e(nT) = \lim_{z \to 1} \frac{T}{z-1}[1 - T(z)]$$

In order to have finite velocity error, $1 - T(z)$ must have a zero at $z = 1$, so let

$$1 - T(z) = (z-1)T_1(z) \qquad (6.3\text{-}14)$$

Then

$$\lim_{n \to \infty} e(nT) = \lim_{z \to 1} T[T_1(z)] \qquad (6.3\text{-}15)$$
$$= T \lim_{z \to 1} T_1(z)$$

Differentiating both sides of (6.3-14) gives

$$-\frac{d}{dz}[T(z)] = (z-1)\frac{d}{dz}[T_1(z)] + T_1(z)$$

and taking the limit as $z \to 1$ we have

$$\lim_{z \to 1}[T_1(z)] = \lim_{z \to 1}\left\{ -\frac{d}{dz}[T(z)] \right\} \qquad (6.3\text{-}16)$$

Therefore (6.3-15) becomes

$$\lim_{n \to \infty} e(nT) = \lim_{z \to 1}\left\{ -T\frac{d}{dz}[T(z)] \right\} \qquad (6.3\text{-}17)$$

From (6.3-11),

$$\lim_{n \to \infty} e(nT) = \frac{1}{K_v}$$

Therefore

$$\frac{1}{K_v} = \lim_{z \to 1}\left\{ -T\frac{d}{dz}[T(z)] \right\} \qquad (6.3\text{-}18)$$

Since we have assumed a finite velocity error, the position error must be zero. This implies that $K_p = \infty$ or $\lim_{z \to 1} T(z) = 1$. Hence using (6.3-8) we can write (6.3-18) as

$$\frac{1}{K_v T} = -\lim_{z \to 1}\frac{(d/dz)[T(z)]}{T(z)}$$
$$= -\lim_{z \to 1}\frac{d}{dz}[\log T(z)]$$

$$= -\lim_{z \to 1} \frac{d}{dz} [\log K + \log (z - z_1) + \cdots + \log (z - z_k)$$
$$- \log (z - p_1) - \cdots - \log (z - p_n)] \quad (6.3\text{-}19)$$

Therefore

$$\frac{1}{K_v T} = -\sum_{i=1}^{k} \frac{1}{1 - z_i} + \sum_{j=1}^{n} \frac{1}{1 - p_j} \quad (6.3\text{-}20)$$

This equation gives us the velocity error constant K_v as a function of the closed-loop poles and zeros.

The Acceleration Error Constant K_a

For acceleration input $r(t) = t^2/2$,

$$R(z) = \frac{T^2 z(z + 1)}{2(z - 1)^3}$$

In this case, from (6.3-1),

$$E(z) = \frac{1}{1 + G(z)} \frac{T^2 z(z + 1)}{2(z - 1)^3}$$

Therefore using (6.3-4)

$$\lim_{n \to \infty} e(nT) = \lim_{z \to 1} (z - 1) \frac{T^2 z(z + 1)/2}{(z - 1)^3 [1 + G(z)]}$$
$$= \lim_{z \to 1} \frac{1}{[(z - 1)^2 / T^2] G(z)} = \frac{1}{K_a} \quad (6.3\text{-}21)$$

where we define

$$\lim_{z \to 1} \frac{(z - 1)^2}{T^2} G(z) \triangleq K_a \triangleq \text{acceleration error constant} \quad (6.3\text{-}22)$$

By using a procedure similar to the one for K_v, it can be shown that

$$\frac{1}{K_a} = -T^2 \lim_{z \to 1} \left\{ \frac{d^2}{dz^2} [T(z)] \right\} \quad (6.3\text{-}23)$$

In terms of poles and zeros of the overall system, we find K_a from (6.3-8) and (6.2-23) as

$$\frac{1}{K_a} = T^2 \left[\sum_{i=1}^{k} \frac{1}{(1 - z_i)^2} - \sum_{j=1}^{n} \frac{1}{(1 - p_j)^2} \right] \quad (6.3\text{-}24)$$

where we have assumed $K_v \to \infty$, i.e., K_a is nonzero. Verification of (6.3-24) is left as an individual exercise.

Higher-order error constants can similarly be defined and obtained, but in general they have little application.

Finally, as in continuous-time systems, we can obtain error constants K_p, K_v, K_a when $T(z)$ is known as a ratio of polynomials in z as

$$T(z) = \frac{a_k z^k + a_{k-1} z^{k-1} + \cdots + a_1 z + a_0}{b_n z^n + b_{n-1} z^{n-1} + \cdots + b_1 z + b_0}, \quad k < n \quad (6.3\text{-}25)$$

In this case if we first obtain $G(z)$ from (6.3-9) as

$$G(z) = \frac{T(z)}{1 - T(z)} \quad (6.3\text{-}26)$$

and then use (6.3-7), (6.3-18), and (6.3-23), we get K_p, K_v, and K_a, respectively, in terms of coefficients of (6.3-25). Note that here, also, we evaluate K_v only if $K_p \to \infty$ and evaluate K_a only if $K_v \to \infty$ and $K_p \to \infty$. Furthermore, the results here are much more complicated algebraically than for the continuous-time case since we now take the limit $z \to 1$ instead of $s \to 0$. Since we do not make use of them, we shall not derive these constants here. The derivation of these constants in terms of the coefficients is again left as an individual exercise.

6.4 FREQUENCY-RESPONSE METHODS

The frequency response is an important tool in the design of SDS. The amplitude and phase characteristics as well as phase margin, gain margin, resonant frequency, and peak frequency overshoot are as useful in the design of SDS as they are in the design of continuous-time systems.

Let us begin by considering the closed-loop frequency response as a function of the open-loop transfer function. The closed-loop transfer function, from (5.2-34), is

$$T(z) = \frac{G(z)}{1 + GH(z)} \quad (6.4\text{-}1)$$

or, using (5.2-4),

$$T^*(s) = \frac{G^*(s)}{1 + GH^*(s)} \quad (6.4\text{-}2)$$

We let $s = j\omega$ and obtain

$$T^*(j\omega) = \frac{G^*(j\omega)}{1 + GH^*(j\omega)} \quad (6.4\text{-}3)$$

The open-loop transfer function is $GH^*(j\omega)$. We wish to plot this as a function of ω in the $GH^*(j\omega)$-plane. We know that $GH^*(j\omega)$ is periodic in ω with a period ω_r, thus if we plot $GH^*(j\omega)$ from $\omega = -\omega_r/2$, to $\omega_r/2$, then all other points on the $j\omega$-axis will fall on top of this plotted curve because of the periodicity. Actually we need only plot $GH^*(j\omega)$ from $\omega = 0$ to $\omega_r/2$ because the plot for $\omega = -\omega_r/2$ to 0 will be the reflection in the real axis of the plot from $\omega = 0$ to $\omega_r/2$. Now we consider the frequency-response plot of the open-loop transfer function.

Frequency-Response Plots of the Open-Loop Transfer Function

EXACT PLOT OF $GH^*(j\omega)$

The exact plot of $GH^*(j\omega)$ is now known because $GH^*(j\omega) = GH(e^{j\omega T})$, the same plot as was developed in the Nyquist criterion when we plotted $GH(z)$ as z goes around the unit circle. This is due to the fact that as ω varies from $-\omega_r/2$ to $\omega_r/2$, $z = e^{j\omega T}$ goes around the unit circle once.

In general,

$$GH^*(j\omega) = GH(z)\bigg|_{z=e^{j\omega T}} = \frac{K(e^{j\omega T} - z_1) \cdots (e^{j\omega T} - z_k)}{(e^{j\omega T} - p_1) \cdots (e^{j\omega T} - p_n)} \qquad (6.4\text{-}4)$$

We note that it is laborious to find magnitude and phase of (6.4-4) by hand methods even though we now have a reduced frequency range to deal with. However, if a computational aid (a digital computer) is available, a frequency response can be calculated as well from (6.4-4) as from a rational function in $j\omega$. Having this frequency response, we may use all the tools used for handling frequency responses that have been developed for analyzing and designing continuous-time systems. In fact, once the frequency response is known there is no need for distinguishing it from a continuous-time system. It is just another frequency response and may be handled as such.

There are, however, several hand methods available for dealing with frequency responses of expressions as in (6.4-4). We consider a few of these here.

APPROXIMATE FREQUENCY-RESPONSE PLOTS

Consider using (5.2-47) on $GH^*(j\omega)$:

$$GH^*(j\omega) = \frac{1}{T} \sum_{k=-\infty}^{\infty} G(\overline{j\omega + k\omega_r}) H(\overline{j\omega + k\omega_r}) \qquad (6.4\text{-}5)$$

$G(j\omega)H(j\omega)$ usually decreases very rapidly as ω increases for the typical control system. If this is the case, then only two or three terms of (6.4-5) will be sufficient to give a good approximation to $GH^*(j\omega)$. Hence we may easily develop a plot of $GH^*(j\omega)$ from a plot of

$G(j\omega)H(j\omega)$. Having $GH^*(j\omega)$, we can determine phase and gain margin exactly as for any other frequency response. If we also determine $G^*(j\omega)$, the closed-loop frequency response can be computed and from it the peak frequency-response overshoot (M_m) and the peak frequency (ω_m) can be determined. (M_m and ω_m are defined in Section 3.4.)

Modifications

It should be noted that the frequency response as given above is based on the Z-transform. So if the $GH^*(j\omega)$ curve does not enclose the $-1+j0$ point, the corresponding system will be stable at the sampling instants. But this does not tell us what happens between samplings. In fact, *hidden oscillations* may exist between samplings.

In order to investigate the behavior in between-sampling instants, we may use the Z_m-transform:

$$T^*(j\omega, m) = \frac{G^*(j\omega, m)}{1 + GH^*(j\omega)} \qquad (6.4\text{-}6)$$

$T^*(j\omega, m)$ constructed for various values of m between 0 and 1 will permit us to make a plot of peak frequency overshoot M_m versus m. From this we will be able to judge behavior between samplings and will also be able to detect hidden oscillations. Applications of this approach of course depend heavily upon the use of automatic computation aids.

Transformation into w-Plane by a Bilinear Transformation

Because of the nonrational form of most transfer functions [(6.4-4) is an example], frequency-response methods are not quite as convenient for SDS as for continuous-time systems, which have rational transfer functions in s. If we had a transformation that would map the unit circle in the z-plane into the imaginary axis of a new plane, resulting in a rational function in the new variable, all the continuous-time system frequency tools (such as Bode plots and Nichols charts) could be used directly. Such a transformation exists and it is the bilinear transformation (as discussed earlier),

$$z = \frac{1+w}{1-w} \qquad (6.4\text{-}7)$$

Under this transformation the open-loop transfer function $GH(z)$ becomes

$$GH(w) = K\frac{(w-z_1') \cdots (w-z_k')}{(w-p_1') \cdots (w-p_n')} \qquad (6.4\text{-}8)$$

which is now an ordinary rational function in the new variable w.

If we plot $GH(w)$ on the imaginary axis, w from $-j\infty$ to $+j\infty$, we actually vary ω (in the s-plane) from $-\omega_r/2$ to $\omega_r/2$. To do this, let $w = jv$, where v is frequency in the w-plane. Then

$$GH(jv) = \frac{(jv - z_1') \cdots (jv - z_k')}{(jv - p_1') \cdots (jv - p_n')} \tag{6.4-9}$$

$GH(jv)$ can be plotted as a Bode plot and the information can be transferred onto the Nichols chart in the usual manner. The phase margin, the gain margin, the peak frequency overshoot (M_m) and the peak frequency (v_m) are seen [in the case $H(s) = 1$] directly from the Nichols chart.

We can find the relationship between v and ω from (6.4-7) as

$$z = e^{j\omega T} = \frac{1 + jv}{1 - jv}$$

or

$$\omega = \frac{2}{T}\tan^{-1}(v) \tag{6.4-10}$$

Equation (6.4-10) enables us to transfer all frequencies from the w-domain to the s-domain.

The bilinear transformation method forms the basis of compensator design for SDS using frequency-response methods. It will be utilized in Chapter 9 for this purpose.

Example 6.4-1

Consider the system shown in Fig. 6.4-1. Let us develop the Bode plots and Nichols chart in the w-plane and determine phase margin, gain margin, peak frequency overshoot, and the peak frequency. Modifications required in the analysis if information on between-sampling instants is desired will also be briefly considered.

We first obtain

$$GH(z) = Z\left\{\frac{2(1 - e^{-sT})}{s^2(s+1)}\right\} = \frac{0.736(z + 0.716)}{(z-1)(z-0.368)}$$

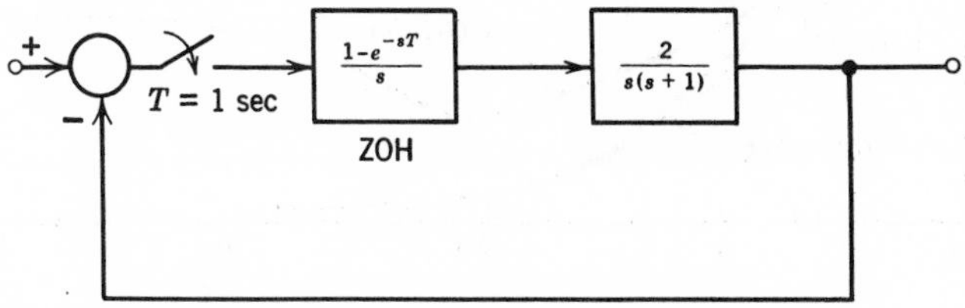

Figure 6.4-1

Putting $z = (1+w)/(1-w)$,

$$GH(w) = \frac{0.736(1-w)(1.716+0.284w)}{2w(0.632+1.368w)}$$

Letting $w = jv$,

$$GH(jv) = \frac{(1-jv)(1+jv/6.05)}{jv(1+jv/0.462)}$$

The magnitude and phase of $GH(jv)$ is plotted in Bode form in Fig. 6.4-2. We note that

$$\text{phase margin} \cong 15°$$
$$\text{gain margin} \cong 3.6 \text{ db}$$

Now we can transfer the magnitude and phase of $GH(jv)$ on the Nichols chart as shown in Fig. 6.4-3. Here

$$\text{peak frequency overshoot} \cong 12 \text{ db} = 4$$
$$\text{frequency } v_m \cong 0.7$$

In order to determine peak frequency, ω_m, in the s-domain, we use (6.4-10) and get

$$\omega_m = 1.22 \text{ rad/sec}$$

This analysis gives us a picture of the behavior of the system in the frequency domain. However, it is based on the information at sampling instants only. In order to see what is happening between samples we consider

$$T(z,m) = \frac{C(z,m)}{R(z)} = \frac{G(z,m)}{1+GH(z)}$$

Since we have unity feedback,

$$G(z,m) = Z_m\{G(s)\}$$

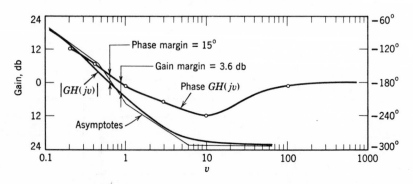

Figure 6.4-2   Bode plot of $GH(jv)$ of Example 6.4-1.

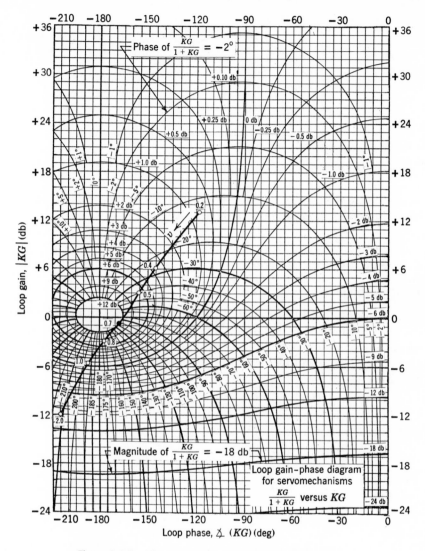

Figure 6.4-3 Nichols chart for $GH(jv)$ of Example 6.4-1.

We now obtain $GH(w)$ and $G(w,m)$. We can plot these on a Bode plot for different values of m. For each value of m we can plot the magnitude and phase of $T(w,m)$ using $GH(w)$ and $G(w,m)$. From these curves we can develop a curve of peak $|T(w,m)|$ versus m. This will enable us to determine the peak of the hidden oscillation and the frequency v at which it occurs if any hidden oscillations do

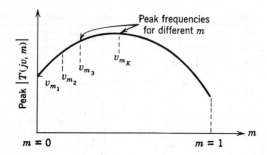

Figure 6.4-4 Peak frequency-response overshoot versus *m* curve for a typical SDS.

occur. This frequency of course can be transferred to s-domain by (6.4-10). A typical peak $|T(jv,m)|$ curve appears in Fig. 6.4-4.

6.5 ROOT LOCUS METHOD

Basic Technique

The root locus method essentially determines the location of the roots of the characteristic equation, i.e., it determines the poles of the closed-loop system from the open-loop pole-zero configuration. In the discrete-time control systems the open-loop transfer function has the form

$$GH(z) = \frac{K(z-z_1) \cdots (z-z_k)}{(z-p_1) \cdots (z-p_n)}, \qquad n \geqslant k \qquad (6.5\text{-}1)$$

The characteristic equation is given by

$$1 + GH(z) = 0 \qquad (6.5\text{-}2)$$

The roots of the characteristic equation and hence the closed-loop poles occur where

$$GH(z) = -1$$

or

$$|GH(z)| = 1 \text{ and } \angle GH(z) = \pm k\,180°, \qquad k = 1, 3, 5, \ldots \quad (6.5\text{-}3)$$

The root locus is then those points in the z-plane where (6.5-3) applies. The K for which the magnitude condition holds is the parameter of the locus.

The definition and the rules for constructing the root locus have been given in Section 4.2. The definition and the rules of construction apply here directly. The only difference is that now we work in the z-plane rather than the s-plane.

In the s-plane, the poles and zeros normally lie in the LHP for consideration of stability. However, in the z-plane, the open-loop poles and zeros usually lie within the unit circle on both the LHS and RHS of the z-plane. This different distribution of open-loop poles and zeros in the z-plane makes the root loci look somewhat different for discrete-time systems. On each root locus plot we usually superimpose the unit circle in order to see whether the root locus goes outside this circle and for what values of K it goes out. The root locus thus instantly indicates for what values of K the system is closed-loop stable.

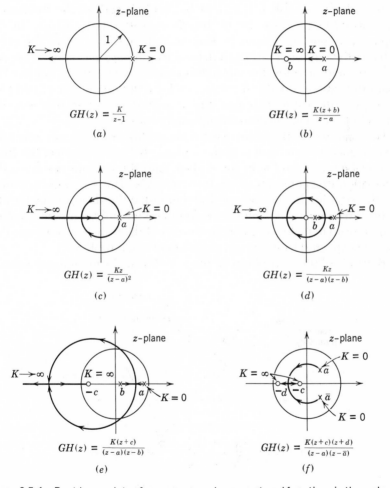

Figure 6.5-1   Root locus plots of some commonly encountered functions in the z-plane.

The root locus plots of some common types of open-loop pole-zero configurations are given in Fig. 6.5-1. These should provide a feel for discrete-time systems' root loci. Note that in most common cases, part of the root locus comes out to be a part of a circle.

Determination of System Behavior from the Root Locus Plot

Just as in the case of continuous-time systems, the transient and stability behavior of SDS can be inferred from the root locus plot. Stability is determined easily since the system will be stable for all those values of K for which the root locus is inside the unit circle.

In order to establish transient behavior based on the location of the closed-loop poles, let us briefly consider the s-plane. In the s-plane, as we have seen in Section 2.3, the location of the poles tells us the damping ratio and natural frequency, because the ζ-lines, σ-lines, and ω-lines are straightforward curves (Fig. 6.5-2). The damping ratio and natural frequency for a dominant pair of complex poles enable us to estimate the transient behavior that may be expected from a given closed-loop configuration.

How can we do the same in the z-plane? Since we already know the significance of ζ, σ, and ω in the s-plane, we can transform the lines of constant ζ, σ, and ω in the s-plane into the z-plane and have the corresponding contours there. If we now superimpose these contours on the root locus, we can use them to judge the damping factor and natural frequency of each closed-loop z-plane pole. This information can be used to make very good estimates of the transient behavior that may be expected from a given closed-loop configuration.

Because of the periodicity of the $z = e^{Ts}$ transformation, we need to

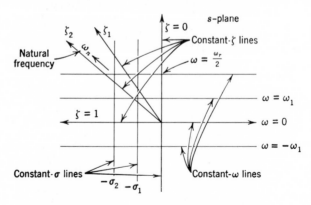

Figure 6.5-2 ζ-, ω-, and σ-lines in the s-plane.

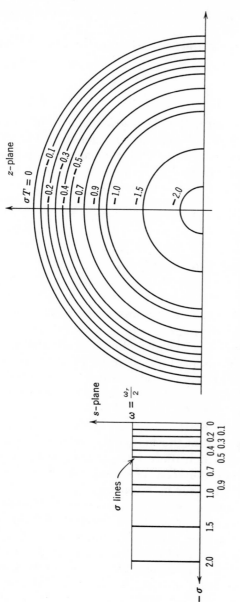

Figure 6.5-3 Transformation of σ-lines from s- to z-plane.

transfer only ζ-, σ-, and ω-lines lying in the primary strip. Furthermore, it is necessary to draw these contours only on the upper half of the z-plane, since the lower half is just a reflection of the upper half in the real axis.

We can write $s = -\sigma + j\omega$, $\sigma > 0$ for s in the LHP. Therefore

$$z = e^{-\sigma T} e^{j\omega T} \tag{6.5-4}$$

or

$$|z| = e^{-\sigma T} \quad \text{and} \quad \angle z = \omega T$$

Each σ-line in the left half s-plane gives a circle of radius $e^{-\sigma T}$ in the z-plane. The constant σT contours are shown in Fig. 6.5-3. Each ω-line in the s-plane gives a line with constant angle ωT in the z-plane. Some lines in terms of sampling frequency are shown in Fig. 6.5-4.

The ζ-lines in the s-plane can be written as

$$s = -\zeta\omega_n + j\omega_n\sqrt{1-\zeta^2} \tag{6.5-5}$$

This gives

$$z = e^{-\zeta\omega_n T} e^{j\omega_n T \sqrt{1-\zeta^2}} \tag{6.5-6}$$

By comparing (6.5-4) with (6.5-6),

$$\zeta\omega_n = \sigma \quad \text{and} \quad \omega_n\sqrt{1-\zeta^2} = \omega \tag{6.5-7}$$

In order to draw constant ζ contours in the z-plane, we can choose a fixed ζ and then vary ω from 0 to $\omega_r/2$ in the primary strips. This enables us to determine $|z|$ and $\angle z$ for each ω in the strip and hence a contour is traced. Contours for $\zeta = 0$ to 1 are shown in Fig. 6.5-5.

It should be noted that any two of the ζ-, σ-, or ω-contours give the complete information, i.e., the location of the corresponding s-plane pole.

From Figs. 6.5-3, 6.5-4, and 6.5-5 we can deduce the transient behavior of a SDS from the knowledge of the damping factor and natural frequency which these curves give us. Once the poles are known for a system, they can be superimposed on these curves and transient behavior can then be estimated.

We can determine σ and ω for a given pole without having these diagrams by joining the origin to the location of the pole. The radius or the length of this line is $e^{-\sigma T}$. The angle of this line with the real axis gives ω. Knowledge of σ and ω of course determines ζ and ω_n from (6.5-7). With a little familiarity, by just looking at the location of the poles within the unit circle, we can visualize the transient behavior, keeping Figs. 6.5-3, 6.5-4, and 6.5-5 in mind. Some of the conclusions are discussed briefly below. (Note that for every pole not on the real axis there is a corresponding conjugate pole.)

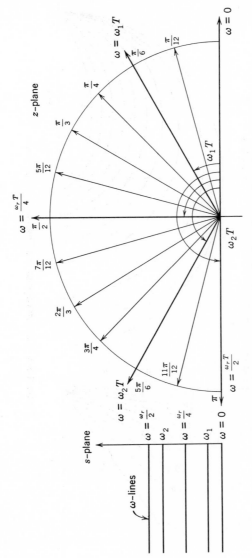

Figure 6.5-4 Transformation of ω-lines from s- to z-plane.

233

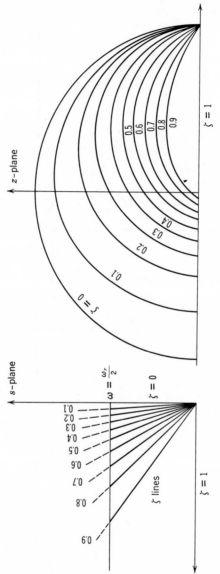

Figure 6.5-5 Transformation of ζ-lines from s- to z-plane.

POLE ON THE POSITIVE REAL AXIS

Here $\omega = 0$, therefore the time response is exponentially decaying. The decay gets faster as the pole moves toward the origin. No decay takes place if the pole is on the unit circle.

POLE IN FIRST QUADRANT

We get decaying sinusoidal oscillation with frequency of oscillation increasing as the pole moves nearer the imaginary axis. Decay is faster near the origin. Stable oscillations occur if the pole is on the unit circle.

POLE IN SECOND QUADRANT

The frequency of oscillation is higher than for a first-quadrant pole. Decay of oscillations will depend on the location — it will be faster near the origin and slower near the unit circle.

Some of these tabulations are shown in Fig. 6.5-6 to familiarize the reader with the correlation of transient behavior with z-plane pole location.

The transient behavior is generally controlled by two dominant poles of the system. Peak overshoot and peak time also can be determined from the location of poles. This is developed in Chapter 9, where the procedure is used to design a discrete compensator. Now we consider an example.

Example 6.5-1

The open-loop transfer function of a unity-feedback SDS is given by

$$GH(s) = \frac{K(1 - e^{-sT})}{s^2(s+1)}, \qquad T = 1 \text{ sec}$$

Plot the root locus in the z-plane, find the closed-loop poles for $K = 1$, and determine ζ, ω, ω_n, and σ for this value of K. Also determine the limiting value of K for which the system is closed-loop stable.

We first obtain $GH(z)$ as

$$GH(z) = Z\{GH(s)\} = \frac{0.368K(z+0.716)}{(z-1)(z-0.368)} = \frac{K_1(z+0.716)}{(z-1)(z-0.368)}$$

In order to plot the root locus, we proceed as follows:

1. We first plot the poles at $z = 1$ and $z = 0.368$ and zero at $z = -0.716$.

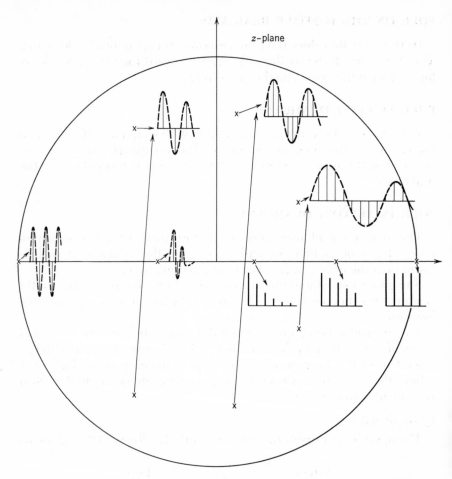

Figure 6.5-6   Time-behavior for various pole locations within the unit circle.

2. The segment of real axis between $z = 1$ and $z = 0.368$ as well as that between $z = -0.716$ and $z = -\infty$ is part of the root locus.

3. Since the root locus starts from poles and goes to the zeros, the locus breaks away from the real axis somewhere between $z = 1$ and $z = 0.368$ and comes back again between $z = -0.716$ and $z = -\infty$.

4. It can be shown that for the pole-zero distribution we have here (two poles and zero), the root locus from the real axis is a circle centered on the zero and with a radius which intersects the axis between the two poles.

5. In order to draw this circle quickly, we can find the breakaway

and break-in points from the real axis in the usual manner. We have

$$\frac{K_1(z+0.716)}{(z-1)(z-0.368)} = -1$$

or

$$K_1 = -\frac{(z-1)(z-0.368)}{(z+0.716)}$$

Differentiating

$$\frac{dK_1}{dz} = \frac{z^2+1.43z-1.348}{(z+0.716)^2} = 0$$

for $z = 0.61$ and -2.05. The roots of the quadratic give $z \cong 0.61$ as the breakaway point and $z \approx -2.05$ as the break-in point.

Using above breakaway and break-in points, we draw the circle as shown in Fig. 6.5-7. Any point on the circle can be checked by adding the angle contributions of each open-loop pole and zero with a spirule.

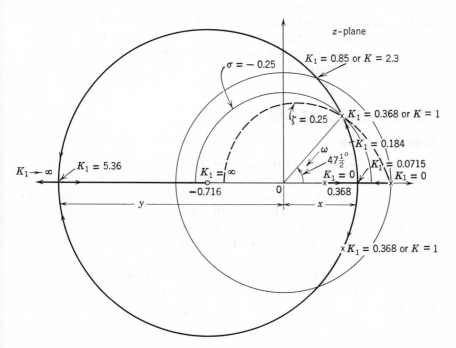

Figure 6.5-7   Root locus plot of $GH(z)$ of Example 6.5-1.

Conclusions

1. The system will become unstable when the root locus goes outside the unit circle. The limiting value of K is obtained as

$$K_1 = 0.368 \times K \approx 0.85 \qquad \text{or} \qquad K \approx 2.3$$

2. $K = 1$ implies $K_1 = 0.368$. This point on the root locus may be found by use of a spirule. In order to determine ζ, σ, and ω, we can superimpose the curves of Figs. 6.5-3, 6.5-4, and 6.5-5. We can also find these quantities if we join the $K_1 = 0.368$ point to the origin as shown.

$e^{-\sigma T}$ = distance between origin and $K_1 = 0.368$ point = 0.8

ωT = angle between this line and real axis = $47\frac{1}{2}° = 0.835$

Since $T = 1$ sec, $\sigma = 0.24$ and $\omega = 0.835$ rad/sec. Now $\zeta \omega_n = \sigma$ and $\omega_n \sqrt{1 - \zeta^2} = \omega$. Therefore $\zeta = 0.275$ and $\omega_n = 0.875$ rad/sec.

However, superimposing the curve gives $\zeta = 0.25$. This type of discrepancy will occur between graphical and numerical work. The knowledge of ζ, ω_n, ω, and σ enables us to determine transient behavior of the system.

The root locus method is a systematic and well-accepted procedure for the analysis and design of SDS. It is widely used for practical work. Computer programs are available for plotting root loci. The root locus method when a factor other than the gain K is the variable is discussed in Chapter 9.

6.6 MITROVIC'S METHOD

Basic Technique

We have developed Mitrovic's method for determining stability and estimating time response for continuous-time control systems in Section 4.3. The technique is based on the consideration of the characteristic equation rather than the open-loop transfer function.

In SDS, a similar procedure can be devised. Consider the LHS of the s-plane and the inside of the unit circle in the z-plane as shown in Fig. 6.6-1.

In Mitrovic's method for continuous-time systems we considered constant ζ-lines in the LHS of the s-plane, which we previously expressed as

$$s = \omega_n e^{j(\pi - \theta)} = -\zeta \omega_n + j \omega_n \sqrt{1 - \zeta^2} \qquad (6.6\text{-}1)$$

where $\zeta = \cos \theta$.

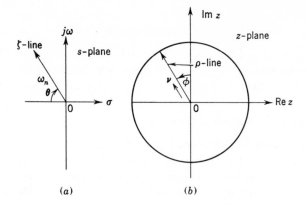

Figure 6.6-1   Zeta-line in s-plane and ρ-line in z-plane.

Of interest in the s-plane are ζ from 0 to 1 and ω_n from 0 to ∞. In this manner any root of the characteristic equation in the LHS of the s-plane will lie on one of the ζ-lines.

In SDS we are concerned with the roots of the characteristic equation within the unit circle. We can express the inside of the unit circle by constant-ρ lines as shown in Fig. 6.6-1b. We let

$$z = \nu e^{j(\pi/2+\phi)} = -\rho\nu + j\nu\sqrt{1-\rho^2} \qquad (6.6\text{-}2)$$

where $\rho = \sin \phi$. As ρ varies from -1 to 1 and ν varies from 0 to 1, we see from (6.6-2) that the whole upper half of the unit circle in the z-plane is covered. This is the region of interest here.

Since (6.6-2) is similar to (6.6-1), the two basic equations for Mitrovic's method in the SDS case will be similar to the two basic equations, (4.3-6), in the method for continuous-time systems. To get the two basic equations for the SDS case, we let the characteristic equation be given by

$$F(z) = a_n z^n + a_{n-1}z^{n-1} + \cdots + a_1 z + a_0, \qquad a_n > 0 \qquad (6.6\text{-}3)$$

By substituting (6.6-2) in (6.6-3), expanding the products, equating both real and imaginary parts to zero, and eliminating common factors, we get the two basic equations which are similar to (4.3-6):

$$-a_1 + a_2\phi_2(\rho)\nu + a_3\phi_3(\rho)\nu^2 + \cdots + a_n\phi_n(\rho)\nu^{n-1} = 0 \qquad (6.6\text{-}4)$$

and

$$a_0 + \nu^2[a_2\phi_1(\rho) + a_3\phi_2(\rho)\nu + \cdots + a_n\phi_{n-1}(\rho)\nu^{n-2}] = 0 \qquad (6.6\text{-}5)$$

where as before [see (4.3-7) and (4.3-8)] we have

$$\phi_k(\rho) = -[2\rho\phi_{k-1}(\rho) + \phi_{k-2}(\rho)], \qquad k \geqslant 2 \qquad (6.6\text{-}6)$$

and

$$\phi_0(\rho) = 0 \quad \text{and} \quad \phi_1(\rho) = -1 \qquad (6.6\text{-}7)$$

Equations (6.6-4) and (6.6-5) are the two basic equations for Mitrovic's method applied to SDS. We may use them to analyze SDS as the two equations of (4.3-6) were used to analyze continuous-time systems.

Development of the Region of Stability

We are interested in the roots inside the unit circle. It is sufficient if we consider just the upper half-circle, because for any complex root there will also be a complex conjugate root in the lower half-circle. On the boundary of the unit circle $\nu = 1$. Thus the boundary of the stable region will be defined by our two basic equations, (6.6-4) and (6.6-5), for the case $\nu = 1$ and $-1 \leq \rho \leq +1$. Let us consider using (6.6-4) and (6.6-5) to determine a stable region in a parameter plane.

In the sampled-data case, we know from (6.2-11) that for stability $F(1) > 0$ and $(-1)^n F(-1) > 0$ must hold. Using (6.6-3), these two conditions become

$$a_0 + a_1 + a_2 + a_3 + \cdots a_n > 0 \qquad (6.6\text{-}8)$$

and

$$(-1)^n[a_0 - a_1 + a_2 - a_3 + \cdots + (-1)^n a_n] > 0 \qquad (6.6\text{-}9)$$

To get another stability condition we now let $\nu = 1$, in (6.6-4) and (6.6-5); then

$$-a_1 + a_2\phi_2(\rho) + a_3\phi_3(\rho) + \cdots + a_n\phi_n(\rho) = 0$$
$$a_0 + a_2\phi_1(\rho) + a_3\phi_2(\rho) + \cdots + a_n\phi_{n-1}(\rho) = 0 \qquad (6.6\text{-}10)$$

Equations (6.6-8) and (6.6-9) give necessary conditions which the coefficients of the characteristic equation must satisfy if the corresponding SDS is to be stable. Equation (6.6-10) also gives two conditions which the coefficients on the limits of stability satisfy. Now, if as in the case of continuous-time systems we consider two of the coefficients, say a_i and a_j, as free parameters and all other coefficients free, then (6.6-8), (6.6-9), and (6.6-10) may be used to find those values of a_i and a_j that result in the corresponding system being stable. If as in Section 4.3 we consider the (a_i, a_j)-plane, (6.6-8), (6.6-9), and (6.6-10) may be used to find the region in this plane for which our system will be stable; i.e., we can use these equations to find the regions of stable values of a_i and a_j in this (a_i, a_j)-plane. Equations (6.6-8) and (6.6-9) are convenient in that they give linear conditions in the coefficients. Thus the boundaries of the linear regions they define will be straight lines in an (a_i, a_j)-plane. Equation (6.6-10), on the other hand, is a function of the variable ρ. As we let ρ go from -1 to $+1$ (and thus consider those values of z on the unit circle), (6.6-10) will give a curve in the (a_i, a_j)-plane which generally will not be linear. Thus more effort usually is required to plot this line, but it gives a line that is on the boundary of the stable region in the parameter plane [i.e., the (a_i, a_j)-plane]. In this way we may construct a region of stability in the parameter plane which tells us at a glance what values of our free parameters we can and cannot use. This generalization of Mitrovic's method is due to Šiljak.[2] Let us consider an example.

Example 6.6-1

The open-loop transfer function of a SDS is given by $G(s)H(s) = 2(1-e^{-sT})/s^2(s+1)$, $T = 1$ sec. Determine whether the system is closed-loop stable.

We find that $GH(z) = 0.736(z+0.716)/(z-1)(z-0.368)$. The characteristic equation of the closed-loop system is then $F(z) = z^2 - 0.632z + 0.896 = 0$. Here $a_1 = -0.632$ and $a_0 = 0.896$; let us use these as our two free parameters. Equations (6.6-8), (6.6-9), and (6.6-10) give us

$$a_0 > -a_1 - 1 \tag{I}$$
$$a_0 > a_1 - 1 \tag{II}$$

and

$$\left.\begin{array}{l} a_1 = \phi_2(\rho) = 2\rho \\ a_0 = -\phi_1(\rho) = 1 \end{array}\right\} \tag{III}$$

Equations I and II give us regions bounded by straight lines in the (a_1, a_0)-plane. These are shown as lines I and II in Fig. 6.6-2. Equation III is also a straight line (as ρ is varied from -1 to 1) in the (a_1, a_0)-plane (also shown in Fig. 6.6-2). The stable region is the shaded area. The point M shown in the figure is $(-0.632, 0.895)$, which corresponds to the values of a_1 and a_0 for the system characteristic equation considered here. It is inside the stable region, so our system will be closed-loop stable. It is, however, very close to the boundary labeled III. Thus we might suspect the system borders on instability. To show this, consider the curve III in Fig. 6.6-2 which is obtained from (6.6-4) and (6.6-5) for $\nu = 1$ and for $-1 \leqslant \rho \leqslant 1$.

If we make $\nu < 1$, curve III will move down and for some value of ν it will pass through the point M. The values of ρ and ν for which this occurs can be determined. This ρ and ν give us the locations of the

[2]D. D. Šiljak, *Nonlinear Systems: The Parameter Analysis and Design*, John Wiley, New York, 1969.

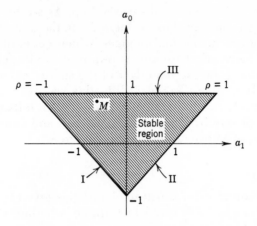

Figure 6.6-2   Stable region in the (a_1, a_0)-plane.

closed-loop poles, which enables us to calculate the corresponding closed-loop damping factor. Knowing this, we can estimate the degree of stability of the systems. In our case, using (6.6-4) and (6.6-5), for $\nu < 1$, we get

$$
\begin{aligned}
a_1 &= 2\rho\nu \\
a_0 &= \nu^2
\end{aligned}
\qquad\qquad (\text{III}^*)
$$

For the point M to be on III^*, we must have

$$\nu^2 = 0.896 \qquad \text{or} \qquad \nu = 0.946$$

Then

$$\rho = \frac{-0.632}{2} \times 0.946 = -0.334$$

Using these values in (6.6-2) gives the corresponding closed-loop poles at $z = -0.315 \pm j0.892$. Using this pole location in Fig. 6.5-5, we find the damping factor $\zeta = 0.04$. The system is underdamped, as we had surmised.

The case where the coefficients of the characteristic equation are functions of two free parameters (say x_1 and x_2) has been discussed for Mitrovic's method applied to continuous-time systems in Section 4.3. The same discussion applies here directly, so we note only that these ideas become very important when we get to the problem of designing compensators for SDS.

The analysis and design of sampled-data control systems using the parameter-plane method have been examined in the literature to a considerable extent.

PROBLEMS

6.1. Use the bilinear transformation method as well as the Jury criterion to determine if the systems with following characteristic equations are stable.

(a) $4z^3 + z^2 - 4z - 1 = 0$
(b) $z^4 + 0.7z^3 + 0.2z - 0.3 = 0$
(c) $z^3 + z^2 + z + 0.5 = 0$
(d) $z^5 + z^4 + 0.6z^3 - 0.2z^2 + 0.3z - 0.8 = 0$

6.2. The open-loop transfer function of a unity-feedback sampled-data system is given by

$$GH(s) = \frac{K(1 - e^{-sT})}{(s+1)(s+a)}, \qquad T = 0.5 \text{ sec}$$

Determine the region of stability in the K-a-plane.

6.3. Use the Nyquist criterion to determine system stability of unity-feedback sampled-data systems with the following open-loop transfer functions.

(a) $\dfrac{2(1 - e^{-sT})}{s(s+1)(s+2)}$, $\qquad T = 1 \text{ sec}$

(b) $\dfrac{1 - e^{-sT}}{s^2(s+1)^2}$, $\qquad T = 0.5 \text{ sec}$

(c) $\dfrac{(1 + sT)}{s^2 T} \dfrac{(1 - e^{-sT})}{(s+1)}$, $\qquad T = 0.5 \text{ sec}$

6.4. The open-loop transfer function of a unity-feedback sampled-data system is given by

$$G(s) = \frac{s^2 + (\pi + 2\alpha)s + \pi(\pi + 1) + \alpha^2}{(s+1)[(s+\alpha)^2 + \pi^2]}, \qquad \alpha = -0.41, \qquad T = 1 \text{ sec}$$

Show that the system shows stability at sampling instants while having an unstable hidden oscillation.

6.5. Find the error constants K_p, K_v, and K_a for a sampled-data system with an overall transfer function.

$$T(z) = \frac{z(z - 0.3)}{(z - 0.1)(z - 0.38 + j0.5)(z - 0.38 - j0.5)}$$

6.6. (a) Make exact plots of $GH^*(j\omega)$ of Problem 6.3.
(b) Make two term plots of these $GH^*(j\omega)$.
(c) Determine the peak overshoot and the resonant frequency by the bilinear transformation method for (a) and (b) of Problem 6.3.

6.7. A sampled-data feedback system is shown in Fig. P6.7. Develop the Bode plots and Nichols chart in the w-plane and determine phase margin, gain margin, peak frequency overshoot, and the resonant frequency.

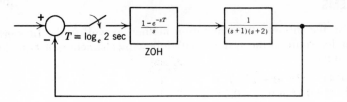

Figure P6.7

Next develop the curve peak $|T^*(j\omega, m)|$ versus m as m varies between 0 and 1. Hence indicate the maximum $|T^*(j\omega, m)|$ and corresponding peak resonant frequency and m.

6.8. (a) Plot the root locus as the system gain K is varied in the loop of a sampled-data system with following open-loop transfer functions:

(i) $\dfrac{K(1-e^{-sT})}{s(s+1)(s+2)}$, $T = 0.5$ sec

(ii) $\dfrac{K}{s(s+4)}$, $T = 0.1$ sec

(b) Find the limiting value of K for stability.
(c) Find ζ and ω_n for the closed loop poles for $K = 1$
(d) Find the value of K that will insure $\zeta = 0.707$.

6.9. Sketch the root locus for a system shown in Fig. P6.9 as K is varied. Find ζ and ω_n for closed-loop poles for $K = 1.5$.

6.10. Plot the region of stability for the system of Fig. P6.9 with $K = 1$ using Mitrovic's method and estimate ζ and ω_n for the closed-loop poles.

6.11. The open-loop transfer function of a feedback sampled-data system is given by

$$GH(s) = \frac{K}{s(s+1)\,(s+2)}, \qquad T = 2 \text{ sec}$$

Plot the region of stability for positive K and determine the range of values of K for which the system is stable.

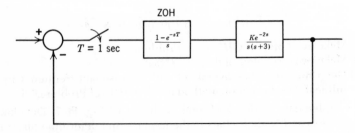

Figure P6.9

6.12. Using $T(z)$ as in (6.3-25), find the error constants K_p, K_v, and K_a in terms of the coefficients of the numerator and denominator polynomials.

6.13. Obtain (6.3-23) from (6.3-22).

6.14. Obtain (6.3-24) from (6.3-8) and (6.3-23).

7

ANALYSIS BY STATE VARIABLE METHODS

7.1 INTRODUCTION[1]

The transfer function approach which we have been using for a given system or subsystem thus far has involved the input, the output, and the corresponding system transfer function. This has the obvious advantage of reducing the problem to its simplest possible form. For many control system problems this is all that is required, and hence the transfer function approach has had and will continue to have great usefulness and popularity. It has been found, however, that many applications in control require something more than this single-input, single-output approach. Many systems of practical importance have many more than one input and output. Moreover, the behavior of individual variables at points between the input-output terminals of a plant is often of vital interest. For instance, it may happen that while the plant output is stable some elements inside the plant are exceeding the specified ratings. In such an event a simultaneous knowledge of the *state* of the variables at some predetermined points along the flow of signal (or information) would be of immense help. These and other considerations of theoretical and practical nature have led to the evolution of what has come to be known as the state variable approach to the analysis and design of control systems. As will be seen shortly, this is a more general approach. In addition, it can be used to handle the multiple-input, multiple-output cases and provides a conceptual framework upon which the analysis of nonlinear systems can be undertaken; these matters will not concern us here but serve to indicate the power of the techniques about to be unfolded.

Our purpose here is not to develop in detail the state variable method of system analysis and design but simply to introduce it for the case of linear time-invariant systems in both the continuous-time and discrete-time cases. We shall attempt to show the connection between the two approaches. We shall be able only to hint at

---

[1]For a detailed view of state-variable methods see L. A. Zadeh and C. A. Desoer, *Linear System Theory, The State Space Approach*, McGraw-Hill, New York, 1963.

the analytical powers available in the state variable approach. The plan of attack is to:

1. Define what is meant by the state variable approach and give the basic formulation for continuous-time and discrete-time systems.
2. Consider how to obtain a state variable formulation from a transfer function and vice versa.
3. Consider stability from the state variable viewpoint.
4. Show how the time response is calculated using the state variable formulation.
5. Evaluate the error constants K_p, K_v, and K_a from the state variable formulation.
6. Introduce the notion of state variable feedback.
7. Show how the Nyquist method, Bode plots, and root locus can be developed from the state variable formulation.

Finally we wish to point out that whereas the Laplace transform is needed for analysis of continuous-time systems and Z-transform for the analysis of discrete-time systems, the state variable approach offers us a way to look at both continuous-time and discrete-time systems with the same formulation. Hence we consider here the state variable formulation for continuous-time and discrete-time systems together. It is indeed an advantage of the state variable approach that the two types of systems can be considered together.

7.2 CONTINUOUS-TIME SYSTEMS

The State Variable Formulation

The state of a system is usually defined as that minimal amount of information required, given the input, to completely determine the response of the system over all future time. Some reflection may be needed to render this definition helpful in system considerations. Let us take, for example, the dynamic systems we have been concerned with heretofore. The dynamical behavior of these systems is governed by an nth-order differential equation that relates the input and the output. It is now easy to see that if the response of this system is to be determined, given the inputs, we must solve this differential equation. To find the solution starting at some time t_0, n independent initial conditions on the dependent variable (the output in this case) valid at time t_0 are needed. Thus for our purposes this set of n initial conditions is the minimal amount of information required to find the future response. Hence this set of n initial conditions may be said to define the state of the system at time t_0. Since there are n independent quantities required to define the state for this case, we say the system is of nth order. The state shall here be denoted by the vector $\mathbf{x}$ where

$$\mathbf{x} = \begin{bmatrix} x_1 \\ x_2 \\ \cdot \\ \cdot \\ \cdot \\ x_n \end{bmatrix} \tag{7.2-1}$$

The x_i, $i = 1, 2, \ldots, n$ are called the components of the *state vector* or simply the *state variables*. Here, rather than considering only one independent variable, we are considering n independent quantities, increasing our resolution on the system considerably.

Let us accept for the moment that for the linear time-invariant systems studied here, suitable state variables and hence a state vector $\mathbf{x}$ may be found. We shall in a moment consider two examples of how this may be done, given the transfer function. First we must ask how the system behavior is characterized. That is, what replaces the transfer function? With the transfer function we have to consider the solution of an nth-order linear, constant-coefficient differential equation. In the state variable approach, the nth-order differential equation is first transformed into n simultaneous, first-order differential equations. These n differential equations are now equations in the n components of the state vector, the state variables, and the output or inputs. For the linear, lumped-parameter, time-invariant systems these differential equations are written in the form of a vector-matrix equation:

$$\dot{\mathbf{x}} = \mathbf{A}\mathbf{x} + \mathbf{B}\mathbf{u} \tag{7.2-2}$$

Here $\mathbf{x}$ is the state vector of form shown in (7.2-1), and $\mathbf{u}$ is the input vector of the form

$$\mathbf{u} = \begin{bmatrix} u_1 \\ u_2 \\ \cdot \\ \cdot \\ \cdot \\ u_m \end{bmatrix} \qquad (7.2\text{-}3)$$

where $u_1, u_2, \ldots, u_m$ are the m inputs to the system. We assume in general that $m \leqslant n$. **A** and **B** are $n \times n$ and $n \times m$ matrices respectively whose elements are all constants.

Let us look for a moment at (7.2-2). The state vector **x** may be considered as an element of n-dimensional space, called the *state space*. We may consider this space to have a Cartesian set of axes. The state variables $x_1, x_2, \ldots, x_n$ will then be the projection of the state vector on the coordinate axes. Assuming an initial value of state vector $\mathbf{x}(t_0)$, if an input control **u** is applied over the time interval $[t_0, t_f]$ then at every instant of time in the time interval of interest $[t_0, t_f]$, the velocity in the state space is given by (7.2-2). The result is a trajectory through state space which starts at $\mathbf{x}(t_0)$. Since the velocity, as seen in (7.2-2), is a function of the input control vector **u**, the shape of this trajectory will in turn be a function of this input control. An example of this is diagramed for the case of a two-dimensional state space in Fig. 7.2-1.

The state variables give the dynamic state of the system. Any practically useful system must, of course, have an output or outputs. The outputs may or may not be the state variables. In the case of the linear lumped-parameter system considered here, the outputs are taken as a linear combination of the state variables and the inputs.

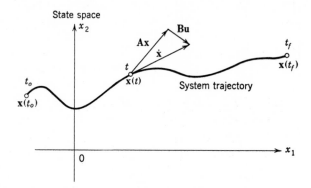

Figure 7.2-1   A trajectory through state space.

Thus if the system has outputs y_1, y_2, ..., and y_p we define the output vector **y** by

$$
\mathbf{y} = \begin{bmatrix} y_1 \\ y_2 \\ \cdot \\ \cdot \\ \cdot \\ y_p \end{bmatrix}
\tag{7.2-4}
$$

If the outputs are then linear combinations of the state variables and the inputs, the output vector may be written as

$$
\mathbf{y} = \mathbf{Cx} + \mathbf{Du}
\tag{7.2-5}
$$

where **C** and **D** are $p \times n$ and $p \times m$ constant matrices, respectively. The general system we have formulated now has m inputs, n state variables, and p outputs (Fig. 7.2-2).

The formulation that has been given here for the state variable approach is for multiple inputs and multiple outputs. Since our main purpose here is to connect the transfer function approach with the state variable approach and since the transfer function approach that has been studied so far has assumed a single-input, single-output system, let us make a special formulation for the state variable

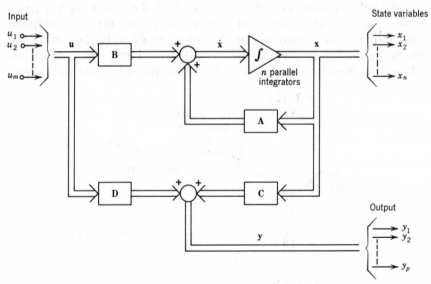

Figure 7.2-2   The state variable formulation for a multi-input, multi-output system.

approach for this case. That is, let us make a special formulation for the case $m = 1, p = 1$. Here (7.2-2) and (7.2-5) become

$$\dot{x} = Ax + bu$$
$$y = c^T x + du \qquad (7.2\text{-}6)$$

where the T denotes the transpose and b and c are just $n \times 1$ vectors; d is just a constant which will normally be zero in our discussion.

Now let us consider how (7.2-6) may be obtained as the formulation for a system whose dynamics are described by an nth-order linear, constant-coefficient differential equation relating the input and the output. If u is the input and y the output of our system, then

$$y^{(n)} + b_{n-1}y^{(n-1)} + \cdots + b_1 y^{(1)} + b_0 y = a_m u^{(m)} + a_{m-1}u^{(m-1)} + \cdots + a_0 u \qquad (7.2\text{-}7)$$

where $y^{(n)} = d^n y / dt^n$, etc. We assume $m < n$ and zero initial conditions. The problem of formulation is reduced to specifying the state variables $x_1, x_2, \ldots, x_n$. There are many possible ways of doing this; a convenient way is to choose the state variable x_1 so that it satisfies the differential equation

$$x_1^{(n)} + b_{n-1}x_1^{(n-1)} + \cdots + b_1 x^{(1)} + b_0 x_1 = u \qquad (7.2\text{-}8)$$

The remaining state variables are then chosen as

$$x_2 = \dot{x}_1 = x_1^{(1)}$$
$$x_3 = \dot{x}_2 = x_1^{(2)}$$
$$\cdot$$
$$\cdot \qquad (7.2\text{-}9)$$
$$\cdot$$
$$x_n = \dot{x}_{n-1} = x_1^{(n-1)}$$

Using (7.2-9) in (7.2-8) and rewriting (7.2-9), we get n first-order differential equations as

$$\begin{aligned}
\dot{x}_1 &= x_2 \\
\dot{x}_2 &= x_3 \\
&\vdots \\
\dot{x}_{n-1} &= x_n \\
\dot{x}_n &= -b_0 x_1 - b_1 x_2 \cdots -b_{n-1}x_n + u
\end{aligned} \qquad (7.2\text{-}10)$$

which may be written in matrix form as

$$
\frac{d}{dt}
\begin{bmatrix}
x_1 \\
x_2 \\
\cdot \\
\cdot \\
\cdot \\
x_{n-1} \\
x_n
\end{bmatrix}
=
\begin{bmatrix}
0 & 1 & 0 & \cdot & \cdot & \cdot & & 0 \\
0 & 0 & 1 & 0 & \cdot & \cdot & \cdot & 0 \\
\cdot & & & & & & & \cdot \\
\cdot & & & & & & & \\
\cdot & & & & & & & \\
0 & 0 & \cdot & \cdot & \cdot & & 0 & 1 \\
-b_0 & -b_1 & \cdot & \cdot & \cdot & \cdot & & -b_{n-1}
\end{bmatrix}
\begin{bmatrix}
x_1 \\
x_2 \\
x_3 \\
\cdot \\
\cdot \\
\cdot \\
x_n
\end{bmatrix}
+
\begin{bmatrix}
0 \\
0 \\
\cdot \\
\cdot \\
\cdot \\
0 \\
1
\end{bmatrix}
u
$$

or

$$
\dot{\mathbf{x}} \quad = \quad \mathbf{A} \quad\quad\quad \mathbf{x} + \mathbf{b}\, u
$$

$$(7.2\text{-}11)$$

To obtain the output y as a linear combination of the input, let us differentiate (7.2-8) w.r.t. time. This gives

$$
\dot{x}_1^{(n)} + b_{n-1}\dot{x}_1^{(n-1)} + \cdots + b_1 \dot{x}_1^{(1)} + b_0 \dot{x}_1 = \dot{u}
$$

Using the first equation of (7.2-9) in this equation gives

$$
x_2^{(n)} + b_{n-1}x_2^{(n-1)} + \cdots + b_1 x_2^{(1)} + b_0 x_2 = u^{(1)}
$$

Differentiating this in turn w.r.t. time and using (7.2-9) repeatedly, we similarly get

$$
x_3^{(n)} + b_{n-1}x_3^{(n-1)} + \cdots + b_0 x_3 = u^{(2)}
$$

$$
x_4^{(n)} + b_{n-1}x_4^{(n-1)} + \cdots + b_0 x_4 = u^{(3)}
$$

$$(7.2\text{-}12)$$

$$
x_{m+1}^{(n)} + b_{n-1}x_{m+1}^{(n-1)} + \cdots + b_0 x_{m+1} = u^{(m)}
$$

If we substitute $u, u^{(1)}, \ldots, u^{(m)}$ as given by (7.2-8) and (7.2-12) into (7.2-7) we get

$$
y^{(n)} + b_{n-1}y^{(n-1)} + \cdots + b_0 y =
$$

$$
a_m x_{m+1}^{(n)} + b_{n-1}a_m x_{m+1}^{(n-1)} + \cdots + b_0 a_m x_{m+1}
$$

$$
+ a_{m-1}x_m^{(n)} + b_{n-1}a_{m-1}x_m^{(n-1)} + \cdots + b_0 a_{m-1}x_m
$$

$$(7.2\text{-}13)$$

$$
+ a_0 x_1^{(n)} + b_{n-1}a_0 x_1^{(n-1)} + \cdots + b_0 a_0 x_1
$$

Comparison of like terms on the left and right sides of (7.2-13) shows that

$$y = a_0x_1 + a_1x_2 + \cdots + a_mx_{m+1} = \mathbf{c}^T\mathbf{x} \qquad (7.2\text{-}14)$$

where

$$\mathbf{c} = \begin{bmatrix} a_0 \\ a_1 \\ a_2 \\ \cdot \\ \cdot \\ \cdot \\ a_m \\ 0 \\ \cdot \\ \cdot \\ 0 \end{bmatrix} \qquad (7.2\text{-}15)$$

Equations (7.2-11) and (7.2-14) are now in the form of (7.2-6) with $d = 0$. We have thus obtained a state variable formulation for a system whose input-output dynamics is described by a linear, constant-coefficient differential equation, (7.2-7). This is a particularly useful formulation because it defines the state variables directly from the input-output relationship. We shall consider obtaining this same formulation directly from a rational transfer function in the next section.

As a final consideration we note that all-zero initial conditions on y and u and their derivatives, in (7.2-7) imply all-zero initial conditions on the state variables, $\mathbf{x}$, in (7.2-11).

Two State Variable Formulations Obtained from the Transfer Function

Let us now consider a system with a transfer function $T(s)$, input u, and output y as diagramed in Fig. 7.2-3. In general for the initially quiescent system, we have the relationship

$$Y(s) = T(s)U(s)$$

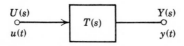

Figure 7.2-3 System with transfer function $T(s)$.

satisfied for all inputs and outputs of the system. Let us consider transfer functions which are rational functions, i.e., those of the form

$$T(s) = \frac{a_m s^m + a_{m-1} s^{m-1} + \cdots + a_1 s + a_0}{s^n + b_{n-1} s^{n-1} + \cdots + b_1 s + b_0}, \qquad m < n \qquad (7.2\text{-}16)$$

The first step toward obtaining a set of first-order differential equations to describe the dynamics of the system and hence the state variable formulation is drawing a block diagram for the system. This diagram is made up of blocks that have individual transfer functions either of the form s^{-1} or constant. To do this, consider $T(s)$ of (7.2-16) written in the form

$$T(s) = \frac{Y(s)}{U(s)} = \frac{a_m s^{m-n} + a_{m-1} s^{m-n-1} + \cdots + a_1 s^{-n+1} + a_0 s^{-n}}{1 + b_{n-1} s^{-1} + \cdots + b_1 s^{-n+1} + b_0 s^{-n}} \qquad (7.2\text{-}17)$$

and let us introduce the new variable $E(s)$. $E(s)$ is related to $U(s)$ by the relationship

$$\frac{E(s)}{U(s)} = \frac{1}{1 + b_{n-1} s^{-1} + \cdots + b_1 s^{-n+1} + b_0 s^{-n}} \qquad (7.2\text{-}18)$$

Comparing (7.2-17) to (7.2-18) we see that $E(s)$ is related to the output $Y(s)$ by

$$Y(s) = a_m s^{m-n} E(s) + a_{m-1} s^{m-n-1} E(s) + \cdots + a_0 s^{-n} E(s) \qquad (7.2\text{-}19)$$

A block diagram which realizes $E(s)$ which satisfies (7.2-18) is shown in Fig. 7.2-4. Note that $E(s)$ is the output of the summation on the LHS of the figure. By using (7.2-19), $Y(s)$ now can be constructed from the nodes available in the block diagram of Fig. 7.2-4. This is also shown in the figure. It is assumed in the figure that $m = n - 1$. If $m < n - 1$, then

$$a_i = 0 \qquad \text{for} \quad i > m \qquad (7.2\text{-}20)$$

Figure 7.2-4 basically shows an analog computer hook-up (where each s^{-1} block represents an integrator) that could be used to realize a block with transfer function $T(s)$.

From the block diagram of Fig. 7.2-4, we obtain directly the differential equations required for the state-variable formulation. To do this, we label the outputs of the s^{-1} blocks by X_i, $i = 1, 2, \ldots, n$ starting on the right as shown in the figure. From the figure,

$$X_i = \frac{1}{s} X_{i-1} \qquad\qquad i = 1, 2, \ldots, n-1 \qquad (7.2\text{-}21)$$

$$X_n = \frac{1}{s}(-b_0 X_1 - b_1 X_2 \cdots - b_{n-1} X_n + U) \qquad (7.2\text{-}22)$$

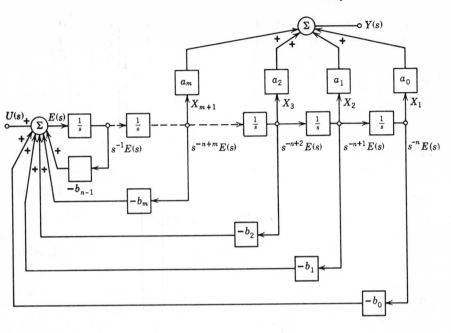

Figure 7.2-4   A state variable realization for a system with a rational transfer function.

or

$$sX_i = X_{i-1} \qquad\qquad i = 1, 2, \ldots, n-1 \qquad (7.2\text{-}23)$$

$$sX_n = -b_0 X_1 - b_1 X_2 \cdots - b_{n-1} X_n + U \qquad (7.2\text{-}24)$$

Now using the inverse Laplace transform on (7.2-23) and (7.2-24),

$$\frac{dx_i}{dt} = x_{i-1}, \qquad i = 1, 2, \ldots, n-1 \qquad (7.2\text{-}25)$$

$$\frac{dx_n}{dt} = -b_0 x_1 - b_1 x_2 \cdots - b_{n-1} x_n + u \qquad (7.2\text{-}26)$$

which is the required system of n first-order differential equations. Further, the output y may be written as a function of the x_i (the state variables) and input u (Fig. 7.2-4). This is then given by.

$$y = a_m x_{m+1} + a_{m-1} x_m + \cdots + a_0 x_1 \qquad (7.2\text{-}27)$$

Equations (7.2-25), (7.2-26) and (7.2-27) may now be written for convenience in matrix form:

$$
\frac{d}{dt}
\begin{bmatrix}
x_1 \\
x_2 \\
x_3 \\
\cdot \\
\cdot \\
\cdot \\
x_{n-1} \\
x_n
\end{bmatrix}
=
\begin{bmatrix}
0 & 1 & 0 & 0 & \cdot & \cdot & \cdot & 0 \\
0 & 0 & 1 & 0 & \cdot & \cdot & \cdot & 0 \\
0 & 0 & 0 & 1 & \cdot & \cdot & \cdot & 0 \\
\cdot & \cdot & \cdot & & & & & \cdot \\
\cdot & \cdot & \cdot & & & & & \cdot \\
\cdot & \cdot & \cdot & & & & & \cdot \\
0 & 0 & 0 & \cdot & \cdot & \cdot & 0 & 1 \\
-b_0 & -b_1 & -b_2 & \cdot & \cdot & \cdot & -b_{n-2} & -b_{n-1}
\end{bmatrix}
\begin{bmatrix}
x_1 \\
x_2 \\
x_3 \\
\cdot \\
\cdot \\
\cdot \\
x_{n-1} \\
x_n
\end{bmatrix}
+
\begin{bmatrix}
0 \\
0 \\
0 \\
\cdot \\
\cdot \\
\cdot \\
0 \\
1
\end{bmatrix}
u
$$

or

$$
\dot{\mathbf{x}} = \mathbf{A} \mathbf{x} + \mathbf{b}\, u. \tag{7.2-28}
$$

and

$$
y = \begin{bmatrix} a_0 & a_1 & a_2 & \cdot & \cdot & \cdot & a_m & 0 & \cdot & \cdot & \cdot & 0 \end{bmatrix}
\begin{bmatrix}
x_1 \\
x_2 \\
\cdot \\
\cdot \\
\cdot \\
x_n
\end{bmatrix}
\tag{7.2-29}
$$

or

$$
y = \mathbf{c}^T \mathbf{x}
$$

which we can now recognize as the same formulation seen in (7.2-11) and (7.2-15). These formulations are good for any rational transfer function.

Let us now consider a more specialized case, where the matrix **A** has a particularly useful form. For this case we assume that the transfer function has only distinct, *real* poles and we may construct the form of **A** as follows.

A FORMULATION FOR WHICH THE MATRIX *A* IS DIAGONAL

The overall transfer function $T(s)$, which is still assumed to be of the form shown in (7.2-16) and to represent the system as shown in Fig. 7.2-3, may be written as

$$
T(s) = \frac{A(s)}{B(s)} = \frac{A(s)}{\displaystyle\prod_{i=1}^{n} (s - p_i)}, \qquad p_i \neq p_j,\ p_i \text{ real} \tag{7.2-30}
$$

$T(s)$ in partial fraction expansion form can be written as

$$
T(s) = \frac{Y(s)}{U(s)} = \frac{K_{11}}{(s - p_1)} + \frac{K_{21}}{(s - p_2)} + \cdots + \frac{K_{n1}}{(s - p_n)} \tag{7.2-31}
$$

And thus

$$Y(s) = \frac{K_{11}U(s)}{(s-p_1)} + \frac{K_{21}U(s)}{(s-p_2)} + \cdots + \frac{K_{n1}U(s)}{(s-p_n)} \qquad (7.2\text{-}32)$$

where now

$$K_{i1} = (s-p_i)T(s)|_{s=p_i}$$

Equation (7.2-32) may be represented by the block diagram of Fig. 7.2-5. The outputs of the first-order blocks in Fig. 7.2-5 have been labeled $x_1, x_2, \ldots, x_n$. From the figure,

$$X_i(s) = \frac{U(s)}{s-p_i}, \qquad i = 1, 2, \ldots, n$$

from which

$$sX_i(s) = p_iX_i(s) + U(s), \qquad i = 1, 2, \ldots, n \qquad (7.2\text{-}33)$$

From (7.2-33) we can obtain the required set of n first-order differential equations by applying the inverse $\mathscr{L}$-transform. Thus

$$\frac{dx_i}{dt} = p_ix_i + u, \qquad i = 1, 2, \ldots, n \qquad (7.2\text{-}34)$$

The output $Y(s)$ is seen from Fig. 7.2-5 to satisfy

$$Y(s) = K_{11}X_1(s) + K_{21}X_2(s) + \cdots + K_{n1}X_n(s)$$

which, after we apply the inverse $\mathscr{L}$-transform, gives

$$y = K_{11}x_1 + K_{21}x_2 + \cdots + K_{n1}x_n \qquad (7.2\text{-}35)$$

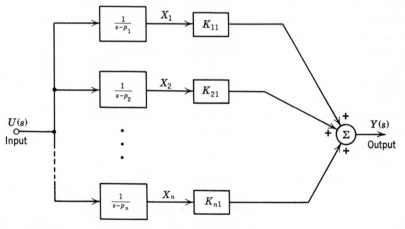

Figure 7.2-5 A realization for a transfer function with real, distinct ples.

Putting (7.2-34) and (7.2-35) into matrix notation,

$$
\frac{d}{dt}
\begin{bmatrix} x_1 \\ x_2 \\ x_3 \\ \cdot \\ \cdot \\ \cdot \\ x_n \end{bmatrix}
=
\begin{bmatrix}
p_1 & 0 & 0 & . & . & . & 0 \\
0 & p_2 & 0 & . & . & . & 0 \\
0 & 0 & p_3 & 0 & . & . & 0 \\
\cdot & & & & & & \cdot \\
\cdot & & & & & 0 & \\
0 & . & . & . & . & 0 & p_n
\end{bmatrix}
\begin{bmatrix} x_1 \\ x_2 \\ x_3 \\ \cdot \\ \cdot \\ \cdot \\ x_n \end{bmatrix}
+
\begin{bmatrix} 1 \\ 1 \\ 1 \\ \cdot \\ \cdot \\ \cdot \\ 1 \end{bmatrix}
u
$$

or

$$
\dot{\mathbf{x}} = \mathbf{A} \qquad \mathbf{x} + \mathbf{b}u
$$

(7.2-36)

and

$$
y = [K_{11} K_{21} \ldots K_{n1}]
\begin{bmatrix} x_1 \\ x_2 \\ \cdot \\ \cdot \\ \cdot \\ x_n \end{bmatrix}
$$

$$
y = \qquad \mathbf{c}^T \qquad \mathbf{x}
$$

Equation (7.2-36) is the state-variable formulation desired. We note that here the matrix $\mathbf{A}$ has nonzero elements only on the diagonal. Since performing matrix operations is generally much easier on a diagonal than on a nondiagonal matrix, this formulation is a popular one when the system of interest has real, distinct poles.

We have obtained two state variable formulations for a system with a given form of transfer function. In fact there is no unique state variable formulation for a given system. In general there are infinitely many state variable formulations for any given system. To see this, let $\mathbf{x}$ be the state vector for a given system such that

$$
\dot{\mathbf{x}} = \mathbf{A}\mathbf{x} + \mathbf{b}u
$$
$$
y = \mathbf{c}^T\mathbf{x} + du
$$

(7.2-37)

Consider now $\mathbf{z} = \mathbf{P}\mathbf{x}$ where $\mathbf{z}$ is now an n component vector and $\mathbf{P}$ is an $n \times n$ nonsingular constant matrix. We may then write

$$
\mathbf{x} = \mathbf{P}^{-1}\mathbf{z}
$$

Substituting this into (7.2-37) gives

$$
\mathbf{P}^{-1}\dot{\mathbf{z}} = \mathbf{A}\mathbf{P}^{-1}\mathbf{z} + \mathbf{b}u
$$

$$
y = \mathbf{c}^T\mathbf{P}^{-1}\mathbf{z} + du
$$

Now multiplying the first of these two equations by $\mathbf{P}$ gives

$$\dot{\mathbf{z}} = \mathbf{PAP}^{-1}\mathbf{z} + \mathbf{P}\mathbf{b}u = \tilde{\mathbf{A}}\mathbf{z} + \tilde{\mathbf{b}}u$$

$$y = \mathbf{c}^T\mathbf{P}^{-1}\mathbf{z} + du = \tilde{\mathbf{c}}^T\mathbf{z} + du$$

which gives another state variable formulation for our system. Thus if $\mathbf{x}$ is a state vector for a given system and $\mathbf{P}$ is a constant, nonsingular, $n \times n$ matrix, then $\mathbf{z} = \mathbf{Px}$ may also be used as a state vector. This allows for a great deal of flexibility in the choice of state variables.

The general method for obtaining a state variable formulation from the transfer function of a given system is to draw up a block diagram whose elemental blocks are constant multipliers and first-order transfer functions.[2] The outputs of the first-order transfer function blocks of type $1/(s+a)$ may then be chosen as the state variables.

Obtaining the Transfer Function for a System from a State Variable Formulation

Let us consider here a system which has a state variable formulation as

$$\dot{\mathbf{x}} = \mathbf{Ax} + \mathbf{b}u$$

$$y = \mathbf{c}^T\mathbf{x} \tag{7.2-38}$$

where now $\mathbf{x}$ is the state vector, u the input, y the output, and $\mathbf{A}$, $\mathbf{b}$, $\mathbf{c}$, constant matrices and vectors. To obtain the transfer function for this system we take the $\mathscr{L}$-transform of (7.2-38):

$$s\mathbf{X}(s) - \mathbf{x}(0) = \mathbf{AX}(s) + \mathbf{b}U(s)$$

$$Y(s) = \mathbf{c}^T\mathbf{X}(s) \tag{7.2-39}$$

where $\mathbf{X}(s) = \mathscr{L}[\mathbf{x}(t)]$, etc. The transfer function is $Y(s)/U(s)$ with zero initial conditions. Thus eliminating $\mathbf{X}(s)$ from the two equations of (7.2-39) gives

$$\mathbf{X}(s) = [s\mathbf{I} - \mathbf{A}]^{-1}\mathbf{x}(0) + [s\mathbf{I} - \mathbf{A}]^{-1}\mathbf{b}U(s) \tag{7.2-40}$$

where $\mathbf{I}$ is the identity matrix having unity as each diagonal element and all other elements zero.

$$Y(s) = \mathbf{c}^T[s\mathbf{I} - \mathbf{A}]^{-1}\mathbf{x}(0) + \mathbf{c}^T[s\mathbf{I} - \mathbf{A}]^{-1}\mathbf{b}U(s) \tag{7.2-41}$$

[2]That is, blocks whose transfer function are of the form $1/s$ or $1/(s+a)$.

From (7.2-41) with $x(0) = 0$ we have directly

$$\frac{Y(s)}{U(s)} = T(s) = c^T[sI - A]^{-1}b \qquad (7.2\text{-}42)$$

which is the transfer function in terms of state variable quantities.

Example 7.2-1

Let us put the transfer function

$$T(s) = \frac{4}{s(s+1)(s+4)}$$

into a state variable formulation so that system matrix A is diagonal.
We begin by making a partial fraction expansion,

$$T(s) = \frac{1}{s} - \frac{4/3}{s+1} + \frac{1/3}{s+4}$$

Comparing with (7.2-31), we note $p_1 = 0$, $p_2 = -1$, $p_3 = -4$, $K_{11} = 1$, $K_{21} = -\frac{4}{3}$ and $K_{31} = \frac{1}{3}$. Hence (7.2-36) gives the state-variable formulation as

$$\frac{d}{dt}\begin{bmatrix} x_1 \\ x_2 \\ x_3 \end{bmatrix} = \begin{bmatrix} 0 & 0 & 0 \\ 0 & -1 & 0 \\ 0 & 0 & -4 \end{bmatrix}\begin{bmatrix} x_1 \\ x_2 \\ x_3 \end{bmatrix} + \begin{bmatrix} 1 \\ 1 \\ 1 \end{bmatrix}u$$

or

$$\dot{x} = \qquad A \qquad x + b\ u$$

and

$$y = [1 - \tfrac{4}{3}\ \tfrac{1}{3}]\begin{bmatrix} x_1 \\ x_2 \\ x_3 \end{bmatrix}$$

or

$$y = \qquad c^T \qquad x$$

Now let us check this formulation by finding the corresponding transfer function by (7.2-42). We have

$$T(s) = [1 - \tfrac{4}{3}\ \tfrac{1}{3}]\begin{bmatrix} s & 0 & 0 \\ 0 & s+1 & 0 \\ 0 & 0 & s+4 \end{bmatrix}^{-1}\begin{bmatrix} 1 \\ 1 \\ 1 \end{bmatrix}$$

$$= [1 - \tfrac{4}{3}\ \tfrac{1}{3}]\begin{bmatrix} \dfrac{1}{s} & 0 & 0 \\ 0 & \dfrac{1}{s+1} & 0 \\ 0 & 0 & \dfrac{1}{s+4} \end{bmatrix}\begin{bmatrix} 1 \\ 1 \\ 1 \end{bmatrix}$$

$$= [1 - \tfrac{4}{3}\tfrac{1}{3}] \begin{bmatrix} \dfrac{1}{s} \\[2ex] \dfrac{1}{s+1} \\[2ex] \dfrac{1}{s+4} \end{bmatrix} = \dfrac{1}{s} - \dfrac{4/3}{s+1} + \dfrac{1/3}{s+4}$$

$$= \dfrac{4}{s(s+1)(s+4)}$$

which checks.

Stability Analysis with a State Variable Formulation

Determining stability is an important problem whatever the formulation we happen to use. So let us examine whether a system is stable when it is presented in a state variable formulation. We assume that the formulation is as in (7.2-38). It has been seen that the transfer function between input and output is given by (7.2-42) as

$$T(s) = \mathbf{c}^T[s\mathbf{I} - \mathbf{A}]^{-1}\mathbf{b} \tag{7.2-43}$$

We know the system will be stable if $T(s)$ has all its poles in the LHP. Let us then look for the poles of $T(s)$. To do this we note that

$$[s\mathbf{I} - \mathbf{A}]^{-1} = \dfrac{[s\mathbf{I} - \mathbf{A}]^*}{\det [s\mathbf{I} - \mathbf{A}]} \tag{7.2-44}$$

where $[s\mathbf{I} - \mathbf{A}]^*$ is the adjoint of the matrix $[s\mathbf{I} - \mathbf{A}]$ and is found by replacing each element of $[s\mathbf{I} - \mathbf{A}]^T$ by its minor with the appropriate sign. Here $\det[s\mathbf{I} - \mathbf{A}]$ is simply the determinant of the matrix $[s\mathbf{I} - \mathbf{A}]$. Now we have

$$T(s) = \dfrac{\mathbf{c}^T[s\mathbf{I} - \mathbf{A}]^*\mathbf{b}}{\det [s\mathbf{I} - \mathbf{A}]} \tag{7.2-45}$$

from which we can see that the denominator of the transfer function or the characteristic polynominal $F(s)$ is given by

$$F(s) = \det[s\mathbf{I} - \mathbf{A}] \tag{7.2-46}$$

The roots of the characteristic equation, i.e. those s for which $F(s) = 0$, are the poles of the system. *Thus the system with state variable formulation*

$$\dot{\mathbf{x}} = \mathbf{A}x + \mathbf{b}u \tag{7.2-47}$$

is stable if det $[sI - A]$ *has all its roots in the LHP*. We call det $[sI - A]$ the *characteristic polynomial* of the matrix A. The roots of a matrix's characteristic equation are called the *characteristic values* or *eigenvalues* of the matrix. Thus *a system will be stable if the eigenvalues of the system matrix A are all in the LHP*. We note that the poles of the transfer functions and the eigenvalues of the system matrix occur at the same points in the complex plane.

If we have determined the characteristic polynomial, det $[sI - A]$, for a given system matrix A, we may of course use the Routh criterion to determine whether or not it has all its roots in the LHP.

Example 7.2-2

Consider the stability of system whose system matrix is given by

$$A = \begin{bmatrix} -7 & 2 & 1 \\ 4 & -5 & -4 \\ 0 & 0 & -6 \end{bmatrix}$$

For this case

$$\det [sI - A] = \begin{bmatrix} s+7 & -2 & -1 \\ -4 & s+5 & 4 \\ 0 & 0 & s+6 \end{bmatrix} = s^3 + 18s^2 + 99s + 162$$

Putting this characteristic polynomial into a Routh table gives

s^3	1	99
s^2	18	162
s^1	1620/18	
s^0	162	

The first column of this Routh table has all positive entries, hence the corresponding polynomial has all LHP roots. Thus in this case the poles of the system which are the eigenvalues of the matrix A, are all in the LHP and the system is stable.

Although this method is straightforward and easy to apply for low-order cases, when n (the dimension of the matrix A) gets to about five and above evaluation of det $[sI - A]$ is a difficult job unless A happens to be near diagonal in form. Most computing centers now have routines that produce the characteristic equation for a matrix and then factor it to give the eigenvalues of the matrix. In this case, determining stability reduces itself to looking to see whether all eigenvalues are in the LHP.

The Solution of $\dot{x} = Ax$ and the matrix e^{At}

The equation

$$\dot{x} = Ax \tag{7.2-48}$$

is the system differential equation with input zero, $[u(t) = 0]$. Taking the $\mathscr{L}$-transform,

$$sX(s) = AX(s) - x(0) \tag{7.2-49}$$

$$X(s) = [sI - A]^{-1}x(0) \tag{7.2-50}$$

The solution for the time response of the state vector $x(t)$, $t \geq 0$ for this case, can be found from (7.2-50) if we can find the inverse $\mathscr{L}$-transform for the matrix $[sI - A]^{-1}$. Let us discuss this briefly. First we write this matrix in the form

$$[sI - A]^{-1} = \frac{1}{s}\left[I - \frac{1}{s}A\right]^{-1} \tag{7.2-51}$$

By expanding the RHS of (7.2-51) into an infinite series, it may be shown that[3]

$$[sI - A]^{-1} = \frac{1}{s}\left[I + \frac{1}{s}A + \frac{1}{s^2}A^2 + \frac{1}{s^3}A^3 + \cdots\right]$$

$$= \frac{1}{s}I + \frac{1}{s^2}A + \frac{1}{s^3}A^2 + \frac{1}{s^4}A^3 + \cdots \tag{7.2-52}$$

for $|s|$ large enough. We consider our abscissa of integration for the inversion integral to be far enough into the RHP to make (7.2-52) valid. Recalling that

$$\mathscr{L}^{-1}\left\{\frac{1}{s^{n+1}}\right\} = \frac{t^n}{n!}\,1(t) \tag{7.2-53}$$

we may invert (7.2-52) to obtain

$$\mathscr{L}^{-1}\{[sI - A]^{-1}\} = \left[I + At + A^2\frac{t^2}{2!} + A^3\frac{t^3}{3!} + \cdots\right]1(t)$$

$$= \sum_{i=0}^{\infty}\frac{(At)^i}{i!}\,1(t) \tag{7.2-54}$$

$$= e^{At}\,1(t)$$

where we define

$$e^{At} \triangleq \sum_{i=0}^{\infty}\frac{(At)^i}{i!} \tag{7.2-55}$$

---

[3] Recall that $1/(1-x) = 1 + x + x^2 + \cdots$ for $|x| < 1$.

Hence the solution of $\dot{\mathbf{x}} = \mathbf{A}\mathbf{x}$ is found by using (7.2-54) in (7.2-50) to get

$$\mathbf{x}(t) = e^{\mathbf{A}t}\mathbf{x}(0), \qquad t > 0 \tag{7.2-56}$$

This result may be checked by using (7.2-54) for $e^{\mathbf{A}t}$ and substituting it into (7.2-48).

The matrix $e^{\mathbf{A}t}$ is called the state-transition matrix. The reason for this name is easily seen from (7.2-56), for multiplying the state $\mathbf{x}(0)$ by $e^{\mathbf{A}t}$ transitions the unforced system to the state at time t, $\mathbf{x}(t)$. This matrix is of paramount importance in all considerations of linear, lumped-parameter, time-invariant systems by state variable methods.

SOME PROPERTIES[4] OF THE MATRIX $e^{\mathbf{A}t}$

1.
$$e^{\mathbf{A}t}\big|_{t=0} = \mathbf{I} \tag{7.2-57}$$

This may be seen directly from (7.2-55).

2.
$$e^{\mathbf{A}(t_1+t_2)} = e^{\mathbf{A}t_1}e^{\mathbf{A}t_2} = e^{\mathbf{A}t_2}e^{\mathbf{A}t_1} \tag{7.2-58}$$

This is similar to the property of the exponential function, $e^{x+y} = e^x e^y$. This may be seen by expanding each term in (7.2-58) into an infinite series by use of (7.2-55) and performing the multiplications indicated and then collecting terms appropriately.

3.
$$[e^{\mathbf{A}t}]^{-1} = e^{-\mathbf{A}t} \tag{7.2-59}$$

To see this we simply apply properties (1) and (2), thus

$$e^{\mathbf{A}(t-t)} = e^{\mathbf{A}t}e^{-\mathbf{A}t} = e^{-\mathbf{A}t}e^{\mathbf{A}t} = \mathbf{I}$$

4.
$$\frac{d}{dt}e^{\mathbf{A}t} = \mathbf{A}e^{\mathbf{A}t} = e^{\mathbf{A}t}\mathbf{A} \tag{7.2-60}$$

This follows directly by differentiating (7.2-55) with respect to t.

It should be noted here that $e^{\mathbf{A}t}$ is conveniently calculated for any given value of t by digital computer and use of the defining equation (7.2-55). It is the type of recurrence relationship which is easily programmed. For a given t the series is continued for N terms until all elements of the matrix $(\mathbf{A}t)^N/(N!)$ are negligible relative to the corresponding sums of the previous terms. This sort of checking is also readily done by digital computer.

---

[4]See S. C. Gupta, *Transform and State Variable Methods in Linear Systems*, John Wiley, New York, 1966, pp. 343–344.

For small dimension matrices (7.2-54) gives a convenient method for determining $\mathbf{e}^{At}$ for a given matrix. To see how this is done, we consider the following example.

Example 7.2-3

Consider

$$\mathbf{A} = \begin{bmatrix} -7 & 2 & 1 \\ 4 & -5 & -4 \\ 0 & 0 & -6 \end{bmatrix}$$

which is the system matrix for the Example 7.2-2. Find $\mathbf{e}^{At}$.

$$[s\mathbf{I} - \mathbf{A}] = \begin{bmatrix} s+7 & -2 & -1 \\ -4 & s+5 & 4 \\ 0 & 0 & s+6 \end{bmatrix}$$

and

$$[s\mathbf{I} - \mathbf{A}]^{-1} = \frac{[s\mathbf{I} - \mathbf{A}]^*}{\det[s\mathbf{I} - \mathbf{A}]}$$

$$= \frac{\begin{bmatrix} (s+5)(s+6) & 2(s+6) & (s-3) \\ 4(s+6) & (s+6)(s+7) & -4(s+6) \\ 0 & 0 & (s+3)(s+9) \end{bmatrix}}{s^3 + 18s^2 + 99s + 162}$$

or

$$[s\mathbf{I} - \mathbf{A}]^{-1} = \begin{bmatrix} \dfrac{s+5}{(s+3)(s+9)} & \dfrac{2}{(s+3)(s+9)} & \dfrac{s-3}{(s+3)(s+6)(s+9)} \\[3mm] \dfrac{4}{(s+3)(s+9)} & \dfrac{s+7}{(s+3)(s+9)} & \dfrac{-4}{(s+3)(s+9)} \\[3mm] 0 & 0 & \dfrac{1}{s+6} \end{bmatrix}$$

Inverting this matrix term by term, from (7.2-54), we have

$$\mathbf{e}^{At} = \begin{bmatrix} (\tfrac{1}{3}e^{-3t} + \tfrac{2}{3}e^{-9t}) & \tfrac{1}{3}(e^{-3t} - e^{-9t}) & (-\tfrac{1}{3}e^{-3t} + e^{-6t} - \tfrac{2}{3}e^{-9t}) \\ \tfrac{2}{3}(e^{-3t} - e^{-9t}) & \tfrac{1}{3}(2e^{-3t} + e^{-9t}) & -\tfrac{2}{3}(e^{-3t} - e^{-9t}) \\ 0 & 0 & e^{-6t} \end{bmatrix}$$

It is now possible to check whether all the properties mentioned above are valid.

The Solution of $\dot{\mathbf{x}} = \mathbf{A}\mathbf{x} + \mathbf{b}u$

To get at this solution let us simply take the $\mathscr{L}$-transform of

$$\dot{\mathbf{x}} = \mathbf{A}\mathbf{x} + \mathbf{b}u \tag{7.2-61}$$

We obtain

$$sX(s) - x(0) = AX(s) + bU(s) \qquad (7.2\text{-}62)$$

which we may now solve for $X(s)$ to get

$$X(s) = [sI - A]^{-1}x(0) + [sI - A]^{-1}bU(s) \qquad (7.2\text{-}63)$$

Now to determine $x(t)$, the desired solution, we need only invert (7.2-63). We already know that

$$\mathcal{L}^{-1}\{[sI - A]^{-1}\} = e^{At}1(t) \qquad (7.2\text{-}64)$$

To invert $[sI - A]^{-1}bU(s)$, we realize (see Table 2.2-1) that

$$\mathcal{L}^{-1}\{F_1(s)F_2(s)\} = \int_0^t f_1(t-\tau)f_2(\tau)\,d\tau \qquad (7.2\text{-}65)$$

and therefore

$$\mathcal{L}^{-1}\{[sI - A]^{-1}bU(s)\} = \int_0^t e^{A(t-\tau)}bu(\tau)\,d\tau \qquad (7.2\text{-}66)$$

Hence the solution of (7.2-61) becomes

$$x(t) = e^{At}x(0) + \int_0^t e^{A(t-\tau)}bu(\tau)\,d\tau \qquad (7.2\text{-}67)$$

which is the desired result. We may check that (7.2-67) is truly the solution to (7.2-61) simply by substituting $x(t)$ as given by (7.2-67) into (7.2-61) and using the properties of e^{At} appropriately. Equation (7.2-67) now gives the state of the system for all $t > 0$ starting from an arbitrary initial state $x(0)$ with an arbitrary control function $u(t)$ applied over the interval $[0, t]$. We can see again that this approach is a more general one than the transfer function approach in that we get a solution here starting from any initial state. The transfer function, of course, assumes that the initial state is always $x(0) = 0$, i.e., the origin. Furthermore, we see that (7.2-67) gives the simultaneous solution for the time response of all the state variables. The transfer function in general gives only the time response of the output. In addition, (7.2-67) is in a particularly convenient form for evaluation by digital computer. e^{At} may be computed to any desired degree of accuracy by (7.2-55). All other operations indicated in (7.2-67) involve only matrix multiplications and additions, all of which the digital computer handles quite readily.

To complete our discussion here, we need to determine the output. Using (7.2-67) in the second equation of (7.2-37),

$$y(t) = c^T e^{At}x(0) + \int_0^t c^T e^{A(t-\tau)}bu(\tau)\,d\tau + du(t) \qquad (7.2\text{-}68)$$

We note that the impulse response of the system is found by using the sifting property of the δ-function in (7.2-68). The result is

$$y_\delta(t) = \mathbf{c}^T e^{A^t}\mathbf{b}, \qquad t > 0 \qquad\qquad (7.2\text{-}69)$$

$$= 0, \qquad\qquad t < 0$$

where the subscript δ indicates the impulse response. The step response, assuming zero initial condition, is found by letting $u(t) = 1(t)$ in (7.2-68):

$$y_{\text{step}}(t) = \int_0^t \mathbf{c}^T e^{A(t-\tau)}\mathbf{b}\, d\tau, \qquad t > 0$$

$$= 0, \qquad t < 0 \qquad\qquad (7.2\text{-}70)$$

To see how these results may be used let us consider an example.

Example 7.2-4

Determine the impulse and step responses of a system whose transfer function is given by

$$T(s) = \frac{(s+2)^2}{s(s+1)(s+4)} \qquad\qquad (7.2\text{-}71)$$

The $\mathscr{L}$-transform of the impulse response $[y_\delta(t)]$ is given by

$$Y_\delta(s) = T(s) = \frac{(s+2)^2}{s(s+1)(s+4)} = \frac{1}{s} - \frac{1/3}{s+1} + \frac{1/3}{s+4}$$

which we may invert term by term to get

$$y_\delta(t) = [1 - \tfrac{1}{3}e^{-t} + \tfrac{1}{3}e^{-4t}]1(t) \qquad\qquad (7.2\text{-}72)$$

We may determine the step response $y_{\text{step}}(t)$ by

$$y_{\text{step}}(t) = \int_0^t y_\delta(\tau)\, d\tau = \mathscr{L}^{-1}\left\{\frac{1}{s}Y_\delta(s)\right\}$$

or

$$y_{\text{step}}(t) = [t - \tfrac{1}{3}(1 - e^{-t}) + \tfrac{1}{12}(1 - e^{-4t})]1(t) \qquad\qquad (7.2\text{-}73)$$

Now let us obtain these results by the state variable methods that we have just developed. We first put the system into a state variable formulation. Multiplying out the numerator and the denominator on the RHS of (7.2-71) gives

$$T(s) = \frac{s^2 + 4s + 4}{s^3 + 5s^2 + 4s} \qquad\qquad (7.2\text{-}74)$$

Equation (7.2-74) may be put directly into the state variable formulation of (7.2-28):

$$\frac{d}{dt}\begin{bmatrix} x_1 \\ x_2 \\ x_3 \end{bmatrix} = \begin{bmatrix} 0 & 1 & 0 \\ 0 & 0 & 1 \\ 0 & -4 & -5 \end{bmatrix}\begin{bmatrix} x_1 \\ x_2 \\ x_3 \end{bmatrix} + \begin{bmatrix} 0 \\ 0 \\ 1 \end{bmatrix} u$$

or

$$\dot{\mathbf{x}} \quad = \quad \mathbf{A} \qquad \mathbf{x} \ + \mathbf{b} \, u$$

$$y = [4 \quad 4 \quad 1]\begin{bmatrix} x_1 \\ x_2 \\ x_3 \end{bmatrix} \tag{7.2-75}$$

or

$$y = \qquad \mathbf{c}^T \qquad \mathbf{x}$$

We now need only determine e^{At} and use this in (7.2-69) and (7.2-70) to determine the impulse and step responses. We have

$$[s\mathbf{I} - \mathbf{A}]^{-1} = \begin{bmatrix} s & -1 & 0 \\ 0 & s & -1 \\ 0 & 4 & s+5 \end{bmatrix}^{-1}$$

$$= \frac{1}{s(s+1)(s+4)}\begin{bmatrix} (s+1)(s+4) & s+5 & 1 \\ 0 & s(s+5) & s \\ 0 & -4s & s^2 \end{bmatrix}$$

which we may invert term by term to give

$$e^{At} = \begin{bmatrix} 1 & (\tfrac{5}{4}-\tfrac{4}{3}e^{-t}+\tfrac{1}{12}e^{-4t}) & (\tfrac{1}{4}-\tfrac{1}{3}e^{-t}+\tfrac{1}{12}e^{-4t}) \\ 0 & (\tfrac{4}{3}e^{-t}-\tfrac{1}{3}e^{-4t}) & (\tfrac{1}{3}e^{-t}-\tfrac{1}{3}e^{-4t}) \\ 0 & (-\tfrac{4}{3}e^{-t}+\tfrac{4}{3}e^{-4t}) & (-\tfrac{1}{3}e^{-t}+\tfrac{4}{3}e^{-4t}) \end{bmatrix} \tag{7.2-76}$$

For this e^{At}, we may compute

$$e^{At}\mathbf{b} = \begin{bmatrix} \tfrac{1}{4}-\tfrac{1}{3}e^{-t}+\tfrac{1}{12}e^{-4t} \\ \tfrac{1}{3}e^{-t}-\tfrac{1}{3}e^{-4t} \\ -\tfrac{1}{3}e^{-t}+\tfrac{4}{3}e^{-4t} \end{bmatrix} \tag{7.2-77}$$

where $\mathbf{b}$ is as defined in (7.2-75). Using $\mathbf{c}$ as in (7.2-75), directly from (7.2-69) we know that

$$y_\delta(t) = \mathbf{c}^T e^{At}\mathbf{b} = 1 - \tfrac{1}{3}e^{-t} + \tfrac{1}{3}e^{-4t}, \qquad t > 0$$

which checks with the same result obtained in (7.2-72). The step response may now be found by using (7.2-70). By integrating (7.2-77),

$$\int_0^t e^{A(t-\tau)}\,\mathbf{b}\,d\tau = \begin{bmatrix} \tfrac{1}{4}t - \tfrac{1}{3}(1-e^{-t}) + \tfrac{1}{48}(1-e^{-4t}) \\ \tfrac{1}{3}(1-e^{-t}) - \tfrac{1}{12}(1-e^{-4t}) \\ -\tfrac{1}{3}(1-e^{-1}) + \tfrac{1}{3}(1-e^{-4t}) \end{bmatrix}$$

which, used in (7.2-70) with c from (7.2-75), gives

$$y_{\text{step}}(t) = t - \tfrac{1}{3}(1 - e^{-t}) + \tfrac{1}{12}(1 - e^{-4t}), \qquad t > 0$$

which checks with (7.2-73).

It has been much more laborious to compute these impulse and step responses by state variable methods using hand calculation. That is, it is more laborious than the transfer-function approach. This is because the transfer-function, s-domain approach is by nature and development well suited to produce impulse and step responses by hand methods. If we wish to use numerical methods on a computer, we will find that the order of difficulty is reversed.

The State Variable Feedback System

The state variable feedback system has been found to have many advantages for stability and response analyses. Hence we consider this system in somewhat more detail.

State variable feedback refers to a function that is a linear combination of the state variables. It takes the form

$$k^T x = \sum_i k_i x_i$$

where x_i is the ith state variable, k_i, $i = 1, 2, \ldots, n$ are n scalar constants. These $k_i x_i$'s are fed back and subtracted from the system input as shown in Fig. 7.2-6 for a single-input, single-output system. We here assume the plant has input u, output y, and state vector x, which are related in a state variable formulation by

$$\dot{x} = Ax + bu \qquad\qquad (7.2\text{-}78)$$
$$y = c^T x$$

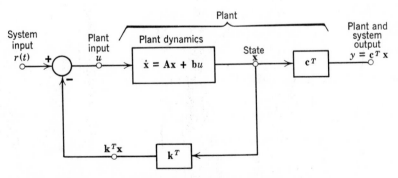

Figure 7.2-6   Single-input, single-output state variable feedback system.

From Fig. 7.2-6 we see that

$$u = r - \mathbf{k}^T \mathbf{x} \tag{7.2-79}$$

where r is the system input. Using (7.2-79) in (7.2-78) gives the state equations for the closed-loop system:

$$\dot{\mathbf{x}} = [\mathbf{A} - \mathbf{b}\mathbf{k}^T]\mathbf{x} + \mathbf{b}r$$
$$y = \mathbf{c}^T\mathbf{x} \tag{7.2-80}$$

which we note is the standard state variable formulation as seen in (7.2-38), except that for the closed-loop, state variable feedback system we replace

$$\mathbf{A} \rightarrow [\mathbf{A} - \mathbf{b}\mathbf{k}^T]$$
$$u \rightarrow r$$

Thus the method of analysis we have been considering for state variable systems may be used for the state variable feedback system. Some insight into the special properties of a state variable feedback system may be gained by considering the closed-loop transfer function for the system of Fig. 7.2-6. This may be found directly by taking the $\mathcal{L}$-transform of (7.2-80) (with zero initial conditions) to get

$$s\mathbf{X}(s) = [\mathbf{A} - \mathbf{b}\mathbf{k}^T]\mathbf{X}(s) + \mathbf{b}R(s)$$

$$Y(s) = \mathbf{c}^T\mathbf{X}(s) = C(s)$$

from which

$$T(s) = \frac{C(s)}{R(s)} = \mathbf{c}^T[s\mathbf{I} - \mathbf{A} + \mathbf{b}\mathbf{k}^T]^{-1}\mathbf{b} \tag{7.2-81}$$

which is the desired result. For another and more familiar form of this closed-loop transfer function, we take the $\mathcal{L}$-transform of (7.2-78) and (7.2-79). This gives

$$s\mathbf{X}(s) = \mathbf{A}\mathbf{X}(s) + \mathbf{b}U(s)$$
$$Y(s) = C(s) = \mathbf{c}^T\mathbf{X}(s)$$
$$U(s) = R(s) - \mathbf{k}^T\mathbf{X}(s)$$

From the first equation just above we get

$$\mathbf{X}(s) = [s\mathbf{I} - \mathbf{A}]^{-1}\mathbf{b}U(s) \tag{7.2-82}$$

and using this in the bottom equation gives

$$U(s) = \frac{R(s)}{1 + \mathbf{k}^T[s\mathbf{I} - \mathbf{A}]^{-1}\mathbf{b}} \tag{7.2-83}$$

Using (7.2-83) in (7.2-82) and the result in the equation for $C(s)$ above,

$$C(s) = \frac{\mathbf{c}^T[s\mathbf{I} - \mathbf{A}]^{-1}\mathbf{b}R(s)}{1 + \mathbf{k}^T[s\mathbf{I} - \mathbf{A}]^{-1}\mathbf{b}}$$

from which we have the closed-loop transfer function for the state variable feedback system in the form

$$T(s) = \frac{C(s)}{R(s)} = \frac{\mathbf{c}^T[s\mathbf{I} - \mathbf{A}]^{-1}\mathbf{b}}{1 + \mathbf{k}^T[s\mathbf{I} - \mathbf{A}]^{-1}\mathbf{b}} \qquad (7.2\text{-}84)$$

Now consider (7.2-84). We note that it is of the form

$$T(s) = \frac{G(s)}{1 + G(s)H(s)} \qquad (7.2\text{-}85)$$

which is the closed-loop transfer function for the system of Fig. 7.2-6. By comparing (7.2-84) and (7.2-85) we see that the forward-path (the plant) transfer function is given by

$$G(s) = \mathbf{c}^T[s\mathbf{I} - \mathbf{A}]^{-1}\mathbf{b} \qquad (7.2\text{-}86)$$

and that the open-loop transfer function is given by

$$G(s)H(s) = \mathbf{k}^T[s\mathbf{I} - \mathbf{A}]^{-1}\mathbf{b} \qquad (7.2\text{-}87)$$

And finally we note that the total effect of state variable feedback may be thought of as a feedback compensator (operating on the plant output) of the form

$$H(s) = \frac{\mathbf{k}^T[s\mathbf{I} - \mathbf{A}]^{-1}\mathbf{b}}{\mathbf{c}^T[s\mathbf{I} - \mathbf{A}]^{-1}\mathbf{b}} \qquad (7.2\text{-}88)$$

From (7.2-88) we see that the practically important case of unity feedback $[H(s) = 1]$ implies that

$$\mathbf{k} = \mathbf{c} \qquad (7.2\text{-}89)$$

We can also see from Fig. 7.2-6 that $\mathbf{k} = \mathbf{c}$ will give a unity-feedback system. Also from (7.2-84) we see that for the state variable feedback system the closed-loop zeros are the same as the forward-path (plant) zeros. The closed-loop poles, on the other hand, are given by the zeros of $(1 + \mathbf{k}^T[s\mathbf{I} - \mathbf{A}]^{-1}\mathbf{b})$, the denominator of (7.2-84).

A root locus for the state variable feedback system may be found by plotting those points in the s-plane where

$$G(s)H(s) = \mathbf{k}^T[s\mathbf{I} - \mathbf{A}]^{-1}\mathbf{b} = -1 \qquad (7.2\text{-}90)$$

By similarly applying frequency response methods, we may make a study of the stability properties of the state variable feedback system. Here we use the Nyquist criterion by making a Nyquist plot of the frequency response of

$$G(j\omega)H(j\omega) = \mathbf{k}^T[j\omega\mathbf{I} - \mathbf{A}]^{-1}\mathbf{b} \qquad (7.2\text{-}91)$$

and counting the appropriate encirclements of the $-1 + j0$ point.

Example 7.2-5

As an example of the use of the preceding analysis on a state variable feedback system, let us consider a plant with transfer function

$$G_f(s) = \frac{K}{s(s+1)(s+4)} \qquad (7.2\text{-}92)$$

This plant with a suitable choice of state variables and the corresponding state variable feedback is shown in Fig. 7.2-7. For the choice of state variables shown there,

$$X_1(s) = \frac{1}{s}X_2(s)$$

$$X_2(s) = \frac{1}{s+1}X_3(s)$$

$$X_3(s) = \frac{K}{s+4}U(s)$$

$$Y(s) = C(s) = X_1(s)$$

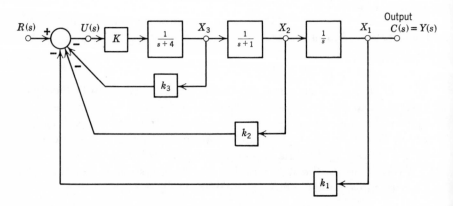

Figure 7.2-7 An example of a state variable feedback system.

which gives

$$sX_1(s) = X_2(s)$$
$$sX_2(s) = -X_2(s) + X_3(s)$$
$$sX_3(s) = \qquad -4X_3(s) + KU(s)$$
$$Y(s) = X_1(s) \qquad (7.2\text{-}93)$$

Inverting (7.2-93),

$$\frac{d}{dt}\begin{bmatrix} x_1 \\ x_2 \\ x_3 \end{bmatrix} = \begin{bmatrix} 0 & 1 & 0 \\ 0 & -1 & 1 \\ 0 & 0 & -4 \end{bmatrix}\begin{bmatrix} x_1 \\ x_2 \\ x_3 \end{bmatrix} + \begin{bmatrix} 0 \\ 0 \\ K \end{bmatrix} u$$

or

$$\dot{\mathbf{x}} = \qquad \mathbf{A} \qquad \mathbf{x} + \mathbf{b}\,u \qquad (7.2\text{-}94)$$

and

$$y = \begin{bmatrix} 1 & 0 & 0 \end{bmatrix}\begin{bmatrix} x_1 \\ x_2 \\ x_3 \end{bmatrix}$$

or

$$y = \qquad \mathbf{c}^T \qquad \mathbf{x}$$

Also from Fig. 7.2-7 we have

$$u = r - \begin{bmatrix} k_1 & k_2 & k_3 \end{bmatrix}\begin{bmatrix} x_1 \\ x_2 \\ x_3 \end{bmatrix} \qquad (7.2\text{-}95)$$

$$u = r - \mathbf{k}^T\mathbf{x}$$

For **A** as given in (7.2-94) we get

$$[s\mathbf{I} - \mathbf{A}] = \begin{bmatrix} s & -1 & 0 \\ 0 & s+1 & -1 \\ 0 & 0 & s+4 \end{bmatrix}$$

which gives

$$[s\mathbf{I} - \mathbf{A}]^{-1} = \begin{bmatrix} \dfrac{1}{s} & \dfrac{1}{s(s+1)} & \dfrac{1}{s(s+1)(s+4)} \\[2mm] 0 & \dfrac{1}{s+1} & \dfrac{1}{(s+1)(s+4)} \\[2mm] 0 & 0 & \dfrac{1}{s+4} \end{bmatrix} \qquad (7.2\text{-}96)$$

Now using **c** and **b** from (7.2-94) and this value of $[s\mathbf{I} - \mathbf{A}]^{-1}$ in (7.2-86), we have

$$G(s) = \mathbf{c}^T[s\mathbf{I} - \mathbf{A}]^{-1}\mathbf{b} = \frac{K}{s(s+1)(s+4)} \qquad (7.2\text{-}97)$$

which checks with (7.2-92). We now use k from (7.2-95) and b and

$[s\mathbf{I}-\mathbf{A}]^{-1}$ from (7.2-94) and (7.2-96), respectively, and from (7.2-87) we get

$$G(s)H(s) = \mathbf{k}^T[s\mathbf{I}-\mathbf{A}]^{-1}\mathbf{b}$$

$$= \frac{Kk_1}{s(s+1)(s+4)} + \frac{Kk_2}{(s+1)(s+4)} + \frac{Kk_3}{(s+4)}$$

Therefore

$$G(s)H(s) = \frac{Kk_3s^2 + (Kk_2+Kk_3)s + Kk_1}{s(s+1)(s+4)} \qquad (7.2\text{-}98)$$

and

$$H(s) = k_3s^2 + (k_2+k_3)s + k_1$$

The results of (7.2-97) and (7.2-98) in (7.2-84) and (7.2-85) give

$$T(s) = \frac{C(s)}{R(s)} = \frac{K}{s^3 + (5+Kk_3)s^2 + (4+Kk_2+Kk_3)s + Kk_1} \qquad (7.2\text{-}99)$$

This result may be checked by use of (7.2-81). This is left as an exercise.

From (7.2-98) we see that the effect of the state variable feedback is to add two open-loop zeros. Since these open-loop zeros [of $H(s)$] are in the feedback path, they do not appear as closed-loop zeros. Also from (7.2-99) we can see that for any given **A**, we can achieve any desired closed-loop pole configuration by properly choosing the feedback coefficients k_1, k_2, and k_3. This facility is helpful in trying to achieve a desirable response for a system.

The root locus for the state variable feedback system under consideration here may be seen to have some very useful properties. The root locus for the system of Fig. 7.2-7 for the case $k_1 = 1$, $k_2 = \frac{3}{8}$, $k_3 = \frac{1}{8}$ used in (7.2-98) is shown in Fig. 7.2-8. We see that for this **k** the root locus has all its branches completely in the LHP. Thus for any value of $K > 0$ the closed-loop system will be stable, which is a desirable feature in a system. The root locus for this system may be brought into a more desirable form (if that of Fig. 7.2-8 is not satisfactory) by putting the closed-loop zeros, which are the roots of $H(s)$ as given by (7.2-98), into different locations. This is easily done if the feedback coefficients, k_1, k_2, and k_3, are free to be chosen.

Steady-State Error Constants for the State Variable Feedback System

In the practically important case of a tracking system, i.e., a system in which the desire is to have the output reproduce the input, the steady-state error constants as defined in Section 3.3 are very convenient for specifying and evaluating the steady-state response. Therefore we

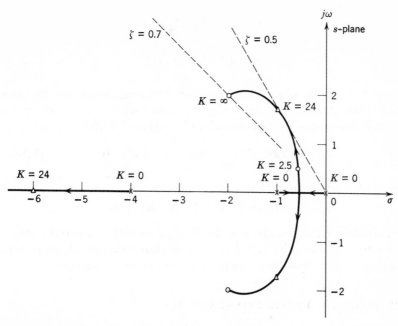

Figure 7.2-8 Root locus for the state variable feedback system of Fig. 7.2-7. $k_1 = 1$, $k_2 = 3/8$, $k_3 = 1/8$.

next consider these constants for the system specified in a state variable formulation. In particular, we shall consider these constants for a state variable feedback system of the form shown in Fig. 7.2-6. The error constants for standard (no-state variable feedback) state variable formulation may be found from the case considered by letting $\mathbf{k} = \mathbf{0}$, where $\mathbf{k}$ is the vector whose components are the feedback coefficients.

Let us define the error for our system as simply the difference between the input and the output:

$$e(t) \overset{\Delta}{=} r(t) - c(t) \qquad (7.2\text{-}100)$$

where e is the error, r the input, and c the output. We are interested only in the steady-state error (e_{ss}):

$$e_{ss} = \lim_{t \to \infty} e(t) \qquad (7.2\text{-}101)$$

To find this steady-state error it is convenient to $\mathcal{L}$-transform the quantities considered and to then use the final-value theorem. Doing this and using (7.2-100) in (7.2-101) gives

$$e_{ss} = \lim_{s \to 0} sE(s) = \lim_{s \to 0} s[R(s) - C(s)]$$

or

$$e_{ss} = \lim_{s \to 0} sR(s)\left[1 - \frac{C(s)}{R(s)}\right]$$

$$= \lim_{s \to 0} sR(s)[1 - T(s)] \qquad (7.2\text{-}102)$$

where $T(s)$ is the closed-loop transfer function. Since we are interested in the steady-state error for the state variable feedback system of Fig. 7.2-6 we use $T(s)$ as given by (7.2-81) in (7.2-102) and get

$$e_{ss} = \lim_{s \to 0} sR(s)(1 - \mathbf{c}^T[s\mathbf{I} - \mathbf{A} + \mathbf{b}\mathbf{k}^T]^{-1}\mathbf{b}) \qquad (7.2\text{-}103)$$

where we have assumed that

$$sE(s) = sR(s)(1 - \mathbf{c}^T[s\mathbf{I} - \mathbf{A} + \mathbf{b}\mathbf{k}^T]^{-1}\mathbf{b})$$

has no poles on the $j\omega$-axis or in the RHP. We shall now use the steady-state error as given by (7.2-103) to determine the steady-state error constants. Let us first consider the position error constant.

THE POSITION ERROR CONSTANT, K_p

The position error constant K_p is defined [from (3.3-6)] by

$$e_{ss}(\text{step}) \overset{\Delta}{=} \frac{1}{1 + K_p}$$

where $e_{ss}(\text{step})$ is the steady-state error to a step input, i.e., for $R(s) = 1/s$. Using $R(s) = 1/s$, we have

$$e_{ss}(\text{step}) = \frac{1}{1 + K_p} = 1 - \mathbf{c}^T[-\mathbf{A} + \mathbf{b}\mathbf{k}^T]^{-1}\mathbf{b} \qquad (7.2\text{-}104)$$

The form of the preceding equation is not really convenient for evaluation since it involves inverting the matrix $[-\mathbf{A} + \mathbf{b}\mathbf{k}^T]$. A more convenient form can be obtained by going back to (7.2-103) with $R(s) = 1/s$ to set

$$e_{ss}(\text{step}) = \lim_{s \to 0}(1 - \mathbf{c}^T[s\mathbf{I} - \mathbf{A} + \mathbf{b}\mathbf{k}^T]^{-1}\mathbf{b}) \qquad (7.2\text{-}105)$$

If we now use the matrix identity

$$[1 + \mathbf{w}^T\mathbf{z}] = \det[\mathbf{I} + \mathbf{z}\mathbf{w}^T] \qquad (7.2\text{-}106)$$

where $\mathbf{w}$ and $\mathbf{z}$ are n-vectors, from (7.2-105) we obtain

$$e_{ss}(\text{step}) = \lim_{s \to 0} \det[\mathbf{I} - [s\mathbf{I} - \mathbf{A} + \mathbf{b}\mathbf{k}^T]^{-1}\mathbf{b}\mathbf{c}^T]$$

If we multiply and divide the determinant on the RHS by $\det [s\mathbf{I} - \mathbf{A} + \mathbf{bk}^T]$ and use the identity

$$(\det [\mathbf{W}])(\det [\mathbf{Z}]) = \det [\mathbf{WZ}] \qquad (7.2\text{-}107)$$

where $\mathbf{W}$ and $\mathbf{Z}$ are square matrices of the same order, we have

$$e_{ss}(\text{step}) = \lim_{s \to 0} \frac{\det [s\mathbf{I} - \mathbf{A} + \mathbf{bk}^T - \mathbf{bc}^T]}{\det [s\mathbf{I} - \mathbf{A} + \mathbf{bk}^T]}$$

from which we have the identity

$$e_{ss}(\text{step}) = \frac{1}{1 + K_p} = \frac{\det [-\mathbf{A} + \mathbf{bk}^T - \mathbf{bc}^T]}{\det [-\mathbf{A} + \mathbf{bk}^T]} \qquad (7.2\text{-}108)$$

which is in a fairly convenient form for evaluation, especially when computational help is available for evaluating the determinants involved.

Example 7.2-6

Consider a unity-feedback system with forward-path transfer function of the form

$$G(s) = \frac{1}{(s+a)(s+b)} = \frac{1}{s^2 + (a+b)s + ab} \qquad (7.2\text{-}109)$$

where we assume $a, b > 0$ so that the open- and closed-loop systems will be stable. We may use (3.3-5) to determine that for this case

$$K_p = \lim_{s \to 0} G(s) = \frac{1}{ab} \qquad (7.2\text{-}110)$$

Now let us do this using (7.2-108). We may use the state variable formulation of (7.2-28) for the transfer function of (7.2-109) to get

$$\frac{d}{dt}\begin{bmatrix} x_1 \\ x_2 \end{bmatrix} = \begin{bmatrix} 0 & 1 \\ -ab & -(a+b) \end{bmatrix}\begin{bmatrix} x_1 \\ x_2 \end{bmatrix} + \begin{bmatrix} 0 \\ 1 \end{bmatrix} u$$

or

$$\dot{\mathbf{x}} \quad = \quad \mathbf{A} \qquad \mathbf{x} \; + \; \mathbf{b}\, u \qquad (7.2\text{-}111)$$

and

$$y \;\; = [1 \quad 0]\begin{bmatrix} x_1 \\ x_2 \end{bmatrix}$$

or

$$y \;\; = \quad \mathbf{c}^T \quad \mathbf{x}$$

For a unity-feedback system, $\mathbf{k} = \mathbf{c}$. Using this and the values $\mathbf{A}$, $\mathbf{b}$, and $\mathbf{c}$ of (7.2-111) in (7.2-108) gives

$$\frac{1}{1+K_p} = \frac{\det\begin{bmatrix} 0 & -1 \\ ab & (a+b) \end{bmatrix}}{\det\begin{bmatrix} 0 & -1 \\ (1+ab) & (a+b) \end{bmatrix}}$$

$$= \frac{ab}{1+ab}$$

from which we have by simple calculation

$$K_p = \frac{1}{ab}$$

which checks with (7.2-110).

Now let us look at the practically more important case of velocity error constant.

THE VELOCITY ERROR CONSTANT, K_v

K_v may be defined from (3.3-9) by

$$\frac{1}{K_v} \overset{\Delta}{=} e_{ss}(\text{ramp}) \tag{7.2-112}$$

where $e_{ss}(\text{ramp})$ is the steady-state error to a unit ramp input, i.e., for $R(s) = 1/s^2$. Using $R(s) = 1/s^2$ in (7.2-103),

$$\frac{1}{K_v} = e_{ss}(\text{ramp}) = \lim_{s \to 0} \frac{(1 - c^T[s\mathbf{I} - \mathbf{A} + \mathbf{b}k^T]^{-1}\mathbf{b})}{s} \tag{7.2-113}$$

Unless $c^T[-\mathbf{A} + \mathbf{b}k^T]^{-1}\mathbf{b} = 1$, $e_{ss}(\text{ramp}) \to \infty$ and $K_v = 0$. So assuming that

$$c^T[-\mathbf{A} + \mathbf{b}k^T]^{-1}\mathbf{b} = 1 \tag{7.2-114}$$

holds, using L'Hospital's rule on the RHS of (7.2-113) gives

$$\frac{1}{K_v} = -\lim_{s \to 0} \frac{d}{ds} c^T[s\mathbf{I} - \mathbf{A} + \mathbf{b}k^T]^{-1}\mathbf{b}$$

Evaluating the derivative,[5] we have

$$\frac{1}{K_v} = \lim_{s \to 0} c^T[s\mathbf{I} - \mathbf{A} + \mathbf{b}k^T]^{-2}\mathbf{b} \tag{7.2-115}$$

---

[5]We know that

$$[s\mathbf{I} - \mathbf{B}][s\mathbf{I} - \mathbf{B}]^{-1} = \mathbf{I}$$

Therefore, taking the derivative,

$$\frac{d}{ds}[s\mathbf{I} - \mathbf{B}][s\mathbf{I} - \mathbf{B}]^{-1} = [\mathbf{I}][s\mathbf{I} - \mathbf{B}]^{-1} + [s\mathbf{I} - \mathbf{B}]\frac{d}{ds}[s\mathbf{I} - \mathbf{B}]^{-1} = 0$$

which gives

$$\frac{d}{ds}[s\mathbf{I} - \mathbf{B}]^{-1} = -[s\mathbf{I} - \mathbf{B}]^{-2}.$$

Again this may be evaluated straightforwardly as

$$\frac{1}{K_v} = \mathbf{c}^T [-\mathbf{A} + \mathbf{b}\mathbf{k}^T]^{-2}\mathbf{b} \tag{7.2-116}$$

which is another not particularly convenient form for evaluation. We can apply (7.2-106) to (7.2-115) to get

$$\frac{1}{K_v} = \lim_{s \to 0} \det\left[\mathbf{I} + [s\mathbf{I} - \mathbf{A} + \mathbf{b}\mathbf{k}^T]^{-2}\mathbf{b}\mathbf{c}^T \right] - 1$$

Multiplying and dividing the determinant by $\det^2 [s\mathbf{I} - \mathbf{A} + \mathbf{b}\mathbf{k}^T]$ and using (7.2-107) gives

$$\frac{1}{K_v} = \lim_{s \to 0} \frac{\det\left[[s\mathbf{I} - \mathbf{A} + \mathbf{b}\mathbf{k}^T]^2 + \mathbf{b}\mathbf{c}^T \right]}{\det^2 [s\mathbf{I} - \mathbf{A} + \mathbf{b}\mathbf{k}^T]} - 1$$

which, after taking the limit, gives

$$\frac{1}{K_v} = e_{ss}(\text{ramp}) = \frac{\det\left[[-\mathbf{A} + \mathbf{b}\mathbf{k}^T]^2 + \mathbf{b}\mathbf{c}^T \right]}{\det^2 [-\mathbf{A} + \mathbf{b}\mathbf{k}^T]} - 1 \tag{7.2-117}$$

which is the desired result. Let us consider an example.

Example 7.2-7

Consider the state variable feedback system of Fig. 7.2-7. We refer here to the state variable formulation for this system given in (7.2-94) with the feedback vector

$$\mathbf{k} = \begin{bmatrix} 1 \\ \frac{3}{8} \\ \frac{1}{8} \end{bmatrix}$$

The root locus for this system is drawn in Fig. 7.2-8. We can see that it is closed-loop stable for all $K > 0$; let us consider the steady-state error constant K_v for this case. Using the formulation of (7.2-94) and $\mathbf{k}$ as above, show that (7.2-114) holds. Now we use $\mathbf{A}$, $\mathbf{b}$, and $\mathbf{c}$ from (7.2-94) and $\mathbf{k}$ as given above in (7.2-117):

$$\frac{1}{K_v} = \frac{\det\left[\begin{bmatrix} 0 & -1 & 0 \\ 0 & 1 & -1 \\ K & \frac{3K}{8} & \left(4 + \frac{K}{8}\right) \end{bmatrix}^2 + \begin{bmatrix} 0 & 0 & 0 \\ 0 & 0 & 0 \\ K & 0 & 0 \end{bmatrix} \right]}{\det^2 \begin{bmatrix} 0 & -1 & 0 \\ 0 & 1 & -1 \\ K & \frac{3K}{8} & \left(4 + \frac{K}{8}\right) \end{bmatrix}} - 1$$

which we may evaluate to get

$$\frac{1}{K_v} = \frac{1}{2} + \frac{4}{K}$$

or

$$K_v = \frac{2K}{K+8}$$

As another exercise, check this result using (7.2-116). We see that for $K > 0$ the maximum for K_v occurs as $K \to \infty$, for which $K_v = 2$.

THE ACCELERATION ERROR CONSTANT, K_a

The acceleration error constant K_a for a state-variable feedback system may be obtained by extension of the methods by which K_v was obtained. We define K_a from (3.3-12) as

$$e_{ss}(\text{accel}) \triangleq \frac{1}{K_a}$$

where $e_{ss}(\text{accel})$ is the steady-state error to a parabolic input, i.e., for $R(s) = 1/s^3$. Using $R(s) = 1/s^3$ in (7.2-103) gives

$$e_{ss}(\text{accel}) = \lim_{s \to 0} \frac{1 - \mathbf{c}^T [s\mathbf{I} - \mathbf{A} + \mathbf{b}\mathbf{k}^T]^{-1}\mathbf{b}}{s^2} \qquad (7.2\text{-}118)$$

Again we note that unless $1 - \mathbf{c}^T [s\mathbf{I} - \mathbf{A} + \mathbf{b}\mathbf{k}^T]^{-1}\mathbf{b}$ has a double zero at $s = 0$ [which also implies that $e_{ss}(\text{ramp}) = 1/K_v = 0$ as given by (7.2-117)], $e_{ss}(\text{accel}) \to \infty$ and $K_a = 0$. Since this is the trivial case, let us consider the case where $1 - \mathbf{c}^T [s\mathbf{I} - \mathbf{A} + \mathbf{b}\mathbf{k}^T]^{-1}\mathbf{b}$ does have the double zero at $s = 0$. For this case, applying L'Hospital's rule to (7.2-118) gives

$$e_{ss}(\text{accel}) = -\lim_{s \to 0} \frac{1}{2} \frac{d^2}{ds^2} \mathbf{c}^T [s\mathbf{I} - \mathbf{A} + \mathbf{b}\mathbf{k}^T]^{-1}\mathbf{b}$$

Performing the differentiation gives

$$e_{ss}(\text{accel}) = -\lim_{s \to 0} \mathbf{c}^T [s\mathbf{I} - \mathbf{A} + \mathbf{b}\mathbf{k}^T]^{-3}\mathbf{b} \qquad (7.2\text{-}119)$$

Now applying (7.2-106) to (7.2-119) and multiplying and dividing by $\det^3 [s\mathbf{I} - \mathbf{A} + \mathbf{b}\mathbf{k}^T]$,

$$e_{ss}(\text{accel}) = \lim_{s \to 0} \left[1 - \frac{\det [[s\mathbf{I} - \mathbf{A} + \mathbf{b}\mathbf{k}^T]^3 + \mathbf{b}\mathbf{c}^T]}{\det^3 [s\mathbf{I} - \mathbf{A} + \mathbf{b}\mathbf{k}^T]} \right]$$

Hence

$$e_{ss}(\text{accel}) = \frac{1}{K_a} = 1 - \frac{\det [[-\mathbf{A} + \mathbf{b}\mathbf{k}^T]^3 + \mathbf{b}\mathbf{c}^T]}{\det^3 [-\mathbf{A} + \mathbf{b}\mathbf{k}^T]} \qquad (7.2\text{-}120)$$

which is the desired result.

Example 7.2-8

As an example of the use of (7.2-120) consider a unity-feedback system with

$$G(s) = \frac{K(s+a)}{s^2(s+b)} \tag{7.2-121}$$

in the forward path. From (3.3-11) we have

$$K_a = \lim_{s \to 0} s^2 G(s) = \frac{aK}{b} \tag{7.2-122}$$

for this $G(s)$. We may use the state variable formulation of (7.2-28) to give

$$\frac{d}{dt}\begin{bmatrix} x_1 \\ x_2 \\ x_3 \end{bmatrix} = \begin{bmatrix} 0 & 1 & 0 \\ 0 & 0 & 1 \\ 0 & 0 & -b \end{bmatrix}\begin{bmatrix} x_1 \\ x_2 \\ x_3 \end{bmatrix} + \begin{bmatrix} 0 \\ 0 \\ 1 \end{bmatrix} u$$

or

$$\dot{\mathbf{x}} = \mathbf{A} \qquad \mathbf{x} + \mathbf{b}\, u$$

and

$$y = \begin{bmatrix} aK & K & 0 \end{bmatrix}\begin{bmatrix} x_1 \\ x_2 \\ x_3 \end{bmatrix}$$

or

$$y = \mathbf{c}^T \qquad \mathbf{x}$$

with unity feedback $\mathbf{k} = \mathbf{c}$. Using these values we have

$$[-\mathbf{A}+\mathbf{b}\mathbf{k}^T] = [-\mathbf{A}+\mathbf{b}\mathbf{c}^T] = \begin{bmatrix} 0 & -1 & 0 \\ 0 & 0 & -1 \\ aK & K & b \end{bmatrix}$$

$$\mathbf{b}\mathbf{c}^T = \begin{bmatrix} 0 & 0 & 0 \\ 0 & 0 & 0 \\ aK & K & 0 \end{bmatrix}$$

Using these values in (7.2-120) we have

$$\frac{1}{K_a} = 1$$

$$-\frac{\det\begin{bmatrix} aK & K & b \\ -abK & -K(b-a) & K-b^2 \\ (-aK^2+aK+ab^2K)(-K^2+K+Kb(b-a))(aK-2bK+b^3) \end{bmatrix}}{\det{}^3\begin{bmatrix} 0 & -1 & 0 \\ 0 & 0 & -1 \\ aK & K & b \end{bmatrix}}$$

$$= \frac{b}{aK}$$

which checks with (7.2-122).

SOME GENERAL REMARKS

1. The error constants K_p, K_v, and K_a as evaluated by (7.2-108), (7.2-117), and (7.2-120), respectively, involve a division by $\det[-\mathbf{A} + \mathbf{bk}^T]$. We next should ask when we have $\det[-\mathbf{A} + \mathbf{bk}^T] = 0$, and what this implies. To answer, let us consider the closed-loop transfer function for the state variable feedback system as given by (7.2-81):

$$T(s) = \mathbf{c}^T[s\mathbf{I} - \mathbf{A} + \mathbf{bk}^T]^{-1}\mathbf{b}$$

By evaluating the inverse matrix on the RHS of this equation we have

$$T(s) = \frac{\mathbf{c}^T[s\mathbf{I} - \mathbf{A} + \mathbf{bk}^T]^*\mathbf{b}}{\det[s\mathbf{I} - \mathbf{A} + \mathbf{bk}^T]} \qquad (7.2\text{-}123)$$

where $[s\mathbf{I} - \mathbf{A} + \mathbf{bk}^T]^*$ is the adjoint matrix of $[s\mathbf{I} - \mathbf{A} + \mathbf{bk}^T]$, i.e., it is the matrix obtained by transposing $[s\mathbf{I} - \mathbf{A} + \mathbf{bk}^T]$ and then replacing each element of the resulting matrix by its minor with the appropriate sign. From (7.2-123) we see that the characteristic polynomial of the closed-loop state variable feedback system (of Fig. 7.2-6) is $\det[s\mathbf{I} - \mathbf{A} + \mathbf{bk}^T]$, and the poles of the closed-loop system are the roots of this polynomial. We then can write

$$\det[s\mathbf{I} - \mathbf{A} + \mathbf{bk}^T] = s^n + b_{n-1}s^{n-1} + \cdots + b_0 \qquad (7.2\text{-}124)$$

and

$$\det[-\mathbf{A} + \mathbf{bk}^T] = b_0$$

and if $\det[-\mathbf{A} + \mathbf{bk}^T] = 0$, we see that the polynomial $\det[s\mathbf{I} - \mathbf{A} + \mathbf{bk}^T]$ has a root at the origin (at $s = 0$). Thus, if $\det[-\mathbf{A} + \mathbf{bk}^T] = 0$, the closed-loop system has a pole at the origin. In general, a closed-loop pole at the origin is not desirable since this means our system is not stable in the bounded input, implying a bounded output sense. Hence the feedback vector $\mathbf{k}$ should in general be chosen so that $\det[-\mathbf{A} + \mathbf{bk}^T] \neq 0$. By looking at (7.2-123) it can be seen that $\mathbf{c}$ and $\mathbf{b}$ may be such that a closed-loop zero also occurs at the origin to cancel out the pole there. If this is the case, the formulae for error constants as given in (7.2-108), (7.2-117), and (7.2-120) will be indeterminate. In this case — a closed-loop pole and zero at the origin cancelling — the error constant may be found by use of (7.2-103). However, now the term $\mathbf{c}^T[s\mathbf{I} - \mathbf{A} + \mathbf{bk}^T]^{-1}\mathbf{b}$ will have to be evaluated as a function of s. After the poles and zeros at the origin have been cancelled out, the steady-state error and the corresponding error constants may then be found directly by going to the limit. It is helpful in doing this to apply (7.2-106) to (7.2-103), which gives

$$e_{ss} = \lim_{s \to 0} sR(s) \det[\mathbf{I} - [s\mathbf{I} - \mathbf{A} + \mathbf{bk}^T]^{-1}\mathbf{bc}^T]$$

$$= \lim_{s \to 0} sR(s) \frac{\det[s\mathbf{I} - \mathbf{A} + \mathbf{b}\mathbf{k}^T - \mathbf{b}\mathbf{c}^T]}{\det[s\mathbf{I} - \mathbf{A} + \mathbf{b}\mathbf{k}^T]} \qquad (7.2\text{-}125)$$

Evaluating the two determinants in (7.2-125) as functions of s will quickly show if cancellation of poles at the origin can be accomplished.

Example 7.2-9

As an example of the use of (7.2-125) consider a unity-feedback system with

$$G(s) = \frac{s}{s^2(s+a)} \qquad (7.2\text{-}126)$$

in the forward path. Using (3.3-8) gives

$$K_v = \lim_{s \to 0} sG(s) = \frac{1}{a} \qquad (7.2\text{-}127)$$

for this $G(s)$. Using the state variable formulation of (7.2-28) on the $G(s)$ of (7.2-126) gives

$$\frac{d}{dt}\begin{bmatrix} x_1 \\ x_2 \\ x_3 \end{bmatrix} = \begin{bmatrix} 0 & 1 & 0 \\ 0 & 0 & 1 \\ 0 & 0 & -a \end{bmatrix}\begin{bmatrix} x_1 \\ x_2 \\ x_3 \end{bmatrix} + \begin{bmatrix} 0 \\ 0 \\ 1 \end{bmatrix} u$$

or

$$\dot{\mathbf{x}} = \mathbf{A} \qquad \mathbf{x} + \mathbf{b}\, u$$

and

$$y = \begin{bmatrix} 0 & 1 & 0 \end{bmatrix}\begin{bmatrix} x_1 \\ x_2 \\ x_3 \end{bmatrix} \qquad (7.2\text{-}128)$$

or

$$y = \mathbf{c}^T \qquad \mathbf{x}$$

For a unity-feedback system $\mathbf{k} = \mathbf{c}$. We see

$$[-\mathbf{A} + \mathbf{b}\mathbf{k}^T] = \begin{bmatrix} 0 & -1 & 0 \\ 0 & 0 & -1 \\ 0 & 1 & a \end{bmatrix}$$

and thus $\det[-\mathbf{A} + \mathbf{b}\mathbf{k}^T] = 0$, hence (7.2-117) will not give any answer for the steady-state error constant K_v. However, using (7.2-125) for this example gives

$$e_{ss} = \lim_{s \to 0} sR(s) \frac{\det\begin{bmatrix} s & -1 & 0 \\ 0 & s & -1 \\ 0 & 0 & (s+a) \end{bmatrix}}{\det\begin{bmatrix} s & -1 & 0 \\ 0 & s & -1 \\ 0 & 1 & (s+a) \end{bmatrix}}$$

$$= \lim_{s \to 0} sR(s) \frac{s(s+a)}{s^2 + as + 1}$$

Now for $R(s) = 1/s^2$ (a ramp input) we have

$$e_{ss}(\text{ramp}) = \frac{1}{K_v} = a$$

which checks with (7.2-127).

2. With a unity-feedback system, unless the forward transfer function has a pole at the origin, the closed-loop system will always have a finite steady-state error to a step-input, i.e., e_{ss} (step) $\neq$ 0. With state variable feedback, however, there exists the possibility of achieving e_{ss}(step) = 0 without a plant pole at the origin. This is done, as we see directly from (7.2-108), by choosing the feedback vector **k** such that

$$\det\left[-\mathbf{A} + \mathbf{bk}^T - \mathbf{bc}^T\right] = 0$$
$$\det\left[-\mathbf{A} + \mathbf{bk}^T\right] \neq 0$$

and the closed-loop system is stable.

7.3 DISCRETE-TIME SYSTEMS

Open-loop Systems

The state variable methods for the analysis of continuous-control systems developed in Section 7.2 can be extended for the analysis of discrete-time or sampled-data systems. Consider first an open-loop system with transfer function $G(s)$ into which we feed a sampled input (Fig. 7.3-1), instead of a continuous input as shown previously. The input either can be passed through a hold circuit first or applied without a hold circuit.

SOLUTION BASED ON THE CONTINUOUS MODEL

Let the input into $G(s)$ be m. Now we can express a state variable formulation for the linear time-invariant system characterized by $G(s)$ as

$$\dot{\mathbf{x}} = \mathbf{Ax} + \mathbf{b}m \qquad (7.3\text{-}1)$$

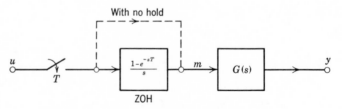

Figure 7.3-1 Open-loop sampled-data system.

and

$$y = \mathbf{c}^T\mathbf{x} \tag{7.3-2}$$

where $\mathbf{x}$ is the state vector (which may be chosen as discussed in Section 7.2) and $\mathbf{A}$ is the system matrix.

The solution of (7.3-1) may be written using (7.2-67) with t_0 as the initial time:

$$\mathbf{x}(t) = \mathbf{e}^{\mathbf{A}(t-t_0)}\mathbf{x}(t_0) + \int_{t_0}^{t} \mathbf{e}^{\mathbf{A}(t-\tau)}\mathbf{b}\,m(\tau)\,d\tau \tag{7.3-3}$$

where, from (7.2-64),

$$\mathbf{e}^{\mathbf{A}t} = \mathscr{L}^{-1}\{[s\mathbf{I} - \mathbf{A}]^{-1}\} \tag{7.3-4}$$

In our case the input is sampled, so we will develop the solution going from one sampling instant $t_0 = nT$ to the next sampling instant $t = (n+1)T$. Substituting these values of t_0 and t in (7.3-3) gives

$$\mathbf{x}(\overline{n+1T}) = \mathbf{e}^{\mathbf{A}T}\mathbf{x}(nT) + \int_{nT}^{(n+1)T} \mathbf{e}^{\mathbf{A}(\overline{n+1T-\tau})}\mathbf{b}\,m(\tau)\,d\tau \tag{7.3-5}$$

The input of $G(s)$, $m(\tau)$, between these two instants will be:

1. No hold circuit, $m(\tau) = u(nT)\,\delta(t-nT)$. $\tag{7.3-6}$
2. With hold circuit $m(\tau) = u(nT)$. $\tag{7.3-7}$

Then:

Without Hold Circuit

$$\mathbf{x}(\overline{n+1T}) = \mathbf{e}^{\mathbf{A}T}\mathbf{x}(nT) + \int_{nT}^{(n+1)T} \mathbf{e}^{\mathbf{A}(\overline{n+1T-\tau})}\mathbf{b}u(nT)\delta(\tau-nT)\,d\tau$$

$$= \mathbf{e}^{\mathbf{A}T}\mathbf{x}(nT) + \mathbf{e}^{\mathbf{A}T}\mathbf{b}u(nT) \tag{7.3-8}$$

With Hold Circuit

$$\mathbf{x}(\overline{n+1T}) = \mathbf{e}^{\mathbf{A}T}\mathbf{x}(nT) + \int_{nT}^{(n+1)T} \mathbf{e}^{\mathbf{A}(\overline{n+1T-\tau})}\mathbf{b}u(nT)\,d\tau$$

$$= \mathbf{e}^{\mathbf{A}T}\mathbf{x}(nT) + \int_{0}^{T} \mathbf{e}^{\mathbf{A}q}\mathbf{b}u(nT)\,dq$$

letting $(n+1)T - \tau = q$. Let[6]

$$\mathbf{h}(T) = \int_{0}^{T} \mathbf{e}^{\mathbf{A}q}\mathbf{b}\,dq \tag{7.3-9}$$

then

$$\mathbf{x}(\overline{n+1T}) = \mathbf{e}^{\mathbf{A}T}\mathbf{x}(nT) + \mathbf{h}(T)u(nT) \tag{7.3-10}$$

[6]If $\mathbf{A}$ is nonsingular then

$$\mathbf{h}(T) = \mathbf{A}^{-1}[\mathbf{e}^{\mathbf{A}T} - \mathbf{I}]\mathbf{b}$$

The solution of (7.3-10) is obtained by realizing that

$$\mathbf{x}(nT) = e^{\mathbf{A}T}\mathbf{x}(\overline{n-1}T) + \mathbf{h}(T)u(\overline{n-1}T)$$
$$\mathbf{x}(\overline{n-1}T) = e^{\mathbf{A}T}\mathbf{x}(\overline{n-2}T) + \mathbf{h}(T)u(\overline{n-2}T)$$

$$\cdot$$
$$\cdot$$
$$\cdot$$

$$\mathbf{x}(T) = e^{\mathbf{A}T}\mathbf{x}(0) + \mathbf{h}(T)u(0)$$

substituting backwards and manipulating limits gives

$$\mathbf{x}(nT) = [e^{\mathbf{A}T}]^n\mathbf{x}(0) + \sum_{k=0}^{n-1} [e^{\mathbf{A}T}]^{n-1-k}\mathbf{h}(T)u(kT) \qquad (7.3\text{-}11)$$

The solution to (7.3-8) is obtained in the same manner, only now $e^{\mathbf{A}T}\,\mathbf{b}$ replaces $\mathbf{h}(T)$. The solution is then

$$\mathbf{x}(nT) = [e^{\mathbf{A}T}]^n\mathbf{x}(0) + \sum_{k=0}^{n-1} [e^{\mathbf{A}T}]^{n-k}\mathbf{b}u(kT) \qquad (7.3\text{-}12)$$

Equations (7.3-11) and (7.3-12) give the state vector at sampling instants in terms of the initial value of the state vector and the sampled values of the input for the case of the system of Fig. 7.3-1. The output at sampling instants is obtained from (7.3-2) as

$$y(nT) = \mathbf{c}^T\mathbf{x}(nT) \qquad (7.3\text{-}13)$$

SOLUTION BETWEEN SAMPLING INSTANTS

The behavior of the state vector and output between sampling instants can be obtained from (7.3-3) by solving for $\mathbf{x}(t)$ starting from $t_0 = nT$.

Assuming a hold circuit, we have $m(\tau) = u(nT)$, for $nT \leqslant t < \overline{n+1}T$. Then (7.3-3) becomes

$$\mathbf{x}(t) = e^{\mathbf{A}(t-nT)}\mathbf{x}(nT) + \int_{nT}^{t} e^{\mathbf{A}(t-\tau)}\mathbf{b}u(nT)d\tau \qquad (7.3\text{-}14)$$

We can now calculate the value of the state vector $\mathbf{x}(t)$ and hence the output $y(t) = \mathbf{c}^T\mathbf{x}(t)$ between any two sampling instants if we know the value of the state vector at the previous sampling instant. We can solve for $\mathbf{x}(nT)$ for any n using (7.3-11) and then using (7.3-14) to determine $\mathbf{x}(t)$ for $nT \leqslant t < \overline{n+1}T$. As is evident, there is a considerable amount of computation involved, which generally requires a digital computer. This approach is an alternate to the modified Z-transform technique which we developed in Chapter 5 to obtain information about between-sampling instants.

SOLUTION BASED ON THE DISCRETE MODEL

A discrete model of a SDS which is valid at sampling instants, can be developed from the z-transfer function for the system. Consider the SDS of Fig. 7.3-2. If there is any hold circuit, it is incorporated in $G(s)$ before $G(z)$ is obtained. In order to obtain the *state-transition equations* which are valid at the sampling instants, we proceed as follows:

1. Let $G(z)$ be in factored form as

$$G(z) = \frac{K(z-z_1)(z-z_2)\cdots(z-z_k)}{(z-p_1)(z-p_2)\cdots(z-p_n)} \tag{7.3-15}$$

We make a state variable diagram as shown in Fig. 7.3-3. A state variable is chosen at the output of each $1/(z-p_i)$ and at the input of each $(z-z_j)$, as shown. Note that the model is good only at sampling instants and $zx_k(nT) = x_k(\overline{n+1T})$. We write the state transition equations from Fig. 7.3-3 as

$$x_1(\overline{n+1T}) = p_1 x_1(nT) + Ku(nT)$$
$$x_2(\overline{n+1T}) = p_2 x_2(nT) + x_1(\overline{n+1T}) - z_1 x_1(nT)$$
$$\vdots$$

$$x_{k+1}(\overline{n+1T}) = p_{k+1} x_{k+1}(nT) + x_k(\overline{n+1T}) - z_k x_k(nT)$$
$$\vdots$$

$$x_n(\overline{n+1T}) = p_n x_n(nT) + x_{n-1}(\overline{n+1T}) - z_{n-1} x_{n-1}(nT)$$

Moreover,

$$y(nT) = x_n(nT) \tag{7.3-16}$$

We can express these equations as

$$\mathbf{x}(\overline{n+1T}) = \mathbf{A}_d \mathbf{x}(nT) + \mathbf{b}\, u(nT) \tag{7.3-17}$$

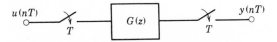

Figure 7.3-2   Discrete model of sampled-data system.

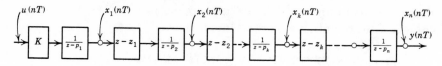

Figure 7.3-3   State variable diagram for $G(z)$ in factored form.

and
$$y(nT) = \mathbf{c}^T \mathbf{x}(nT)$$
where
$$\mathbf{x}^T = [x_1 \cdots x_n], \quad \mathbf{c}^T = [0 \cdots 1], \quad \mathbf{b}^T = [K \quad K \quad \cdots \quad K]$$
and
$$\mathbf{A}_d = \begin{bmatrix} p_1 & 0 & \cdot\ \cdot & 0 \\ (p_1 - z_1) & p_2 & \cdot\ \cdot\ \cdot & 0 \\ (p_1 - z_1) & (p_2 - z_2) & \cdot\ \cdot\ \cdot & 0 \\ \cdot & \cdot & & \\ \cdot & \cdot & & \\ \cdot & \cdot & & \\ (p_1 - z_1) & (p_2 - z_2) & \cdot\ \cdot\ \cdot & p_n \end{bmatrix} \tag{7.3-18}$$

$\mathbf{A}_d$ is the *discrete-system matrix*.

Note that the form of (7.3-17) is similar to (7.3-10). Therefore the solution can be written as

$$\mathbf{x}(nT) = \mathbf{A}_d^n \mathbf{x}(0) + \sum_{k=0}^{n-1} \mathbf{A}_d^{n-1-k} \mathbf{b}u(kT) \tag{7.3-19}$$

2. Consider $G(z)$ of the form

$$G(z) = \frac{a_k z^k + a_{k-1} z^{k-1} + \cdots + a_0}{z^n + b_{n-1} z^{n-1} + \cdots + b_0}, \quad k \leqslant n \tag{7.3-20}$$

This can be described as in Fig. 7.3-4, which is similar to the block diagram for the continuous case shown in Fig. 7.2-4. The difference in the two figures is seen to be that z^{-1} replaces s^{-1}. The selection of state variables is also shown in Fig. 7.3-4. The state transition equations, then, are

$$x_1\overline{(n+1T)} = x_2(nT)$$
$$x_2\overline{(n+1T)} = x_3(nT)$$
$$\cdot$$
$$\cdot \tag{7.3-21}$$
$$\cdot$$
$$x_{n-1}\overline{(n+1T)} = x_n(nT)$$

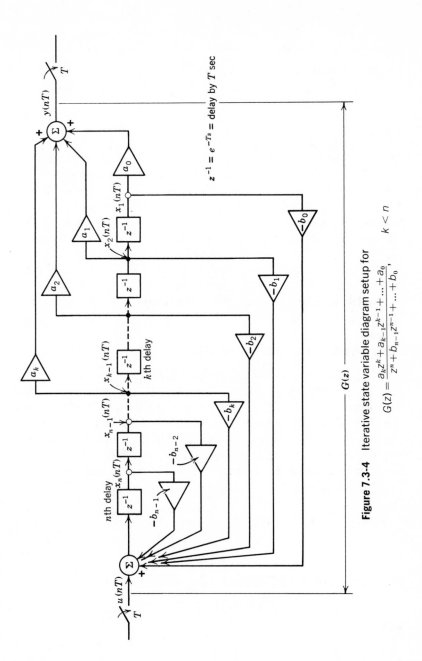

Figure 7.3-4 Iterative state variable diagram setup for

$$G(z) = \frac{a_k z^k + a_{k-1} z^{k-1} + \dots + a_0}{z^n + b_{n-1} z^{n-1} + \dots + b_0}, \qquad k < n$$

$z^{-1} = e^{-Ts} = $ delay by T sec

$$x_n\overline{(n+1T)} = -b_0 x_1(nT) - b_1 x_2(nT) - \cdots - b_{n-1}x_n(nT) + u(nT)$$

and

$$y(nT) = a_0 x_1(nT) + \cdots + a_k x_{k+1}(nT) \qquad (7.3\text{-}22)$$

These equations can be written as

$$\mathbf{x}\overline{(n+1T)} = \mathbf{A}_d\mathbf{x}(nT) + \mathbf{b}u(nT)$$
$$y(nT) = \mathbf{c}^T\mathbf{x}(nT) \qquad (7.3\text{-}23)$$

where $\mathbf{A}_d$, $\mathbf{b}$, and $\mathbf{c}$ are exactly the same as $\mathbf{A}$, $\mathbf{b}$, and $\mathbf{c}$, respectively, in (7.2-28) and (7.2-29). Equation (7.3-23) is exactly the same as (7.3-17) and the solution [for $\mathbf{x}(nT)$] is again as in (7.3-19).

Note that the discrete model develops the solution only at sampling instants, whereas the continuous model can also be used to obtain between-sampling information. There are a number of different choices of state variables and corresponding state variable diagrams possible for a given configuration of $G(z)$. Further, the state variables for the discrete model are different from the state variables of the continuous model. Also, functions like $[e^{AT}]^n$ and $[\mathbf{A}_d]^n$ can be computed by standard digital computer routines from knowledge of $\mathbf{A}$ and $\mathbf{A}_d$ respectively.

State Variable Feedback Sampled-Data Systems

The notion of state variable feedback (SVF) is as powerful in the case of sampled-data systems as it is in the case of continuous-time systems. The advantages of SVF are the same in either case. So let us consider the SVF SDS as shown in Fig. 7.3-5. We consider the case with a ZOH after the sampler. The case without the hold circuit can be found from our results here by the substitution $\mathbf{h}(T) = e^{AT}\mathbf{b}$ as we have already seen. We begin by finding the z-transfer function for the system of Fig. 7.3-5. We have from the figure

$$\dot{\mathbf{x}} = \mathbf{A}\mathbf{x} + \mathbf{b}u \qquad (7.3\text{-}25)$$

$$u = r - \mathbf{k}^T\mathbf{x} \qquad (7.3\text{-}26)$$

$$y = \mathbf{c}^T\mathbf{x} \qquad (7.3\text{-}27)$$

The solution of (7.3-25) from (7.3-10) is

$$\mathbf{x}\overline{(n+1T)} = e^{AT}\mathbf{x}(nT) + \mathbf{h}(T)u(nT) \qquad (7.3\text{-}28)$$

where

$$\mathbf{h}(T) = \int_0^T e^{A\tau}\mathbf{b}\,d\tau$$

$$= \mathbf{A}^{-1}[e^{AT} - \mathbf{I}]\mathbf{b} \qquad (7.3\text{-}29)$$

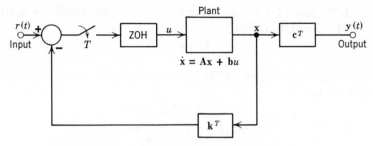

Figure 7.3-5   State variable feedback sampled-data system.

where the last equation applies if $\mathbf{A}$ is nonsingular. We now let

$$\mathbf{A}_d = \mathbf{e}^{AT} \tag{7.3-30}$$

in (7.3-28) and get

$$\mathbf{x}(\overline{n+1}T) = \mathbf{A}_d\mathbf{x}(nT) + \mathbf{h}(T)u(nT) \tag{7.3-31}$$

Taking the Z-transform of this and (7.3-26) and (7.3-27) with zero initial conditions gives

$$z\mathbf{X}(z) = \mathbf{A}_d\mathbf{X}(z) + \mathbf{h}(T)U(z) \tag{7.3-32}$$

$$U(z) = R(z) - \mathbf{k}^T\mathbf{X}(z) \tag{7.3-33}$$

$$Y(z) = \mathbf{c}^T\mathbf{X}(z) \tag{7.3-34}$$

From (7.3-32) we have

$$\mathbf{X}(z) = [z\mathbf{I} - \mathbf{A}_d]^{-1}\mathbf{h}(T)U(z) \tag{7.3-35}$$

which substituted into (7.3-33), gives

$$U(z) = \frac{R(z)}{1 + \mathbf{k}^T[z\mathbf{I} - \mathbf{A}_d]^{-1}\mathbf{h}(T)}$$

This is substituted into (7.3-35) and the result in (7.3-34) gives

$$\frac{Y(z)}{R(z)} = T(z) = \frac{\mathbf{c}^T[z\mathbf{I} - \mathbf{A}_d]^{-1}\mathbf{h}(T)}{1 + \mathbf{k}^T[z\mathbf{I} - \mathbf{A}_d]^{-1}\mathbf{h}(T)} \tag{7.3-36}$$

This is of the form $T(z) = G(z)/[1 + GH(z)]$, which is the transfer function of the SDS in Fig. 5.2-4c. Thus the forward-path (open-loop) transfer function is

$$G(z) = \mathbf{c}^T[z\mathbf{I} - \mathbf{A}_d]^{-1}\mathbf{h}(T) \tag{7.3-37}$$

and the open-loop transfer function is

$$GH(z) = \mathbf{k}^T[z\mathbf{I} - \mathbf{A}_d]^{-1}\mathbf{h}(T) \tag{7.3-38}$$

From (7.3-37) and (7.3-38) we note that for a unity-feedback system $G(z) = GH(z)$ and therefore

$$k = c \qquad (7.3\text{-}39)$$

We gain insight into the SVF SDS by using (7.2-106) on the denominator of (7.3-36) to get

$$T(z) = \frac{c^T[zI - A_d]^{-1}h(T)}{\det[I + [zI - A_d]^{-1}h(T)k^T]}$$

Multiplying the numerator and denominator by $\det[zI - A_d]$ gives

$$T(z) = \frac{c^T[zI - A_d]^*h(T)}{\det[zI - A_d + h(T)k^T]} \qquad (7.3\text{-}40)$$

where $[zI - A_d]^*$ is the adjoint matrix of $[zI - A_d]$. Equation (7.3-40) shows that the SVF SDS has the same zeros as the forward-path transfer function $G(z)$ in (7.3-37) and the closed-loop poles are where $\det[zI - A_d + h(T)k^T]$ has its roots, i.e., they are the eigenvalues of the matrix $[A_d - h(T)k^T]$.

An alternate expression for the closed-loop transfer function of the SVF SDS of Fig. 7.3-5 is found by evaluating $u(nT) = r(nT) - k^T x(nT)$ from (7.3-26) and using this in (7.3-28) to get

$$x\overline{(n+1T)} = [A_d - h(T)k^T]x(nT) + h(T)r(nT) \qquad (7.3\text{-}41)$$

with $A_d = e^{AT}$ from (7.3-30). Taking the Z-transform of (7.3-41) gives directly

$$T(z) = \frac{Y(z)}{R(z)} = c^T[zI - A_d + h(T)k^T]^{-1}h(T) \qquad (7.3\text{-}42)$$

which is the desired result. We note that (7.3-36) and (7.3-42) are similar to the equivalent results for the continuous-time SVF case as seen in (7.2-84) and (7.2-81), respectively. We also note that the root locus for the SVF SDS occurs where $GH(z) = -1$ with $GH(z)$ as given by (7.3-38). The Nyquist criterion may be applied to this type of system by taking the frequency response $GH(e^{j\omega T})$ where $GH(z)$ is again given by (7.3-38).

To reiterate, the unity-feedback transfer functions are found by letting $k = c$, the open-loop transfer functions are found by letting $k = 0$, and for the case of no-hold circuit after the sampler our results apply with $h(T) = e^{AT}b$.

Evaluation of A_d^n

Consider the equation

$$\overline{x(n+1T)} = A_d x(nT) \tag{7.3-43}$$

Starting with $x(T) = A_d x(0)$ and working forward iteratively to nT we have

$$x(nT) = A_d^n x(0) \tag{7.3-44}$$

z-transforming (7.3-43) gives

$$zX(z) - zx(0) = A_d X(z)$$

or

$$X(z) = z[zI - A_d]^{-1} x(0)$$

Taking the inverse Z-transform,

$$x(nT) = Z^{-1}\{z[zI - A_d]^{-1}\}x(0)$$

Comparing this result with (7.3-44) gives

$$A_d^n = Z^{-1}\{z[zI - A_d]^{-1}\} \tag{7.3-45}$$

Similarly,

$$[e^{AT}]^n = Z^{-1}\{z[zI - e^{AT}]^{-1}\} \tag{7.3-46}$$

where

$$e^{AT} = \mathscr{L}^{-1}\{[sI - A]^{-1}\}|_{t=T} \tag{7.3-47}$$

Alternately,

$$[e^{AT}]^n = e^{nAT} = \mathscr{L}^{-1}\{[sI - A]^{-1}\}|_{t=nT}$$

Note that

$$[zI - A_d]^{-1} = \frac{[zI - A_d]^*}{\det [zI - A_d]}$$

which can be computed directly given A_d. The preceding development on the utilization of state variable methods enables us to obtain the transient response of a SDS.

Again A_d^n and $[e^{AT}]^n$ can be computed by standard digital computer routines.

Example 7.3-1

Consider a feedback SDS as shown in Fig. 7.3-6.

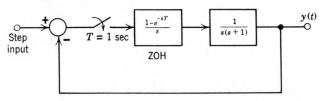

Figure 7.3-6 Example of feedback sampled-data system.

1. Set up the state variable diagram for the continuous model and develop the corresponding state transition equations. Find the solution of the state vector $x(nT)$ and the closed-loop transfer function.

2. Set up the state variable diagram of the discrete model and repeat part 1.

Continuous Model

The states can be chosen as shown in Fig. 7.3-7. The differential equations for the state variables may be found, as shown in Example 7.2-5, to be

$$\dot{x}_1 = x_2$$

$$\dot{x}_2 = -x_2 + e(nT), \qquad nT \leqslant t < \overline{n+1}\,T$$

and from Fig. 7.3-7,

$$e(nT) = r(nT) - y(nT)$$

$$y = x_1 = c^T x, \qquad c^T = [1 \quad 0]$$

we get

$$\dot{x} = \begin{bmatrix} 0 & 1 \\ 0 & -1 \end{bmatrix} x + \begin{bmatrix} 0 \\ 1 \end{bmatrix} e(nT) \qquad (7.3\text{-}48)$$

or

$$\dot{x} = \quad A \quad x + b \ e(nT)$$

We first find e^{At}:

$$e^{At} = \mathcal{L}^{-1}\{[sI - A]^{-1}\}$$

Now

$$[sI - A] = \begin{bmatrix} s & -1 \\ 0 & s+1 \end{bmatrix}$$

therefore

$$[sI - A]^{-1} = \begin{bmatrix} 1/s & 1/s(s+1) \\ 0 & 1/(s+1) \end{bmatrix}$$

or, inverting, we get

$$e^{At} = \begin{bmatrix} 1 & 1 - e^{-t} \\ 0 & e^{-t} \end{bmatrix}$$

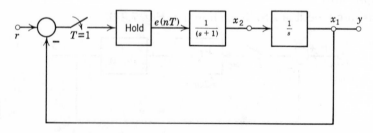

Figure 7.3-7 Continuous model state variable diagram for system of Example 7.3-1.

and

$$\mathbf{h}(T) = \int_0^T \mathbf{e}^{\mathbf{A}\tau}\mathbf{b}\, d\tau = \int_0^T \begin{bmatrix} 1-e^{-\tau} \\ e^{-\tau} \end{bmatrix} d\tau = \begin{bmatrix} T+e^{-T}-1 \\ 1-e^{-T} \end{bmatrix}$$

For $T = 1$ sec,

$$\mathbf{e}^{\mathbf{A}T} = \begin{bmatrix} 1 & 0.632 \\ 0 & 0.368 \end{bmatrix} \quad \text{and} \quad \mathbf{h}(T) = \begin{bmatrix} 0.368 \\ 0.632 \end{bmatrix} \qquad (7.3\text{-}49)$$

Hence the state-transition equation is

$$\begin{aligned} \mathbf{x}(n+1) &= \mathbf{e}^{\mathbf{A}T}\mathbf{x}(n) + \mathbf{h}(T)e(n), \qquad T = 1 \text{ sec} \\ &= \mathbf{e}^{\mathbf{A}T}\mathbf{x}(n) + \mathbf{h}(T)[r(n) - x_1(n)] \\ &= \mathbf{Q}(T)\mathbf{x}(n) + \mathbf{h}(T)r(n) \end{aligned}$$

where

$$\begin{aligned} \mathbf{Q}(T)\mathbf{x}(n) &= \mathbf{e}^{\mathbf{A}T}\mathbf{x}(n) - \mathbf{h}(T)x_1(n) \\ &= \begin{bmatrix} 1 & 0.632 \\ 0 & 0.368 \end{bmatrix}\begin{bmatrix} x_1(n) \\ x_2(n) \end{bmatrix} - \begin{bmatrix} 0.368 \\ 0.632 \end{bmatrix}x_1(n) \\ &= \begin{bmatrix} 0.632 & 0.632 \\ -0.632 & 0.368 \end{bmatrix}\mathbf{x}(n) \end{aligned}$$

Since the input is a step, $r(n) = 1 \ \forall \ n \geqslant 0$. Therefore

$$\mathbf{x}(n+1) = \mathbf{Q}(T)\mathbf{x}(n) + \mathbf{h}(T)$$

The solution of this equation is given in (7.3-19) as

$$\mathbf{x}(n) = [\mathbf{Q}(T)]^n\mathbf{x}(0) + \sum_{k=0}^{n-1}[\mathbf{Q}(T)]^{n-1-k}\mathbf{h}(T)$$

where

$$[\mathbf{Q}(T)]^n = Z^{-1}\{z[z\mathbf{I} - \mathbf{Q}(T)]^{-1}\} = Z^{-1}\left\{\begin{bmatrix} \dfrac{z(z-0.368)}{z^2-z+0.632} & \dfrac{0.632z}{z^2-z+0.632} \\[3mm] \dfrac{-0.632z}{z^2-z+0.632} & \dfrac{z(z-0.632)}{z^2-z+0.632} \end{bmatrix}\right\}$$

Inverting term by term in the matrix on the RHS gives

$$[\mathbf{Q}(T)]^n = (0.795)^{n-1}\begin{bmatrix} \begin{matrix} 0.632\cos(n-1)51° \\ -0.511\sin(n-1)51° \end{matrix} & \begin{matrix} 0.632\cos(n-1)51° \\ +0.511\sin(n-1)51° \end{matrix} \\[3mm] \begin{matrix} -0.632\cos(n-1)51° \\ -0.511\sin(n-1)51° \end{matrix} & \begin{matrix} 0.368\cos(n-1)51° \\ -0.725\sin(n-1)51° \end{matrix} \end{bmatrix}$$

In order to obtain $[\mathbf{Q}(T)]^{n-1-k}$, we just replace n by $n-1-k$ in the preceding expression. The overall transfer function is given in (7.3-37) as

$$T(z) = \mathbf{c}^T [z\mathbf{I} - \mathbf{Q}(T)]^{-1}\mathbf{h}(T)$$

$$= \begin{bmatrix} 1 & 0 \end{bmatrix} \begin{bmatrix} z-0.632 & -0.632 \\ 0.632 & z-0.368 \end{bmatrix}^{-1} \begin{bmatrix} 0.368 \\ 0.632 \end{bmatrix}$$

$$= \frac{0.368\,(z-0.368) + 0.632^2}{(z-0.632)\,(z-0.368) + 0.632^2}$$

$$= \frac{0.368z + 0.264}{z^2 - z + 0.632} \tag{7.3-50}$$

This checks with the standard Z-transform results. Information on the state $x(t)$ between sampling instants may be obtained using (7.3-14).

We may obtain these same results using the state variable feedback SDS results. The state variable feedback SDS is shown in Fig. 7.3-5. For the unity-feedback case, the case of interest, we recall that $\mathbf{k} = \mathbf{c}$ applies. Using $\mathbf{A}_d = e^{\mathbf{A}T}$ and $\mathbf{h}(T)$ from (7.3-40) and $\mathbf{k}^T = \mathbf{c}^T = \begin{bmatrix} 1 & 0 \end{bmatrix}$ in (7.3-41), we get

$$\mathbf{x}(\overline{n+1T}) = \left\{ \begin{bmatrix} 1 & 0.632 \\ 0 & 0.368 \end{bmatrix} - \begin{bmatrix} 0.368 \\ 0.632 \end{bmatrix} \begin{bmatrix} 1 & 0 \end{bmatrix} \right\} \begin{bmatrix} x_1(nT) \\ x_2(nT) \end{bmatrix} + \begin{bmatrix} 0.368 \\ 0.632 \end{bmatrix} r(nT)$$

or

$$\mathbf{x}(\overline{n+1T}) = \begin{bmatrix} 0.632 & 0.632 \\ -0.632 & 0.368 \end{bmatrix} \begin{bmatrix} x_1(nT) \\ x_2(nT) \end{bmatrix} + \begin{bmatrix} 0.368 \\ 0.632 \end{bmatrix} r(nT)$$

Using the same values for $\mathbf{A}_d$, $\mathbf{h}(T)$, and $\mathbf{k}$ in (7.3-42) gives

$$T(z) = \begin{bmatrix} 1 & 0 \end{bmatrix} \left[\begin{bmatrix} z-1 & -0.632 \\ 0.632 & z-0.368 \end{bmatrix} + \begin{bmatrix} 0.368 \\ 0.632 \end{bmatrix} \begin{bmatrix} 1 & 0 \end{bmatrix} \right]^{-1} \begin{bmatrix} 0.368 \\ 0.632 \end{bmatrix}$$

which may be evaluated to give (7.3-50).

Discrete Model

In order to develop the discrete model, we first obtain

$$Z\left\{ \frac{1-e^{-sT}}{s} \frac{1}{s(s+1)} \right\} = \frac{0.368(z+0.717)}{(z-1)(z-0.368)}$$

This is now shown in Fig. 7.3-8. We note that x_2 is not the same now as it was in the case of the continuous model.

We can now write the state-transition equations as

$$x_1(\overline{n+1T}) = 0.368x_1(nT) + x_2(\overline{n+1T}) + 0.717x_2(nT)$$

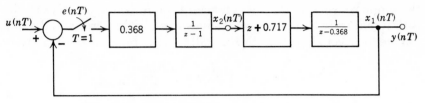

Figure 7.3-8   Discrete model of system of Example 7.3-1.

$$x_2(\overline{n+1}T) = x_2(nT) + 0.368e(nT)$$
$$e(nT) = r(nT) - x_1(nT)$$

and

$$y(nT) = x_1(nT)$$

Solving these equations for $x_1(\overline{n+1}T)$ and $x_2(\overline{n+1}T)$ gives

$$x_1(\overline{n+1}T) = 1.717x_2(nT) + 0.368r(nT)$$
$$x_2(\overline{n+1}T) = -0{\cdot}368x_1(nT) + x_2(nT) + 0.368r(nT)$$

or

$$\mathbf{x}(\overline{n+1}T) = \begin{bmatrix} 0 & 1.717 \\ -0.368 & 1 \end{bmatrix} \mathbf{x}(nT) + \begin{bmatrix} 0.368 \\ 0.368 \end{bmatrix} r(nT)$$

$$= \tilde{\mathbf{A}}_d\mathbf{x}(nT) + \mathbf{b}r(nT)$$

and

$$y(nT) = \begin{bmatrix} 1 & 0 \end{bmatrix}\mathbf{x}(nT) = \mathbf{c}^T\mathbf{x}(nT) = x_1(nT)$$

Since the input is a step $r(nT) = 1$ for $n \geqslant 0$, the solution is obtained again using (7.3-19):

$$\mathbf{x}(nT) = \tilde{\mathbf{A}}_d^n\mathbf{x}(0) + \sum_{k=0}^{n-1} \tilde{\mathbf{A}}_d^{n-1-k}\mathbf{b}$$

where

$$\tilde{\mathbf{A}}_d^n = Z^{-1}\{z(z\mathbf{I} - \tilde{\mathbf{A}}_d)^{-1}\}$$

$$= Z^{-1}\left\{\begin{bmatrix} \dfrac{z(z-1)}{z^2 - z + 0.632} & \dfrac{1.717z}{z^2 - z + 0.632} \\ \dfrac{-0.368z}{z^2 - z + 0.632} & \dfrac{z^2}{z^2 - z + 0.632} \end{bmatrix}\right\}$$

$$= (0.795)^{n-1}\begin{bmatrix} \begin{array}{l} -1.023\cos{(n-1)51°} \end{array} & \begin{array}{l} 1.717\cos{(n-1)51°} \\ +1.389\sin{(n-1)51°} \end{array} \\ \begin{array}{l} -0.318\cos{(n-1)51°} \\ -0.298\sin{(n-1)51°} \end{array} & \begin{array}{l} \cos{(n-1)51°} \\ -0.213\sin{(n-1)51°} \end{array} \end{bmatrix}$$

We emphasize again that the choice of states for the discrete model is different from that of the continuous model. The error and the output are the same; however, the overall transfer function is given by

$$T(z) = \mathbf{c}^T [z\mathbf{I} - \tilde{\mathbf{A}}_d]^{-1}\mathbf{b} = \frac{0.368z + 0.264}{z^2 - z + 0.632}$$

This checks with the result in (7.2-50). In the discrete model case we do not get information on the response in between-sampling instants.

In the solution obtained above, the major difficulty is a numeric one, associated with the evaluation of $\mathbf{e}^{\mathbf{A}T}$ or $\tilde{\mathbf{A}}_d^n$. Rather than using Laplace and Z-transforms, these can be obtained numerically by use of the computer.

Stability and Related Topics

We have seen that the closed-loop transfer function of a SDS can be represented in terms of the system matrix, input and output vectors as shown in (7.3-36) and (7.3-42). The closed-loop characteristic equation can be seen from the denominator of (7.3-40), or from (7.3-42) directly, to be

$$\det [z\mathbf{I} - \mathbf{A}_d + \mathbf{h}(T)\mathbf{k}^T] = 0 \qquad (7.3\text{-}51)$$

If the system has no feedback, we have $k = 0$ and (7.3-51) simplifies to

$$\det [z\mathbf{I} - \mathbf{A}_d] = 0 \qquad (7.3\text{-}52)$$

After evaluating the determinant in (7.3-51), we can apply the standard bilinear transformation method or the Jury criterion to determine if there are any roots outside the unit circle. This determines whether or not the system is stable.

NYQUIST CRITERION, BODE PLOTS, AND THE ROOT LOCUS METHODS

All these methods depend on the knowledge of the open-loop transfer function. The open-loop transfer function in terms of system matrix, input and output vectors has been given for the state variable feedback SDS of Fig. 7.3-5 in (7.3-38) as

$$GH(z) = \mathbf{k}^T [z\mathbf{I} - \mathbf{A}_d]^{-1}\mathbf{h}(T) \qquad (7.3\text{-}53)$$

where $\mathbf{A}_d$, $\mathbf{h}(T)$, and k are defined in (7.3-30), (7.3-29), and (7.3-26), respectively. With the open-loop transfer function as given in (7.3-53), standard procedures can be used to apply the Nyquist criterion for stability using the frequency response in the form of Bode plots or Nichols charts, or to make root locus plots.

Error Constants K_p, K_v, K_a for Unity-Feedback Sampled-Data System

In the development of the relationship of K_p, K_v, K_a to the system matrix, input and output vectors, we shall consider the unity-feedback case first.

For unity-feedback systems under consideration here, we can define

$$K_m \triangleq \lim_{z \to 1} \frac{(z-1)^m}{T^m} [1 + G(z)] \tag{7.3-55}$$

where

$$K_0 = K_p + 1, \qquad K_1 = K_v, \qquad \text{and} \qquad K_2 = K_a$$

Here, K_p, K_v, and K_a are the error constants as defined in (6.3-7), (6.3-12), and (6.3-22), respectively. For unity feedback, $\mathbf{k} = \mathbf{c}$ in (7.3-36), and from (7.3-37)

$$G(z) = \mathbf{c}^T [z\mathbf{I} - \mathbf{A}_d]^{-1} \mathbf{h}(T) \tag{7.3-56}$$

Hence

$$K_m = \lim_{z \to 1} \frac{(z-1)^m}{T^m} [1 + \mathbf{c}^T [z\mathbf{I} - \mathbf{A}_d]^{-1} \mathbf{h}(T)] \tag{7.3-57}$$

Using (7.2-106) and multiplying and dividing by $\det [z\mathbf{I} - \mathbf{A}_d]$,

$$[1 + \mathbf{c}^T [z\mathbf{I} - \mathbf{A}_d]^{-1} \mathbf{h}(T)] = \frac{\det [z\mathbf{I} - \mathbf{A}_d + \mathbf{h}(T)\mathbf{c}^T]}{\det [z\mathbf{I} - \mathbf{A}_d]} \tag{7.3-58}$$

Hence

$$K_m = \lim_{z \to 1} \frac{(z-1)^m}{T^m} \frac{\det [z\mathbf{I} - \mathbf{A}_d + \mathbf{h}(T)\mathbf{c}^T]}{\det [z\mathbf{I} - \mathbf{A}_d]} \tag{7.3-59}$$

If the SDS is of type 0, then the position constant can be determined by letting $m = 0$ and $z = 1$ in (7.3-59). We get

$$K_0 = 1 + K_p = \frac{\det [\mathbf{I} - \mathbf{A}_d + \mathbf{h}(T)\mathbf{c}^T]}{\det [\mathbf{I} - \mathbf{A}_d]} \tag{7.3-60}$$

When the system is of type 1, we have a factor $(z-1)$ in the expansion of $\det [z\mathbf{I} - \mathbf{A}_d]$ which must be removed before letting $z = 1$. Thus we get

$$K_v = \lim_{z \to 1} \frac{(z-1)}{T} \frac{\det [\mathbf{I} - \mathbf{A}_d + \mathbf{h}(T)\mathbf{c}^T]}{\det [z\mathbf{I} - \mathbf{A}_d]} \tag{7.3-61}$$

Similarly, if the system is of type 2, we will have a factor $(z-1)^2$ in the expansion $\det [z\mathbf{I} - \mathbf{A}_d]$. So we get

$$K_a = \lim_{z \to 1} \frac{(z-1)^2}{T^2} \frac{\det [\mathbf{I} - \mathbf{A}_d + \mathbf{h}(T)\mathbf{c}^T]}{\det [z\mathbf{I} - \mathbf{A}_d]} \tag{7.3-62}$$

Thus the procedure for determining the type of system and the finite nonzero error constant is first to expand $\det[z\mathbf{I} - \mathbf{A}_d]$ and to check for roots at $z = 1$. The number of roots at $z = 1$ gives the type of the system. Once the type is known, we can determine the error constant by (7.3-60), (7.3-61), or (7.3-62).

In order to determine the number of roots of $\det[z\mathbf{I} - \mathbf{A}_d]$ at $z = 1$, we determine the rank of matrix $[\mathbf{I} - \mathbf{A}_d]$. Then

$$(\text{roots at } z = 1) = (\text{order} - \text{rank})$$

Example 7.3-2

The state-transition equations of a SDS are given by

$$x(\overline{n+1T}) = \begin{bmatrix} 1 & 0 \\ 0 & e^{-aT} \end{bmatrix} x(nT) + \begin{bmatrix} \dfrac{\beta T}{a} \\ \dfrac{\beta(1 - e^{-aT})}{a^2} \end{bmatrix} u(nT)$$

and

$$y(nT) = [1 \quad 1]x(nT)$$

Find the system type and evaluate the nonzero finite error constant if we use the given system in a unity-feedback configuration.

$$[\mathbf{I} - \mathbf{A}_d] = \begin{bmatrix} 0 & 0 \\ 0 & 1 - e^{-aT} \end{bmatrix}$$

The rank of the preceding matrix is one. Therefore the number of roots of $\det[z\mathbf{I} - \mathbf{A}_d]$ at $(z = 1) = (\text{order} - \text{rank}) = 1$. Hence the system is of type 1. The nonzero finite error constant is then the velocity error constant K_v given by

$$K_v = \lim_{z \to 1} \frac{z-1}{T} \frac{\det[\mathbf{I} - \mathbf{A}_d + \mathbf{h}(T)\mathbf{c}^T]}{\det[z\mathbf{I} - \mathbf{A}_d]}$$

$$= \frac{\det \begin{bmatrix} \beta T/a & \beta T/a \\ \beta(1 - e^{-aT})/a^2 & 1 - e^{-aT} + \beta(1 - e^{-aT})/a^2 \end{bmatrix}}{T(1 - e^{-aT})}$$

$$= \frac{\beta}{a}$$

The Steady-State Error Constants for the State Variable Feedback Sampled-Data System

For the SVF SDS we may define the error, as in the continuous-time case, by

$$e(t) \triangleq r(t) - y(t) \qquad (7.3\text{-}63)$$

where $r(t)$ and $y(t)$ are defined as in Fig. 7.3-5. Then, at the sampling instants

$$e(nT) = r(nT) - y(nT)$$

which, after taking the Z-transform, gives

$$E(z) = R(z) - Y(z)$$

$$= R(z)\left[1 - \frac{Y(z)}{R(z)}\right] \qquad (7.3\text{-}64)$$

Defining steady-state error now by

$$e_{ss} \triangleq \lim_{n \to \infty} e(nT) \qquad (7.3\text{-}65)$$

we may apply the final-value theorem of the Z-transform to (7.3-64) to get

$$e_{ss} = \lim_{z \to 1} (z-1)E(z)$$

$$= \lim_{z \to 1} (z-1)R(z)\left[1 - \frac{Y(z)}{R(z)}\right] \qquad (7.3\text{-}66)$$

Now using $Y(z)/R(z)$ from (7.3-42) in (7.3-66) and by a development that parallels step by step the development for continuous-time systems in Section 7.2, we may determine the steady-state error and the corresponding error constants for unit step, unit ramp, and unit acceleration inputs as

$$e_{ss}(\text{step}) \triangleq \frac{1}{1+K_p} = \frac{\det[\mathbf{I} - \mathbf{A}_d + \mathbf{h}(T)\mathbf{k}^T - \mathbf{h}(T)\mathbf{c}^T]}{\det[\mathbf{I} - \mathbf{A}_d + \mathbf{h}(T)\mathbf{k}^T]} \qquad (7.3\text{-}67)$$

$$e_{ss}(\text{ramp}) \triangleq \frac{1}{K_v} = T\left[\frac{\det[[\mathbf{I} - \mathbf{A}_d + \mathbf{h}(T)\mathbf{k}^T]^2 + \mathbf{h}(T)\mathbf{c}^T]}{\det^2[\mathbf{I} - \mathbf{A}_d + \mathbf{h}(T)\mathbf{k}^T]} - 1\right] \qquad (7.3\text{-}68)$$

$$e_{ss}(\text{accel}) \triangleq \frac{1}{K_a} = T^2\left[1 - \frac{\det[[\mathbf{I} - \mathbf{A}_d + \mathbf{h}(T)\mathbf{k}^T]^3 + \mathbf{h}(T)\mathbf{c}^T]}{\det^3[\mathbf{I} - \mathbf{A}_d + \mathbf{h}(T)\mathbf{k}^T]}\right] \qquad (7.3\text{-}69)$$

By comparing (7.3-67), (7.3-68), and (7.3-69), with similar results for the continuous-time case as seen in (7.2-108), (7.2-117), and (7.2-120), respectively, we see that the only real difference is that $\mathbf{I} - \mathbf{A}_d$ in the SDS results replaces $-\mathbf{A}$ in the continuous-time results. The extra $\mathbf{I}$ arises from the fact that in the final-value theorem for the Z-transform we take $\lim z \to 1$, whereas for the final-value theorem for the $\mathcal{L}$-transform we take $\lim s \to 0$. Equations (7.3-67), (7.3-68), and (7.3-69) are in a fairly convenient form for computation in that all that is involved is matrix and vector multiplication and addition and the taking of determinants. All these operations are readily accomplished by digital computer, especially since none of the determinants need be expanded in terms of the variable z.

Example 7.3-3

As an example of the use of these formulas to find an error constant let us use them on the system equations of Example 7.3-2. For this case

$$\mathbf{A}_d = \begin{bmatrix} 1 & 0 \\ 0 & e^{-aT} \end{bmatrix}$$

$$\mathbf{h}(T) = \begin{bmatrix} \beta T/a \\ \beta(1-e^{-aT})/a^2 \end{bmatrix}$$

$$\mathbf{c} = \begin{bmatrix} 1 \\ 1 \end{bmatrix}$$

Since the system here is assumed to be a unity-feedback one, $\mathbf{k} = \mathbf{c}$. Using this fact in (7.3-67), we have

$$e_{ss}(\text{step}) = \frac{1}{1+K_p} = \frac{\det[\mathbf{I}-\mathbf{A}_d]}{\det[\mathbf{I}-\mathbf{A}_d+\mathbf{h}(T)\mathbf{c}^T]}$$

We know from Example 7.3-2 (and by inspection) that $\det[\mathbf{I}-\mathbf{A}_d]=0$ and $\det[\mathbf{I}-\mathbf{A}_d+\mathbf{h}(T)\mathbf{c}^T] \neq 0$; therefore $e_{ss}(\text{step})=0$, which implies $K_p \to \infty$. Since K_p is infinite, we may use (7.3-68) to determine K_v [if K_p were finite, (7.3-68) and (7.3-69) would be invalid for this system]. Equation (7.3-68) gives

$$e_{ss}(\text{ramp}) = \frac{1}{K_v} = T\frac{\det\left[\begin{bmatrix} \beta T/a & \beta T/a \\ \beta X/a^2 & \beta X/a^2+X \end{bmatrix}^2 + \begin{bmatrix} \beta T/a & \beta T/a \\ \beta X/a^2 & \beta X/a^2 \end{bmatrix}\right]}{\det^2\begin{bmatrix} \beta T/a & \beta T/a \\ \beta X/a^2 & \beta X/a^2+X \end{bmatrix}} - T$$

where $X = 1-e^{-aT}$. From routine computation,

$$e_{ss}(\text{ramp}) = \frac{1}{K_v} = \frac{a}{\beta}$$

from which $K_v = \beta/a$, which checks with the result in Example 7.3-2. Since $e_{ss}(\text{ramp})$ is nonzero, we now know that $K_a = 0$ and (7.3-69) does not apply for this system. In general, for an arbitrary system for which we wish to find the applicable error constant, we start evaluating

(7.3-67), (7.3-68), and (7.3-69) in order. We stop at the first one that gives a nonzero steady-state error. Steady-state error for the higher-order inputs will be infinite, and the corresponding higher-order error constants hence will be zero. Equations (7.3-67), (7.3-68), and (7.3-69) appear more complicated than the equivalent results as given by (7.3-62); however, they generally do not involve expanding a determinant as a function of the variable z. This usually means less effort in evaluating the expressions involved. Again, if a computer is available, evaluation of (7.3-67), (7.3-68), and (7.3-69) is routine where the terms involved are known in numerical form.

PROBLEMS

7.1. Give three state variable formulations for systems whose transfer functions are given by the following:

$$\text{(a) } T(s) = \frac{K}{(s+a)^3}$$

$$\text{(b) } T(s) = \frac{\omega}{s^2+\omega^2}$$

$$\text{(c) } T(s) = \frac{\omega_n^2}{s^2+2\zeta\omega_n s+\omega_n^2}$$

$$\text{(d) } T(s) = \frac{(s+a)\omega_n^2/a}{s^2+2\zeta\omega_n s+\omega_n^2}$$

$$\text{(e) } T(s) = \frac{(s+a)(s+b)(s+c)}{(s+d)(s+e)(s+f)(s+g)}$$

7.2. Check all the four properties of the exponential matrix as developed in the text for e^{At} of Example 7.2-3.

7.3. Show that
$$\mathbf{x}(t) = e^{At}\mathbf{x}(0) + \int_0^t e^{A(t-\tau)}\mathbf{b}u(\tau)\,d\tau$$
satisfies $\dot{\mathbf{x}} = A\mathbf{x} + \mathbf{b}u$.

7.4. Show that
$$e^{A(t_1+t_2)} = e^{At_1}e^{At_2} = e^{At_2}e^{At_1}$$

7.5. Show that $e^{x+y} \neq e^x e^y$ unless $XY = YX$.

7.6. Determine the transfer function for the following systems. Also draw the state variable diagrams for each system.

$$\text{(a) } \frac{d}{dt}\begin{bmatrix} x_1 \\ x_2 \\ x_3 \end{bmatrix} = \begin{bmatrix} -1 & -4 & -1 \\ -1 & -6 & -2 \\ -1 & -2 & -3 \end{bmatrix}\begin{bmatrix} x_1 \\ x_2 \\ x_3 \end{bmatrix} + \begin{bmatrix} 0 \\ 1 \\ 1 \end{bmatrix}u, \quad y = \begin{bmatrix} 1 & 1 & 1 \end{bmatrix}\begin{bmatrix} x_1 \\ x_2 \\ x_3 \end{bmatrix}$$

$$\text{(b) } \dot{x}_1 = Ku, \quad \dot{x}_2 = x_1 - x_2, \quad y = x_2$$

(c) $\begin{bmatrix} \dot{x}_1 \\ \dot{x}_2 \end{bmatrix} = \begin{bmatrix} -7 & 3 \\ -14 & 6 \end{bmatrix} \begin{bmatrix} x_1 \\ x_2 \end{bmatrix} + \begin{bmatrix} K \\ 3K \end{bmatrix} u,$ $y = [3/2 \quad -1/2] \begin{bmatrix} x_1 \\ x_2 \end{bmatrix}$

(d) $\dfrac{d}{dt} \begin{bmatrix} x_1 \\ x_2 \\ x_3 \end{bmatrix} = \begin{bmatrix} -2 & 0 & 0 \\ -1 & -4 & 0 \\ -1 & -1 & -5 \end{bmatrix} \begin{bmatrix} x_1 \\ x_2 \\ x_3 \end{bmatrix} + \begin{bmatrix} 1 \\ 1 \\ 1 \end{bmatrix} u,$ $y = \dfrac{40}{18} x_3 + \dfrac{40}{18} u$

(e) $\dfrac{d}{dt} \begin{bmatrix} x_1 \\ x_2 \\ x_3 \end{bmatrix} = \begin{bmatrix} -2 & 0 & 0 \\ -1 & -4 & 0 \\ -1 & -1 & -5 \end{bmatrix} \begin{bmatrix} x_1 \\ x_2 \\ x_3 \end{bmatrix} + \begin{bmatrix} 1 \\ 1 \\ 1 \end{bmatrix} u,$ $y = \dfrac{40}{3} x_3$

(f) $\dfrac{d}{dt} \begin{bmatrix} x_1 \\ x_2 \end{bmatrix} = \begin{bmatrix} 0 & \omega \\ -\omega & 0 \end{bmatrix} \begin{bmatrix} x_1 \\ x_2 \end{bmatrix} + \begin{bmatrix} 1 \\ 1 \end{bmatrix} u,$ $y = x_1$

7.7. Determine e^{At} for the system matrices of Problem 7.6.

7.8. Determine the stability of the systems, using the **A** matrices as given in Problem 7.6.

7.9. Find expression for $x(t)$ for input $u(t)$, initial condition $x(0)$ for the systems of Problem 7.6.

7.10. For a unity negative-feedback connection ($k = c$) determine the steady-state error constants from the state variable formulations of Problem 7.6.

7.11. For the multivariable system shown in Fig. P7.11 write the state equations and hence solve for the state vector **x** and output vector. Assume step inputs and zero initial conditions.

7.12. Make at least one discrete state variable diagram and write the state transition equation for following systems:

(a) $\dfrac{z(2z^2 - 3z + 7)}{z^3 + 2z^2 + 3z + 0.5}$

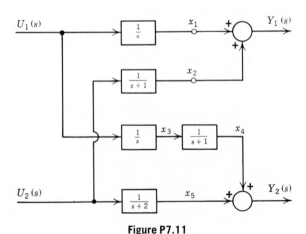

Figure P7.11

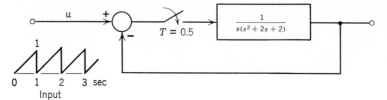

Figure P7.12

(b) $\dfrac{z(z-1)}{(z-0.7)(z+0.9)}$

(c) $\dfrac{0.2(z-0.3)}{z^2+0.5z+0.1}$

    (d) See system of Fig. P7-12.

7.13. For the sampled-data system shown in Fig. P7.13:
    (a) Construct the state variable diagram based on the continuous model and establish the state vector at sampling instants and between sampling instants. Determine also the overall transfer function.
    (b) Construct the discrete model and find the state vector at sampling instants. Find the overall transfer function.
    (c) Establish stability using the system matrix from (b).
    (d) Find the velocity error constant.

7.14. Given the sampled-data system shown in Fig. P7.14. Develop the state-transition equations from the continuous model and find the state vector $x(nT)$. Plot the sampled output. Also establish stability and find the finite error constant.

7.15. The open-loop transfer function of a unity-feedback sampled-data system is given by

$$G(s) = \frac{2(1-e^{-sT})}{s(s+1)(s+2)}, \qquad T = 1\ \text{sec}$$

Determine the output at sampling instants and also between sampling instants and determine the finite error constant. Assume zero initial conditions, step input and use the state variable approach.

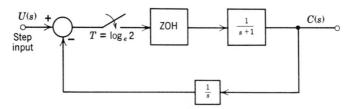

Figure P7.13

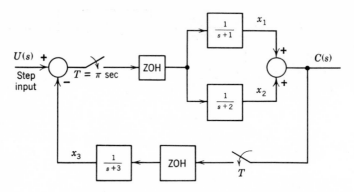

Figure P7.14

7.16. The discrete model of a sampled-data feedback system leads to the following set of equations:

$$x_1(n+1) = x_2(n)$$
$$x_2(n+1) = x_1(n) + 3x_2(n) + u(n)$$
$$c(n) = x_1(n) + x_2(n)$$

Given zero initial conditions and step input, find the output at discrete instants of time.

8

CLASSICAL COMPENSATION OF CONTINUOUS-TIME CONTROL SYSTEMS

8.1 INTRODUCTION TO CASCADE COMPENSATION

Basic Considerations

The notions of damping factor, overshoot, phase and gain margins, and resonant frequency were introduced in Chapters 2 and 3. These have a direct bearing on the stability and other specifications of a dynamic system. The object of providing compensation is to prevent any abnormal values of these factors from adversely affecting the overall efficiency of the system. For an analysis of such compensations, one visualizes the complete time response of the system to be broken up into two parts, the transient response and the steady-state response. Any network or subsystem designed to correct system behavior may be cascaded with the plant in question to achieve the compensation.

The transient state of a physical dynamic system exists when energy conditions of one steady-state are being changed to those of a second steady-state. Such a state has no response in the limit as time goes to infinity. In contrast to this, a dynamic system is in steady-state when variables describing its behavior are either time invariant or periodic functions of time. Classical methods of compensation involve different design considerations for the improvement of these two responses. We begin with the transient response.

TRANSIENT RESPONSE   *Lead Comp.*

The transient response is judged on the basis of how well the output of the system tracks a unit step input during the transient portion of the response. Ideally, the output of the system would instantaneously change from its quiescent value to its steady-state or final value when a step input was applied. Real systems generally do not exhibit this type of response. Realistically, the output requires some time to come up to the final value when excited by a step input, and it may not settle to this final value at once on reaching it. Thus, as far as the transient response of a real system is concerned, it has been found that a

310

very satisfactory and realizable alternative to this ideal response is the step response for a second-order system with damping factor ζ in the range $0.3 \leq \zeta \leq 0.8$. These responses are seen in Fig. 2.3-3. This means that we generally wish to have an overshoot somewhere between 5% and 35% which decays to the final value quickly after the overshoot. More overshoot than this usually means that the system is too oscillatory and that more damping is required; less overshoot or no overshoot at all means that the system is sluggish and the response may be speeded up by decreasing the damping.

The second-order system response with damping in the desired range gives the general form of the step response desired. The speed of the response is measured by the time to the first peak, t_p, or the settling time, t_s. Settling time is most conveniently taken as <u>four time constants</u>, when a time constant can be defined for the system's response, or the time it takes after the step input is applied for the output to settle to within 2% at its final value.[1] Of course, how fast the system has to respond and hence what t_p or t_s has to be depends entirely upon the particular system application, so no general guidelines on these can be set. We assume that the system designer knows by some means what the speed of response has to be and can from this determine suitable values for t_p and t_s.

STEADY-STATE RESPONSE   2 pg Comp.

The steady-state response is judged entirely on the basis of the magnitude of the steady-state error constants K_p, K_v, and K_a, which were discussed in Section 3.3. We recall that for a unity-feedback closed-loop system, the steady-state error to a unit step, unit ramp,

---

[1] We note that for a second-order system, four time constants take the output to 1.8% of its final value.

and unit acceleration input are given by

$$e_{ss}(\text{step}) = \frac{1}{1+K_p}$$

$$e_{ss}(\text{ramp}) = \frac{1}{K_v}$$

$$e_{ss}(\text{accel}) = \frac{1}{K_a}$$

for a type 0, 1, or 2 system, respectively. It is readily seen that the larger the steady-state error constant K_p, K_v, or K_a for a type 0, 1, or 2 system, respectively, the smaller will be the respective error to a unit step, ramp, or parabolic input. Thus the larger the error constant, the better the system will perform insofar as the steady-state portion of the response. Our efforts in trying to improve the steady-state portion of the response are hence directed at increasing the magnitude of the error constant of the particular system.

Why do we choose to judge the response of systems solely on the basis of the step response and the magnitude of the steady-state error constant? Empirically, it has been found that if the step response of a system is acceptable and the steady-state error constant is brought up high enough, the system generally will behave well to any input. By behaving well, we mean that the system will track almost any input well. Moreover, the step response and the steady-state error constants provide a criterion we can relate to the compensator parameters fairly easily. That is, our design methods work well with the step response and the steady-state error constants as measures, so we use them. Intuitively, this should not be too surprising.

We shall also see that in the case of the design of compensators by frequency-response methods our basic method is to shape the overall response of the system in such a way that it looks like the frequency response of a system which has a good second-order step response. Thus our methods are indirect to this extent. Again we find that a system with a frequency response approximately equal to that of a second-order system with a good step response generally has a time response that is satisfactory for any input.

The Basic Idea in Classical Compensation

The basic unity-feedback system of the general form shown in Fig. 8.1-1 is termed the *compensated system*. The plant, with transfer function $G_f(s)$, is operated in closed-loop fashion with the compensator,

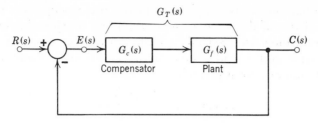

Figure 8.1-1 The compensated system.

with transfer function $G_c(s)$, in cascade and preceding the plant. This is not the only way that systems may be compensated. For some applications it is desirable, if not necessary, to put the compensator in the feedback path. However, forward-path cascade compensation very well illustrates what may be accomplished with simple classical compensation, and since that is the primary objective here we shall not consider feedback compensation.

We note that the overall open-loop transfer function, $G_T(s)$, for the system of interest here (Fig. 8.1-1) is given by

$$G_T(s) = G_c(s)G_f(s) \tag{8.1-1}$$

and that the closed-loop transfer function is given by

$$T(s) = \frac{C(s)}{R(s)} = \frac{G_c(s)G_f(s)}{1 + G_c(s)G_f(s)} \tag{8.1-2}$$

The open-loop poles (zeros) of our system are the poles (zeros) of the compensator, $G_c(s)$, and the plant, $G_f(s)$. The characteristic polynomial (the denominator of closed-loop transfer function) for the closed-loop system is seen from (8.1-2) to be given by the numerator of $1 + G_c(s)G_f(s)$ after this expression is rationalized.

For classical compensation, a compensator of the general form

$$G_c(s) = A\frac{(s-z)}{(s-p)} \tag{8.1-3}$$

is assumed and the problem reduces itself to determining parameters A, z, and p. As will be seen later, z is usually set by rule of thumb so that the problem is further reduced to determining only A and p. In general, if $z > p$, i.e., if the zero is to the right of the pole on the real axis, then the compensator is termed a lead compensator. We will see later that a lead compensator is used to improve the transient response of the system. On the other hand, if $p > z$, i.e., if the pole is to the right of the zero on the real axis, the compensator is termed a lag compensator.

Lag compensators are used to improve the steady-state response characteristics of the system. In addition, we consider lag-lead compensators which are basically simple series combinations of a lag compensator and a lead compensator. The transfer function of the lag-lead compensator is given by

$$G_c(s) = A \frac{(s-z_1)\ (s-z_2)}{(s-p_1)\ (s-p_2)} \tag{8.1-4}$$

where now $z_1 < p_1$ and $p_2 < z_2$, so that we have both a lag and a lead section. For reasons of ease of realizability we also shall require that $z_1 z_2 = p_1 p_2$. Use of the lag-lead compensator usually improves both the transient response and the steady-state performance of the system.

In all three types of compensator it is, in general, desirable that the compensator itself be a stable one, so all compensator poles will fall in the LHP. This means, first, that the compensators may be physically realized with a passive network (in series with an amplifier with a variable gain) and, second, that if the plant is open-loop stable, then the whole system will be open-loop stable. If the system is operated closed-loop, it must of course be closed-loop stable. In fact, this is a primary requirement. In most systems it is also desirable to have open-loop stability, so that if the loop is accidentally opened perhaps through some malfunction or failure, the system will not "run away". This may or may not be important, depending upon the individual case; however, it occurs often enough so that it is reasonable to require that the compensator have only LHP poles.

Let us now look in somewhat more detail at just how it is that a lead compensator can be expected to improve the transient response while a lag compensator improves the steady-state error characteristics of the system.

LEAD COMPENSATION

To see how we accomplish improvement of the transient response through the use of lead compensation, consider the root-locus diagram for a three-pole open-loop system shown in Fig. 8.1-2a. We see a root locus with two branches which go off into the RHP as $K \to \infty$. Thus for high values of K, the closed-loop system will be unstable. Also shown in the figure is the line in the LHP along which the damping factor $\zeta = 0.707$. This value of ζ is approximately what one might desire for a control system, so let us assume that this is the value that we want. For this value of ζ, if K is adjusted properly (K^1 as shown in the figure), we may achieve a resonant frequency ω_n, shown as ω_{n_1}

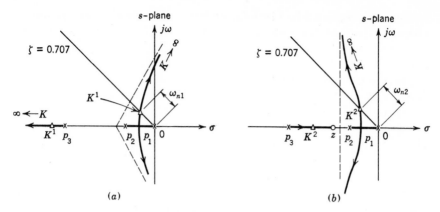

Figure 8.1-2 (a) A typical uncompensated root locus. (b) Root locus compensated through the addition of a single zero.

in the figure. For this value of $K(K^1)$, the step response of the system may be expected to be acceptable since the third, closed-loop pole is far into the LHP relative to the two complex poles and hence the complex poles will dominate. In fact, the step response may be expected to be very close to that of the two-pole, no-zero system which is seen in Fig. 2.3-3. The only question that could be expected to arise about the transient (step) response is whether or not the value of ω_{n_1} is large enough to give a system that responds fast enough. Recall that the time constant (τ) for the response is given by

$$\tau = \frac{1}{\zeta \omega_n}$$

Thus the larger the value of ω_n, the faster the response of the system. If we find that the value of ω_{n_1} as shown in Fig. 8.1-2a is sufficiently large to give a transient response that is fast enough, then the decision would be made, as far as the transient response is concerned, that no compensator is really required. All that is required is that the loop gain (and hence K) be adjusted properly.

Let us assume, though, that the closed-loop system response is not fast enough for the purposes for which the system is to be used, but that the overshoot characteristics are just about what they should be. Thus to improve the response of the system we want to increase ω_n of the dominant poles (this will make for a smaller $\tau = 1/\zeta\omega_n$) and keep $\zeta (=0.707)$ the same.[2] To do this, let us consider adding a

[2]For the system (Fig. 8.1-2a) we see that a larger ω_n could be obtained by increasing K. This would, however, result in a smaller value of $\zeta\omega_n$ for the dominant poles with the result that $\tau = 1/\zeta\omega_n$ would actually increase.

single zero in the midst of the plant poles. This would correspond to adding a compensator of the form

$$G_c(s) = A(s - z) \tag{8.1-5}$$

in cascade (as in Fig. 8.1-1) with the plant for which the root locus is shown in Fig. 8.1-2a. The resulting root locus is shown in Fig. 8.1-2b. The zero z has been chosen so that it falls approximately midway between the poles at the ends of the range of plant poles. Just precisely where it falls is not critical to its general effect on the root locus. The resulting root locus is drawn in Fig. 8.1-2b, where it can be seen that the effect on the root locus has been quite radical. First, all branches of the root locus now lie completely in the LHP, so we can adjust K to any positive value (no matter how large) and the system will still be closed-loop stable. The system thus has an added margin of stability with the additional zero. Moreover, the two branches of the root locus that lie in the complex part of the plane (off the real axis) have been bent away from the $j\omega$-axis. Now if K is adjusted to give the two complex poles with $\zeta = 0.707$ ($K = K^2$ in Fig. 8.1-2b), we can see in Fig. 8.1-2b that ω_n (ω_{n_2} in the figure) is increased. It may be judged from the figure that ω_n (and hence $\zeta\omega_n$) has been increased by a factor of approximately two. Thus $\tau = 1/\zeta\omega_n$ has been decreased by a factor of approximately one-half and we may expect that the overall closed-loop system with the compensator zero added will respond approximately twice as fast as the uncompensated system. In this way, the transient response has been speeded up.

Something else we might note in Fig. 8.1-2b is that besides the two complex poles, we have a third pole, real and closed-loop, lying on the branch between the open-loop pole at p_3 and the zero at z. This pole is, however, sufficiently far out into the LHP (relative to the two complex, closed-loop poles) and near enough the zero at z, which is also a closed-loop zero, that no detrimental effects[3] due to this pole may be expected. Since the zero at z is nearer the $j\omega$-axis than this far LHP pole, one may in fact expect the step response to show a slightly higher overshoot than the uncompensated system (root locus of Fig. 8.1-2a). Again the closed-loop pole is near enough so that this should be a negligible amount, and so the compensation with $G_c(s)$ as given in (8.1-5) should be acceptable.

With lead compensation we try to achieve the effect of adding a simple zero. That is, we try to increase the stability and speed of response of the system. It is, however, difficult to physically realize

---

[3]See the discussion in Section 2.3 on the effect of an additional pole or zero on the real axis on the response of a system with two complex poles.

a compensator which has a simple zero as its transfer function and no pole. Thus for the lead compensator we add an additional compensator pole, putting it far enough into the LHP so that it has a relatively negligible effect upon the root locus in the region where the two dominant closed-loop, complex poles are to occur. For the single-zero compensated system of Fig. 8.1-2*b* we might add a compensator pole (to make the compensator physically realizable) as in Fig. 8.1-3.

Here we see that the root locus has been little changed in the region where the two dominant closed-loop poles lie, hence so far as the transient response is concerned the system will behave approximately the same as with the single-zero compensation proposed in Fig. 8.1-2*b*. However, now we do have the two branches of the root locus bending over into the RHP for *K* large. In this case the value of *K* for which a RHP pole may be had is much higher than the value of *K* for which the uncompensated system (see Fig. 8.1-2*a*) would have a RHP pole. Thus the margin of stability has been increased, and ω_n, the natural frequency, for the compensated system is larger in magnitude so that the response is faster. Hence the addition of the lead compensator is seen to have speeded up the transient response and also increased the margin of stability.

We see that the lead compensator has a pole far into the LHP and a zero closer to the *jω*-axis (to the right of the pole). The usual way of writing the lead compensator transfer function is

$$G_c(s) = A \frac{(s+1/\tau)}{(s+\alpha/\tau)}, \qquad \alpha > 1, \quad A, \tau > 0 \qquad (8.1\text{-}6)$$

We note that $\alpha > 1$ insures that the pole is to the left of the zero at $-1/\tau$.

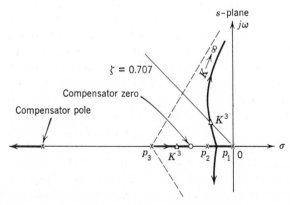

Figure 8.1-3 Root locus of Fig. 8.1-2*a* compensated with a lead compensator.

LAG COMPENSATION

With a lag compensator we may increase the magnitude of the error constant K_p, K_v, or K_a depending upon the type of system at hand. We have seen earlier in this section that steady-state error varies inversely with the error constant. Hence increasing the magnitude of the steady-state error constant decreases the steady-state error. To see how a lag compensator may be used to increase the steady-state error constant, consider the system of Fig. 8.1-1, which is to be compensated. Let us consider the error constant for this system for the uncompensated case, i.e., where $G_c(s) = 1$. Let us assume also that loop gain K has been adjusted to give a satisfactory transient response. The error constant (see Section 3.3) given by

$$K_m = \lim_{s \to 0} s^m G_f(s) \qquad (8.1\text{-}10)$$

where the system $[G_f(s)]$ is of type m.[4] Here $K_m = K_p$ if the system is of type 0 ($m = 0$), $K_m = K_v$ if the system is of type 1 ($m = 1$), and $K_m = K_a$ if the system is of type 2 ($m = 2$). Now let us say that steady-state error is large enough to be unsatisfactory, so that it is desirable to increase the magnitude of the error constant K_m. If we now add a compensator of the form

$$G_c(s) = \frac{A}{s} \qquad (8.1\text{-}11)$$

we will have a compensated error constant ($K_m{}^c$) given by

$$K_m{}^c = \lim_{s \to 0} s^m G_c(s) G_f(s) = \lim_{s \to 0} A \frac{K_m}{s} \to \infty$$

where K_m is as given by (8.1-10). Thus by adding the compensator of (8.1-11) the type of the resulting system has been increased by one and the compensated error constant has been increased to infinity, so the system will have a better steady-state error characteristics. This is of course all very helpful; however, the pole added at the origin means that the resulting system will almost assuredly be closed-loop unstable. To see why this is so, let us consider the case where $G_f(s)$ is such that the root locus for it is as shown in Fig. 8.1-2a. If we add a compensator as in (8.1-11), the resulting $G_T(s) = G_c(s) G_f(s)$ will have a root locus as shown in Fig. 8.1-4. We see that the result is two complete branches of the root locus in the RHP, making the system closed-loop unstable for any value of loop gain. Thus the advantages we thought we had obtained in steady-state error characteristics are

---

[4]Recall that a system is of type m if the corresponding $G(s)$ has m poles at the origin.

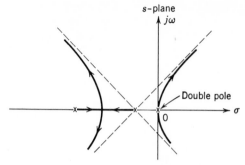

Figure 8.1-4 The root locus of a type 1 system compensated with a compensator $G_c(s) = A/s$.

lost due to instability. The addition of a compensator pole at the origin will not affect so drastically the stability of all such systems, in general it will shift the root locus to the right and will tend to destabilize the system. In any case, the transient response will be degraded.

To get around these detrimental effects of a single compensator pole at the origin, we might add an additional compensator zero very near the pole at the origin and in the LHP. That is, we might add a compensator of the form

$$G_c(s) = \frac{A(s+z)}{s}, \qquad A > 0, \quad z > 0 \tag{8.1-12}$$

where z is a small positive number. For a compensator such as this and with a $G_f(s)$ with root locus as in Fig. 8.1-2a, the resulting root locus for $G_T(s) = G_c(s)G_f(s)$ will be as in Fig. 8.1-5.

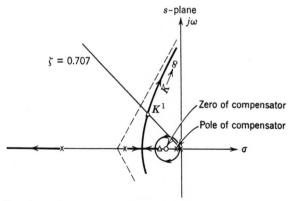

Figure 8.1-5 Root locus for a type 1 system compensator with a lag compensator pole-zero near the origin.

The general character of the root locus has not been changed greatly, although it has been shifted slightly to the right. For the closed-loop system adjusted now so as to have two complex poles with $\zeta = 0.707$ (K^1 in the figure) there will be a closed-loop pole-zero dipole near the origin. The pole and zero here are so close together that the pole is almost cancelled out by the zero and hence the pole will not contribute significantly to the response of the closed-loop system. The response thus still will be dominated by the two complex poles, as desired. Since the pole-zero dipole for the closed-loop system has the zero nearer the origin than the pole (see Fig. 8.1-5), the step-response will be slightly more peaked[5] – there will be more overshoot – than will the same system uncompensated. This can be made a negligible amount though.

The effect on the error constant of a compensator of the form given in (8.1-12) is readily seen by determining the error constant for the compensated system of Fig. 8.1-1. We then have

$$K_m{}^c = \lim_{s\to0} s^m \, G_T(s) = \lim_{s\to0} s^m G_c(s)\, G_f(s) = \lim_{s\to0} \frac{A(s+z)}{s} K_m \to \infty$$

where K_m is the uncompensated error constant as determined by (8.1-10). Again the error constant is infinite, the type of the system has been increased by one through the addition of the compensator pole at the origin.

In actual practice it is difficult to realize a compensator with a single pole at the origin. Consequently, the lag compensator pole is put not at the origin but just slightly to the left of it on the real axis (see Section 8.2). The usual lag compensator then has a transfer function of the form

$$G_c(s) = A \frac{s+1/\tau}{s+1/\alpha\tau}, \qquad A, \tau > 0, \quad \alpha > 1 \qquad (8.1\text{-}13)$$

which then has a zero at $-1/\tau$ and a pole at $-1/\alpha\tau$. Since $\alpha > 1$ we see that the pole is to the right of the zero and nearer the origin than the zero. With a lag compensator as in (8.1-13), the error constant for the system of Fig. 8.1-1 is given by

$$K_m{}^c = \lim_{s\to0} G_c(s)\, G_f(s) = \lim_{s\to0} \frac{A(s+1/\tau)}{s+1/\alpha\tau}\, G_f(s) = A\alpha K_m$$

where K_m is as given in (8.1-10). For the usual case $A \cong 1$, hence

$$K_m{}^c \cong \alpha K_m \qquad (8.1\text{-}14)$$

---

[5]See Section 2.3 for consideration of the effect of a zero on the real axis near the origin on the response of a two complex-pole system.

α gives the approximate factor by which the error constant is improved through the addition of the lag compensator. An α in the range $3 < \alpha \le 10$ is readily accomplished for a realizable compensator, so this is the range of improvement in error constant which can be readily achieved through the use of a lag compensator. We should note that for the compensator of (8.1-13),

$$\frac{z}{p} = \frac{1/\tau}{1/\alpha\tau} = \alpha$$

LAG-LEAD COMPENSATION

As the name implies, the lag-lead compensator is basically just a lag compensator and a lead compensator in series. It has one pole-zero pair corresponding to the lag compensator section and another pole-zero pair corresponding to the lead section of the compensator. A lag-lead compensator is used if both the transient response, which is compensated with the lead section, and the steady-state response, which is compensated with the lag section, require improvement. The transfer function for the lag-lead compensator is of the form

$$G_c(s) = A \underbrace{\frac{(s + 1/\tau_1)}{(s + \alpha/\tau_1)}}_{\substack{\text{Lead} \\ \text{section}}} \underbrace{\frac{(s + 1/\tau_2)}{(s + 1/\alpha\tau_2)}}_{\substack{\text{Lag} \\ \text{section}}}, \qquad A, \tau_1, \tau_2 > 0, \quad \alpha > 1 \qquad (8.1\text{-}15)$$

Although it is not necessary to have the lag section and the lead section with the same value of α, for ease of realization this is convenient.

8.2 REALIZATION OF COMMON COMPENSATORS

Lead Compensator Realization

We shall give here a simple realization for the three types of fixed-form compensators used in this chapter. The realizations suggested here are by no means unique. Although we consider only network realizations, these compensators may be realized by mechanical means through the use of duality relationships. The purpose of our discussion is simply to show that these compensators are easily realizable and to illustrate how a practical realization might look.

To begin, consider the lead compensator. Recall that a lead compensator has a single pole and single zero, with the pole lying to left of the zero on the real axis. The general form of the transfer function for the lead compensator block will be assumed to be of the form

$$G_c(s) = A\frac{(s+1/\tau)}{(s+\alpha/\tau)}, \qquad \alpha > 1, \quad A, \tau > 0 \qquad (8.2\text{-}1)$$

We thus have three parameters that have to be determined to specify the compensators, A, τ, and α. Our realization here for (8.2-1) is shown in Fig. 8.2-1. For this block we see that

$$\frac{E_o(s)}{E_i(s)} = G_c(s) = \frac{AR_2}{R_2 + \dfrac{R_1/sC}{R_1 + 1/sC}} \qquad (8.2\text{-}2)$$

$$= \frac{A(s+1/R_1C)}{s + [(R_1+R_2)/R_2](1/R_1C)} \qquad (8.2\text{-}3)$$

Now by comparing (8.2-3) and (8.2-1) we have

$$\tau = R_1C, \qquad \alpha = \frac{R_1+R_2}{R_2}, \qquad A = A \qquad (8.2\text{-}4)$$

Thus given T, α, and A for a given lead compensator we may use (8.2-4) to find values for R_1, R_2, C, and A of the compensator of Fig. 8.2-1. We note that the realization of Fig. 8.2-1 has four free parameters while the lead compensators (8.2-1) has only three parameters. We thus have an additional degree of freedom in the choice of the compensator for our realization. Usually this extra degree of freedom is used to set the impedance level of the circuit so that it matches the circuit which produces the input voltage [$E_i(s)$ in Fig. 8.2-1] of the compensator.

SOME PRACTICAL CONSIDERATIONS IN THE CHOICE OF α

In choosing values for the compensator it may be seen from (8.2-4) that τ depends solely upon R_1 and C. The range of resistance and

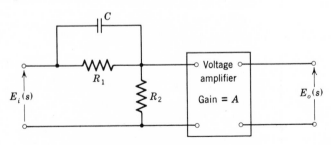

Figure 8.2-1   Lead compensator realization.

capacitance available allows the designer to choose practically any value of τ. One does not have quite as free a choice with the quantity α. Usually α is chosen so that $3 < \alpha \leq 10$. The reasons for limiting α to something equal to or less than 10 are based mainly upon the problem of noise in the control loop. It is a matter of practical reality that every control system has a source of noise somewhere within it. This noise is generally of a high-frequency nature relative to the frequencies of the control signals that appear in the loop. The capacitor C of the lead compensator of Fig. 8.2-1 thus appears almost like a short circuit to the noise, whereas to the control signals it is almost an open circuit. Hence to a fair degree of approximation the lead compensator of Fig. 8.2-1 amplifies the noise by the gain A, while the control signals are amplified by the factor A/α. This means the lead compensator boosts the noise signal level in the loop relative to the control signal and the higher the value of α, the higher this boost. Thus if noise is a problem in a particular system one must be careful about choosing a high α for a lead compensator, hence the upper limit of 10. An α lower than 3 generally will not give enough improvement in the response to warrant the use of a compensator.

Lag Compensator Realization

We recall that a lag compensator is one that has a simple pole and zero on the real axis with the pole to the right of the zero. In general we are interested only in stable compensators, so the pole must be in the LHP. Thus the pole is nearer the origin than the zero. A simple realization for such a compensator is shown in Fig. 8.2-2. For this compensator one may write

$$\frac{E_o(s)}{E_i(s)} = G_c(s) = \frac{A(R_2 + 1/sC)}{R_1 + R_2 + 1/sC}$$

$$= \frac{AR_2}{(R_1 + R_2)} \frac{(s + 1/R_2C)}{[s + R_2/(R_1 + R_2)R_2C]} \qquad (8.2\text{-}5)$$

We wish the lag compensator transfer function to be of the form

$$G_c(s) = A'\frac{(s + 1/\tau)}{(s + 1/\alpha\tau)}, \qquad \alpha > 1, \quad \tau > 0 \qquad (8.2\text{-}6)$$

We compare (8.2-5) and (8.2-6) to get

$$\tau = R_2C, \qquad \alpha = \frac{R_1 + R_2}{R_2}, \qquad A' = \frac{A}{\alpha} = \frac{AR_2}{R_1 + R_2} \qquad (8.2\text{-}7)$$

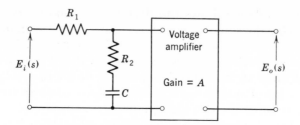

Figure 8.2-2   Lag compensator realization.

Thus given the pole-zero location for a lag compensator, we put the zero at $-1/\tau$ and determine τ. We put the pole at $-1/\alpha\tau$ and determine α and then use (8.2-7) to determine the component values to be used in the realization of Fig. 8.2-2. Again we have four free parameters in the compensator realization and only three parameters in the compensator transfer function, (8.2-6). We now may use the remaining degree of freedom to match the impedance in the circuit in which the compensator is to be used.

The lag compensator, in contrast to the lead compensator, attenuates high-frequency noise signals in the control loop. This may be seen from Fig. 8.2-2 and (8.2-5) if we again take the noise components to be very high in frequency and the control signals of a relatively low frequency. Looking at (8.2-5) we see that the noise signal is amplified approximately by the factor A/α while the control signals are amplified by the factor A. Since $\alpha > 1$ we have an improvement in the ratio of control signal to noise signal in the loop due to the addition of the lag compensator. In fact, if we have a control loop in which noise is a problem, the addition of a lag compensator will attenuate the noise.

In choosing values for τ and α for the lag compensator, considerations are approximately the same as for the lead compensator. τ may be chosen almost arbitrarily; however, α is again usually chosen in the range $3 < \alpha \leqslant 10$.

Lag-Lead Compensator Realization

The lag-lead compensator, as the name implies, is a combination of a lag compensator and a separate lead compensator. The lag section is one real pole and one real zero with the pole to the right of the zero. The lead section is one pole and one zero with the zero to the right of the pole. The lag-lead compensator is thus of second order.

A simple realization for the lag-lead compensator is as shown in

Fig. 8.2-3. To see that this is a lag-lead compensator, we consider the transfer function for the compensator;

$$G_c(s) = \frac{E_o(s)}{E_i(s)} = A\frac{R_2 + 1/sC_2}{R_2 + 1/sC_2 + \dfrac{R_1}{sC_1} \Big/ \Big(R_1 + \dfrac{1}{sC_1}\Big)} \qquad (8.2\text{-}8)$$

After some clearing, this becomes

$$G_c(s) = \frac{A(sR_1C_1 + 1)(sR_2C_2 + 1)}{(sR_1C_1 + 1)(sR_2C_2 + 1) + sR_1C_2}$$

$$= \frac{A(s\tau_1 + 1)(s\tau_2 + 1)}{(s\tau_1 + 1)(s\tau_2 + 1) + s\tau_{12}} \qquad (8.2\text{-}9)$$

where

$$\tau_1 = R_1C_1, \qquad \tau_2 = R_2C_2, \qquad \tau_{12} = R_1C_2 \qquad (8.2\text{-}10)$$

Equation (8.2-9) may now be written into the form

$$G_c(s) = \frac{A(s + 1/\tau_1)(s + 1/\tau_2)}{(s + \alpha/\tau_1)(s + 1/\alpha\tau_2)}, \qquad \alpha > 1, \quad \tau_1, \tau_2 > 0 \quad (8.2\text{-}11)$$

where α is determined as the number that is greater than one which satisfies

$$\frac{\alpha}{\tau_1} + \frac{1}{\alpha\tau_2} = \frac{1}{\tau_1} + \frac{1}{\tau_2} + \frac{\tau_{12}}{\tau_1\tau_2} \qquad (8.2\text{-}12)$$

Equation (8.2-12) is obtained by multiplying out the denominators of (8.2-9) and (8.2-11) and equating the coefficients. We see from (8.2-11) that the lead section of the lag-lead compensator has a zero at $-1/\tau_1$ and pole at $-\alpha/\tau_1$ ($\alpha > 1$ so the pole is to the left of the zero). The lag section has a zero at $-1/\tau_2$ and pole at $-1/\alpha\tau_2$ ($\alpha > 1$ so the pole is to the right of the zero). It may be seen by comparing (8.2-1), (8.2-6), and (8.2-11) that the lag-lead compensator looks exactly like an ordinary lag compensator followed by an ordinary lead compensator

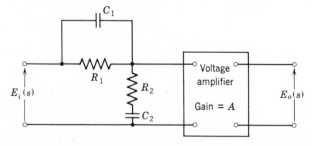

Figure 8.2-3 Lag-lead compensator realization.

(or vice versa) where the lag and lead compensators have the same value of α.

In this case, as in the case of the lead and lag compensators, we have an additional degree of freedom of the choice of circuit parameter values in the realization, so the impedance level may be chosen to match the impedance level of the circuit in which it will have to operate.

With the lag-lead compensator there is in general no decided amplification or attenuation of noise signals relative to the control signals. Thus in this respect nothing will be gained or lost by adding one. As a practical matter, however, α is usually chosen in the range $3 < \alpha \leqslant 10$. τ_1 and τ_2 are limited only by the availability of resistance and capacitance values to be used in the realization.

8.3 ROOT LOCUS COMPENSATION

Lead Compensation

As has been pointed out earlier, lead compensation generally is used to improve the transient response of a system. Now we consider just how the root locus may be used to design a compensator so as to realize a desired improvement in transient response. The basic idea in root locus compensation is to add the compensator so that the resulting closed-loop transfer function has a pair of complex poles positioned in the LHP with a desired damping factor ζ and natural frequency ω_n and with all other poles either near zeros or relatively far into the LHP. This, as will be shown, insures that the two complex

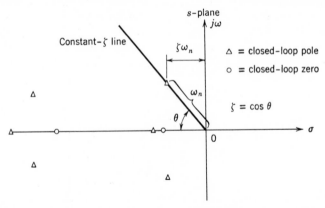

Figure 8.3-1   A desirable closed-loop pole-zero configuration for a feedback control system.

poles with the desired ζ and ω_n will dominate the response: all other poles will contribute a negligible amount to the output of the system. Thus the closed-loop system should respond essentially like the second-order system considered in Section 2.3. To obtain the desired response, we simply must choose ζ and ω_n of the dominant two poles properly. Shown in Fig. 8.3-1 is an example of what we try to achieve in the way of a closed-loop pole-zero configuration for a feedback control system.

We can see that poles far into the LHP and near zeros contribute a negligible amount to the output response by considering a closed-loop transfer function of the form

$$T(s) = \frac{K_T \prod_{i=1}^{m} (s - z_i)}{\prod_{i=1}^{n} (s - p_i)}, \qquad m < n \qquad (8.3\text{-}1)$$

Thus we have m zeros, z_i, and n poles, p_i. Let us consider the step response of this system, which is given by

$$C_{\text{step}}(s) = \frac{1}{s} T(s) = \frac{K_T \prod_{i=1}^{m} (s - z_i)}{s \prod_{i=1}^{n} (s - p_i)} \qquad (8.3\text{-}2)$$

For the sake of simplicity let us assume that the poles are all distinct, so that the RHS of (8.3-2) may be expanded into a partial fraction expansion as follows:

$$C_{\text{step}}(s) = \frac{K_{01}}{s} + \frac{K_{11}}{s + \zeta\omega_n - j\omega_n\sqrt{1-\zeta^2}} + \frac{\overline{K_{11}}}{s + \zeta\omega_n + j\omega_n\sqrt{1-\zeta^2}} + \frac{K_{31}}{s - p_3}$$

$$\cdots + \frac{K_{n1}}{s - p_n} \qquad (8.3\text{-}3)$$

where

$$K_{i1} = \frac{K_T \prod_{j=1}^{m} (p_i - z_j)}{p_i \prod_{\substack{j=1 \\ j \neq i}}^{n} (p_i - p_j)}$$

The pair of complex poles that should dominate the response have been taken to be poles p_1 and p_2 [of (8.3-2)], which here have the real and imaginary parts written out in detail in terms of ζ and ω_n. Now the time response is what we are really interested in, so inverting (8.3-3) term by term we have

$$c_{\text{step}}(t) = [K_{01} + 2|K_{11}|e^{-\zeta\omega_n t} \cos(\omega_n\sqrt{1-\zeta^2}t + \angle K_{11}) +$$
$$K_{31}e^{p_3 t} + \cdots + K_{n1}e^{p_n t}] 1(t) \quad (8.3\text{-}4)$$

If we want this response to be dominated by the complex poles $(-\zeta\omega_n \pm j\omega_n\sqrt{1-\zeta^2})$, we must in general arrange to have

$$2|K_{11}|e^{-\zeta\omega_n t} \gg |K_{i1} e^{p_i t}| = |K_{i1}|e^{\text{Re } p_i t} \quad (8.3\text{-}5)$$

for $i = 3,4\ldots, n$. Since we assume a stable system, Re $p_i < 0$. If we have Re $p_i \ll -\zeta\omega_n$, then as t increases the RHS of (8.3-5) will decay much more quickly than the LHS, and the inequality in (8.3-5) will hold almost regardless of the relative magnitude of $|K_{11}|$ and $|K_{i1}|$. This means that if we have the poles other than the dominant ones far into the LHP, then (8.3-5) will be satisfied. The contribution of these far LHP poles in (8.3-4) will thus be insignificant and the two complex poles will dominate. If, however,

$$\text{Re } p_i \cong -\zeta\omega_n$$

then it can be seen that the only way that (8.3-5) can be satisfied is that

$$|K_{11}| \gg |K_{i1}| \quad (8.3\text{-}6)$$

be satisfied. Thus the object is to make $|K_{i1}|$ small. From (8.3-3) we have

$$|K_{i1}| = \frac{K_T \prod\limits_{j=1}^{m} |p_i - z_j|}{|p_i| \prod\limits_{\substack{j=1 \\ j \neq i}}^{n} |p_i - p_j|} \quad (8.3\text{-}7)$$

Thus if pole p_i is very near a zero (say z_j), the term $|p_i - z_j|$ in the numerator of (8.3-7) will be small and consequently $|K_{i1}|$ will be small. $|K_{i1}|$ will be much smaller than $|K_{11}|$ if the pair of dominant complex poles are not near any zeros. This is generally easy to arrange, since the zeros are usually all down on the real axis. In conclusion, if all poles besides the complex ones are either far into the LHP or near zeros, the complex poles will dominate.

SPECIFYING ζ AND ω_n FOR THE DOMINANT COMPLEX POLES

The method of specifying ζ and ω_n basically is to assume initially that the system will behave like the second-order system of Section

2.3. That is, it will be completely dominated by the two complex poles. We then choose the damping factor ζ so that the step response of the closed-loop system will have the desired or tolerable amount of overshoot. From Fig. 2.3-3 we see that the percent overshoot to the second-order case is a function of ζ alone. From Fig. 2.3-3 we can see directly that damping factor will usually be in the range $0.3 \leqslant \zeta \leqslant 0.8$. A popular value is $\zeta = 0.707$, which gives an overshoot of 6%. Once a value of ζ is decided upon, ω_n is determined from the desired speed of response. If one can determine a desired value of time to the first peak (t_p) for the step response, then Fig. 2.3-4 may also be used to determine a suitable value of ω_n. If it is easier to determine a desired settling time t_s, then one may use

$$t_s = 4\tau = \frac{4}{\zeta \omega_n} \tag{8.3-8}$$

to determine ω_n.

Once we have chosen values of ζ and ω_n, the compensator is designed to give the dominant complex poles with these values. Usually the sum total of the contributions of the poles which have been placed far into the LHP or near zeros will not be completely negligible. Hence adjustment will have to be made, and this is where the trial and error part of the design process enters. *In general, if the speed of response for the compensated system is too slow, one increases* ω_n; *if the overshoot is excessive, ζ is increased; etc.* With these new values of ζ and ω_n the compensator is redesigned and the step response is checked again. Alternatively, one may attempt to position the compensator pole and zero to reduce the detrimental contribution of the nondominant poles. More will be said about this at the end of this section. Adjustments on ζ and ω_n may then be made again if the desired time response is not achieved. The entire process is repeated until a compensator with an acceptable response is found.

DESIGN OF A LEAD COMPENSATOR BY ROOT LOCUS

Having decided where, we want the dominant two poles to be located, we next must add the compensation so that:

1. A root locus branch of $G_T(s)$ passes through the desired closed-loop pole position.
2. For a gain which puts a closed-loop pole at the desired dominant pole position, all other closed-loop poles are far into the LHP or near zeros. All closed-loop poles of course have to be in the LHP.

The form of the compensator will be

$$G_c(s) = A\frac{(s+1/\tau)}{(s+\alpha/\tau)}, \qquad A, \tau > 0, \quad \alpha > 1 \qquad (8.3\text{-}9)$$

which is to be placed in series with a plant $G(s)$ and operated in a unity-feedback closed-loop system as in Fig. 8.1-1. The open-loop transfer function will be given by

$$G_T(s) = G_c(s)G_f(s) \qquad (8.3\text{-}10)$$

The closed-loop poles will now appear on the root locus of $G_T(s)$. Our compensator has three quantities to be determined, A, τ, and α. We first determine the location of the compensator pole and zero. This is done by making a pole-zero map for the transfer function $G_f(s)$. These are the open-loop plant poles and zeros. At the position of the desired closed-loop pole (let us call it p_d) we determine $\angle G_f(s)$, i.e., we determine $\angle G_f(p_d)$. This may be done with the spirule. Now

$$\angle G_T(s) = \angle G_f(s) + \angle G_c(s) \qquad (8.3\text{-}11)$$

from (8.3-10). If the root locus of $G_T(s)$ is to pass through p_d, we must have

$$\angle G_T(p_d) = \pm 180° \qquad (8.3\text{-}12)$$

and therefore

$$\angle G_c(p_d) = \pm 180° - \angle G_f(p_d) \qquad (8.3\text{-}13)$$

Since we have already determined $G_f(p_d)$ from the pole-zero map of $G_f(s)$, (8.3-13) gives the angle the compensator pole-zero pair must subtend at the desired closed-loop pole position. The angle contribution of a pole-zero pair at a given position p_d in the s-plane is illustrated in Fig. 8.3-2.

For a lead compensator, $\angle G_c(p_d)$ as shown in Fig. 8.3-2 will be positive. Thus using (8.3-13) we determine $\angle G_c(p_d)$ and all that

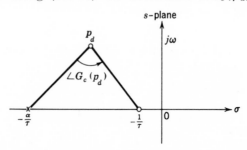

Figure 8.3-2 The angle contribution at a desired closed-loop pole position due to a lead compensator pole-zero pair.

remains to be done is to determine the zero at $-1/\tau$. The zero generally is placed to meet the requirement that closed-loop poles near the $j\omega$-axis be near zeros. This usually means that the first open-loop plant pole on the real axis to the left of the origin is cancelled out. To illustrate this process consider designing a lead compensator for a plant

$$G_f(s) = \frac{1}{s(s+1)}$$

The pole-zero map for this transfer function is shown in Fig. 8.3-3.

Now let us assume that the specifications for the transient response are such that $\zeta = 0.707$, $\omega_n = 2$ for the desired closed-loop pole will give a satisfactory response. We locate the corresponding point in the LHP on the pole-zero map of Fig. 8.3-3 (shown as p_d). We determine straightforwardly from angles of the two poles at -1 and the origin that

$$\angle G_f(p_d) = -135° - 106° = -241°$$

We now use (8.3-13) to determine that if the root locus is to pass through p_d, then we must have

$$\angle G_c(p_d) = \pm 180° - \angle G_f(p_d)$$

$$= -180° - (-241°) = 61°$$

Having the angle that the compensator pole-zero pair must subtend at the point p_d, we now only need locate the zero. Placing it so that the open-loop pole at -1 is cancelled results in a root locus that will have no closed-loop pole nearer the $j\omega$-axis than the desired dominant closed-loop pole at p_d. Now with the compensator zero at -1, the compensator pole is adjusted so that the angle subtended at p_d is $\angle G_c(p_d) = 61°$ (Fig. 8.3-3). The result is a compensator pole at -2.83.

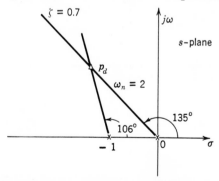

Figure 8.3-3   Pole-zero map for $G_f(s) = 1/s(s+1)$.

The final step required to complete the design of a compensator of the form of (8.3-9) is to determine the value of the gain constant A. This is determined so that the closed-loop system will have a pole at p_d. The root locus for the compensated open-loop transfer function $[G_T(s)]$ is shown in Fig. 8.3-4. For the root locus drawn, a value of $K = 4$ will result in a closed-loop pole at p_d. Thus we have

$$G_T(s) = \frac{A(s+1)}{s+2.83} \frac{1}{s(s+1)} = \frac{A}{s(s+2.83)} \tag{8.3-14}$$

and we find that $A = K = 4$ will result in a closed-loop pole in the desired position. In this case there are no other closed-loop poles and hence the complex poles at p_d and $\bar{p}_d$ will of course dominate. The step response of the resulting system will be exactly of the form of the second-order system step response (with $\zeta = 0.707$, $\omega_n = 2$) discussed in Section 2.3. This is what we were trying to achieve.

Before going on to a full-scale example, it will be helpful to discuss in detail the steps in lead compensator design by root locus.

Step 1. Determine damping factor ζ to give the desired overshoot and resonant frequency ω_n to give the desired speed of response to the closed-loop system.

Step 2. At the desired pole position (p_d) determined by the ζ and ω_n of Step 1 determine $\angle G_f(p_d)$. Then

$$\angle G_c(p_d) = \pm 180° - \angle G_f(p_d) \tag{8.3-15}$$

Step 3. Add the compensator pole-zero pair so that $\angle G_c(p_d)$ is as

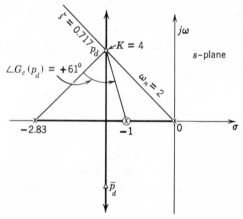

Figure 8.3-4 Root locus for $G_f(s) = 1/s(s+1)$ compensated to give a closed-loop pole with $\zeta = 0.707$, $\omega_n = 2$.

determined in Step 2 and place the compensator zero such that the resulting root locus will have all poles beside p_d and $\bar{p}_d$ either far into the LHP or near zeros. Usually this means cancelling out the plant pole nearest (but not on) the $j\omega$-axis on the negative real axis.

Step 4. With the spirule determine K, which for the compensated closed-loop system will give a pole at p_d. Determine compensator gain A to give this K.

Step 5. Check the time response to see that the desired overshoot and speed of response have been obtained. If not, go back to Step 3 and adjust the position of the compensator zero so that the desired overshoot and speed of response have been obtained. If this adjustment does not result in the desired overshoot and speed of response, return to Step 1 and adjust ζ and ω_n in the direction required to give a more desirable response.

Let us now apply these steps to a detailed example.

Example 8.3-1

Let us assume that we want to operate a motor in a feedback position control system as diagramed in Fig. 8.3-5. The input is the voltage e_{in}, the output shaft position is θ, and the output potentiometer voltage, which is directly proportional to shaft position, is e_o. For our purposes let us assume that the motor and potentiometer parameters are such that we have a transfer function between the armature voltage and the output potentiometer given by

$$G_f(s) = \frac{E_o(s)}{V_c(s)} = \frac{1}{s(s+1)(s+4)} \qquad (8.3\text{-}16)$$

The system of Fig. 8.3-5 may then be block-diagramed as in Fig. 8.3-6.

The problem now is to design the compensator, $G_c(s)$ of Fig. 8.3-6, so that the system will be a suitable position control device. Before going into the problem of designing the compensator, however, let us consider what in the way of a system response can be obtained

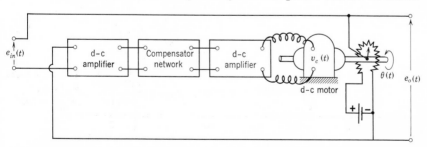

Figure 8.3-5   A schematic diagram of a feedback position control system.

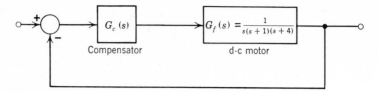

Figure 8.3-6 Block diagram for the feedback position control system of Fig. 8.3-5.

without any additional compensator, i.e., by adjusting the loop gain only. This can be done by adjusting the gain of the amplifier driving the d-c motor to obtain the uncompensated system response.

The Uncompensated System Response

To analyze the closed-loop response of our system (uncompensated), we begin by constructing the root locus for the plant transfer function of (8.3-16). This root locus is drawn in Fig. 8.3-7. For our purposes, let us say that an overshoot of approximately 16% is satisfactory. From Fig. 2.3-3 it can be seen that $\zeta = 0.5$ gives 16% overshoot for a second-order system. The line along which $\zeta = 0.5$ is drawn in the second quadrant of the s-plane of Fig. 8.3-7. Where it crosses the root

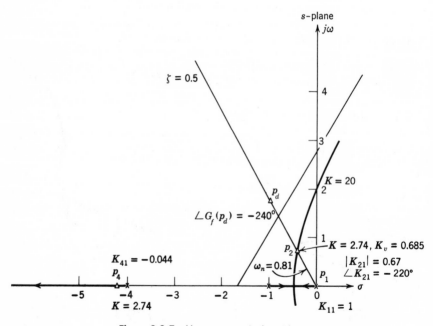

Figure 8.3-7 Uncompensated root locus.

locus we find, by use of the spirule, that $K = 2.74$. Thus if we adjust the amplifier gain in the loop such that

$$G_T(s) = G_c(s)G_f(s) = \frac{2.74}{s(s+1)(s+4)} \qquad (8.3\text{-}17)$$

then a closed-loop pole will result at position p_2 in Fig. 8.3-7; i.e., we will have a pair of complex poles with damping factor $\zeta = 0.5$. A third closed-loop pole will occur on the branch of the root locus that goes out on the negative real axis from the pole at -4 to $-\infty$. This pole is found to occur at -4.2, relatively far out in the LHP relative to the pair of complex closed-loop poles (p_2 and p_3). Hence it should contribute a negligible amount to the transient response of the system. This is verified by determining the step response:

$$C_{\text{step}}(s) = \frac{1}{s}T(s) = \frac{1}{s}\frac{G_T(s)}{1+G_T(s)}$$

from the closed-loop system of Fig. 8.3-6. Using the $G_T(s)$ of (8.3-17),

$$C_{\text{step}}(s) = \frac{2.74}{s(s^3 + 5s^2 + 4s + 2.74)}$$

$$= \frac{2.74}{s(s+4.2)(s+0.4-j0.7)(s+0.4+j0.7)} \qquad (8.3\text{-}18)$$

The denominator is factored easily since we already know where the closed-loop poles are from the root locus of Fig. 8.3-7. Now the RHS of (8.3-18) may be expanded into a partial fraction expansion of the form

$$C_{\text{step}}(s) = \frac{K_{11}}{s} + \frac{K_{21}}{s+0.4-j0.7} + \frac{\bar{K}_{21}}{s+0.4+j0.7} + \frac{K_{41}}{s+4.2} \qquad (8.3\text{-}19)$$

The magnitude and angle of these partial fraction expansion coefficients may be determined by use of the spirule (see Section 4.2). The values obtained for these coefficients are shown adjacent to the corresponding poles in Fig. 8.3-7. Use of these values in (8.3-19) and inverting the $\mathscr{L}$-transform as in (4.2-41) gives a step time response:

$$c_{\text{step}}(t) = [1 + 1.35e^{-0.4t}\cos(0.7t - 220°) - 0.044e^{-4.2t}]1(t) \qquad (8.3\text{-}20)$$

We see that the coefficient of the term $e^{-4.2t}$ (the term due to the pole at -4.2) is very small relative to the coefficients of the other terms. This term hence contributes an insignificant amount to the output time response, which is as desired. The time response of (8.3-20), the uncompensated step response, is plotted in Fig. 8.3-8. The overshoot

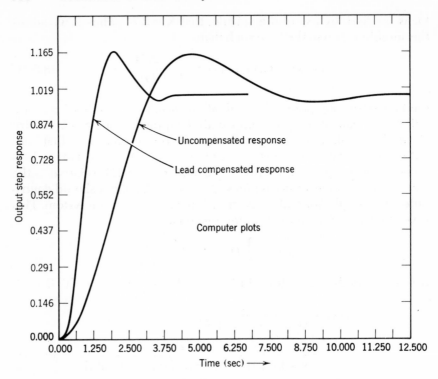

Figure 8.3-8   Lead compensated and uncompensated step response.

is about 16.5%, which is slightly higher than the 16% that we were trying for, but well within the spirule accuracy. The peak is reached in about 5 sec and there is a settling time of approximately 10 sec.

We can also consider the steady-state response of the closed-loop system. We note from (8.3-16) that the plant is a type 1 system (one pole at the origin). The total forward transfer function of (8.3-17) is hence also type 1; therefore the closed-loop system will have a finite, nonzero velocity error constant K_v. In this case we determine K_v by (3.3-8):

$$K_v = \lim_{s \to 0} [sG_T(s)] = \frac{2.74}{4} = 0.685 \qquad (8.3\text{-}21)$$

and the system will hence track a unit ramp with steady-state error given by

$$e_{ss}(\text{ramp}) = \frac{1}{K_v} = 1.46 \qquad (8.3\text{-}22)$$

The response of the system is well detailed in Fig. 8.3-8 and

(8.3-22). The form of the step response of Fig. 8.3-8 is very good. We can see from the root locus of Fig. 8.3-7 that increasing the loop gain will decrease the damping factor. The settling time will also increase with increased gain since it is seen that this decreases $\zeta\omega_n(\text{Re } p_2)$. On the other hand, increasing the gain will increase K_v and hence decrease e_{ss}(ramp). Lowering the gain increases the damping factor and hence decreases the overshoot. However, very little improvement in settling time will be realized. Decreasing the gain of course decreases K_v and correspondingly increases e_{ss}(ramp). We thus see what can be realized with system response by simply adjusting the loop gain. If the speed of response as seen in the step response of Fig. 8.3-8 is adequate and if K_v of (8.3-21) is sufficiently high for the application intended, the system may well be satisfactory as is. However, our effort here now will be to speed up the response through lead compensation and to increase K_v through lag compensation.

Design of a Lead Compensator for the System

We begin with Step 1 in the lead compensator procedure as outlined earlier in the section, i.e., we determine ζ and ω_n for the two complex poles that we want to have dominate the response of the system.

Step 1. The uncompensated response has been seen (in Fig. 8.3-8) to give an overshoot of approximately 16% for a damping factor $\zeta = 0.5$. This is a very reasonable amount of damping, so let us stay with this value of damping for the compensated system. For the uncompensated system we see (Fig. 8.3-7) that for $\zeta = 0.5$, the dominant pole has $\omega_n = 0.81$ and a settling time then of 10 sec (see Fig. 8.3-8), the peak of the step response occurs about 5 sec after the initial instant. Let us now assume that for our purposes we would like a response about twice this fast: we want a settling time of about 5 sec, peaking in about 2.5 sec. For good measure and to make sure we achieve this with the compensator we get, let us take as our design goals a settling time t_s and time to the first peak t_p as

$$t_s = 4\text{ sec}, \qquad t_p \cong 2\text{ sec} \qquad (8.3\text{-}23)$$

Since $t_s = 4\tau = 4/\zeta\omega_n$, we have for $\zeta = 0.5$ and t_s as in (8.3-23) that $\omega_n = 2$. Thus for our desired dominant pole p_d we have

$$\zeta = 0.5, \qquad \omega_n = 2 \qquad (8.3\text{-}24)$$

which gives

$$p_d = -\zeta\omega_n + j\omega_n\sqrt{1-\zeta^2} = -1 + j1.73 \qquad (8.3\text{-}25)$$

Step 2. We make a pole-zero plot of the plant $G_f(s)$ as given in (8.3-16) and determine $\angle G_f(p_d)$. This may be done readily with the pole-zero plot of Fig. 8.3-7. We find

$$\angle G_f(p_d) = -240°$$

Using (8.3-15) we find that in order for the root locus to go through p_d the angle of the compensator must be

$$\angle G_c(p_d) = \pm 180° - (-240°) = 60°$$

Step 3. Locating the compensator pole-zero pair so that an angle of 60° is subtended at p_d and locating the zero so that the plant pole nearest the $j\omega$-axis on the real axis (at $s = -1$) is cancelled, we find that the compensator pole must go at $s = -4$. Thus the form of the lead compensator is

$$G_c(s) = A\frac{s+1}{s+4} \qquad (8.3\text{-}26)$$

Step 4. The root locus for the compensated system, i.e., for

$$G_T(s) = G_c(s)G_f(s) = \frac{A}{s(s+4)^2} \qquad (8.3\text{-}27)$$

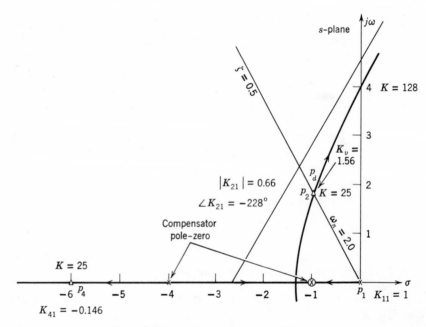

Figure 8.3-9   Lead compensated root locus.

is drawn in Fig. 8.3-9. It of course passes right through p_d. With the spirule we determine that $K = 25$ gives a closed-loop pole at p_d. From (8.3-27) we see that $K = A$; thus $A = 25$ realizes a closed-loop pole at the desired position. The lead compensator is thus determined. We have

$$G_c(s) = 25\frac{s+1}{s+4} \qquad (8.3\text{-}28)$$

Comparing this compensator with the standard form lead compensator of (8.2-1), we see that here

$$\tau = 1, \qquad \alpha = 4$$

For $\alpha = 4$, the lead compensator is very easily realized as shown in Section 8.2, so the compensator should be an acceptable one from this standpoint.

Step 5. The compensated root locus (Fig. 8.3-9) has a third closed-loop pole (for $K = 25$, which realizes a closed-loop pole at p_d) at $s = -6$. This is relatively far into the LHP for the two complex poles, so that the two complex poles should definitely dominate the response of the system. We check this by determining the step response of the closed-loop system. The partial fraction expansion coefficients, for use in (4.2-41) to give the step time response, have been determined with the spirule as shown in Section 4.2 and are shown adjacent to the corresponding poles. Using these in (4.2-41) gives

$$c_{\text{step}}(t) = [1 + 1.32e^{-t}\cos(1.73t - 228°) - 0.416e^{-6t}]1(t)$$
$$(8.3\text{-}29)$$

We see that the coefficient of the term corresponding to the far LHP pole (the one at $s = -6$) is much larger than it was in the uncompensated case of (8.3-20); however, the pole is farther into the LHP this time (-6 as compared to -4.2) so its contribution should still be negligible. The lead compensated time response is shown plotted in Fig. 8.3-8. Again we have about 16% overshoot as in the uncompensated response case seen in the same figure. The peak occurs at $t \cong 2.0$ sec and $t_s \cong 4$ sec. Thus the design specifications have been met as closely as could be expected. The general form of the lead compensated response is almost exactly that of the uncompensated step response. Hence adding the lead compensator has speeded up the response, as we would expect. In addition to this improvement in transient response we may check the new velocity error constant

(K_v^c) to see what has happened to the steady-state response characteristics. We have

$$K_v^c = \lim_{s \to 0} [sG_T(s)] = \lim_{s \to 0} s\,\frac{25}{s(s+4)^2} = 1.56 \qquad (8.3\text{-}30)$$

Comparing this with the K_v for the uncompensated system in (8.3-21), we see that adding the lead compensator has more than doubled the velocity error constant. Thus the lead compensator has also improved the steady-state response characteristics of the system. We begin to see the value of a simple lead compensator.

Lag Compensation

As mentioned in Section 8.1, a lag compensator is used to improve the steady-state response characteristics of the system. It does this by increasing the magnitude of the steady-state error constant K_p, K_v, or K_a depending on whether the system is type 0, 1, or 2, respectively. The lag compensator is of the form

$$G_c(s) = A\,\frac{(s+1/\tau)}{(s+1/\alpha\tau)}, \qquad A, \tau > 0, \quad \alpha > 1 \qquad (8.3\text{-}31)$$

We see the pole is nearer the origin (to the right) than the zero. For the usual case, $A \cong 1$ so that, as shown in (8.1-14), the error constant is increased in magnitude approximately by the factor α. The lag compensator is usually added only to a system which already has a satisfactory transient response, hence the compensator pole and zero have to be added so that they will not disturb the general form of the root locus. This is accomplished by putting the compensator pole and zero very near the origin relative to the other open-loop poles and zeros. The usual rule of thumb is to put the compensator zero about 10% of the distance between the origin and the first pole on the negative real axis out from the origin (Fig. 8.3-10).

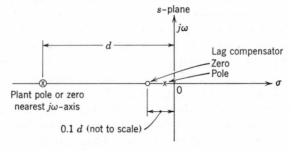

Figure 8.3-10   Placing the lag compensator zero.

Designing a lag compensator is much simpler than designing a lead compensator. The steps of the procedure for a system where the transient response is satisfactory but the steady-state response characteristics are not, follow:

Step 1. Determine by what factor the error constant K_p, K_v, or K_a, depending on the type of the open-loop system, must be increased in order to give satisfactory steady-state response characteristics. For a compensator of the form of (8.3-31), choose α slightly greater than the factor by which the error constant is to be increased. As a rule of thumb choose α greater by 10–20%.

Step 2. Choose τ of (8.3-31) such that the zero at $-1/\tau$ is approximately 10% of the way from the origin to the first open-loop singularity on the negative real axis.

Step 3. Choose A such that the dominant closed-loop poles are as near as possible to where the dominant closed-loop poles were before compensation.

Step 4. Check the transient response of the compensated system. If necessary, adjust loop gain to correct any degradation of the transient response due to the addition of the lag compensator. Check to see if the desired increase in magnitude of the error constant has been obtained.

Let us now use this procedure in an example.

Example 8.3-2

Let us consider here the system discussed in Example 8.3-1. With $G_c(s) = A = 2.74$ and for

$$G_f(s) = \frac{1}{s(s+1)(s+4)} \qquad (8.3\text{-}32)$$

the uncompensated (step) response is shown in Fig. 8.3-8. Let us assume for the moment that this transient response is satisfactory and that the location of the dominant closed-loop pole with $\zeta = 0.5$ and $\omega_n = 0.81$ (see Fig. 8.3-7 for the uncompensated root locus) is acceptable. We have $K_v = 0.685$ for this case from (8.3-21). For this value of velocity error constant $e_{ss}(\text{ramp}) = 1.46$. For the sake of argument let us assume that this is not satisfactory. Let us say that an $e_{ss}(\text{ramp}) = 0.5$ is the most that can be tolerated in the way of steady-state error. Thus K_v must in turn be increased by a factor of 3. We turn to the lag-compensation procedure developed earlier to satisfy this condition.

Step 1. We choose the α of the compensator, (8.3-31), to be slightly larger than 3. Though a factor of 10–20% larger is suggested, let us choose

$$\alpha = 4 \qquad (8.3\text{-}33)$$

since 4 is the next largest integer value. We thus avoid fractional values that would be inconvenient.

Steps 2, 3. We choose τ [see (8.3-31)] so that the compensator zero is about 10% of the way from the origin toward the first open-loop (plant) singularity on the negative real axis. The plant has a pole at $s = -1$; we thus choose our zero to be at $s = -0.1$. Since the zero is at $s = -(1/\tau)$, this gives

$$\tau = 10 \qquad (8.3\text{-}35)$$

By (8.3-31) we see that the compensator pole is at $s = -0.025$. The root locus for the plant compensated with a pole and zero at $s = -0.025$ and $s = -0.1$, respectively, is shown in Fig. 8.3-11. For a closed-loop dominant pole with $\zeta = 0.5$, the spirule has been used to determine that $K = 2.52$ is required. The compensated, open-loop transfer function $[G_T(s)]$ is given by

$$
\begin{aligned}
G_T(s) &= G_c(s)G_f(s) \\
&= \frac{A(s+0.1)}{(s+0.025)} \frac{1}{s(s+1)(s+4)}
\end{aligned}
\qquad (8.3\text{-}36)
$$

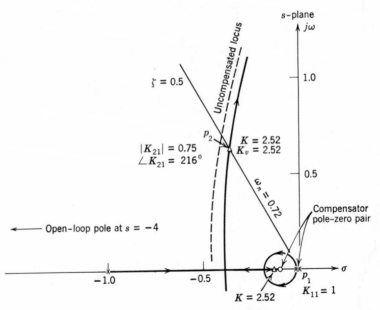

Figure 8.3-11 Lag compensated root locus.

where $G_f(s)$ is as given in (8.3-32). We see from (8.3-36) that $K = A$ for the compensated open-loop transfer function. Thus to realize a $K = 2.52$ to give the dominant closed-loop pole with $\zeta = 0.5$, we must then have

$$A = 2.52 \tag{8.3-37}$$

The lag compensator may now be written out completely using (8.3-31), (8.3-33), (8.3-35), and (8.3-37) as

$$G_c(s) = 2.52 \frac{s + 0.1}{s + 0.025} \tag{8.3-38}$$

With this compensation we immediately determine that the compensated velocity error constant is given by

$$K_v{}^c = \lim_{s \to 0} [sG_T(s)] = 2.52 \tag{8.3-39}$$

for which we have

$$e_{ss}(\text{ramp}) = 0.39$$

which is well within the $e_{ss}(\text{ramp}) = 0.5$ that we were trying to realize.

Step 4. We plot the lag-compensated step response (Fig. 8.3-12). We see that the addition of the lag compensator has indeed affected the transient response. This is seen by comparing the uncompensated step response of Fig. 8.3-8 with the lag-compensated response of Fig. 8.3-12. The overshoot is now up to 27, almost 28%. The reason for this can be seen in the root locus of Fig. 8.3-11. The compensator pole-zero pair has added another branch to the root locus and the result is a closed-loop pole near the compensator zero (which is also a closed-loop zero). The closed-loop transfer function thus has a pole-zero pair near the origin. Since the zero is nearest, its effect will dominate. In Section 2.3 it was seen that a zero on the real axis caused a peaking in the step response. This has occurred here in our lag-compensated system. In addition to this increase in overshoot, the settling time has been increased from approximately 10 sec to approximately 11 sec. This effect can also be predicted from the root locus of Fig. 8.3-11. This is seen from the fact $\omega_n = 0.72$ for the dominant, complex poles of the lag-compensated system while the uncompensated system has an $\omega_n = 0.81$, as seen in Fig. 8.3-7. We hence would expect about 10% increase in settling time.

These detrimental effects on the step response may be alleviated slightly by dropping the gain A of the compensator (and hence K_v) slightly. Looking at the root locus of Fig. 8.3-11, we see that this will increase the damping ζ and also slightly increase $\zeta\omega_n$. Since K_v in this case is slightly higher than the specifications actually

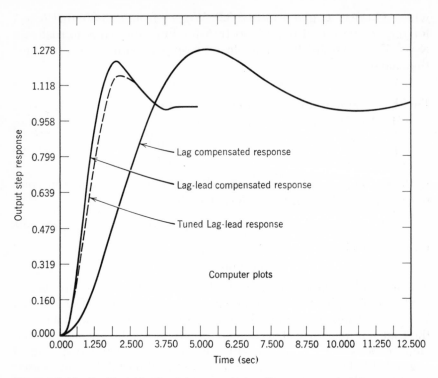

Figure 8.3-12   Root locus lag-lead compensated and lag compensated step response.

called for, we can do this here if the degradation in the lag-compensated transient response as seen in Fig. 8.3-12 is not tolerable.

Lag-Lead Compensation

A lag-lead compensator is used when the system under consideration requires improvement in both transient and steady-state response characteristics. We have seen in (8.1-4) and in (8.2-11) that the transfer function for the lag-lead compensator is of the form

$$G_c(s) = A \underbrace{\frac{(s + 1/\tau_1)}{(s + \alpha/\tau_1)}}_{\substack{\text{Lead} \\ \text{section}}} \underbrace{\frac{(s + 1/\tau_2)}{(s + 1/\alpha\tau_2)}}_{\substack{\text{Lag} \\ \text{section}}}, \qquad \tau_1, \tau_2, A > 0, \quad \alpha > 1 \qquad (8.3\text{-}40)$$

We thus have a lag section and a lead section, both with the same value of α. The reason for both sections having the same value of α is that this makes physical realization of the compensator simpler.

For α the same in both sections there is in general no real handicap. Except for the same value of α, the poles and zeros of either section can be placed as desired. The basic method of designing the lag-lead compensator is then to design the lead section precisely in the manner as described earlier in the section. The lag-compensator section is then designed, using the same value of α as in the lead section, also described earlier in this section. The lag section is designed for the already lead-compensated system. The lag-lead compensator transfer function is then found by multiplying together the lead and lag compensator transfer functions obtained.

There remains the question of how to determine the value of α for the compensator. A reasonable procedure here is to go ahead and design the lead section using whichever α (say α_l) is required to realize the required ζ and ω_n for the dominant complex poles. The error constant (K_p, K_v, or K_a) is then determined for the lead compensated system, say it is $K_m{}^l$. If $K_m{}^l$ is sufficiently high to give the desired steady-state response, we need go no farther. Lead compensation is all that is required. If, however, $K_m{}^l$ is not large enough in magnitude, we consider adding the lag section, which should result in an error constant

$$K_m{}^{ll} \cong \alpha_l K_m{}^l \qquad (8.3\text{-}41)$$

since the addition of a lag compensator increases the magnitude of the error constant by approximately the factor α.

If $K_m{}^{ll}$ so obtained is large enough in magnitude to meet the steady-state specifications, the lag section is designed using this value of α. If, however, $K_m{}^{ll}$ as given in (8.3-41) is not large enough to meet steady-state specifications, we choose a value of α (say α_{ll}) such that $\alpha_{ll} K_m{}^l$ is large enough. We then redesign both the sections using α_{ll} as the value of α for both the lead and lag sections. Usually the same value of ζ used in the first designed lead compensator is chosen; next, whatever value of ω_n, the larger value of $\alpha(\alpha_{ll})$ is taken. Then the lag section is designed with $\alpha = \alpha_{ll}$. The result should be a response that gives a faster acting transient response than required and a steady-state response which more closely meets the required steady-state specifications.

Let us now set down step by step the lag-lead compensator design procedure.

Step 1. From the transient and steady-state performance requirements determine ζ and ω_n for the desired dominant poles to meet transient specifications and the error constant (let us call it K_{md}) to meet the steady-state specifications.

Step 2. Design the lead section to meet the desired transient specifications.

Step 3. If the lead compensated error constant ($K_m{}^l$, say) satisfies

(a) $K_m{}^l \geq K_{md}$, then only a lead section is required; go to Step 5.

(b) If $\alpha_l K_m{}^l \geq K_{md}$ (α_l used in Step 1), then go to Step 4.

(c) If $\alpha_l K_m{}^l < K_{md}$, then choose an α_{ll} such that $\alpha_{ll} K_m{}^l > K_{md}$ by about 10 or 20%. Go back to Step 1 and design the lead compensator there using $\alpha = \alpha_{ll}$.

Step 4. Design a lag-section using the α of the lead section in Step 1.

Step 5. Determine the compensator gain A (of 8.3-40) so as to realize a loop gain K so that the dominant closed-loop poles are as near as possible to the dominant poles realized with the lead compensator of Step 1.

Step 6. Determine the step response and steady-state error constant to see that all specifications are met. Make adjustments (usually in the gain A) in the compensator parameters required if the specifications are not met. Let us now use this procedure in an example.

Example 8.3-3

Let us again consider the position control system described in Example 8.3-1 and block diagramed in Fig. 8.3-6. The uncompensated step response for this system with $G_c(s) = 2.74$, i.e., a constant gain, is shown in Fig. 8.3-8 where we see an overshoot of 16% with a 10-sec settling time, T_s. The open-loop system is type 1, for which $K_v = 0.685$ as determined in (8.3-21). Thus we have $e_{ss}(\text{ramp}) = 1/K_v = 1.46$. Let us now assume, as was the case in the lead compensator design example, that the overshoot is acceptable but that we want a faster time response, about twice as fast, or $T_s \cong 5$ sec. In addition, let us seek a steady-state error to a unit ramp $e_{ss}(\text{ramp}) \leq 0.5$.

Step 1. With these system response specifications we can design the compensator. The uncompensated response of Fig. 8.3-8 has a damping factor ζ of the dominant, complex poles of $\zeta = 0.5$. This is satisfactory. To be sure to meet the settling time requirement, let us choose $T_s = 4$ sec. Letting $T_s = 4\tau = 4/\zeta\omega_n$, we have

$$\zeta = 0.5, \qquad \omega_n = 2$$

for the dominant complex poles. To meet the steady-state error specification $e_{ss}(\text{ramp}) \leq 0.5$, we must have $K_v = 1/e_{ss}(\text{ramp}) = 2$. Thus we have $K_{md} = K_v \geq 2$ for the steady-state specification.

Step 2. The lead compensator required to meet the transient specifications of Step 1 has been designed in Example 8.3-1. The required result here is then

$$G_c(s) = 25\frac{s+1}{s+4} \qquad \frac{\tau}{\alpha\tau} \qquad (8.3\text{-}42)$$

which we have from (8.3-28). This lead compensator has

$$\alpha = 4$$

Step 3. We saw in (8.3-30) and also in Fig. 8.3-9 that for the lead-compensated system $K_v = 1.56$. This does not meet the $K_v \geq 2$ requirement. Hence a lag section is required to boost the magnitude of K. We note that with a lag section using $\alpha = 4$ (the α of the lead section of Step 2) we should obtain

$$K_v \cong 4 \times 1.56 = 6.24 \geq 2$$

Thus a lag-lead compensator with $\alpha = 4$ should meet both transient and steady-state specifications.

Step 4. We design a lag section with the same α (4) as the lead section of Step 2. In Example 8.3-2 a lag compensator has already been designed with an α of 4. We may use this same compensator as the lag section of our lag-lead compensator. Using the lag-compensator pole and zero from (8.3-38) and the lead compensator pole and zero from (8.3-42), we see that the lag-lead compensator required for our system is given by

$$G_c(s) = A\frac{(s+1)(s+0.1)}{(s+4)(s+0.025)} \qquad (8.3\text{-}44)$$

Step 5. The root locus for our compensated system, $G_c(s)$ as in (8.3-44) and $G_f(s)$ as in (8.3-16) and Fig. 8.3-6, has been drawn in Fig. 8.3-13. We see from the figure that $K = 25$ gives a pair of dominant complex poles with $\zeta = 0.5$. For this case $K = A$ of the compensator, hence we choose $A = 25$ and the lag-lead compensator for this system [from (8.3-44)] is then

$$G_c(s) = \frac{25(s+1)(s+0.1)}{(s+4)(s+0.025)} \qquad (8.3\text{-}45)$$

It is easy to calculate the steady-state error constant:

$$K_v = 6.24 \qquad (8.3\text{-}46)$$

which adequately satisfies the steady-state error specification $K_v \geq 2$.

Step 6. The step response for the system with lag-lead compensation of (8.3-45) is shown in Fig. 8.3-12. We see an overshoot of slightly more than 22% with settling time of just over 4 sec. Thus the settling

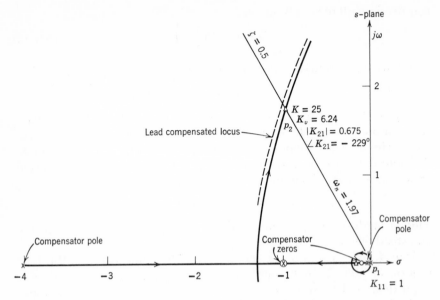

Figure 8.3-13   Lag-lead compensated root locus.

time specification ($T_s \le 5$ sec) is met; however, there is slightly more overshoot than was originally intended (16%). This excess overshoot has been seen in the lag compensated response of Fig. 8.3-12. It appears here for the same reason that it appears there, i.e., the closed-loop pole zero pair near the origin, as seen in the root locus of Fig. 8.3-13. Reducing K will result in increased damping and since the velocity error constant (K_v) from (8.3-46) is much higher than the steady-state error specifications requires, reducing A (and hence also K_v) might be appropriate if this excess overshoot is a problem. In the next section we shall consider another alternative for reducing the overshoot. In any case we see from the lag-lead compensated response of Fig. 8.3-12 that the addition of the lag-lead compensator has materially speeded up the response of the system. This is seen by comparing the lag-lead step response of Fig. 8.3-12 with the just lag compensated step response of the same figure. K_v has been increased significantly, as seen in (8.3-46), thus the compensator has improved both the transient and steady-state response.

Tuning the Transient Response

The step response of the lag-lead compensated system seen in Fig. 8.3-12 has a slightly higher overshoot (22%) than desired (16%). It has already been noted that reducing K for the loop (and hence also reducing K_v) should cure the excess overshoot if it is a problem. Since the lag-lead compensator has a lead section that is used to control the shape of the transient response, there is an alternative here of positioning the lead section pole and zero to remove the excess overshoot. We can now begin the complete redesign of the compensator, again choosing a slightly higher value of damping factor to lower the overshoot. A simpler way of doing this can be seen by considering the source of the excess overshoot. This is the closed-loop pole-zero on the real axis very near the origin, seen on the root locus of Fig. 8.3-13. This closed-loop pole-zero pair is due to the addition of the pole and zero of the lag section of the compensator. Since the zero is nearer the origin than the pole, its effect dominates. It has been seen in Section 2.3 that a zero on the real axis has a severe peaking effect on the step response of the two-pole system (see Fig. 2.3-6). This is what is happening here. Since the closed-loop pole is very near the zero, this peaking is not great; however, there is enough of it to show up in the excess overshoot, as seen in Fig. 8.3-12.

Now if in some way we could obtain another closed-loop pole-zero pair near the origin with the pole nearer the origin than the zero, the effect of the pole-zero pair causing the excess overshoot could be nullified and hence the excess overshoot could be reduced. This we can do by changing the location of the lead section zero, which is cancelling out the plant pole at $s = -1$ (see Fig. 8.3-13). By not cancelling out this pole we introduce another branch to the root locus and upon this branch will appear another closed-loop pole. If the lead section zero is near the plant pole at $s = -1$, this additional closed-loop pole will fall near the zero and hence its effect will be minor. However, by placing the zero properly, so that

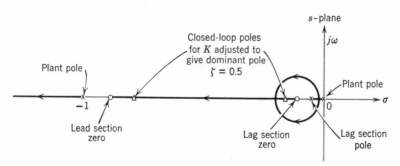

Figure 8.3-14 The root locus with the lead section zero placed to the right of the plant pole at $s = -1$.

the closed-loop pole is nearer the origin than the zero, the effect will be in such a direction as to reduce the excess overshoot. If the lead section zero is placed to the right of the plant pole at $s = -1$, the approximate shape of the root locus on the negative real axis for $-1 \leq \sigma \leq 0$ will be as shown in Fig. 8.3-14. We see that the effect has been an added closed-loop pole near the lead section zero on the negative real axis. For this closed-loop pole-zero pair the pole is now nearest the origin. By choosing the location of the lead section zero properly we should in fact be able to almost completely cancel out the effect of the closed-loop pole-zero pair nearer the origin which is causing the excess overshoot.

To proceed on this assumption, we try a lead section of the form

$$\frac{s+0.9}{s+3.6} \tag{8.3-47}$$

(a zero at -0.9 and pole at -3.6). The pole was chosen so that the α of the lead section was 4, which matched the α of the already designed lag section. Any other value of α would have meant a redesign of the lag section. Using the lead section of (8.3-47) and leaving the lag section as before in (8.3-45), $K = 22.6$ gives dominant complex poles with $\zeta = 0.5$. The new lag-lead compensator was hence

$$G_c(s) = \frac{22.6(s+0.9)(s+0.1)}{(s+3.6)(s+0.025)} \tag{8.3-48}$$

With this compensator, the tuned lag-lead step response is shown in Fig. 8.3-12. Overshoot has been brought down to just less than the 16% that was the original design objective. We calculate immediately that $K_v = 5.65$ with the compensator of (8.3-48); thus there has been some reduction in K_v. Hence reducing the excess overshoot has cost something by way of steady-state error performance. This might have been expected since we did not raise α at all.

It should be noted that if the opposite effect was required, i.e., if more overshoot was desired, the lead section zero could have been pushed over to the left of the plant pole at $s = -1$. In this manner the response may be tuned as desired. It should also be noted that this sort of tuning is possible not only with lag-lead compensators but also with lead compensators.

Root Contours

In root locus design the gain K is always the parameter on the root locus. This is usually the most critical parameter and the one that has to be controlled since the steady-state error constant as well as the transient response of the system is directly related to it. Frequently, however, some other parameter is open to the choice of the designer and in this case it is desirable to know how the closed-loop poles behave as a function of this other parameter. For example, if one is compensating with a standard lag or lead compensator it would be convenient to see what the loci of the closed-loop poles are with α or τ of the compensator as the free parameter. These non loop-gain parameter loci are termed root contours. If the coefficients of the open-loop transfer function are linear functions of the parameter of interest, the standard root locus methods may be used to determine the root contours for this parameter. To see how this may be done let us consider an example.

Example 8.3-3

Let us consider the closed-loop system we have been studying, the system of Fig. 8.3-6. For this case

$$G_f(s) = \frac{1}{s(s+1)(s+4)} \tag{8.3-49}$$

Let us assume that we wish to increase the steady-state error constant K_v and hence we use the lag compensator

$$G_c(s) = \frac{A(s+1/\tau)}{(s+1/\alpha\tau)} \tag{8.3-50}$$

and let us consider finding the root contours as a function of the parameter α. In Example 8.3-2, $A = 2.52$ and $\tau = 10$ were chosen and found to give a satisfactory response for the case $\alpha = 4$. So let us choose these same values for A and τ and find the root contours as α is allowed to vary.

The closed-loop poles and hence the root contours occur for those values of s where

$$1 + G_c(s)G_f(s) = 0 \tag{8.3-51}$$

using (8.3-49) and (8.3-50) in this gives

$$1 + \frac{2.52(s+0.1)}{s(s+1/10\alpha)(s+1)(s+4)} = 0$$

Letting $p = 1/10\alpha$ this may be written as

$$s^4 + 5s^3 + 4s^2 + 2.52s + 0.252 + p(s^3 + 5s^2 + 4s) = 0$$

from which we get

$$1+\frac{ps(s+1)(s+4)}{(s+4.18)(s+0.11)(s+0.35-j0.62)(s+0.35+j0.62)} \tag{8.3-52}$$

Equation (8.3-52) is in the form

$$1+G(s)H(s) = 1+\frac{pA(s)}{B(s)} = 0$$

which is the standard form for plotting root loci as discussed earlier only here p replaces the parameter K.

The root contour corresponding to (8.3-52) is plotted in Fig. 8.3-15. We see that for α in the practical range of $1 \le \alpha \le 10$ $(0.01 \le p \le 0.1)$ the closed-loop poles are very insensitive to α. This is exactly what we should expect since we put the compensator zero (at $-1/\tau$) very close to the origin and the compensator pole between it and the origin. This choice was made so that adding the compensator would have negligible effect upon the transient response, i.e., on the dominant complex poles.

The root contour for values of $\alpha < 1$ $(p > 0.1)$ correspond to lead compensation of our system with the compensator zero set at $s = -0.1$. We thus see that to have a large effect upon the dominant, complex closed-loop poles we must use lead compensation, which is exactly what we concluded in Section 8.1.

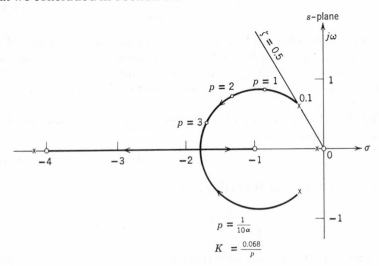

Figure 8.3-15 Root contour for the system of Fig. 8.3-6 with $G_c(s) = 2.52(s+0.1)/(s+p)$.

We may also find the steady-state error constant K_v as a function of the root contour parameter p. We have

$$K_v = \lim_{s \to 0} G_c(s)\, G_f(s)$$

which, using (8.3-49) and (8.3-50) with $A = 2.72$, $\tau = 10$, and $p = 1/10\alpha$. gives

$$K_v = \frac{0.068}{p} \tag{8.3-53}$$

We see, as expected, that high values of K_v occur for small p, i.e., for large α.

We should note that it is possible to plot these root contours using different values of τ and A. From such a plot we see the potential of simple lag or lead compensation so far as closed-loop pole location is concerned.

8.4 FREQUENCY-RESPONSE COMPENSATION

Some Basic Considerations

Frequency-response compensation methods enjoy a great popularity in practice because of the ease with which the frequency response for a real system can be measured. This has been shown in Section 3.4. The frequency response for a real plant may be measured and, as will be seen below, a compensator may be designed solely using this measured frequency response. The result in most cases gives a very satisfactory time response. The ease with which this can be done is completely independent of the complexity of the plant under consideration. All that is required is that the plant be linear and time invariant. In fact, there is no need that the plant have a rational transfer function as in the case of the root locus. Distributed parameter systems with nonrational transfer functions are handled with equal ease, once the frequency response is known. Besides all these advantages there is a variety of equipment now available for measuring and working with frequency responses. Knowledge of frequency-response methods is now widespread and their use is deeply ingrained throughout the electrical and mechanical (and even chemical) industries. So these methods are important in control system design. However, the frequency response does not give much direct information about the time response of the system. The connection of the frequency response with the time response is by no means as direct as its connection with the root locus. With the root locus the designer sees very clearly where the closed-loop poles and zeros are and with a little practice he can extract a wealth of information concerning the time response as some of the parameters change. In this regard, the frequency-response methods do not do as well and consequently there is usually more trial and error necessary to design a compensator by frequency response methods.

In this section we shall work with a system with the same configuration as in the previous sections of this chapter, i.e., the system shown in Fig. 8.1-1. We assume that the frequency response of the plant, $G_f(j\omega)$, is known over the range of frequency (ω) in which compensation is required. The basic idea in frequency-response transient-response compensation is to add the compensator, a lead compensator here, with frequency response $G_c(j\omega)$ such that the closed-loop frequency response looks like that of a second-order system with a desirable damping factor. These second-order system frequency responses are seen in Fig. 3.4-4 with the damping factor in the range $0.3 \leqslant \zeta \leqslant 0.8$. We choose ζ in this range since, as we saw in Fig. 2.3-3, this gives a desirable step response. As in root locus compensation, the error constant (K_p, K_v, or K_a) is increased to improve the steady-state response. We may interpret the error constant definition [see (8.1-10)] as

$$K_m = \lim_{\omega \to 0} (j\omega)^m G(j\omega) \tag{8.4-1}$$

where K_m is K_p, K_v, or K_a depending on whether m, the type of the system, is 0, 1, or 2, respectively. Thus to compensate our system in order to improve the steady-state response we choose a lag compensator frequency response, $G_c(j\omega)$, so that

$$G_T(0) = G_c(0)\, G_f(0)$$

is such that the corresponding error constant has the desired magnitude. This specifies the lag-compensator frequency response in the low-frequency range. In the high-frequency range the lag-compensator frequency response, $G_c(j\omega)$, is chosen so that

$$|G_c(j\omega)| \ll |G_f(j\omega)|$$

$$|\angle G_c(j\omega)| \ll |\angle G_f(j\omega)|$$

In root locus compensation this corresponds to choosing the pole and zero of the lag compensator very near the origin.

We shall consider here the design of lead and lag compensators by use of the Nichols chart. There are many different approaches to the design of compensators, depending upon the presentation of the frequency-response data. Although each approach has its own advantages, we limit ourselves here to the Nichols chart method since it is practical and it illustrates basic ideas in frequency-response compensation. Nichols chart compensation has the additional advantage that both the open-loop and closed-loop frequency responses can be seen simultaneously on the same chart. The Bode plot does not have this advantage.

Lead Compensation by Nichols Chart

As in root locus compensation, lead compensation is used to improve the transient response of a system. The system that we wish to compensate is that of Fig. 8.1-1. The object is to add a compensator, $G_c(s)$, such that the closed-loop frequency response looks like the frequency response of a second-order system as shown in Fig. 3.4-4. It will be recalled that in the root locus compensation the transient-response specifications were given in terms of the damping factor ζ and undamped natural frequency ω_n of the dominant, closed-loop poles we wished to achieve. In Nichols chart compensation the specifications must be made in terms of the quantities that are meaningful from the standpoint of the closed-loop frequency response. The quantities most convenient for use with Nichols chart are the peak magnitude of the frequency response, M_m, and the frequency, ω_m, at which this peak occurs (see Fig. 3.4-5). From (3.4-18), for a second-order system

$$\omega_m = \omega_n \sqrt{1 - 2\zeta^2} \qquad (8.4\text{-}2)$$

and

$$M_m = -20 \log_{10} 2\zeta \sqrt{1 - \zeta^2} \text{ db} \qquad (8.4\text{-}3)$$

These two relationships give M_m and ω_m in terms of the ζ and ω_n that we used in root locus compensation. We see, for the case of a second-order system, that M_m is a function of ζ alone. From Fig. 2.3-4 we see that step-response overshoot is also a function of ζ only. Thus step-response and frequency-response overshoot can be related directly. This has been done in Fig. 8.4-1. Equation (8.4-2) relates ω_m and ω_n. We note that ω_m is in general a lower frequency than ω_n. Figure 8.4-2 shows ω_m/ω_n versus ζ. Using Figs. 8.4-1 and 8.4-2 and the relationship for the settling time, $t_s = 4\tau = 4/\zeta\omega_n$, we can quickly relate step response and the corresponding s-plane characteristics to the frequency-response characteristics M_m and ω_m.

In lead compensation we determine M_m and ω_m from the desired step response. We then seek a frequency response to shape the open-loop frequency response on the Nichols chart so that it is tangent to the M-circle, which gives the desired M_m. The frequency at the point

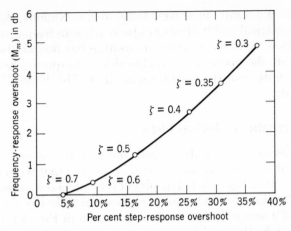

Figure 8.4-1 Step response versus frequency-response overshoot for a second-order system.

of tangency has then to be adjusted to be the desired ω_m. As an example of what is meant here, see Fig. 3.4-12. An open-loop frequency response that is reasonably smooth in the immediate neighborhood of the point of tangency will give a closed-loop frequency response that has the general shape of the second-order frequency response of Fig. 3.4-4. This also is shown by the frequency response of Fig. 3.4-12. The Bode plot of the closed-loop frequency response for this curve is shown in Fig. 3.4-13.

Now to see more clearly what we require the lead compensator to

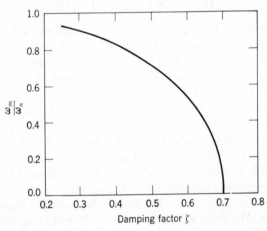

Figure 8.4-2 Ratio of peak to corner frequency (ω_m/ω_n) versus damping factor (ζ) for a second-order system.

do, consider Fig. 8.4-3. This shows the frequency response of a system that might require lead compensation. Also shown on the figure are the M-circle corresponding to the desired M_m and the desired ω_m. The lead compensator must now supply the required phase shift, shown in the figure, and required gain (at frequency ω_m), also shown in the figure, so that the compensated frequency response will be tangent to the desired M-circle at the frequency of the desired ω_m. To accomplish the desired compensation, essentially we have to shift the frequency-response curve of Fig. 8.4-3 over to the right (which means positive phase shift, hence a lead compensator) and upward (which means increased gain) by the amounts shown in the figure. For our system with the plant frequency response $G_f(j\omega)$ and the compensator frequency response $G_c(j\omega)$, we then have the compensated open-loop frequency response $G_T(j\omega)$ given by

$$G_T(j\omega) = G_c(j\omega)G_f(j\omega)$$

from which

$$20\log_{10}|G_T(j\omega)| = 20\log_{10}|G_c(j\omega)| + 20\log_{10}|G_f(j\omega)|$$

$$\angle G_T(j\omega) = \angle G_c(j\omega) + \angle G_f(j\omega) \tag{8.4-4}$$

Therefore

$$20\log_{10}|G_c(j\omega)| = 20\log_{10}|G_T(j\omega)| - 20\log_{10}|G_f(j\omega)|$$

$$\angle G_c(j\omega) = \angle G_T(j\omega) - \angle G_f(j\omega) \tag{8.4-5}$$

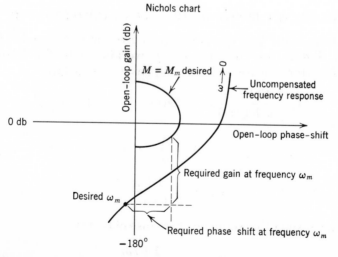

Figure 8.4-3　A typical uncompensated frequency response on a Nichols chart.

We see from (8.4-4) the compensated open-loop frequency response is obtained on the Nichols chart by simply adding together the gain (in db) and phase of the compensator and the plant. By (8.4-5) the compensator gain and phase are simply the difference between the desired compensated open-loop frequency and the plant response. Since (8.4-5) must hold for all ω, it must of course hold for the desired ω_m, hence

$$20\log_{10}|G_c(j\omega_m)| = 20\log_{10}|G_T(j\omega_m)| - 20\log_{10}|G_f(j\omega_m)| \quad (8.4\text{-}6)$$
$$\angle G_c(j\omega_m) = \angle G_T(j\omega_m) - \angle G_f(j\omega_m)$$

Thus, returning to Fig. 8.4-3, we have

$$20\log_{10}|G_c(j\omega_m)| = \text{Required gain at frequency } \omega_m$$
$$\angle G_c(j\omega_m) = \text{Required phase shift at frequency } \omega_m \quad (8.4\text{-}7)$$

Hence using the Nichols chart we can see immediately from the frequency response of the plant how much gain and how much phase shift the compensator must provide to achieve a desired M_m and ω_m. The problem in lead compensator design is to choose the compensator parameters so that the required gain and phase shift are obtained. Before we can do this it is necessary to consider the frequency-response of a lead compensator.

THE FREQUENCY RESPONSE OF A LEAD COMPENSATOR

The form of the lead compensator as given in (8.1-6) and (8.2-1) is altered slightly for convenience in working with the frequency responses. Here we use the lead compensator in the form

$$G_c(s) = \frac{A}{\alpha}\frac{1+s\tau}{1+s\tau/\alpha} = A'G_c{}^o(s), \qquad \alpha > 1, \quad A, \tau > 0 \quad (8.4\text{-}8)$$

where

$$A' = \frac{A}{\alpha}$$

$$G_c{}^o(s) = \frac{1+s\tau}{1+s\tau/\alpha} \quad (8.4\text{-}9)$$

We shall consider the frequency response of $G_c{}^o(s)$

$$G_c{}^o(j\omega) = \frac{1+j\omega\tau}{1+j\omega\tau/\alpha} \quad (8.4\text{-}10)$$

For $G_c{}^o(j\omega)$ we have then that

$$20 \log_{10} |G_c{}^o(j\omega)| = 20 \log_{10}|1+j\omega\tau| - 20 \log_{10}\left|1+\frac{j\omega\tau}{\alpha}\right| \tag{8.4-11}$$

$$\angle G_c{}^o(j\omega) = \tan^{-1}\omega\tau - \tan^{-1}\frac{\omega\tau}{\alpha}$$

By using (8.4-11), the frequency response of $G_c{}^o(j\omega)$ has been plotted in Fig. 8.4-4 on $20 \log_{10}$ magnitude versus phase graphs. The compensator curves as they are drawn may be used to compensate frequency-response curves drawn on Nichols chart of the same scale.

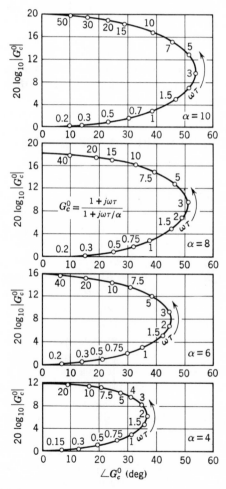

Figure 8.4-4   Lead compensator curves for various values of α.

It should be noted here that the maximum phase shift available with a given lead compensator is a function of the α of the compensator. Maximum phase shift (ϕ_{max}) occurs at a frequency

$$\omega_{max} = \frac{\sqrt{\alpha}}{\tau} \tag{8.4-12}$$

Using this in (8.4-11), we have

$$\phi_{max} = \tan^{-1}\frac{\alpha - 1}{2\sqrt{\alpha}} \tag{8.4-13}$$

A plot of this is shown in Fig. 8.4-5. We see here a reason for not using an α much greater than 10. Above this the curve is very flat so that increasing α results in a very little increase in the available phase shift.

Now that the frequency response curves for the compensators are available, let us consider how we obtain the compensated frequency response for the addition on the Nichols chart of the compensator and plant frequency responses.

COMPENSATED SYSTEM FREQUENCY RESPONSE (ON THE NICHOLS CHART)

For a plant frequency response drawn on the Nichols chart and by use of the compensator frequency response curves, as drawn in Fig. 8.4-4, for example, the compensated system frequency response can be graphically drawn quickly. Let us assume that we have chosen a lead compensator and we wish now to obtain the compensated frequency response. From (8.4-5) we know that the compensator frequency response is simply added to the plant frequency response to obtain the compensated system's frequency response. This can be done graphically. The values of α and τ for the compensator are assumed given so that the compensator curve can be calibrated with ω as the parameter (using the compensator curves of Fig. 8.4-4 we simply have $\omega = \omega\tau/\tau$). Now for each point on the plant frequency response curve corresponding to a given value of ω we must add the compensator gain and phase at this same value of ω. If the compensator curve is drawn to the same scale as the Nichols chart on which it is to be used and is then cut out so as to form a template, it may be slid easily along the plant frequency response. The graphical addition of the plant and compensator frequency response may in this way be done quite quickly. This is shown for two points in Fig. 8.4-6. This graphical addition must be done for enough points to give the compensated curve in the range of frequency of interest. Since the curve

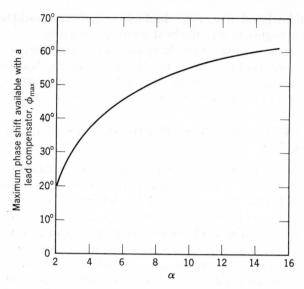

Figure 8.4-5   Maximum phase shift versus α.

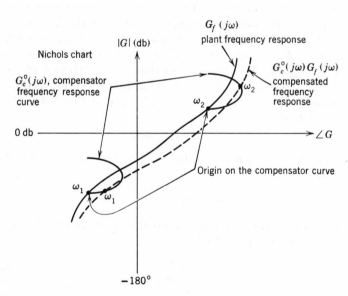

Figure 8.4-6   Graphical construction of the compensated frequency response on the Nichols chart.

in the neighborhood of ω_m is of chief interest, there should be enough points in this region to give the best accuracy possible.

We have now seen what the frequency-response curves for a lead compensator are and how these may be added on the Nichols chart to a plant frequency response to obtain the compensated frequency response. We next consider compensator parameters.

CHOOSING THE COMPENSATOR PARAMETERS IN LEAD COMPENSATION

We see from Fig. 8.4-3 and (8.4-7) how to determine the compensator gain and phase shift required at frequency ω_m to realize a desired ω_m and M_m. We now need to determine the compensator parameters so as to realize (8.4-7). First we note that (8.4-7) gives us two conditions and the lead compensator as seen in (8.4-8) has three parameters. We thus have one degree of freedom here in the choice of the parameters. It is logical and desirable to use this degree of freedom to minimize the value of α of the compensator. Minimizing α generally tends to maximize the signal to noise ratio in the control loop if noise is a problem. It also tends to minimize the amplitude of the signal at the output of the compensator, which is the input to the plant. We thus choose α so that the compensator provides just the minimum amount of phase shift required. This minimum amount of phase shift is just the amount required to give the desired ω_m and M_m as specified in (8.4-7). Thus we choose α of the compensator so that the maximum phase shift provided by the compensator is just that required by (8.4-7). We choose τ so that this maximum phase shift occurs at the desired ω_m. To do this, after determining the phase shift required as in (8.4-7), α is chosen by use of Fig. 8.4-5. We thus choose α such that

$$\angle G_c(j\omega_m) = \phi_{max}$$

Once we have α, τ is chosen so that ϕ_{max} occurs at ω_m. From (8.4-12) we then have

$$\omega_m = \omega_{max} = \frac{\sqrt{\alpha}}{\tau} \tag{8.4-14}$$

where ω_{max} is the frequency of the compensator at which maximum phase shift occurs. From (8.4-14),

$$\tau = \frac{\sqrt{\alpha}}{\omega_m} \tag{8.4-15}$$

To complete the design of the compensator, the parameter A [see (8.4-8)] must be determined so as to give the desired M_m. We may

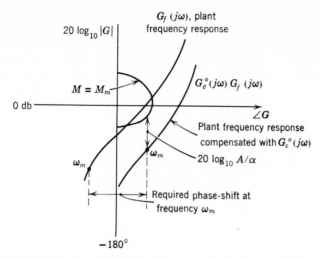

Figure 8.4-7 Determination of lead compensator gain A on the Nichols chart.

determine A from $20 \log_{10}|G_c(j\omega_m)|$ as given in (8.4-7). Since from (8.4-8)

$$20 \log_{10}|G_c(j\omega)| = 20 \log_{10}\frac{A}{\alpha} + 20 \log_{10}|G_c^{\,o}(j\omega)|$$

Now using (8.4-7), we get

$$20 \log_{10}\frac{A}{\alpha} = 20 \log_{10}|G_c(j\omega_m)| - 20 \log_{10}|G_c^{\,o}(j\omega_m)| \quad (8.4\text{-}16)$$

Since $G_c^{\,o}(s)$ involves only the parameters α and τ [see (8.4-9)], $|G_c^{\,o}(j\omega_m)|$ may be determined once α and τ have been determined as described above. A may be determined from (8.4-16) or it may be determined graphically. The frequency response of the compensator, $G_c^{\,o}(j\omega)$, is added to the plant frequency response $G_f(j\omega)$, to obtain the frequency response of $G_c^{\,o}(j\omega)G_f(j\omega)$. A/α (in db) then has to be that gain which, on the Nichols chart, is required to lift the curve of $G_c^{\,o}(j\omega)G_f(j\omega)$ up to tangency on the desired M-circle, i.e., to give the desired M_m. This is shown in Fig. 8.4-7. We are now in a position to consider, in order, the step by step design of a lead compensator using this method.

STEPS IN THE DESIGN OF A LEAD COMPENSATOR BY NICHOLS CHART

Step 1. Determine ω_m and M_m to give the desired transient response.

Step 2. Plot the plant frequency response on the rectangular axes of the Nichols chart and determine the required phase shift to give the desired ω_m.

Step 3. Determine the α of the compensator from Fig. 8.4-5 to give the required phase shift as determined in Step 2. Determine τ of the compensator so as to give this phase shift at frequency ω_m. From (8.4-15),

$$\tau = \frac{\sqrt{\alpha}}{\omega_m}$$

Step 4. Graphically add $G_c{}^0(j\omega)$ to the plant frequency response which was plotted on the Nichols chart in Step 2. The compensator curves of Fig. 8.4-4 may be used, or another may be drawn using (8.4-11). From the plot of $G_c{}^0(j\omega)G_f(j\omega)$ determine A/α such that the desired M_m is achieved. The lead compensator is then

$$G_c(s) = \frac{A}{\alpha}\frac{1+s\tau}{1+s\tau/\alpha}$$

Step 5. Determine the step response. If it does not meet specifications, adjust A to meet specifications. If this fails, readjust ω_m and M_m in the direction required to give the desired time response. Return to Step 2.

Let us now turn to an example.

Example 8.4-1

We here consider compensating the same system compensated by root locus methods in Section 8.3. The system is shown in Fig. 8.3-5 and is block diagramed in Fig. 8.3-6. From Fig. 8.3-6 the plant is seen to have the transfer function

$$G_f(s) = \frac{1}{s(s+1)(s+4)} \tag{8.4-17}$$

Before designing a lead compensator for this plant, let us use the Nichols chart to see what type of response can be achieved by simply adjusting the loop gain properly.

As a first try let us choose a damping factor $\zeta = 0.5$. From Fig. 8.4-1 we see that for $\zeta = 0.5$ we should have $M_m = 1.25$ db. By trial it is found that $G_c(s) = A = 2.72$ gives a forward transfer function

$$G_T(s) = G_c(s)\,G_f(s) = \frac{2.72}{s(s+1)(s+4)} \tag{8.4-18}$$

which has a frequency response, as seen in Fig. 8.4-8, that is just tangent to the $M = 1.25$ db circle on the Nichols chart. We see also

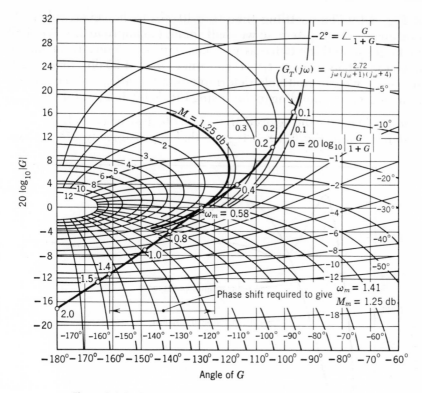

Figure 8.4-8 The uncompensated frequency response.

in Fig. 8.4-8 that with this M_m, $\omega_m = 0.58$. Using Fig. 8.4-2, we have for $\omega_m = 0.58$, $\omega_n = 0.83$ (recall $\zeta = 0.5$). For this case then the settling time is given by

$$t_s = 4\tau = \frac{4}{\zeta\omega_n} = 9.6 \text{ sec}$$

From Fig. 8.4-1 it is seen that for $M_m = 1.25$ db the step-response overshoot should be approximately 16.5%. The uncompensated step response for the closed-loop system with $G_T(s)$ as in (8.4-18) is shown in Fig. 8.4-9 and it is seen that the predicted overshoot and settling time have been very closely met.

Now let us turn to the design of a lead compensator for this system. Since the overall form of the step response is in general very good and since the overshoot is almost exactly what the second-order curves predicted, it is reasonable to keep the same frequency-response overshoot (M_m) that gave the uncompensated step response of Fig. 8.4-9. The settling time ($t_s = 10$ sec) is, however, relatively long for a control system, so we shall attempt to improve this. We shall try to get something twice as fast, i.e., with $t_s \cong 5$ sec. To be on the safe side let us choose $t_s = 4$ sec. Since $M_m = 1.25$ db corresponds to $\zeta = 0.5$, we can routinely determine that $t_s = 4/\zeta\omega_n = 4$ sec implies an $\omega_n = 2$. Corresponding to this, using Fig. 8.4-2, is $\omega_m = 1.41$.

Step 1. We have the desired $M_m = 1.25$ db, desired $\omega_m = 1.41$.
Step 2. From Fig. 8.4-8 it is seen that approximately 37° of phase

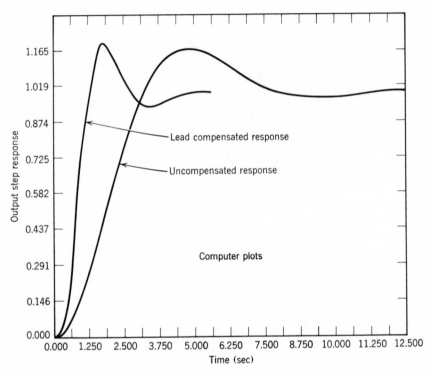

Figure 8.4-9 Nichols chart lead compensated and uncompensated system response.

shift at the frequency of the desired ω_m is required to give ω_m and M_m as determined in Step 1.

Step 3. Figure 8.4-5 shows that this amount of phase shift requires a minimum α of 4; thus we choose $\alpha = 4$. We set τ to give the maximum phase shift at the desired ω_m, thus, using (8.4-15),

$$\tau = \frac{\sqrt{\alpha}}{\omega_m} = \frac{\sqrt{4}}{1.41} = 1.41$$

Hence we have

$$G_c^o(s) = \frac{1+1.41s}{1+0.353s} \qquad (8.4\text{-}19)$$

Step 4. We use the lead compensator curve for $\alpha = 4$ from Fig. 8.4-4 and calibrate the frequency thereon to correspond to $\tau = 1.41$; now the open-loop frequency response of the uncompensated system of Fig. 8.4-8 has been compensated with the lead compensator just designed. The result is shown in Fig. 8.4-10. We see from the figure

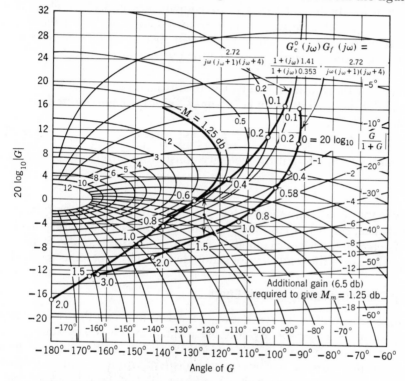

Figure 8.4-10   Lead compensation to achieve $\omega_m = 1.4$, $M_m = 1.25$.

that 6.5 db of additional gain is required to achieve the desired $M_m = 1.25$ db. We incidentally see that with this additional 6.5 db, $\omega_m = 1.4$ as desired. Since 6.5 db corresponds to a gain of 2.11, we have

$$G_T(s) = 2.11 \frac{1+1.41s}{1+0.353s} \frac{2.72}{s(s+1)(s+4)} \qquad (8.4\text{-}20)$$

which will have a frequency response with $\omega_m = 1.4$, $M_m = 1.25$ db.

Thus

$$G_c(s) = \frac{G_T(s)}{G_f(s)} = 23\frac{s+0.708}{s+2.83} \qquad (8.4\text{-}21)$$

is the desired lead compensator, using (8.4-17). If the compensated frequency-response curve of Fig. 8.4-10 is lifted 6.5 db at all frequencies, the closed-loop frequency response may be read off the closed-loop nonrectangular set of coordinates.

Step 5. The step response for the lead compensated closed-loop system has been computed and is shown in Fig. 8.4-9. We see that the design objectives $(t_s = 4\text{ sec}$, overshoot $= 16.5\%)$ have been fairly closely met. The form of the response is very close to that of the lead compensated step response by root locus method as seen in Fig. 8.3-8.

As a further point of interest with this lead compensator, we may consider the velocity error constant $(K_v{}^c)$ for the compensated system. Using (8.4-20),

$$K_v{}^c = \lim_{s\to 0} sG_T(s) = 1.43 \qquad (8.4\text{-}22)$$

Using (8.4-18), we get the velocity constant before compensation to be

$$K_v = 0.68$$

Thus the lead compensator has not only given us a faster transient response, but it has resulted in an improved steady-state response as reflected in the higher velocity error constant. This was also observed for the case of the lead compensation designed by root locus methods.

Lag Compensation by Nichols Chart

Lag compensation, as we know, is used to improve the steady-state response of a system. By this we mean that the steady-state error constant is increased in magnitude. We assume it is applied to a system that already has adequate transient response characteristics. As has been seen in Section 8.1, adding a lag compensator increases the magnitude of error constant by the factor α of the compensator. So when designing a lag compensator, we first determine the factor by which the error constant has to be increased and then choose the α for the compensator slightly larger, say 10 or 20%. The addition of the lag compensator must not detract materially from good transient characteristics of the system. Since the transient characteristics of the system are determined by the frequencies, in the neighborhood of ω_m the lag compensator must have little effect on the open-loop and closed-loop frequency responses of the system in that frequency range. Our method of doing this is to design the compensator so that its main effects fall in the very low-frequency range. To do this we use the rule of thumb that the compensator should contribute only approximately 5° of phase shift to the open-loop frequency response at frequency ω_m. It will then contribute less phase shift at the higher frequencies, as will be seen. This corresponds in root locus design to putting the pole and zero of the lag compensator very near the origin on the negative real axis of the s-plane so as not to affect the root locus significantly in the neighborhood of the dominant complex poles.

THE FORM OF THE LAG COMPENSATOR FREQUENCY RESPONSE

The lag compensator as given in (8.1-13) and (8.2-6) is rearranged for convenience in working with frequency responses into the form

$$G_c(s) = A\alpha \frac{1+s\tau}{1+s\alpha\tau}, \qquad A, \tau > 0, \quad \alpha > 1 \qquad (8.4\text{-}23)$$

This is written in the form

$$G_c(s) = A'G_c{}^o(s) \qquad (8.4\text{-}24)$$

where

$$G_c{}^o(s) = \frac{1+s\tau}{1+s\alpha\tau}, \qquad A' = A\alpha \qquad (8.4\text{-}25)$$

We now consider the frequency response of $G_c{}^o(s)$. From (8.4-25),

$$20 \log_{10} |G_c{}^o(j\omega)| = 20 \log_{10} |1+j\omega\tau| - 20 \log_{10} |1+j\alpha\omega\tau|$$
$$\angle G_c{}^o(j\omega) = \angle(1+j\omega\tau) - \angle(1+j\alpha\omega\tau) \qquad (8.4\text{-}26)$$
$$= \tan^{-1} \omega\tau - \tan^{-1} \alpha\omega\tau$$

The frequency response of the lag compensator of (8.4-23) is then given by

$$20 \log_{10} |G_c(j\omega)| = 20 \log_{10} A' + 20 \log_{10} |G_c{}^o(j\omega)|$$
$$\angle G_c(j\omega) = \angle G_c{}^o(j\omega) \qquad (8.4\text{-}27)$$

The frequency response of $G_c{}^o(s)$ for various values of α has been computed from (8.4-26) and is plotted in Fig. 8.4-11 on a log magnitude versus phase plane suitable for use in compensating on a Nichols chart. We note that they are the same form as the lead compensator frequency-response curves as seen in Fig. 8.4-4. In fact, the lag compensator curves may be seen to be simply the lead compensator

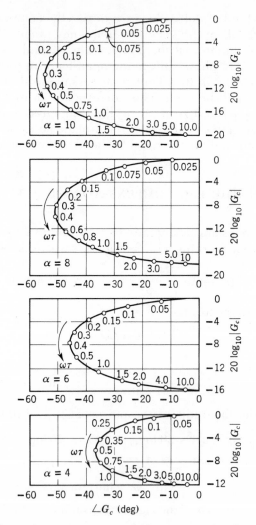

Figure 8.4-11   Frequency-response curves for lag compensators for various values of α.

curves turned over. The maximum phase shift available (though it is not important in the design of the compensator) for a given α is the same as for the lead compensator. However, with a lag compensator, this is negative phase shift.

Of interest on the compensator frequency-response curves of Fig. 8.4-11 is the value of τ that gives 5° of phase shift at the high-frequency end of the curve. We note that for every value of α, $\omega\tau = 10$ gives $\angle G_c{}^o(j\omega) \cong -5°$. Thus if we desire a lag compensator to give

$5°$ of phase shift at frequency ω_m, we might simply choose the τ of the compensator so that $\omega_m\tau \cong 10$.

THE STEPS IN THE DESIGN OF A LAG COMPENSATOR BY NICHOLS CHART

Step 1. Determine by what factor the error constant (K_p, K_v, or K_a depending on the type of the system) has to be increased to give the steady-state response required.

Step 2. Choose the α of the compensator so that it is 10 or 20% (or the nearest round number) larger than the factor of increase in error constant determined in Step 1.

Step 3. Draw the open-loop frequency response on a Nichols chart. Determine ω_m. Choose τ of the compensator so that

$$\omega_m\tau \cong 10 \qquad\qquad (8.4\text{-}28)$$

This should give approximately $5°$ of phase shift at ω_m.

Step 4. With α and τ as determined in Steps 2 and 3, add the frequency response of $G_c{}^o(j\omega)$ as given by (8.4-26) to the open-loop frequency response drawn in Step 3. Determine the gain $A' = A\alpha$ such that the compensated frequency response will have the same M_m as the uncompensated system had. Then the lag compensator is

$$G_c(s) = A'G_c{}^o(s) = A\alpha\frac{1+s\tau}{1+s\alpha\tau} \qquad\qquad (8.4\text{-}29)$$

Step 5. Determine the time response and the compensated system error constant. Make adjustments as required to meet transient and steady-state specifications.

As with root locus compensation, the lag compensator pole-zero pair is very near the origin. The closed-loop zero will be nearer the origin than the pole, so some peaking of the step response, i.e., an increased overshoot, may be expected.

Example 8.4-2

Let us lag compensate the system we have been considering, i.e., the system block diagramed in Fig. 8.3-6 and for the plant transfer function as given in (8.4-17). Let us assume that the loop gain has been adjusted to give us an open-loop transfer function as in (8.4-18), for which the frequency response is plotted on the Nichols chart of Fig. 8.4-8 and for which the resulting uncompensated step response is shown in Fig. 8.4-9.

We assume the transient response as seen in Fig. 8.4-9 is satisfactory. Next, consider the steady-state response. From (8.4-18) it is seen that the system is type one and the corresponding

$$K_v = \lim_{s \to 0} sG_T(s) = 0.68 \qquad (8.4\text{-}30)$$

Thus the system will track a unit-ramp input with steady-state error $e_{ss}(\text{ramp}) = 1/K_v = 1.47$. As in the case of the root locus lag compensation example let us try for $e_{ss}(\text{ramp}) = 0.5$.

Step 1. The factor that K_v must be increased by is $1.47/0.5 \cong 3$.

Step 2. Since K_v must be increased by at least 3, let us choose $\alpha = 4$.

Step 3. The open-loop frequency response has been drawn on the Nichols chart of Fig. 8.4-12. We see from the figure that $\omega_m \cong 0.58$. Thus we choose τ so that $\omega_m \tau \cong 10$ to get 5° of phase shift at ω_m. This gives $\tau = 17.3$. We thus have

$$G_c{}^o(s) = \frac{1 + s17.3}{1 + s69.2} \qquad (8.4\text{-}31)$$

Step 4. For the $G^o(s)$ as determined in Step 3, the corresponding frequency response curve of Fig. 8.4-11 is added to the uncompensated

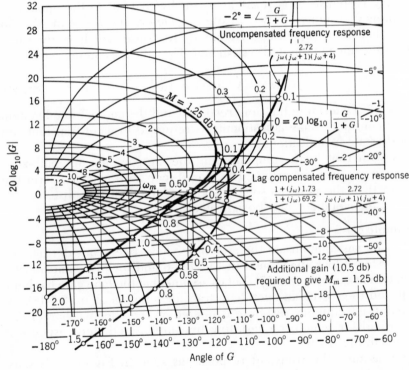

Figure 8.4-12   Lag compensated frequency response.

open-loop frequency response of Fig. 8.4-12. The result is shown in Fig. 8.4-12. We note that the lag-compensated frequency response may be obtained in the same manner as the lead-compensated frequency response curve—by adding graphically the frequency response of the uncompensated system to that of the compensator directly on the Nichols chart. We see that 10.5 db of additional gain is required to give $M_m = 1.25$ db. This 10.5 db gives a gain of 3.35. Thus the lag compensated open-loop frequency response

$$G_T(j\omega) = 3.35 \frac{1 + (j\omega)17.3}{1 + (j\omega)69.2} \frac{2.72}{j\omega(1 + j\omega)(4 + j\omega)} \qquad (8.4\text{-}32)$$

will give $M_m = 1.25$ db. For a fixed plant then as given by (8.4-17), the required lag compensator is

$$G_c(s) = \frac{G_T(s)}{G_f(s)} = 2.28 \frac{(s + 0.0577)}{(s + 0.0145)}$$

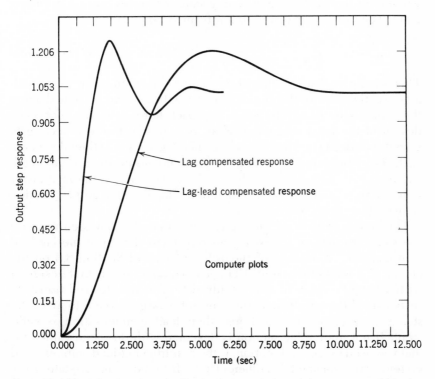

Figure 8.4-13 Nichols chart lag-lead compensated and lag compensated step response.

Step 5. From (8.4-32) we see that the compensated velocity error constant $K_v = 2.28$. Thus $e_{ss}(\text{ramp}) = 0.438$, which more than fulfills the requirement of $e_{ss}(\text{ramp}) = 0.5$, our design goal. The step response for the lag compensated system is shown in Fig. 8.4-13. As mentioned under Step 5 of the design procedure, the addition of the lag compensator may be expected to increase the overshoot somewhat. This is seen clearly by comparing the uncompensated response in Fig. 8.4-9 and the lag compensated response in Fig. 8.4-13. The settling times are approximately the same, although one might expect a slightly longer settling time for the compensated system since it has a slightly lower ω_m. This may be noted in the slightly longer time that it takes the lag compensated step response to reach its peak value.

Lag-Lead Compensation by Nichols Chart

The lag-lead compensator, as was mentioned in Section 8.1, is used when both the transient and steady-state responses require improvement. The form of the compensator from (8.1-15) is

$$G_c(s) = A \frac{s + 1/\tau_1}{s + \alpha/\tau_1} \cdot \frac{s + 1/\tau_2}{s + 1/\alpha\tau_2}, \qquad A, \tau_1, \tau_2 > 0, \quad \alpha > 1 \quad (8.4\text{-}33)$$

This is simply a lead section of the standard form in conjunction with a lag section also of the standard form, with the two sections having the same value of α. The basic approach here, as in the root locus design of the lag-lead compensator, is to design the lead section to give the desired transient response. The lag section is then designed, with the same value of α, so as to lag compensate the already lead-compensated system. The desired lag-lead compensator transfer function thus is simply the product of the lead section and lag section transfer functions.

The choice of an α for the lag-lead compensator is made in exactly the same manner as it was in the root locus design of the lag-lead compensator. (A review of the relevant part of Section 8.3 on the choice of α might be helpful.) The lead compensator is first designed with whatever value of α is required to give the desired transient response. If the resulting error constant is large enough to meet the steady-state error specifications, then lead compensation is sufficient. If not, then a lag section is designed with the same value of α that was used in the lead section. If the resulting lag-lead compensated system error constant is sufficiently high to meet steady-state error specifications, the design is complete. If not, α is increased by whatever factor (plus a margin of say 10 to 20%) is required to bring the error constant up to the required magnitude and a new lead and lag section is redesigned using this new value of α. This will now be outlined.

THE STEPS IN THE NICHOLS CHART DESIGN OF A LAG-LEAD COMPENSATOR

Step 1. Determine M_m and ω_m to give the desired transient response and steady-state error constant (K_p, K_v, or K_a, depending on the type of system) to give the desired steady-state response. Let us call it $K_m{}^d$.

Step 2. Design the lead compensator as described earlier in this section to realize the desired M_m and ω_m. This gives $G_c^{\text{lead}}(s)$.

Step 3. If the lead compensated system error constant ($K_m{}^l$) satisfies
(a) $K_m{}^l \geq K_m{}^d$, then go to Step 5.
(b) $\alpha K_m{}^l \geq K_m{}^d$, where α is as determined in Step 2, then go to Step 4.
(c) $\alpha K_m{}^l < K_m{}^d$, then choose an α such that $\alpha K_m{}^l \cong 1.1 K_m{}^d$ and return to Step 2 and use this value of α for the lead compensator there.

Step 4. Design a lag compensator for the lead compensated system of Step 2 using the value of α used in Step 2. This will give $G_c^{\text{lag}}(s)$. Then

$$G_c^{\text{lag-lead}}(s) = G_c^{\text{lag}}(s) G_c^{\text{lead}}(s)$$

Step 5. Determine the transient response and the error constant of the lag-lead compensated system. Make adjustments as required to meet the transient and steady-state specifications.

Example 8.4-3

As an example of lag-lead compensation let us again look at the system for which we have designed lead and lag compensators in the previous two sections, i.e., the system that is block diagramed in Fig. 8.3-6 and whose frequency response is drawn on the Nichols chart of Fig. 8.4-8 with uncompensated step response as given in Fig. 8.4-9. Let the compensation objective for the transient and steady-state responses be those that were chosen to be realized by lead compensation and lag compensation separately. These were

$$\text{settling time } t_s = 4 \text{ sec}$$
$$\text{step overshoot} \cong 16\%$$
$$e_{ss}(\text{ramp}) \leq 0.5$$

Step 1. From Fig. 8.4-1 we see that a step overshoot of 16% implies an $M_m = 1.25 \, \text{db}$ and a damping factor $\zeta = 0.5$ for a second-order system. Since $t_s = 4\tau = 4/\zeta\omega_n$, we have $\omega_n = 2$ for $t_s = 4$. From Fig.

8.4-2 we see that for $\omega_n = 2$, $\omega_m = 1.41$. Since $e_{ss}(\text{ramp}) = 1/K_v$ where K_v is the velocity error constant, $e_{ss}(\text{ramp}) \leq 0.5$ implies $K_v \geq 2$. Thus our compensator design objectives are

$$M_m = 1.25 \text{ db}$$
$$\omega_m = 1.41 \text{ rad/sec}$$
$$K_v \geq 2$$

Step 2. A lead compensator which realizes the transient response specifications of Step 1 has been designed in Example 8.4-1. From (8.4-19) we know the pole and zero of this compensator are given by

$$\frac{1 + 1.41s}{1 + 0.353s}$$

The lead-compensated frequency response is shown in Fig. 8.4-14. The compensator has $\alpha = 4$.

Step 3. It has been determined in Example 8.4-1 that the lead-compensated system has $K_v = 1.43$ [see (8.4-22)]. This does not meet the specification $K_v \geq 2$ of Step 1. Hence additional lag compensation is required. In this case $K_v = 4 \times 1.43 = 5.72 \geq 2$.

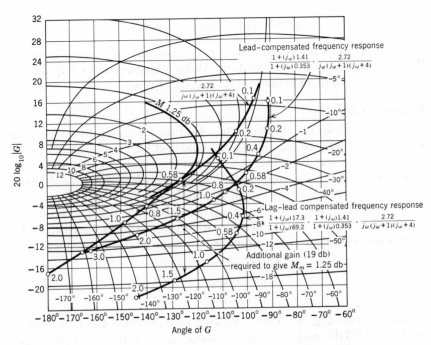

Figure 8.4-14   Lag-lead compensation by Nichols chart.

Step 4. We wish to design a lag compensator section for our system with $\alpha = 4$, the same α as has the lead compensator of Step 2. This has been done for this system in Example 8.4-2. The lag compensator there was found in (8.4-31) to be

$$G_c^0(s) = \frac{1+s17.3}{1+s69.2}$$

The frequency response for this lag compensator section (the $\alpha = 4$ curve of Fig. 8.4-11) was added to the lead compensated frequency response of Fig. 8.4-14. The result is also shown in Fig. 8.4-14. We see from the figure that the lag-lead compensated frequency response requires 19 db of additional gain to give a closed-loop $M_m = 1.25$ db. This we make up by proper choice of the factor A of the compensator [see (8.4-33)]. Now 19 db gives a gain of 8.82. Therefore

$$G_T(j\omega) = 8.82\,\frac{1+j\omega 17.3}{1+j\omega 69.2} \cdot \frac{1+j\omega 1.41}{1+j\omega 0.353} \cdot \frac{2.72}{j\omega(j\omega+1)(j\omega+4)}$$

gives an $M_m = 1.25$ db. Therefore the lag-lead compensator of the form of (8.4-33) is given by

$$G_c(s) = \frac{G_T(s)}{G_f(s)} = 23.9\,\frac{(s+0.708)(s+0.0578)}{(s+2.84)(s+0.0144)} \qquad (8.4\text{-}34)$$

Step 5. The lag-lead compensated step response using the compensator of Step 4 is shown in Fig. 8.4-13. Again the overshoot is slightly higher than the 16% we were seeking. Using $G_T(s)$ as determined in Step 4, we determine immediately that

$$K_v = \lim_{s\to 0} sG_T(s) = \frac{8.82 \times 2.72}{4} = 5.9$$

which is quite a bit higher than the $K_v = 2$ required in Step 1. Thus, if the excess overshoot over 16% was undesirable, we might lower the A of the compensator from the value of 23.9 shown in (8.4-34). This would give a lower K_v and also a lower M_m and consequently a lower step-response overshoot. Precisely how much it would have to be lowered to give an overshoot of 16% can be determined by trial and error. Another method of lowering this step response overshoot would be to vary the position of the zero of the lead section of our lag-lead compensator. Just which way its position should be changed to produce a lower overshoot cannot readily be told from the Nichols chart that we are working with here. Hence this would also have to be determined by trial and error.

8.5 COMPENSATION BY MITROVIC'S METHOD

Basic Considerations

With Mitrovic's method we essentially use the two basic equations, (4.3-6), to get two conditions that the compensator parameters must satisfy. Of course our compensators have more than two free parameters, so additional conditions must be found to set all the parameters. Our method here will be to simply set the position of the compensator zero or zeros by rule of thumb. If we look closely, it will be seen that this is exactly what has been done already in the root locus and frequency-response methods of compensator design. Setting the zeros will leave two free parameters (A and α) to be determined in the case of the lead, lag, and lag-lead compensators and these are now determined by solution of the two basic equations (4.3-6). A simple plot in the (A, α)-plane will show just what can be realized in the way of transient and steady-state response by varying A and α.

Pursuing this, consider the two basic equations of the method (4.3-6). Here we write in a slightly more compact form:

$$a_0 + \sum_{j=2}^{n} a_j \omega_n{}^j \phi_{j-1}(\zeta) = 0$$

$$\sum_{j=1}^{n} a_j \omega_n{}^{j-1} \phi_j(\zeta) = 0 \qquad (8.5\text{-}1)$$

where a_j, $j = 0, 1, \ldots, n$ are the coefficients of the characteristic equation of the system. In our case it is the characteristic equation of the closed-loop system that we consider. The undamped natural frequency ω_n and damping factor ζ corresponding to a pair of closed-loop poles; and the functions $\phi_j(\cdot)$, $j = 1, 2, \ldots, n$ are given by (4.3-7) (some typical and most used values of ϕ are given in Table 4.3-1). Now let us look at the case where the coefficients $(a_j, j = 0, 1, 2, \ldots, n)$ of the characteristic equation are linear functions of two parameters x_1 and x_2:

$$a_j = a_{j1} x_1 + a_{j2} x_2 + a_{j0}, \qquad j = 0, 1, 2, \ldots, n \qquad (8.5\text{-}2)$$

For our use, x_1 and x_2 will usually be the A and α of the compensator which is being designed, but they need not necessarily be these two particular parameters. Substituting (8.5-2) into (8.5-1) gives

$$\left[a_{01} + \sum_{j=2}^{n} a_{j1} \omega_n{}^j \phi_{j-1}(\zeta) \right] x_1 + \left[a_{02} + \sum_{j=2}^{n} a_{j2} \omega_n{}^j \phi_{j-1}(\zeta) \right] x_2$$

$$= -\left[a_{00} + \sum_{j=2}^{n} a_{j0} \omega_n{}^j \phi_{j-1}(\zeta) \right]$$

$$\left[\sum_{j=1}^{n} a_{j1}\omega_n^{\,j-1}\phi_j(\zeta) \right] x_1 + \left[\sum_{j=1}^{n} a_{j2}\omega_n^{\,j-1}\phi_j(\zeta) \right] x_2 = - \sum_{j=1}^{n} a_{j0}\omega_n^{\,j-1}\phi_j(\zeta)$$

$$(8.5\text{-}3)$$

which are two linear equations in the parameters x_1 and x_2. Given the coefficients of the characteristic equation as functions of the two parameters x_1 and x_2, we may routinely use (8.5-3) to solve for values of x_1 and x_2 that will give a pair of complex poles with any desired ω_n and ζ. This is very useful, for it allows us to solve for the compensator parameters directly with a minimum of effort. Equations (8.5-3) also permit us to construct contours in the (A, α)-plane which very explicitly demonstrate what values of ζ and ω_n can be achieved for A and α in specific ranges. Now let us show in detail how (8.5-3) may be used to design compensators of the type we have been considering in this chapter.

Lead Compensation by Mitrovic's Method

As ever, the lead compensator is used to improve the transient response of the system. Let us consider the basic system as shown in Fig. 8.1-1 and assume the compensator is a lead compensator of the form

$$G_c(s) = A\frac{s+1/\tau}{s+\alpha/\tau}, \qquad A, \tau > 0, \qquad \alpha > 1 \qquad (8.5\text{-}4)$$

and the fixed plant has the form

$$G_f(s) = \frac{N_f(s)}{D_f(s)} \qquad (8.5\text{-}5)$$

Now for the system considered here (Fig. 8.1-1) we have the overall transfer function $T(s)$ given by

$$T(s) = \frac{G_c(s)G_f(s)}{1+G_c(s)G_f(s)} \qquad (8.5\text{-}6)$$

Using (8.5-4) and (8.5-5) in (8.5-6),

$$T(s) = \frac{A(s+1/\tau)N_f(s)}{(s+\alpha/\tau)D_f(s)+A(s+1/\tau)N_f(s)}$$

The characteristic polynomial [$F(s)$, the denominator of the overall transfer function] is given by

$$F(s) = \left(s+\frac{\alpha}{\tau}\right)D_f(s) + A\left(s+\frac{1}{\tau}\right)N_f(s) \qquad (8.5\text{-}7)$$

Since the plant is assumed given, the coefficients of $N_f(s)$ and $D_f(s)$ are known. If the position of the compensator zero (at $-1/\tau$) is determined by rule of thumb or other means, then (8:5-7) shows that the coefficients of the characteristic equation $[F(s)]$ are linear functions of the two compensator parameters A and α. Hence we may let A and α take the place of x_1 and x_2 in (8.5-3) and use that set of equations to solve for an A and α that will give a closed-loop pole with the desired ω_n and ζ. This assures us of a pair of closed-loop poles with the desired ζ and ω_n. It says nothing about where the other closed-loop poles fall. In fact, one might fall into the RHP and we would hence have an unstable closed-loop system. Or there might be other closed-loop poles very near the $j\omega$-axis relative to the poles with the desired ω_n and ζ. These would dominate the response of the system and the resulting response certainly would be undesirable. Here the success of the method depends upon how well the compensator zero is placed. Brief consideration of the root locus will generally be useful here to determine just how the zero should be placed to come up with values of A and α that will make the poles with the desired ω_n and ζ dominant. As always, application of trial and error is helpful with the problem of locating the compensator zero.

THE STEPS IN THE DESIGN OF A LEAD COMPENSATOR BY MITROVIC'S METHOD

Step 1. Determine ζ and ω_n for a pair of dominant poles which will give the desired transient response.

Step 2. Using (8.5-7), determine the coefficients of the characteristic polynomial as functions of the compensator parameters τ, A, and α [see (8.5-4)].

Step 3. Determine the zero of the compensator (at $-1/\tau$) so that the poles with ζ and ω_n as determined in Step 1 will be dominant. Usually, cancelling out the plant pole nearest the origin on the negative real axis will suffice to locate the zero. This determines the value of τ. Thus, using the results of Step 2, we have the coefficients (a_j) of the characteristic polynomial as functions of A and α:

$$a_j = a_{j1}A + a_{j2}\alpha + a_{j0}, \qquad j = 0, 1, 2, \ldots, n \qquad (8.5-8)$$

Using this in (8.5-3), we have

$$\left[a_{01} + \sum_{j=2}^{n} a_{j1}\omega_n{}^j\phi_{j-1}(\zeta) \right] A + \left[a_{02} + \sum_{j=2}^{n} a_{j2}\omega_n{}^j\phi_{j-1}(\zeta) \right] \alpha$$
$$= -\left[a_{00} + \sum_{j=2}^{n} a_{j0}\omega_n{}^j\phi_{j-1}(\zeta) \right]$$

$$\left[\sum_{j=1}^{n} a_{j1}\omega_n{}^{j-1}\phi_j(\zeta)\right]A + \left[\sum_{j=1}^{n} a_{j2}\omega_n{}^{j-1}\phi_j(\zeta)\right]\alpha = -\left[\sum_{j=1}^{n} a_{j0}\omega_n{}^{j-1}\phi_j(\zeta)\right]$$

$$(8.5\text{-}9)$$

Using ζ and ω_n as determined in Step 1, solve (8.5-9) for A and α. This determines $G_c(s)$ as in (8.5-4).

Step 4. Determine the step response using the compensator as determined in Step 3. Adjust minor discrepancies from the desired transient response by adjusting A of the compensators. If this fails or if the response is too far from the desired transient response, return to Step 2.

Let us next consider an example.

Example 8.5-1

We consider the same system for compensation described in the previous sections of this chapter. From the block diagram of the system (Fig. 8.3-6), we see that

$$G_f(s) = \frac{1}{s(s+1)(s+4)}$$

However, before trying to design a lead compensator for this plant, let us see how we might use the basic equations of Mitrovic's method to adjust the loop gain so as to achieve the best possible response without the use of a compensator. To do this we let $G_c(s) = A$ in Fig. 8.3-6 and the overall transfer function $T(s)$ is given by

$$T(s) = \frac{G_c(s)G_f(s)}{1 + G_c(s)G_f(s)} = \frac{A}{s^3 + 5s^2 + 4s + A} \qquad (8.5\text{-}10)$$

From (8.5-10) we see that the coefficients of the characteristic polynomial are

$$a_3 = 1, \quad a_2 = 5, \quad a_1 = 4, \quad \text{and} \quad a_0 = A$$

Using these values in (8.5-1) gives

$$a_0 = -5\omega_n{}^2\phi_1(\zeta) - \omega_n{}^3\phi_2(\zeta)$$

$$a_1 = 5\omega_n\phi_2(\zeta) + \omega_n{}^2\phi_3(\zeta) \qquad (8.5\text{-}11)$$

We already have the value of a_1; however, it is included in (8.5-11) so that we can draw up an (a_1, a_0)-plane plot showing constant-ζ contours for some of the most desirable values of ζ. This has been done in Fig. 8.5-1 for $\zeta = 0.3$, 0.5, and 0.7. Since our plant has fixed $a_1 = 4$, we look along the $a_1 = 4$ line in the figure and see what ζ and ω_n are available to us by varying $a_0 = A$. Now let us assume that the

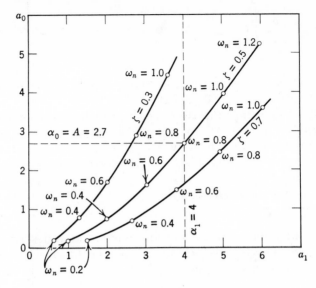

Figure 8.5-1 (a_1, a_0)-plane for the system of Fig. 8.3-6.

overshoot of approximately 16% which goes along with $\zeta = 0.5$ is about what we want. Looking along the $a_1 = 4$ line of Fig. 8.5-1, we see that $\zeta = 0.5$, $\omega_n = 0.8$ for $a_0 = A = 2.7$. Using $A = 2.7$ in (8.5-10), we have the corresponding closed-loop transfer function. The step response for this closed-loop transfer function is shown in Fig. 8.5-2, where it is labeled the uncompensated response. We see what looks almost exactly like a second-order response with $\zeta = 0.5$, $\omega_n = 0.8$.

From Fig. 8.5-2, the uncompensated settling time is seen to be approximately 10 sec. Let us now choose as our design objective the reduction of this settling time to 5 sec while maintaining the overshoot to that shown in the uncompensated response of Fig. 8.5-2. Thus choose $\zeta = 0.5$ and $t_s = 4\tau = 4/\zeta\omega_n = 4$ sec which gives us 20% margin on the desired settling time of 5 sec. This implies $\omega_n = 2$.

Step 1. We thus have for the desired dominant two poles $\zeta = 0.5$, $\omega_n = 2$.

Step 2. From Fig. 8.3-6, we have

$$T(s) = \frac{G_c(s)G_f(s)}{1+G_c(s)G_f(s)} = \frac{A(s+1/\tau)}{(s+\alpha/\tau)(s)(s+1)(s+4)+A(s+1/\tau)}$$

from which

$$T(s) = \frac{A(s+1/\tau)}{s^4+(5+\alpha/\tau)s^3+(5\alpha/\tau+4)s^3+(4\alpha/\tau+A)s+A/\tau}$$

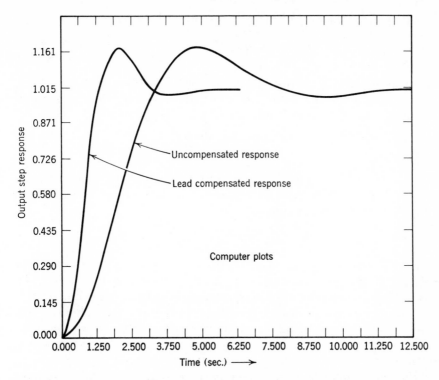

Figure 8.5-2   Mitrovic's method lead compensated and uncompensated step response.

This gives as the coefficients of the characteristic polynomial

$$a_4 = 1, \quad a_3 = 5 + \alpha/\tau, \quad a_2 = 5\alpha/\tau + 4, \quad a_1 = 4\alpha/\tau + A, \quad a_0 = A/\tau$$
$$(8.5\text{-}12)$$

Step 3. We must now set the zero of the compensator so that if we achieve poles with ζ and ω_n as specified in Step 1, they will dominate. From the root locus we know that this usually can be accomplished by cancelling out the plant pole nearest the origin on the negative real axis. This pole, for the plant we have here, is at $s = -1$. Hence setting the compensator zero at $s = -1/\tau = -1$ we have $\tau = 1$. Using this in (8.5-12) and these coefficients in (8.5-1) gives

$$A + (5\alpha + 4)\omega_n{}^2\phi_1(\zeta) + (5 + \alpha)\omega_n{}^3\phi_2(\zeta) + \omega_n{}^4\phi_3(\zeta) = 0$$
$$-4\alpha - A + (5\alpha + 4)\omega_n\phi_2(\zeta) + (5 + \alpha)\omega_n{}^2\phi_3(\zeta) + \omega_n{}^3\phi_4(\zeta) = 0$$

which we may rearrange to give

$$A + [5\omega_n{}^2\phi_1(\zeta) + \omega_n{}^3\phi_2(\zeta)]\alpha = -4\omega_n{}^2\phi_1(\zeta) - 5\omega_n{}^3\phi_2(\zeta) - \omega_n{}^4\phi_3(\zeta)$$

$$-A + [-4 + 5\omega_n\phi_2(\zeta) + \omega_n{}^2\phi_3(\zeta)]\alpha = -4\omega_n\phi_2(\zeta) - 5\omega_n{}^2\phi_3(\zeta) - \omega_n{}^3\phi_4(\zeta)$$

$$(8.5\text{-}13)$$

Equations (8.5-13) correspond to the two equations of (8.5-9). Now using $\omega_n = 2$, $\zeta = 0.5$ (from Step 1) and evaluating the $\phi_j(\zeta)$, coefficients $j = 1, 2, 3, 4$, from Table 4.3-1, (8.5-13) becomes

$$A - 12\alpha = -24$$
$$-A + 6\alpha = 0$$

From these two equations it is routine to determine that $A = 24$, $\alpha = 4$. From this we use (8.5-4) and τ as determined just above to find that the desired compensator is

$$G_c(s) = 24\frac{s+1}{s+4} \qquad (8.5\text{-}14)$$

Step 5. The lead-compensated step response using the compensator of (8.5-14) in the system of Fig. 8.3-6 is shown in Fig. 8.5-2. The overshoot is now down to just slightly more than 15% (as compared to 16% in the uncompensated response). Settling time is almost 5 sec, so the 20% margin of safety we took for the settling time objective, $t_s = 4$ sec, was a good choice. The response meets the design objectives.

For the compensator of (8.5-14) and the plant of Fig. 8.3-6 we have $K_v = 1.5$. For the uncompensated system step response of Fig. 8.5-2, we can determine $K_v = 0.68$. Thus the lead compensator here has also improved the steady-state response of the system.

Lag Compensation by Mitrovic's Method

Lag compensation by Mitrovic's method proceeds along the same basic lines as lead compensation. We again determine the coefficients of the closed-loop system characteristic polynomial in terms of the compensator parameters A, α, and τ. The position of the compensator zero is set by rule of thumb (which fixes τ) and then A and α are determined by use of the two basic equations, (8.5-1). With lag compensation, however, we are not trying to realize a fixed ζ and ω_n for the dominant poles but rather some value of steady-state error constant (K_p, K_v, or K_a). Hence the basic two equations are used to fix lines of constant ζ in the (A, α)-plane. Since the error constant for the system will also be a function of A and α, we may also draw lines of constant error constant in the (A, α)-plane. With such a parameter plane plot one can see directly what values of error constant may be realized for a given choice of A and α and what the trade-off is so far as the transient response and error constant are concerned. A great deal of helpful design information can be collected on such a parameter plane. This will be seen in the next example.

To set up the equations for making the plots of constant ζ and constant error constant in the parameter plane proceed as follows. The lag-compensator has the transfer function

$$G_c(s) = A\frac{s+1/\tau}{s+1/\alpha\tau}, \qquad A, \tau > 0, \quad \alpha > 1 \tag{8.5-15}$$

If we again take $G_f(s) = N_f(s)/D_f(s)$, we have as the overall transfer function of the system (Fig. 8.1-1):

$$T(s) = \frac{A(s+1/\tau)N_f(s)}{(s+1/\alpha\tau)D_f(s) + A(s+1/\tau)N_f(s)}$$

from which we see that the characteristic polynomial is given by

$$F(s) = \left(s+\frac{1}{\alpha\tau}\right)D_f(s) + A\left(s+\frac{1}{\tau}\right)N_f(s) \tag{8.5-16}$$

Now we select the position of the compensator zero (at $-1/\tau$) and consequently fix τ. With τ fixed, it can be seen from (8.5-16) that the coefficients of the characteristic equation will be linear functions of A and $1/\alpha$, i.e., the coefficients may be written in the form

$$a_j = a_{j1}A + a_{j2}\frac{1}{\alpha} + a_{j0}, \qquad j = 0, 1, 2, \ldots, n \tag{8.5-17}$$

Thus the basic equations of Mitrovic's method may be written in the form of (8.5-3) with $x_1 = A$, $x_2 = 1/\alpha$. This gives

$$\left[a_{01} + \sum_{j=2}^{n} a_{j1}\omega_n^{j}\phi_{j-1}(\zeta)\right]A + \left[a_{02} + \sum_{j=2}^{n} a_{j2}\omega_n^{j}\phi_{j-1}(\zeta)\right]\frac{1}{\alpha}$$

$$= -\left[a_{00} + \sum_{j=2}^{n} a_{j0}\omega_n^{j}\phi_{j-1}(\zeta)\right]$$

$$\left[\sum_{j=1}^{n} a_{j1}\omega_n^{j-1}\phi_j(\zeta)\right]A + \left[\sum_{j=1}^{n} a_{j2}\omega_n^{j-1}\phi_j(\zeta)\right]\frac{1}{\alpha} = -\left[\sum_{j=1}^{n} a_{j0}\omega_n^{j-1}\phi_j(\zeta)\right]$$

$$\tag{8.5-18}$$

Now if the coefficients of the characteristic polynomial are determined as in (8.5-17), we may determine $\phi_j(\zeta), j = 1, 2, \ldots, n$ for the values of desired ζ from Table 4.3-1. We may then solve (8.5-18) for A and $1/\alpha$ for a range of values of ω_n to give the lines of constant ζ in the (A, α)-plane.

Lines of constant error constant in the (A, α)-plane may be obtained even more simply. If the plant is of type m [$G_f(s)$ has m poles at the origin], then the steady-state error constant (K_m) for the compensated system may be determined by

$$K_m = \lim_{s \to 0} s^m G_c(s) G_f(s) = A \alpha K_{mf} \qquad (8.5\text{-}19)$$

where $G_c(s)$ has been taken to be the lag compensator of (8.5-15) and

$$K_{mf} = \lim_{s \to 0} s^m G_f(s)$$

From (8.5-19) we may write directly

$$\alpha = \frac{K_m}{K_{mf}} \frac{1}{A} \qquad (8.5\text{-}20)$$

which may be used directly to plot lines of constant K_m in the (A, α)-plane.

Again to apply our method we must set the position of the compensator zero by rule of thumb. The pole and zero of the lag compensator must of course fall very close together and very near the origin if the compensator is not to have a large detrimental effect upon the transient response of the system. It has been seen in discussing lag compensator design by root locus (Section 8.3) that placing the lag compensator zero approximately 10% of the way from the origin and toward the first plant pole on the negative real axis will usually have negligible effect on the transient response. Thus we may use the same rule here. In lag compensator design by Nichols chart it was also seen that the lag compensator had negligible effect on the system if we chose the τ of the compensator so that $\omega_m \tau = 10$ where $\omega_m = \omega_n \sqrt{1 - 2\zeta^2}$. This gave 5° of compensator phase shift at frequency ω_m. Thus we might alternately choose $\tau = 10/\omega_n \sqrt{1 - 2\zeta^2}$ if we know the ζ and ω_n of the poles dominating the transient response of the system. Either rule may be used, and there are others that may be used as well. As long as the lag compensator has negligible effect on the transient response, the position of the compensator zero is not really critical. This condition is usually easy to meet.

THE STEPS IN THE DESIGN OF A LAG COMPENSATOR BY MITROVIC'S METHOD

Step 1. Determine what value of steady-state error constant is required to give the steady-state response desired. Also, determine ζ and ω_n for the dominant poles of the system.

Step 2. Determine the coefficients of the characteristic polynomial as functions of the compensator parameters A, α, and τ by using (8.5-16).

Step 3. Determine τ by setting the zero of the compensator close enough to the origin so that the compensator will have a negligible effect upon the transient response. This may be done by setting the compensator zero approximately 10% of the way between the origin and first plant singularity on the negative real axis or, alternatively, by letting $\omega_m \tau = 10$ where $\omega_m = \omega_n \sqrt{1 - 2\zeta^2}$.

Step 4. Determine contours of constant ζ in the (A, α)-plane for the value of ζ of interest by use of (8.5-18). Values for ω_n which are slightly lower (10–20%) than the ω_n as determined in Step 1 should be considered.

Step 5. Plot in the (A, α)-plane of Step 4 lines of constant steady-state error constant. Equation (8.5-20) may be used to accomplish this. Values of the error constant as determined in Step 1 are of interest here.

Step 6. Choose values of A and α that are seen from the (A, α)-plane of Steps 4 and 5 to give the desired or greater value of steady-state error constant (as determined in Step 1) and also ζ and ω_n which will give a satisfactory transient response. The desired compensator is then

$$G_c(s) = A\left(\frac{s + 1/\tau}{s + 1/\alpha\tau}\right)$$

Step 7. Determine the transient response for the system using the compensator of Step 6. Make adjustments in A and α by using the (A, α)-plane constructed in Step 4; use that of Step 5 if the steady-state response is not satisfactory. Let us now consider an example.

Example 8.5-2

Again let us consider the feedback system of Fig. 8.3-6, for which the plant is given by

$$G_f(s) = \frac{1}{s(s + 1)(s + 4)} \tag{8.5-21}$$

It was determined in Example 8.5-1 by use of Fig. 8.5-1 that a fixed gain $G_c(s) = A = 2.7$ gave an uncompensated transient response as in

Fig. 8.5-2. We see that we have an overshoot of approximately 16% and settling time of 10 sec. Let us assume that this is an adequate transient response for our needs. For $G_c(s) = 2.7$ the velocity error constant is determined to be $K_v = 0.68$, which results in steady-state error to a ramp, $e_{ss}(\text{ramp}) = 1.48$. Let us now design a lag compensator that will give $e_{ss}(\text{ramp}) \leq 0.5$ and correspondingly $K_v \geq 2$.

Step 1. We want $K_v \geq 2$. From Fig. 8.5-1 we see that $\zeta = 0.5$, $\omega_n = 0.8$ give the satisfactory uncompensated transient response of Fig. 8.5-2.

Step 2. Using the plant as given by (8.5-21) and the compensator as in (8.5-15), we have

$$T(s) = \frac{G_c(s)G_f(s)}{1 + G_c(s)G_f(s)} = \frac{A(s+1/\tau)}{(s+1/\alpha\tau)(s)(s+1)(s+4) + A(s+1/\tau)}$$

$$= \frac{A(s+1/\tau)}{s^4 + (5+1/\alpha\tau)s^3 + (4+5/\alpha\tau)s^2 + (A+4/\alpha\tau)s + A/\tau} \tag{8.5-22}$$

from which we have the coefficients of the characteristic polynomial:

$$a_4 = 1, \quad a_3 = 5 + \frac{1}{\alpha\tau}, \quad a_2 = 4 + \frac{5}{\alpha\tau}, \quad a_1 = A + \frac{4}{\alpha\tau}, \quad a_0 = A/\tau$$

Step 3. The plant pole on the negative real axis nearest the origin (but not at the origin itself) is at $s = -1$. We choose the compensator zero to fall 10% of the distance between the origin and this pole, i.e., we choose $-1/\tau = -0.1$. Therefore we have $\tau = 10$.

Step 4. With $\tau = 10$ as determined in Step 3 and the characteristic polynomial coefficients as determined in Step 2, using (8.5-1) we have

$$0.1A + \left(4 + \frac{0.5}{\alpha}\right)\omega_n^2\phi_1(\zeta) + \left(5 + \frac{0.1}{\alpha}\right)\omega_n^3\phi_2(\zeta) + \omega_n^4\phi_3(\zeta) = 0 \tag{8.5-23}$$

$$-A - \frac{0.4}{\alpha} + \left(4 + \frac{0.5}{\alpha}\right)\omega_n\phi_2(\zeta) + \left(5 + \frac{0.1}{\alpha}\right)\omega_n^2\phi_3(\zeta) + \omega_n^3\phi_4(\zeta) = 0$$

We saw in Step 1 that the desired value of damping factor is $\zeta = 0.5$. We can now consider plotting a constant-ζ contour in the (A, α)-plane for $\zeta = 0.5$. Using Table 4.3-1 to determine values of the $\phi_j(\zeta)$, $j = 1, 2, 3, 4$ in (8.5-23) gives

$$0.1A + (-0.5\omega_n^2 + 0.1\omega_n^3)\frac{1}{\alpha} = 4\omega_n^2 - 5\omega_n^3$$

$$\tag{8.5-24}$$

$$-A + (-0.4 + 0.5\omega_n)\frac{1}{\alpha} = -4\omega_n + \omega_n^3$$

Equation (8.5-24) may be seen to be (8.5-18) evaluated for this example with $\zeta = 0.5$. By using (8.5-24) the $\zeta = 0.5$ contour was constructed in the (A, α)-plane for ω_n in the range $0.72 \leqslant \omega_n \leqslant 0.76$ (Fig. 8.5-3). This range of ω_n was chosen since it covered the range of α that was thought desirable for this compensator. A little trial and error work was necessary here; however, it was minor since $\omega_n = 0.8$ was known from Step 1 as the ω_n for the uncompensated system.

Step 5. Using (8.5-21) and the compensator transfer function of (8.5-15), we have

$$K_v = \lim_{s \to 0} s G_c(s) G_f(s) = \frac{A\alpha}{4}$$

from which we have

$$\alpha = \frac{4K_v}{A} \tag{8.5-25}$$

which may be used to draw lines of constant K_v in the (A, α)-plane. Lines with $K_v = 1.5$, 2.0, 2.5, and 3.0 were computed and drawn in the (A, α)-plane of Fig. 8.5-3. These values of K_v were chosen because it was known from Step 1 that we were interested in K_v in the neighborhood of $K_v = 2$.

Step 6. By looking at the (A, α)-plane of Fig. 8.5-3 we can see easily the kind of trade-off in transient response that will have to be made if a higher K_v is desired. A higher K_v means a lower ω_n. We wanted to get $K_v \geqslant 2$ while maintaining $\zeta = 0.5$. So, looking at the (A, α)-plane of Fig. 8.5-3, let us choose a $K_v \cong 2.5$. This will give us a little margin so far as $e_{ss}(\text{ramp})$ is concerned and will not cost too much

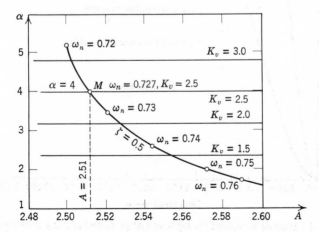

Figure 8.5-3 (A, α)-Plane for the system of Fig. 8.3-6 lag compensated with $\tau = 10$.

so far as a reduced ω_n is concerned. It can be seen that ω_n decreases fairly slowly for increasing K_v. Once having chosen $K_v = 2.5$, we have directly from Fig. 8.5-3 that $\alpha = 4$, $A = 2.51$ are the desired parameter values. Since we know from Step 3 that $\tau = 10$ we have as our desired lag compensator,

$$G_c(s) = 2.51 \frac{s + 0.1}{s + 0.025} \qquad (8.5\text{-}26)$$

Step 7. From (8.5-25) we may calculate more exactly that $K_v = 2.51$ for $A = 2.51$, $\alpha = 4$. The lag compensated transient response using the compensator of (8.5-26) in our system (Fig. 8.3-6) has been computed and is shown in Fig. 8.5-4. Again as in all the previous cases of lag compensator design, the addition of the lag compensator has increased the overshoot. As has been mentioned before, this happens because the compensator pole-zero pair near the zero introduces a closed-loop pole-zero pair near the origin. The zero of this pair is

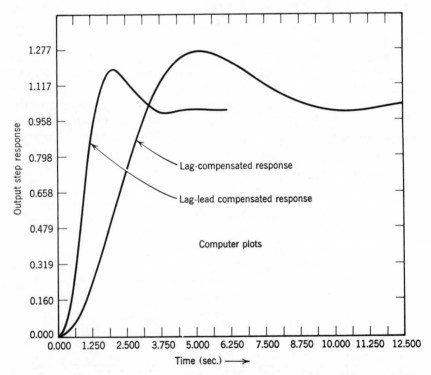

Figure 8.5-4 Mitrovic's method for lag-lead compensated and lag compensated step response.

nearer the origin than the pole, hence the step response is slightly peaked, resulting in additional overshoot. Since K_v is slightly higher than it need be ($K_v = 2.5$ as compared to the design specification $K_v \geq 2$), we might reduce the gain A slightly to correct this excess overshoot if it is a problem. If we did not wish to reduce K_v, we might consider a higher damping factor ζ for the transient, dominating, closed-loop poles. To do this we could simply draw another constant-ζ contour in the (A, α)-plane of Fig. 8.5-3 for a slightly higher value of ζ and then on this contour look for values of A and α that would give the desired K_v ($= 2.5$) and a satisfactory ω_n.

Settling time has also been increased somewhat, as seen by comparing the uncompensated and the lag compensated responses. Reducing A will also decrease the settling time somewhat since it will increase the damping factor.

Lag-Lead Compensation by Mitrovic's Method

Lag-lead compensation is used whenever both the transient response and the steady-state response of a system are unsatisfactory and require improvement. A simple lead compensator, while improving transient response, will also give some increase in steady-state error constant. Thus if improvement is required in both the transient and steady-state response, as a first try a lead compensator should be designed to give the desired improvement in transient response. The steady-state response then may be checked to see whether it also has been improved enough. If not, a lag-lead compensator is called for.

The design of a lag-lead compensator by Mitrovic's method does not require design of a lead compensator to improve the transient response as required followed by design of a lag compensator with the same value of α, the method used in both root locus and frequency-response design. Although this approach may be used with Mitrovic's method, it is more direct to proceed from system specifications to the lag-lead compensator parameters. Using this procedure, we gain even more insight into what is possible with the compensator.

The design of a lag-lead compensator by Mitrovic's method differs only in minor details from the design method for the lag compensator. We still find the coefficients of the characteristic equation as functions of the compensator parameters, the zeros of the compensator are set by rule of thumb (we now have two zeros to set) and then by use of the two basic equations of Mitrovic's method, constant-ζ contours are drawn in the (A, α) parameter plane. The error constant contours may also be drawn in this plane. With this parameter plane the values of A and α (and hence the compensator itself) may be determined so as to give the damping factor ζ, the undamped natural frequency ω_n and the error constant that will give the desired transient and steady-state response.

The transfer function of the compensator of interest from (8.1-15) is

$$G_c(s) = A\,\frac{(s+1/\tau_1)\,(s+1/\tau_2)}{(s+\alpha/\tau_1)\,(s+1/\alpha\tau_2)}\;, \qquad A, \tau_1, \tau_2 > 0, \quad \alpha > 1$$

$$\text{(8.5-27)}$$

$$= \frac{A[s^2 + (1/\tau_1 + 1/\tau_2)s + 1/\tau_1\tau_2]}{s^2 + (\alpha/\tau_1 + 1/\alpha\tau_2)s + 1/\tau_1\tau_2}$$

If $G_f(s) = N_f(s)/D_f(s)$, we have as the closed-loop transfer function of the system of interest (Fig. 8.1-1) $T(s)$ as

$$\frac{A[s^2 + (1/\tau_1 + 1/\tau_2)s + 1/\tau_1\tau_2]N_f(s)}{[s^2 + (\alpha/\tau_1 + 1/\alpha\tau_2)s + 1/\tau_1\tau_2]D_f(s) + A[s^2 + (1/\tau_1 + 1/\tau_2)s + 1/\tau_1\tau_2]N_f(s)}$$

$$\text{(8.5-28)}$$

from which the characteristic polynomial is given by

$$\left(s^2 + \alpha's + \frac{1}{\tau_1\tau_2}\right)D_f(s) + A\left[s^2 + \left(\frac{1}{\tau_1} + \frac{1}{\tau_2}\right)s + \frac{1}{\tau_1\tau_2}\right]N_f(s) \quad \text{(8.5-29)}$$

where

$$\alpha' = \frac{\alpha}{\tau_1} + \frac{1}{\alpha\tau_2} \tag{8.5-30}$$

We see that the coefficients of the characteristic equation are linear functions of the parameters A and α'. If the two compensator zeros are set, which sets τ_1 and τ_2, then we may calculate α directly from (8.5-30) if α' is known. Using (8.5-29), we may write the coefficients of the characteristic polynomial in the form

$$a_j = a_{j1}A + a_{j2}\alpha' + a_{j0}, \qquad j = 0, 1, 2, \ldots, n$$

Substituting this in (8.5-3) gives the two basic equations with A and α' replacing x_1 and x_2, respectively, as

$$\left[a_{01} + \sum_{j=2}^{n} a_{j1}\omega_n{}^j\phi_{j-1}(\zeta)\right]A + \left[a_{02} + \sum_{j=2}^{n} a_{j2}\omega_n{}^j\phi_{j-1}(\zeta)\right]\alpha'$$

$$= -\left[a_{00} + \sum_{j=2}^{n} a_{j0}\omega_n{}^j\phi_{j-1}(\zeta)\right]$$

$$\left[\sum_{j=1}^{n} a_{j1}\omega_n{}^{j-1}\phi_j(\zeta)\right]A + \left[\sum_{j=1}^{n} a_{j2}\omega_n{}^{j-1}\phi_j(\zeta)\right]\alpha' = -\sum_{j=1}^{n} a_{j0}\omega_n{}^{j-1}\phi_j(\zeta)$$

$$\text{(8.5-31)}$$

Equations (8.5-31) and (8.5-30) may now be used to draw constant-ζ contours in the (A, α)-plane. Error constant contours may be drawn even more easily in this case than in the case of the lag compensator. We simply note that the error constant, K_m, for our system (Fig. 8.1-1) is given (assuming the plant is type m) by

$$K_m = \lim_{s \to 0} s^m G_c(s) G_f(s) = A K_{mf} \qquad (8.5\text{-}32)$$

where $G_c(s)$ is given by (8.5-27) and

$$K_{mf} = \lim_{s \to 0} s^m G_f(s)$$

From (8.5-32) we see that the error constant is independent of α in this case and hence the constant error constant contours will be vertical lines in the (A, α)-plane.

The zeros of the compensator are set in the same way as are those of the lead and lag compensators. The lead section zero [at $-1/\tau_1$ in (8.5-27)] is set as described earlier for the lead compensator zero. It should be used so that the complex poles with the desired ζ and ω_n are the dominant poles. Usually this can be assured by using the lead section zero to cancel out the plant pole nearest the origin (but not at the origin) on the negative real axis. The lag section zero [at $-1/\tau_2$ in (8.5-27)] is placed so that the lag section pole-zero pair will have negligible effect upon the transient response of the system. Usually this can be assured simply by setting the lag section zero about 10% of the way from the origin between the origin and the first plant singularity on the negative real axis. Or alternately, this can be done by using a frequency-response approach, choosing τ_2 so that $\omega_m \tau_2 = 10$ where $\omega_m = \omega_n \sqrt{1 - 2\zeta^2}$ and ω_n and ζ are taken for the complex poles dominating the transient response.

THE STEPS IN THE DESIGN OF A LAG-LEAD COMPENSATOR BY MITROVIC'S METHOD

Step 1. Determine ζ and ω_n for the dominant complex poles, which will give the desired transient response. Determine the steady-state error constant, K_m^d, which will give the desired steady-state response.

Step 2. Determine the coefficients of the characteristic polynomial as functions of the compensator parameters A, α, τ_1, and τ_2 by using (8.5-29).

Step 3. Determine τ_1 by setting the lead section zero at $-1/\tau_1$ so that the complex poles with the desired ζ and ω_n (as determined in Step 1) will be dominant. This is done as explained in Step 1.

Step 4. Determine contours of constant ζ in the (A, α)-plane by use of (8.5-31) where

$$\alpha' = \frac{\alpha}{\tau_1} + \frac{1}{\alpha\tau_2}$$

Values of ζ and ω_n should be chosen in the neighborhood of the ζ and ω_n determined in Step 1.

Step 5. Plot, in the (A, α)-plane obtained in Step 4, lines of constant steady-state error constant by use of (8.5-32). These contours should fall in the same region of the (A, α)-plane as the constant-ζ contours obtained in Step 4.

Step 6. In the (A, α)-plane constructed in Steps 4 and 5 choose values of A and α that are seen to give the ζ, ω_n, and K_m^d as determined in Step 1. Since all three values in general will not be satisfied exactly, the vales of ζ, ω_n, and K_m^d of Step 1 should be used as minimum values to be achieved. With these values of A, α, τ_1, and τ_2 as determined in Step 3, the desired compensator is

$$G_c(s) = A\frac{(s+1/\tau_1)(s+1/\tau_2)}{(s+\alpha/\tau_1)(s+1/\alpha\tau_2)}$$

Step 7. Determine the step response. If adjustments in this transient response are necessary return to Step 6 and choose new values of A and α which alter the step response in the direction required.

Example 8.5-3

Again let us consider the example of Fig. 8.3-6, where

$$G_f(s) = \frac{1}{s(s+1)(s+4)} \tag{8.5-33}$$

We may use as our design goals the goals of Examples 8.5-1 and 8.5-2. Here for the transient response we take as the objective an overshoot of approximately 16% and a settling time of 4 sec. This implies $\zeta = 0.5$, $\omega_n = 2$. The steady-state error design goal was $e_{ss}(\text{ramp}) \leq 0.5$, which implies $K_v \geq 2$. We see that with the lead compensator of Example 8.5-1 [the compensator is given in (8.5-14)] the compensated velocity error constant $K_v = 1.5$. This is not high enough to satisfy our design objective here of $K_v \geq 2$. Hence lead compensation is not enough and so we turn to the design of a lag-lead compensator.

Step 1. We have $\zeta = 0.5$, $\omega_n = 2$, $K_v \geq 2$.

Step 2. To determine the coefficients of the characteristic equation as functions of the compensator parameters we make use of (8.5-29) to get as the characteristic polynomial

$$\left(s^2 + \alpha's + \frac{1}{\tau_1\tau_2}\right)(s)(s+1)(s+4) + A\left[s^2 + \left(\frac{1}{\tau_1} + \frac{1}{\tau_2}\right)s + \frac{1}{\tau_1\tau_2}\right]$$

which we may multiply out to give

$$s^5 + (5+\alpha')s^4 + \left(4+5\alpha' + \frac{1}{\tau_1\tau_2}\right)s^3 + \left(4\alpha' + A + \frac{5}{\tau_1\tau_2}\right)s^2$$

$$+ \left[A\left(\frac{1}{\tau_1} + \frac{1}{\tau_2}\right) + \frac{4}{\tau_1\tau_2}\right]s + \frac{A}{\tau_1\tau_2}$$

where we have as the coefficients of the characteristic equation

$$a_5 = 1, \quad a_4 = 5+\alpha', \quad a_3 = 4+5\alpha' + \frac{1}{\tau_1\tau_2}, \quad a_2 = 4\alpha' + A + \frac{5}{\tau_1\tau_2},$$

$$a_1 = A\left(\frac{1}{\tau_1} + \frac{1}{\tau_2}\right) + \frac{4}{\tau_1\tau_2}, \quad a_0 = \frac{A}{\tau_1\tau_2} \tag{8.5-34}$$

where α' is given by (8.5-30).

Step 3. The plant pole nearest the origin and on the negative real axis is at $s = -1$. We choose the lead section zero (at $-1/\tau_1$) to cancel this pole. This gives $\tau_1 = 1$. We choose the lag section zero (at $-1/\tau_2$) 10% of the way from the origin to this first plant pole at $s = -1$, therefore $-1/\tau_2 = -0.1$ and $\tau_2 = 10$.

Step 4. Using τ_1 and τ_2 as determined in Step 3 and substituting the coefficients of (8.5-34) into (8.5-31) gives

$$(0.1 - \omega_n^2)A + [-4\omega_n^2 + 5\omega_n^3\phi_2(\zeta) + \omega_n^4\phi_3(\zeta)]\alpha'$$
$$= 0.5\omega_n^2 - 4.1\omega_n^3\phi_2(\zeta) - 5\omega_n^4\phi_3(\zeta) - \omega_n^5\phi_4(\zeta) \tag{8.5-35}$$

$$[-1.1 + \omega_n\phi_2(\zeta)]A + [4\omega_n\phi_2(\zeta) + 5\omega_n^2\phi_3(\zeta) + \omega_n^3\phi_4(\zeta)]\alpha'$$
$$= 0.4 - 0.5\omega_n\phi_2(\zeta) - 4.1\omega_n^2\phi_3(\zeta) - 5\omega_n^3\phi_4(\zeta) - \omega_n^4\phi_5(\zeta)$$

Using (8.5-35) and Table 4.3-1 we plot the constant-ζ contours for $\zeta = 0.5$, 0.6, and 0.7 in the (A, α)-plane (Fig. 8.5-5). The range of ω_n was chosen as $1.6 \leqslant \omega_n \leqslant 2.4$ since this bracketed the $\omega_n = 2$, which was determined as the design objective in Step 1.

Step 5. Using (8.5-32) for the lag-lead compensation of $G_f(s)$ as in (8.5-33) gives

$$K_v = \frac{A}{4}, \tag{8.5-36}$$

which may be used to draw constant K_v lines in the (A, α)-plane plotted in Step 4. The constant-K_v lines are also shown in Fig. 8.5-5.

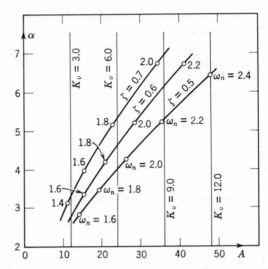

Figure 8.5-5 (A, α)-Plane for a lag-lead compensation of the system of Fig. 8.3-6.

Step 6. From Step 1 we want $\zeta = 0.5$, $\omega_n = 2.0$, and $K_v \geqslant 2$. From the (A, α)-plane of Fig. 8.5-5 we see that the point where $\zeta = 0.5$, $\omega_n = 2.0$ gives $K_v \cong 7$. Thus we may use the A and α values of this point for our compensator since the steady-state specification $K_v \geqslant 2$ is also satisfied there. We see $A = 26.3$, $\alpha = 4.25$ at this point. Using these values with τ_1 and τ_2 as determined in Step 3 gives as the desired compensator:

$$G_c(s) = 26.3 \frac{(s+1)(s+0.1)}{(s+4.25)(s+0.0235)} \qquad (8.5\text{-}37)$$

Using (8.5-36), we find $K_v = 6.57$.

Step 7. The lag-lead compensated step response using the compensator of (8.5-37) is shown in Fig. 8.5-4. There is more overshoot here than we originally intended to have (20% as compared to the desired 16%). This is again due to the introduction of the pole-zero pair near the origin in the lag section of our compensator. The overshoot is, however, not very far off; thus since the settling time is very near the 4-sec value that was desired, the lag-lead compensated step response (Fig. 8.5-4) is likely to be considered a very satisfactory response. In fact, the step response we have here is about as close to the design objectives as anyone could expect to come on the first pass. Now we may use the (A, α)-plane of Fig. 8.5-5 to tune the response further. Since K_v is much higher than the $K_v \geqslant 2$ required, we can easily see in the figure that A can be reduced slightly to give us a higher damping factor, which will reduce the overshoot if this is desired. We see, for instance, that taking the same α used in Step 6 ($\alpha = 4.25$) but reducing A to 21 will give $\zeta = 0.6$. From Fig. 2.3-4 this can be seen to give a drop in overshoot of about 5%, bringing us back down near 16% overshoot. We also see from Fig. 8.5-5 that by having $A = 21$ we have also reduced ω_n to 1.8. If this reduction in ω_n is undesirable, again from the (A, α)-plane plot we can maintain $\omega_n = 2$ and $\zeta = 0.6$ by taking $A = 28$, $\alpha = 5.2$. In general, it can be seen that with an (A, α)-plane plot as shown in Fig. 8.5-5 it is not at all difficult to choose compensator parameters so as to tune the transient response to any desirable response. Moreover, K_v is seen simultaneously, so that the steady-state response of the system may be handled as well.

It should be appreciated that a parameter plane such as the one in Fig. 8.5-5 with ζ and error constant contours drawn upon it gives much immediate design information. The possible responses with a simple lag-lead compensator are immediately apparent. We can also quickly answer the question of whether the type of compensation being employed will achieve our design objectives. Furthermore, the construction of an (A, α)-plane as in Fig. 8.5-5, though perhaps somewhat laborious by hand, is a strictly routine job if a digital computer is available. Mechanizing equations of the form of (8.5-3) for solution is very straightforward. This is one advantage the Mitrovic's method of compensation schemes has over frequency response and root locus methods. It is basically analytic rather than graphic in character and hence lends itself well to implementation on the computer.

8.6 DESIGN OF COMPENSATORS BY S-PLANE SYNTHESIS[6]

Basic Technique

In the compensators we have designed heretofore, the approach has been to assume the form of the compensator and to then choose the parameters of the compensator so that a satisfactory closed-loop transfer function results. In classical lead, lag, or lag-lead compensation we start with the plant transfer function and work out a satisfactory closed-loop transfer function, i.e., we go from the inside out. With s-plane synthesis, this process is reversed. We start with

[6]The presentation here is essentially as that of J. G. Truxal, *Automatic Control System Synthesis*, McGraw-Hill, New York, 1955, Chapter 6.

a closed-loop transfer function that we know gives satisfactory transient and steady-state responses and then proceed to design a compensator that will force the system to give this closed-loop transfer function. The general method is, at least on the surface, a more direct one than the methods we have been considering. However, in the process we find that part of this directness is lost, for unless we exercise considerable care in the selection of the desired closed-loop transfer function, the resulting compensator may be unduly complex or undesirable for other reasons. Hence trial and error is involved in the choice of the closed-loop transfer function and the solution for the corresponding compensators and then some adjustment of the closed-loop transfer function in the light of the obtained compensator. We assume now that the compensator transfer function is not fixed in form but does have at least as many poles as zeros and is rational. If the compensator is not to be fixed in form, as has been the case thus far, the question of the realizability of the compensator arises. This question shall not be answered here; here we merely assume that a network, either passive or active, can be found by the methods of network synthesis that will realize the required compensators. This is not a totally unrealistic assumption. We use as our guide the fact that in general the lower the order of the compensator transfer function, the easier it is to realize. Also, if the poles of the compensator are all on the negative real axis, it may be realized as an *RC* network in series with an amplifier of adjustable gain. This is usually a simple, practical realization.

Let us consider the design of a compensator that will realize a given open-loop transfer function. Our system is as diagrammed in Fig. 8.1-1. The closed-loop transfer function is given by

$$T(s) = \frac{G_c(s)G_f(s)}{1 + G_c(s)G_f(s)}$$

which we solve for the compensator transfer function $G_c(s)$ to get

$$G_c(s) = \frac{T(s)}{[1 - T(s)]G_f(s)} \tag{8.6-1}$$

If we put $T(s)$ in the form

$$T(s) = \frac{N(s)}{F(s)} \tag{8.6-2}$$

then

$$G_c(s) = \frac{N(s)}{[F(s) - N(s)]G_f(s)} \tag{8.6-3}$$

A $G_c(s)$ as in (8.6-3) will realize the closed-loop transfer function as in (8.6-2). From (8.6-1), we see that the compensator here cancels out the plant and then substitutes in an open-loop transfer function which will give the desired closed-loop transfer function $T(s)$. Let us look at the limitations which stability and physical realizability place upon this cancelling out of the plant.

1. THE CASE OF A RHP PLANT POLE

We have an unstable plant. Looking at (8.6-3), we see that if the polynomial $[F(s) - N(s)]$ does not have a zero at the unstable pole of $G_f(s)$, then the compensator $G_c(s)$ will have a zero at this point. Thus in effect we are trying to cancel out an unstable plant pole with a RHP compensator zero. It is a fact of life in the design of compensators that perfect cancellation of an unstable pole can never be realized. The result is always a RHP closed-loop pole and hence the closed-loop system is unstable, an unacceptable result. The conclusion thus is that *if the plant* $G_f(s)$ *has a RHP pole, then* T(s) = N(s)/F(s) *must be chosen so that* [F(s) − N(s)] *has a zero at the location of this plant RHP pole.* We see from (8.6-1) that this RHP pole of $G_f(s)$ must thus be a zero of $1 - T(s)$.

2. THE CASE OF A RHP PLANT ZERO

Looking at (8.6-3), we can see that if $G_f(s)$ (the plant) has a zero at some point in the RHP and $N(s)$ [therefore also $T(s)$] does not have a zero at this same point, then the compensator will have a RHP pole at this point. We will again be trying to cancel out a RHP plant zero with a compensator RHP pole. Again, imperfect cancellation insures that the resulting closed-loop system will have a RHP pole and will thus be unstable. Since this is again unacceptable, the conclusion is that *if the plant* [G_f(s)] *has a RHP zero, then the* T(s) *we choose must have this same RHP zero.* We can see that if we have a plant RHP zero, then we really have no choice but to live with it. Indeed, this is no unique failing of the *s*-plane method. Whatever the choice of compensation method, RHP plant zeros have to be accepted where they are. This method simply points this fact out more strongly than the others.

3. THE EXCESS POLE RESTRICTION ON THE CHOICE OF THE CLOSED-LOOP TRANSFER FUNCTION $T(s)$

In order to be physically realizable, a compensator must have at least as many poles as zeros, i.e., must have no excess zeros. From (8.6-3) we see that this means the factor $N(s)/[F(s)-N(s)]$ must have as many or more excess poles as $G_f(s)$ has excess poles:

$$\left\{ \begin{array}{c} \text{excess poles of} \\ \dfrac{N(s)}{[F(s)-N(s)]} \end{array} \right\} \geq \left\{ \begin{array}{c} \text{excess poles of} \\ G_f(s) \end{array} \right\}$$

But $N(s)/[F(s)-N(s)]$ has the same number of excess poles as $T(s) = N(s)/F(s)$. We thus see that physical realizability requires that

$$\left\{ \begin{array}{c} \text{excess poles of} \\ T(s) \end{array} \right\} \geq \left\{ \begin{array}{c} \text{excess poles of} \\ G_f(s) \end{array} \right\} \qquad (8.6\text{-}4)$$

In words, the chosen $T(s)$ must have at least as many excess poles as the plant has. This is by far the most important of the restrictions we face, since most plants will have all LHP poles and zeros, and hence the first two restrictions seldom apply. *Our method of choosing a desired* T(s) *is then in general to choose a pair of dominant complex poles that will give the desired transient response. In addition to these two, far LHP poles are chosen in sufficient number to satisfy (8.6-4).* It will be seen later in this section that an additional pole-zero pair may be chosen near the origin if a higher steady-state error constant is required.

We now summarize the restrictions that stability and physical realizability requirements place upon the choice of a closed-loop transfer function $T(s) = N(s)/F(s)$:

1. If $G_f(s)$ has a RHP pole, then $[F(s)-N(s)]$ must have a zero at the same point as this RHP pole.

2. If $G_f(s)$ has a RHP zero, then $T(s)$ must have this same RHP zero.

3. $$\left\{ \begin{array}{c} \text{excess poles of} \\ T(s) \end{array} \right\} \geq \left\{ \begin{array}{c} \text{excess poles of} \\ G_f(s) \end{array} \right\}$$

We now proceed to consider a $T(s)$ that satisfies the three restrictions above and to obtain a desired transient response. Following this we consider adding a pole-zero pair near the origin of $T(s)$ to increase the steady-state error constant of the system to a desired specification.

Design of a Compensator to Achieve a Desired Transient Response

The desired closed-loop pole-zero configuration in s-plane synthesis is exactly the same as in the case of root locus compensator design. As was seen in Section 8.3 and Fig. 8.3-1, the basic requirement is two dominant complex poles with ζ and ω_n chosen to give a step response (and hence a transient response) with desirable overshoot, settling time, time to the first peak (t_p), etc. The ζ and ω_n must be chosen in the light of the requirement that the plant and the closed-loop transfer function must have the same RHP zeros. All other poles and zeros should be far into the LHP. If a pole is not far into the LHP, it should be very near a zero, i.e., it should be very nearly cancelled by a zero. Those closed-loop poles and zeros other than the dominant two poles must be chosen to satisfy the three requirements on the closed-loop transfer function considered earlier. These are dictated by the plant. We consider these in the following order:

1. *RHP plant poles.* To simplify the problem we shall assume that the plant is a stable one. This assumption holds in the majority of practical cases; otherwise one may attack the problem by first designing a subsidiary loop around the plant which has the sole function of stabilizing it.

2. *RHP plant zeros.* If the plant has a RHP zero, it has been seen that this same zero must appear in the chosen closed-loop transfer function. If it is nearer the $j\omega$-axis relative to the desired dominant two poles, then it will have an effect on transient (i.e., step) response, which is of the same form but opposite in sign to the effect of a LHP zero near the $j\omega$-axis. It was seen in Section 2.3 that a LHP zero near the $j\omega$-axis produced early peaking and excess overshoot in the step response of a two-pole system. For a RHP zero, the effect may be described as a tendency toward a later first peak in the step response and also increased overshoot. If the closed-loop transient response which we are seeking must have a given RHP zero, its effects must be taken into account in choosing ζ and ω_n for the dominant poles. This ζ and ω_n must be chosen with the understanding that the first peak will be somewhat delayed and will have a somewhat increased overshoot. Trial and error methods are usually necessary to insure a proper choice of ζ and ω_n if the step response requirements are critical. One simply computes the step response for a two-pole system with various values of ζ and ω_n and with a RHP zero at the point of the plant RHP zero. This is not difficult by conventional means since the system for which the step response is sought is only of second order.

3. *The excess poles requirement.* As seen in (8.6-4), the chosen

$T(s)$ must have as many excess poles as the plant. Lacking RHP zeros, we can see that the two dominant complex poles give us an excess of two poles. If the plant has no more than two excess poles, we need go no further. If, however, the plant should have a RHP zero or, not having a RHP zero, if it should have three or more excess poles, we must choose additional poles so that (8.6-4) is satisfied. These additional poles are chosen far (relative to the two dominant complex poles) into the LHP so that their effect on the transient response will be negligible. We may, of course, choose these far LHP poles at random as long as they are far enough to the left. In general, if the far LHP pole is p_i and we choose $|\text{Re } p_i| \geq 6\zeta\omega_n$ where ζ and ω_n are of the dominant complex poles, then the contribution due to p_i to the transient response will truly be negligible (see Fig. 2.3-8). These far LHP poles should be chosen with some care if the resulting compensator as given by (8.6-3) is to be of minimum complexity. We can see from (8.6-3) that if a pole of $G_f(s)$ (say p_j) is also a zero of $[F(s) - N(s)]$, the resulting $G_c(s)$ will be of order one lower than if this coincidence did not occur, i.e., if p_j were not a zero of $[F(s) - N(s)]$. It is thus desirable to choose the far LHP poles of $T(s) = N(s)/F(s)$ so that the far LHP poles of $G_f(s)$ are cancelled by zeros of $[F(s) - N(s)]$. We choose to have the latter coincide with the zeros of $[F(s) - N(s)]$ since this will result in the near LHP poles of $G_f(s)$ being cancelled by zeros of the compensator.

From the root locus discussion in Section 8.3 we know that this is usually desirable from the standpoint of the transient response. Besides, it is logical to choose the far LHP poles of $T(s)$ [zeros of $F(s)$] near the far LHP zeros of $[F(s) - N(s)]$. We can choose the far LHP poles of $T(s) = N(s)/F(s)$ [zeros of $F(s)$] so that zeros of $[F(s) - N(s)]$ will coincide with the far LHP poles of $G_f(s)$. We do this by using synthetic division to find conditions for this coincidence to occur. For example, consider

$$G_f(s) = \frac{1}{s(s+1)(s+4)(s+6)} \tag{8.6-5}$$

We have four excess plant poles, hence the chosen $T(s)$ must have at least four excess poles. Let us say that we wish $\zeta = 0.5$ and $\omega_n = 2$ for the two dominant poles. We hence want $T(s)$ of the form

$$T(s) = \frac{4c}{(s^2 + 2s + 4)(s^2 + bs + c)} = \frac{N(s)}{F(s)} \tag{8.6-6}$$

The factor $(s^2 + 2s + 4)$ in $F(s)$ is chosen to give the two dominant poles with the desired ζ and ω_n. The factor $(s^2 + bs + c)$ is chosen to give the required two additional poles so that the excess pole condition of (8.6-4) is satisfied. $N(s) = 4c$ is chosen so that $T(0) = 1$, which will insure that $e_{ss}(\text{step}) = 0$ for the closed-loop system. This means the open-loop system will be of at least type 1. Since the plant, (8.6-5), is of type 1, this is desirable. In order to design the compensator of minimum complexity we must choose $[F(s) - N(s)]$ with zeros at the two far LHP plant poles, i.e., at $s = -4$ and $s = -6$. In other words we want $[F(s) - N(s)]$ to have the factor $(s^2 + 10s + 24)$. Using synthetic division and $[F(s) - N(s)]$ with $N(s)$ and $F(s)$ from (8.6-6), we have

$$
\begin{array}{r}
s^2 + (b-8)s \\
s^2 + 10s + 24 \overline{\smash{\big)}\ s^4 + (2+b)s^3 + (4+2b+c)s^2 + (4b+2c)s} \\
\underline{s^4 + 10s^3 + 24s^2} \\
(b-8)s^3 + (-20+2b+c)s^2 + (4b+2c)s \\
\underline{(b-8)s^3 + (10b - 80)s^2 + (24b - 192)s} \\
(60 - 8b + c)s^2 + (192 - 20b + 2c)s
\end{array}
$$

$$(8.6\text{-}7)$$

From (8.6-7) we see that if $s^2 + 10s + 24 = (s+6)(s+4)$ is to be a factor of $[F(s) - N(s)]$ for this case, then

$$60 - 8b + c = 0$$
$$192 - 20b + 2c = 0$$

$$(8.6\text{-}8)$$

must be satisfied. We have from (8.6-8) that $b = 18$, $c = 84$. Using these values in (8.6-6) gives

$$T(s) = \frac{336}{(s^2 + 2s + 4)(s^2 + 18s + 84)} \qquad (8.6\text{-}9)$$

Using (8.6-9) and (8.6-5) in (8.6-3),

$$G_c(s) = \frac{336(s+1)}{(s+10)} \qquad (8.6\text{-}10)$$

The compensator here is a simple lead compensator with $\alpha = 10$, a not unreasonable one to realize.

There are a few points to be noticed in this example. The first is the fact that the equations (8.6-8), which the undetermined coefficients of the factor of $F(s)$ must satisfy and which were obtained from the synthetic division, are linear. Solution of these equations is hence routine. Second, the factor of $F(s)$ (the characteristic polynomial) from the solution of (8.6-8) is $(s^2 + 18s + 84) = (s + 9 - j\sqrt{3})(s + 9 + j\sqrt{3})$. The closed-loop poles which we have added to satisfy the excess pole requirements are hence at $s = -9 \pm j\sqrt{3}$, which are relatively far into the LHP compared to the dominant two poles with $\zeta = 0.5$, $\omega_n = 2$, i.e., with $\zeta\omega_n = 1$. The closed-loop response should hence be dominated completely by the two complex poles with $\zeta = 0.5$ and $\omega_n = 2$. The response should be almost exactly of the form of the two-pole step response as given for $\zeta = 0.5$ in Fig. 2.3-3 with time to the first peak (t_p) as given by Fig. 2.3-4 for these values of ζ and ω_n.

STEPS IN THE DESIGN OF A COMPENSATOR TO ACHIEVE A DESIRED TRANSIENT RESPONSE

Step 1. Choose ζ and ω_n for the dominant two poles, which will give the desired transient response. Figures 2.3-3 and 2.3-4 may be used for this if the plant has no RHP zero. If the plant has a RHP zero, ζ and ω_n must be chosen so that a second-order system with this same RHP zero will give a satisfactory transient response.

Step 2. Choose a closed-loop transfer function of the form

$$T(s) = \frac{N(s)}{F(s)} = \frac{(s+z)a_0\omega_n^2/z}{(s^2 + 2\zeta\omega_n + \omega_n^2)(s^m + a_{m-1}s^{m-1} + \cdots + a_0)} \quad (8.6\text{-}11)$$

where z is the RHP zero of the plant $G_f(s)$. If $G_f(s)$ has no RHP zero, choose the factor $(s+z)/z = 1$. Now ζ and ω_n are as determined in Step 1. The degree m of the undetermined factor of $F(s)$ is then determined so that (8.6-4) is satisfied, usually with equality. The coefficients a_i, $i = 0, 1, 2, \ldots, m-1$ are determined so that the far LHP poles of $G_f(s)$ coincide with zeros of $[F(s) - N(s)]$. This is accomplished by dividing $[F(s) - N(s)]$ by the factor whose roots are the far LHP poles of $G_f(s)$ and then determining the coefficients a_i, $i = 0, 1, \ldots, m-1$ so that the synthetic division has no remainder.

Step 3. Determine the compensator by

$$G_c(s) = \frac{N(s)}{[F(s) - N(s)]G_f(s)}$$

Note that once Step 2 has been used to determine the coefficients a_i, $i = 0, 1, \ldots, m-1$, the roots of the polynomial $s^m + a_{m-1}s^{m-1} + \cdots + a_0$ are poles of the closed-loop system. This we see from (8.6-11). We should now check to see that these roots do indeed lie far into the LHP relative to the two complex poles with ζ and ω_n as determined in Step 1. This usually does occur, since these roots will be near the far LHP poles of $G_f(s)$, which we are trying to cancel in the denominator of $G_c(s)$ above. It may turn out though that these roots of $s^m + a_{m-1}s^{m-1} + \cdots + a_0$ are not far enough into the LHP. This is most reasonably checked by determining the step response using the $T(s)$ of (8.6-11). If the step response is close to the desired one, we can make the appropriate adjustments in ζ and ω_n of Step 1 and proceed with the redesign. If the step response is completely unsatisfactory, a higher-order compensator will be required. Proceeding under this assumption, we can do one of two things. Either we increase the value of m in (8.6-11) or we keep it fixed and reduce the number of far LHP poles of $G_f(s)$ which are to cancel with zeros of $[F(s) - N(s)]$. In either case ingenuity is called for to use the increased freedom provided by the higher-order compensator in such a way that the resulting compensator is of minimum complexity and the additional closed-loop poles (the roots of the factor $s^m + a_{m-1}s^{m-1} + \cdots + a_0$) are all far enough into the LHP. An example may help clarify this.

Example 8.6-1

Let us again consider the design of the compensator $G_c(s)$ for the system of Fig. 8.3-6. In this case

$$G_f(s) = \frac{1}{s(s+1)(s+4)} \qquad (8.6\text{-}12)$$

Let us take as the design objectives an overshoot of 16% and a settling time (t_s) of 4 sec. From Fig. 8.3-6 we find that $\zeta = 0.5$ will give 16% of overshoot for a second-order system. $t_s = 4/\zeta\omega_n = 4$ sec implies $\omega_n = 2$ for this value of ζ. We proceed by the steps outlined earlier.

Step 1. We choose $\zeta = 0.5, \omega_n = 2$.

Step 2. We see from $G_f(s)$ of (8.6-12) that we have no RHP zeros, thus $T(s)$ need have none. Also it is seen that $G_f(s)$ has three excess poles, hence we choose a $T(s)$ of the form

$$T(s) = \frac{\omega_n{}^2 a_0}{(s^2 + 2\zeta\omega_n s + \omega_n{}^2)(s + a_0)}$$

Comparing this with (8.6-11) we see that $m = 1$ has been chosen so that $T(s)$ will have the same number (three) of excess poles as $G_f(s)$. Using ζ and ω_n as found in Step 1,

$$T(s) = \frac{4a_0}{(s^2 + 2s + 4)(s + a_0)} = \frac{N(s)}{F(s)}$$

We have one free parameter (a_0) here, thus we can choose this parameter so that one zero of $[F(s) - N(s)]$ will coincide with a pole of $G_f(s)$. Let us choose a_0 so that $[F(s) - N(s)]$ will have a zero at $s = -4$, the location of the left-most pole of $G_f(s)$. To do this we divide $[F(s) - N(s)]$ by the factor $(s + 4)$ by synthetic division.

$$
s+4 \overline{\left)\begin{array}{l} s^2+(a_0-2)s \\ s^3+(a_0+2)s^2+(4+2a_0)s \\ \underline{s^3+4s^2} \end{array}\right.}
$$

$$
\begin{array}{c}
(a_0-2)s^2+(4+2a_0)s \\
\underline{(a_0-2)s^2+(-8+4a_0)s} \\
(-2a_0+12)
\end{array}
$$

We see from this division that if $(s+4)$ is to be a factor of $[F(s) - N(s)]$, we must have $-2a_0+12 = 0$, i.e., $a_0 = 6$. We thus have

$$
T(s) = \frac{24}{(s^2+2s+4)(s+6)} \tag{8.6-13}
$$

Using this in (8.6-1) and $G_f(s)$ of (8.6-12) gives

$$
G_c(s) = \frac{24(s+1)}{(s+4)} \tag{8.6-14}
$$

We see from (8.6-13) that the far LHP pole which has been added here to satisfy the excess pole requirement is at $s = -6$. This is relatively far into the LHP relative to the two complex poles with $\zeta = 0.5$, $\omega_n = 2$. The complex poles should dominate this response very well. The compensator obtained in (8.6-14) is now an ordinary lead compensator with $\alpha = 4$. This is in fact the same compensator that was obtained in Section 8.5 in (8.5-14) by Mitrovic's method. The lead compensated step response using this compensator is shown in Fig. 8.5-2. The overshoot of 16% and settling time of 4 sec have been closely met.

The Addition of a Pole-Zero Pair to the Closed-Loop Transfer Function to Increase the Magnitude of the Steady-State Error Constant

We now assume that the system under consideration has already been compensated in such a manner that the transient response is satisfactory but the steady-state response is not satisfactory. The problem, as in the cases considered before, is reduced to increasing the magnitude of the error constant. Let us now consider a forward plant of type 1 to illustrate our approach. What we say and the methods we use will apply equally well to type 0 and type 2 systems. However, the approach here is best seen in the case of a type 1 system; also this is the most important case from the practical standpoint. It has been seen in (3.3-24) that for a unity-feedback system as in Fig. 8.1-1 with a type 1 forward plant and with closed-loop poles p_i, $i = 1, 2, \ldots, n$

and closed-loop zeros z_i, $i = 1, 2, \ldots, m$, the velocity error constant, K_v, is nonzero and finite and is given by

$$\frac{1}{K_v} = \sum_{i=1}^{m} \frac{1}{z_i} - \sum_{i=1}^{n} \frac{1}{p_i} \tag{8.6-15}$$

Let us now consider increasing the magnitude of the velocity error constant by adding a pole p_j and a zero z_j to the closed-loop transfer function. It will be understood that the pole p_j and the zero z_j will be so close together that no appreciable effect upon the transient response will be noted due to their addition. The pole p_j will of course be located in the LHP. Now if K_v is the velocity error constant before the addition of the pole-zero pair p_j and z_j and is given by (8.6-15), the velocity error constant ($K_v{}^c$) after the addition will be given by

$$\frac{1}{K_v{}^c} = \frac{1}{K_v} + \frac{1}{z_j} - \frac{1}{p_j} \tag{8.6-16}$$

Let us choose $p_j < z_j < 0$. We will thus have

$$\frac{1}{z_j} - \frac{1}{p_j} < 0$$

and the result will be

$$K_v{}^c > K_v$$

i.e., we will have increased the magnitude of the error constant. If we now wish a large improvement in error constant and hence large $|1/z_j - 1/p_j|$, we must choose $|z_j|$ and $|p_j|$ small. That is, we must choose the pole p_j and zero z_j near the origin. This is what we did in the case of lag compensation: we added a pole-zero pair very close together and very near the origin. It should be recalled that in lag compensation we added a pole-zero pair to the open-loop (forward) transfer function in which the pole was nearer the origin than the zero. To increase K_v we add a pole-zero pair to the closed-loop transfer function with the zero nearer the origin than the pole.

Now given a system for which we wish to increase K_v, we could use (8.6-16) and simply set the zero z_j (or pole p_j) near the origin relative to the two dominant complex poles and then solve for p_j (or z_j) to give the desired improvement in K_v. Since we are working with the closed-loop transfer function we can do better than this. In fact, we can predict very closely the effect of addition of this pair on the step response. This information makes effective use of the freedom available in the location of the pole and zero. To see how this is done consider a given closed-loop transfer function of the form

$$
T(s) = \frac{\left[\displaystyle\prod_{i=1}^{n} (-p_i) \right] (s - z_1)(s - z_2) \cdots (s - z_m)}{\left[\displaystyle\prod_{i=1}^{m} (-z_i) \right] (s - p_1)(s - p_2) \cdots (s - p_n)}
$$

where now the multiplying constant has been adjusted so that $T(0) = 1$. The resulting system will then have a nonzero K_v. We assume $p_i \neq 0$, $i = 1, 2, \ldots, n$ and $z_i \neq 0$, $i = 1, 2, \ldots, m$. Further consider adding a pole-zero pair p_j, z_j to this transfer function to increase K_v as in (8.6-16). The compensated closed-loop transfer function will then be

$$
T^c(s) = \frac{p_j(s - z_j) \left[\displaystyle\prod_{i=1}^{n} (-p_i) \right] \left[\displaystyle\prod_{i=1}^{m} (s - z_i) \right]}{z_j(s - p_j) \left[\displaystyle\prod_{i=1}^{m} (-z_i) \right] \left[\displaystyle\prod_{i=1}^{n} (s - p_i) \right]}
\tag{8.6-17}
$$

where we have again adjusted the multiplying constant so that $T^c(0) = 1$. Let us now consider the step response for the system of (8.6-17).

$$
C_{\text{step}}(s) = \frac{1}{s} T^c(s) = \frac{1}{s} \left[\frac{\displaystyle\prod_{i=1}^{n} (-p_i) \prod_{i=1}^{m} (s - z_i)}{\displaystyle\prod_{i=1}^{m} (-z_i) \prod_{i=1}^{n} (s - p_i)} \right] \frac{p_j(s - z_j)}{z_j(s - p_j)}
\tag{8.6-18}
$$

We may expand the RHS of (8.6-18) into a partial fraction expansion as

$$
C_{\text{step}}(s) = \frac{1}{s} + \frac{K_{11}}{(s - p_1)} + \frac{K_{21}}{(s - p_2)} + \cdots + \frac{K_{n1}}{s - p_n} + \frac{K_{j1}}{s - p_j}
\tag{8.6-19}
$$

where we have assumed that all poles are distinct. Inverting (8.6-19) term by term we have

$$
c_{\text{step}}(t) = 1 + \sum_{i=1}^{n} K_{i1} e^{p_i t} + K_{j1} e^{p_j t}, \qquad t > 0
\tag{8.6-20}
$$

We see that the contribution to the step response due to the added pole at p_j is the term

$$
K_{j1} e^{p_j t}
$$

We may determine K_{j1} from (8.6-19) as

$$K_{j1} = \frac{\prod_{i=1}^{n} (-p_i) \prod_{i=1}^{m} (p_j - z_i) \, (p_j - z_j)}{\prod_{i=1}^{m} (-z_i) \prod_{i=1}^{n} (p_j - p_i) \quad (z_j)} \tag{8.6-21}$$

In (8.6-21) the pole p_j is very near the origin compared to the other poles (i.e., $|p_j| \ll |p_i|$, $i = 1, 2, \ldots, n$) and zeros (i.e., $|p_j| \ll |z_i|$, $i = 1, 2, \ldots, m$); therefore

$$K_{j1} \cong \frac{p_j - z_j}{z_j} = -1 + \frac{p_j}{z_j} \tag{8.6-22}$$

Thus using (8.6-20),

(contribution to step response by pole at p_j) $\cong \left(\frac{p_j}{z_j} - 1\right) e^{p_j t}$ $\qquad$ (8.6-23)

We have placed the pole at p_j so that it is in the LHP but much closer to the origin than all the other poles of $T^c(s)$, i.e., p_i, $i = 1, 2, \ldots, n$. Therefore Re $p_i \ll p_j$ and thus the term due to the pole at p_j will decay much more slowly than the terms due to all other poles including the two dominant complex poles. In fact at the time of the first peak, t_p, of the step response we will have $e^{p_j t_p} \cong 1$. Using this in 8.6-23 we see that the magnitude of the first peak is increased by an amount

$$K_{j1} \cong \frac{p_j}{z_j} - 1$$

We recall that $p_j/z_j > 1$ if we wish to increase K_v as seen in (8.6-16), hence we have

$$\left[\frac{p_j}{z_j} - 1\right] > 0$$

We thus conclude that *by adding the pole-zero pair (at* p_j *and* z_j*) to increase the magnitude of* K_v*, the effect on the step response is to increase the percent overshoot by the factor* $(p_j/z_j - 1) \times 100\%$. If $X\%$ increase in step overshoot can be tolerated to increase the error constant, then

$$\frac{X}{100} \cong \frac{p_j}{z_j} - 1$$

or

$$\frac{p_j}{z_j} \cong 1 + \frac{X}{100} \tag{8.6-24}$$

Equation (8.6-24) may be used with (8.6-16) to calculate p_j and z_j, i.e., to locate the desired pole-zero pair.

As an example let us consider a system with $K_v = 2$ and step response overshoot of 6%. Suppose we wish a velocity error constant $K_v^c = 4$ and can tolerate 10% of step overshoot. Thus $10 - 6 = 4\%$ increase in step overshoot can be tolerated, or $X = 4$. Using this value of X in (8.6-24) and the K_v and K_v^c we have in (8.6-16) gives

$$\frac{1}{4} = \frac{1}{2} + \frac{1}{z_j} - \frac{1}{p_j}$$

$$\frac{p_j}{z_j} = 1 + \frac{4}{100}$$

whence we have $p_j = 0.16$, $z_j = 0.16/1.04$. With p_j and z_j determined by (8.6-16) and (8.6-24) and used in the closed-loop transfer function of (8.6-17), the required compensator may be determined by (8.6-1).

It should be noted that with the closed-loop $T^c(s)$ as given in (8.6-17),

$$T^c(s) = T(s) \frac{p_j}{z_j} \frac{(s - z_j)}{(s - p_j)}$$

$$= \frac{N(s)}{F(s)} \frac{p_j}{z_j} \frac{(s - z_j)}{(s - p_j)} \qquad (8.6\text{-}25)$$

$$= \frac{N^c(s)}{F^c(s)}$$

If now $T(s) = N(s)/F(s)$ has been chosen so that $[F(s) - N(s)]$ has zeros which coincide with the far LHP poles of $G_f(s)$, adding the pole-zero pair (p_j, z_j) as in (8.6-25) will destroy this coincidence when the corresponding compensator is calculated as in (8.6-1) or (8.6-3). Since the added pole and zero of the pair are very close together, the far LHP poles of $G_f(s)$ and the zeros of $[F_c(s) - N_c(s)]$ will be close together but not coincident. One may check this and if coincidence in the resulting compensator is close enough, the cancellation and corresponding reduction in the order of $G_c(s)$ may be carried out. A negligible effect may be expected due to this pole-zero cancellation in the compensator in the face of imperfect coincidence. If this does not turn out to be the case, then the far LHP poles of $T_c(s) = N_c(s)/F_c(s)$ may be reset so that the zeros of $[F_c(s) - N_c(s)]$ will coincide with the desired far LHP poles of $G_f(s)$. This again may be done by synthetic division in the manner described earlier in this section.

THE STEPS IN DESIGN OF AN S-PLANE SYNTHESIS COMPENSATION TO INCREASE THE MAGNITUDE OF THE VELOCITY ERROR CONSTANT

Step 1. Choose a $T(s)$ that will result in a satisfactory transient response and realizable compensator, i.e., design a compensator as outlined earlier in the section.

Step 2. Determine K_v by (8.6-15) for the system with $T(s)$ as found in Step 1. Determine what velocity error constant $(K_v{}^c)$ is required to give the steady-state response desired. If $K_v \geqslant K_v{}^c$ the compensator as determined in Step 1 is sufficient. If $K_v < K_v{}^c$ proceed to Step 3.

Step 3. Determine the increase in percent step-response overshoot (X) that can be tolerated. Determine z_j and p_j to satisfy

$$\frac{1}{K_v{}^c} = \frac{1}{K_v} + \frac{1}{z_j} - \frac{1}{p_j}$$

and

$$\frac{p_j}{z_j} = 1 + \frac{X}{100}$$

We have then

$$T^c(s) = \frac{p_j}{z_j} \frac{(s - z_j)}{(s - p_j)} T(s) = \frac{N^c(s)}{F^c(s)}$$

Step 4. Determine the compensator by

$$G_c(s) = \frac{N^c(s)}{[F^c(s) - N^c(s)] G_f(s)} \qquad (8.6\text{-}26)$$

Accomplish any cancellation of poles and zeros of $G_c(s)$ that are near coincidence. If the far LHP poles of $G_f(s)$ and the zeros of $[F^c(s) - N^c(s)]$ are not near enough to permit this cancellation and a reduction in the order of $G_c(s)$ is desirable, the far LHP zeros of $[F^c(s) - N^c(s)]$ may be redetermined so as to coincide with the far LHP poles of $G_f(s)$. The steps required to do this are described earlier in the section.

Step 5. Determine step response so as to check whether or not the transient response is as desired.

Example 8.6-2

Let us again consider the example of Fig. 8.3-6 (considered in Example 8.6-1). The plant $G_f(s)$ is as given in (8.6-12). In Example 8.6-1 a compensator was designed to give an overshoot of 16% and a settling time of 4 sec. The result for Step 1 is found in (8.6-13) to be

$$T(s) = \frac{24}{(s^2 + 2s + 4)(s + 6)} \qquad (8.6\text{-}27)$$

Step 2. By using (8.6-15), K_v for this system is found to be

$$\frac{1}{K_v} = -\frac{1}{-6} - \frac{1}{-1+j\sqrt{3}} - \frac{1}{-1-j\sqrt{3}}$$

and therefore

$$K_v = 1.5 \qquad (8.6\text{-}28)$$

Let us assume that $K_v{}^c = 6$ is a satisfactory value of velocity error constant for the system. We thus require an increase in the magnitude of the velocity error constant. We can get this increase by adding a pole-zero pair near the origin to the closed-loop transfer function of (8.6-27) that was determined in Step 1.

Step 3. The step response for $T(s)$ as shown in (8.6-27) has been computed and is shown in Fig. 8.5-2. We have 16% of overshoot. Say that for our purposes an additional 5% overshoot can be tolerated, i.e., 21% of overshoot is permissible. Thus $X = 5$. Using this value in (8.6-24) and K_v and $K_v{}^c$ as determined in Step 2 in (8.6-16), we have

$$\frac{1}{6} = \frac{1}{1.5} + \frac{1}{z_j} - \frac{1}{p_j}$$

$$\frac{p_j}{z_j} = 1.05$$

Thus

$$p_j = 0.1, \quad z_j = \frac{0.1}{1.05} = 0.09524$$

Using these values,

$$T^c(s) = \frac{25.2\,(s+0.09524)}{(s^2+2s+4)\,(s+6)\,(s+0.1)} = \frac{N_c(s)}{F_c(s)} \qquad (8.6\text{-}29)$$

Step 4. The corresponding compensator, using (8.6-26), is

$$G_c(s) = \frac{25.2\,(s+0.09524)\,(s+1)\,(s+4)}{(s+0.0241)\,(s^2+8.076s+16.6)} \qquad (8.6\text{-}30)$$

The factor $(s^2+8.076s+16.6)$ gives complex poles at $s = -4.04 \pm j0.505$, which are very close to zero at $s = -4$. It will be recalled that the pole of $T(s)$ at $s = -6$ in (8.6-27) was chosen so that the corresponding $G_c(s)$ [as seen in (8.6-14)] would have this zero at $s = -4$ exactly cancelled by a pole at the same point. So far as the response of the system is concerned, the effect of assuming that these two complex poles (at $s = -4.04 \pm j0.505$) is a double pole at $s = -4$ and accomplishing the corresponding cancellation to get

$$G_c(s) = \frac{25.2\,(s+0.09524)\,(s+1)}{(s+0.0241)\,(s+4)} \qquad (8.6\text{-}31)$$

will be negligible for almost any application. The compensator obtained is then not only of lower order, but also has all its poles on the negative real axis. It can hence be realized as an *RC* network (followed by an amplifier of the appropriate gain). The compensator with the complex pole cannot be realized so easily. However, let us assume that the simplification as indicated by (8.6-31) is not a satisfactory substitute for the compensator of (8.6-30). To design a compensator in which the zero at $s = -4$ (due to the plant pole at this same point) is cancelled exactly by a pole at this same point we may relocate the far LHP pole of $T^c(s)$ of (8.6-29) so that this cancellation will occur. Looking at (8.6-29) we see that this pole is at $s = -6$. Let us put a new pole at $s = -a_0$ and determine a_0 so that the desired cancellation will occur. Then

$$T^c(s) = \frac{4.2a_0(s+0.09524)}{(s^2+2s+4)(s+a_0)(s+0.1)} = \frac{N^c(s)}{F^c(s)}$$

We want $[F^c(s) - N^c(s)]$ to have a zero at $s = -4$. We determine a_0 now so that $[F^c(s) - N^c(s)]$ will have a factor of $(s+4)$. This is accomplished by synthetic division as indicated in Example 8.6-1. The result is $a_0 = 6.16$. The resulting closed-loop transfer function with this value of a_0 is then

$$T^c(s) = \frac{25.9(s+0.09524)}{(s^2+2s+4)(s+6.16)(s+0.1)} \tag{8.6-32}$$

The corresponding compensator using (8.6-26) with this $T^c(s)$ is

$$G_c(s) = \frac{25.9(s+0.09524)(s+1)}{(s+0.0236)(s+4.236)} \tag{8.6-33}$$

We thus have a second-order compensator as desired. It is very nearly a lag-lead compensator, having a lag section and a lead section. The two sections do not, however, have the same value of α, so it is not truly a lag-lead compensator as defined in Section 8.1. It should be noted that this compensator, (8.6-33), is very nearly the lag-lead compensator found for the same system by Mitrovic's method in (8.5-37). The response using the compensator of (8.6-33) was computed and found to be virtually indistinguishable from the response using the compensator of (8.5-37). This lag-lead compensated response is shown in Fig. 8.5-2. We see that the overshoot is 20%. Thus by comparing this with the lead-compensated response obtained earlier (Fig. 8.5-2), it is seen that the addition of the pole-zero pair near the origin has increased the overshoot by only 4%, rather than the 5% which was assumed tolerable and used in the calculation of z_j and p_j

in Step 3 above. Using $T^c(s)$ as found in (8.6-32) in (8.6-15), we determine immediately that

$$K_v^c = 6.16 \qquad (8.6\text{-}34)$$

We note that here $K_v^c = 6.16$ instead of 6 as used in the calculations to determine z_j and p_j in Step 3. This discrepancy occurs because in Step 4 the far LHP pole of $T(s)$ [of (8.6-27) in Step 1] that was at -6 was relocated to have the zeros of $[F^c(s) - N^c(s)]$ coincide with the far LHP poles of $G_f(s)$. This relocation resulted in the increased K_v as seen in (8.6-34).

8.7 A COMPARISON OF THE VARIOUS METHODS OF COMPENSATION

Let us now consider the compensation methods we have been discussing in the light of their usefulness to a design engineer. Let us first make a few general comments about all the methods and then consider each in somewhat more detail.

First, each method tends to highlight one or more aspects of the response of linear systems. Each method tends to bring out aspects of the response of the system that are not found out by the other methods. Also, for a given system, usually one method leads more easily to a satisfactory response than another. With experience the designer should be able to judge which of the various methods will work best for a given system. And, of course, no single compensation method is best for all systems, so the designer should have several methods at his disposal. Understanding of what is required and the strengths and weaknesses of each method should lead to the most fruitful approach. The different methods may be thought of as viewing the problem from different angles; together they can bring an understanding to a given problem which one alone cannot give.

It should be appreciated also that with the example that was considered in this chapter, the results obtained differed insignificantly from one method to another. This was thought desirable here, for having one method give better results than another might tend to bias the reader to one particular method. It should be pointed out, however, that this uniformity of results should not be expected for every case.

Now considering the methods separately, we see that they are basically of two types: the frequency-response methods and those we may describe as the analytic methods.

Frequency-Response Methods

The frequency-response methods are unique in that they do not require explicit knowledge of the plant transfer function. All that is required is a plant frequency response. Once the frequency response is known, the frequency-response compensation methods proceed unaffected by the order or complexity of the plant being compensated. In fact, the plant transfer function need not even be a rational function, which the other methods require. Coupled with the fact that the frequency response can be measured readily by using sinusoidal inputs, this makes the frequency-response methods highly attractive if one is faced with the problem of controlling a complicated plant, the parameters and dynamics of which are not too well known or understood.

The disadvantages of these frequency-response methods are also very real, however. First, in working with the frequency response, the quantities one deals with are not really directly related to the time-response quantities, which are usually important in control system design specifications. Thus much trial and error and skill can be required to do satisfactory compensations. Also, given a compensated system frequency response, the amount and direction of the change in the time response due to changes in plant or compensator parameters are not evident. It should be noted further that the frequency-response compensation methods as described in Section 8.4 assume that the plant has no RHP zeros. If the plant does happen to have a RHP zero, frequency-response methods may be used; however, the additional phase shift due to the fact that a zero is over in the RHP has to be carefully considered before proceeding.

Another aspect of the frequency-response methods is the fact that they are graphical by nature and hence work very well when all the analysis work has to be done by hand. This is an advantage; however, it is more and more the case that the engineer has a digital or analog computer available to help with the analysis. It is usually found that these aids are not of much help, other than to check the result obtained, if one uses a strict frequency-response approach.

The Analytic Design Methods

By the analytic design methods, we mean the root locus, Mitrovic's, and s-plane synthesis methods. Looking at all three methods, it can be seen that they all require a knowledge of the transfer function of the plant before a compensator can be designed. Also they all require that the transfer function be rational. With such a transfer function for the plant, it generally may be said that one of the analytic design methods will lead more easily and quickly to a satisfactory compensator than the frequency-response methods. Otherwise one tends to use the frequency-response methods, since the plant frequency-response can usually be measured experimentally without too great an effort. Let us now look at these methods in some detail.

THE ROOT LOCUS METHOD

The root locus approach to compensator design has the overwhelming advantage that once a compensator has been selected and the root locus drawn, the closed-loop poles and zeros are immediately available on the root locus. Using the spirule, as described in Section 4.2, the time response can be obtained quickly for the closed-loop system. Thus correlation of the time response with the root locus is very good. Moreover, once the root locus is drawn, one usually can see very well, qualitatively at least, what the effect of changing the compensator parameters (such as gain and pole-zero location) will be on the system's time response. This is usually exactly what the system designer wants, for it allows him to see just what sort of trade-off can be made between compensator complexity and system time response.

Also from the root locus the designer can get limits of closed-loop stability for various compensator parameters, the most important being loop gain. All these advantages have led to the growing popularity of the root locus in industrial applications in the face of the entrenched position of the frequency-response methods and the obvious advantage these methods have if the problem includes an unidentified plant.

The root locus has a disadvantage in the fact that the complexity of its use increases as the order of the system under consideration increases. In fact, for a system of higher than sixth or seventh order, some computational aid (e.g., a digital computer) is almost mandatory. Of course if a computer must be used, some of the flexibility which makes the method attractive in the first place is lost.

MITROVIC'S METHOD

This method in general demands a fair amount of numerical effort to construct the contours in the parameter plane. This numerical work lends itself well to solution on the digital computer, but it can be laborious if done by hand. Once such a parameter plane is constructed the designer has before him a lot of quantitative information on just what can be accomplished in the way of system response and just what the compensator parameters have to be to accomplish a specific response. The designer can quickly design a compensator to give a desired response. If after trying the compensator it is found that the response is still not quite what it should be, the parameter plane shows how much the parameters will have to be changed to get the desired response. This is, of course, very attractive from the designer's standpoint.

On the other hand, it should be noted that Mitrovic's method works solely with the characteristic equation of the closed-loop system and therefore ignores the presence of any zeros. It is well known that closed-loop zeros can have a decided effect upon the response of a given system and thus some provisions usually have to be made to avoid or to compensate for the effects of undesired zeros. A little manipulation with the root locus is usually helpful in this case.

S-PLANE SYNTHESIS

In s-plane synthesis, we start from a desired closed-loop transfer function and then solve for the compensator. This is a very direct and straightforward method which has the advantage that we know from the beginning what the closed-loop transfer function will be. The disadvantages are, first, that the form of the compensator is not fixed beforehand and hence the result is usually a complicated compensator; further, once we have solved for the compensator, if it is found too complicated, we get very little insight into what to do to get a simpler one or what the cost in performance degradation will be if we do insist upon a simpler compensator. And even if the compensator obtained is a reasonable one, if the resulting system response is unsatisfactory (possibly because the plant transfer function does not fit the actual plant as well as it should or because one of the far LHP closed-loop poles that has been added to satisfy the excess pole requirements has gotten too close in to the $j\omega$-axis), then the design method provides very little insight in how to tune the compensator parameters to get a more desirable response. In general it can be said that the method is a simple and direct one that gives a quick answer with a minimum of effort. Since the method is a simple one, it can be used in an initial attempt at compensator design. If the result is a reasonable compensator and this compensator gives a satisfactory response, we are through. If not, we go on to another method.

SUMMARY

We have seen that as far as the analytic methods of compensation are concerned, the s-plane synthesis method should be tried first. If this works, i.e., if a satisfactory response is obtained with the compensator obtained, then "well and good." Otherwise, if the system is of low order, the root locus approach is probably most suitable. If the system is of high order, then a combination of the root locus and Mitrovic's method used with a digital computer will probably give the best results.

PROBLEMS

8.1. Given $G(s) = 110/s(s+5)$ operated in a unity negative-feedback configuration, determine a cascade compensator so that the dominant closed-loop poles will have $\zeta = 0.6$, $\omega_n = 23$ rad/sec. The addition of the compensator should not affect the steady-state error characteristics (i.e., it should keep K_v constant).

8.2. Given the control system of Fig. P8.2, what type of controller transfer function will give zero steady-state error to a ramp input with $K, a > 0$?

8.3. Given

$$G(s) = \frac{1}{s^5 + 10s^4 + 45s^3 + 114s^2 + 166s + 136}, \qquad H(s) = 1$$

operated in a standard closed-loop configuration. Find the cascade compensator such that the closed-loop system has step-response overshoot of 16% and settling time of 4 sec.

8.4. Given

$$G(s) = \frac{K(s+16)}{s(s+7)(s+25)(s+33)}$$

(a) For this plant operated in a unity, negative-feedback, closed-loop configuration find the value of K that will give an overshoot (to a step input) of approximately 15%. What is the value of K_v and the settling time for this value of K? Plot the output step response for this value of K out to the settling time.

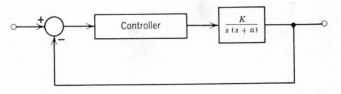

Figure P8.2

(b) Design a cascade compensator that will give the same step-response overshoot as in (a) but will give a settling time of approximately 0.32 sec. Assume that K is fixed at the value found in (a). What is K_v for this compensator? Plot the output step response out to the settling time to see if the design specifications have been met.

(c) Design a cascade compensator that will give the same damping factor as in (a) and will increase K_v by a factor of 6. How much has the overshoot to a step been increased by the addition of this compensator?

(d) Design a compensator that will meet the transient specifications as in (a) and the steady-state error specification as in (c). Plot the output step response out to settling time. Are the specifications met?

(e) Design a compensator, either cascade or feedback, which will meet the transient and steady-state specifications as in (a) and (c).

8.5. Given

$$G(s) = \frac{K}{s(s+3)(s+9)}$$

to be operated in a unity, negative-feedback, closed-loop system.

(a) Make a frequency-response plot of $G(j\omega)$ on the Nichols chart for ω in the range $0.1 \leq \omega \leq 10$ or thereabout. What value should be chosen for K if 15% overshoot to a step input is desired for the closed-loop system response?

(b) What settling time can be expected with K adjusted to give 15% step-response overshoot? What is K_v for this case?

(c) Design a cascade compensator that will give approximately 15% step overshoot, a factor of 2.5 decrease in settling time, and $K_v \geq 20$.

(d) Make a Bode plot (magnitude and phase) of the closed-loop frequency response for the uncompensated and compensated systems. Make them on the same graph so that you can see the difference. The plots should be over sufficient range in ω to show the salient features of the response.

8.6. The frequency response plotted on the Nichols chart of Fig. P8.6 is that of a gun in an anti-aircraft installation. For an aircraft coming straight overhead at constant velocity and within the usable range of the gun, it was found that the gun would track to within 1.5° of the target's projected position as given by the computer. This was too much error. With the proximity fuse in use and with the range of the gun taken into consideration, something of the order of 15 min of arc tracking error was calculated to be the maximum tolerable. Design a cascade compensator to be added in the forward path of the control loop of the gun, that will reduce this tracking error to a tolerable amount and not materially affect the transient response of the gun to evasive action of the target.

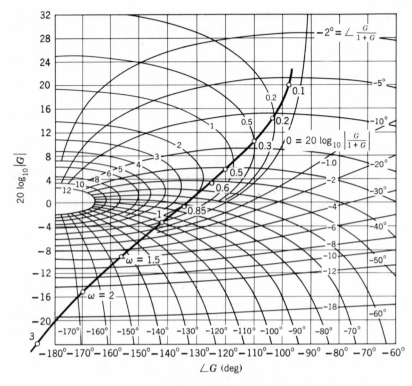

$$-2° = \angle \frac{G}{1+G}$$

$$0 = 20 \log_{10} \left| \frac{G}{1+G} \right|$$

Figure P8.6

8.7. Given

$$G(s) = \frac{10(s+5)(s+13)}{s(s+4)(s+9)(s^2+22s+122)}$$

Design a cascade compensator to give step-response overshoot of approximately 10%, time to peak = 0.7 secs, and $K_v \geq 15$.

8.8. Given the antenna positioning system as shown in Fig. P8.8.
   (a) Determine the open-loop transfer function.
   (b) What is the steady-state error constant for this system $[G_c(s) = 1]$?
   (c) Determine the unit step response for this system $[G_c(s) = 1]$; plot.
   (d) Determine a cascade compensator $[G_c(s)]$ which will give approximately 10% step-response overshoot and a settling time of about 2.0 sec.
   (e) Determine and plot the step response for the system with compensation as determined in (d). Carry out the plot past the settling time and show clearly the overshoot and time of peak on the plot. What is K_v?

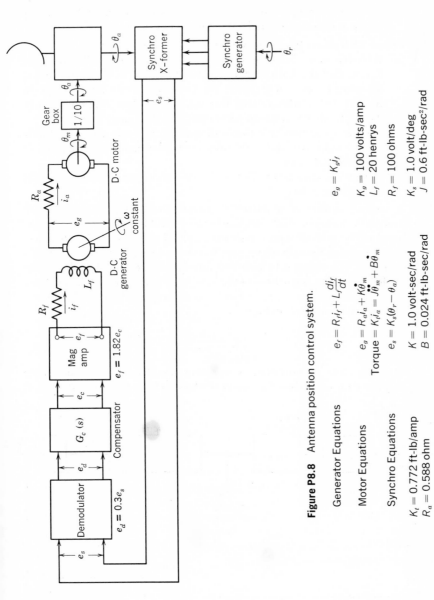

Figure P8.8 Antenna position control system.

Generator Equations

$$e_f = R_f i_f + L_f \frac{di_f}{dt}$$

$$e_g = K_g i_f$$

$K_g = 100$ volts/amp
$L_f = 20$ henrys
$R_f = 100$ ohms

Motor Equations

$$e_g = R_a i_a + K\dot{\theta}_m$$
$$\text{Torque} = K_t i_a = J\ddot{\theta}_m + B\dot{\theta}_m$$

Synchro Equations

$$e_s = K_s(\theta_r - \theta_a)$$

$K = 1.0$ volt-sec/rad
$B = 0.024$ ft-lb-sec/rad

$K_s = 1.0$ volt/deg
$J = 0.6$ ft-lb-sec²/rad

$K_t = 0.772$ ft-lb/amp
$R_a = 0.588$ ohm

421

(f) Determine a compensation that will meet the transient requirements as in (d) but with $K_v \geqslant 5$.

(g) Determine and plot the step response for the system with compensation as determined in (f). What is the new K_v?

8.9. Figure P8.9 shows the pitch axis control system for an aircraft.

(a) With the loop closed by feeding back pitch rate, as shown in the
figure, find the gain A [for $G_c(s) = A$] which will give maximum
damping to the pitch rate $\dot{\theta}$. Find $\dot{\theta}(t)$ for a unit step change in
control stick position using this value of A. How long does it take
the pitch rate $\dot{\theta}$ to settle to within 2% of its final value?

(b) Design a compensator that will give a pitch-rate damping factor
$\zeta \cong 0.7$ (approximately 10% step overshoot) and a settling time of
approximately 1 sec.

(c) Design a compensator that will give a transient response as in (b)
but with $K_p \geqslant 20$.

8.10. A control system to stabilize a radar antenna against yaw aircraft
motion is shown in block diagram form in Fig. 8P.10. The yaw displacement of the aircraft with respect to a spatial reference is θ_2. Displacement of the antenna with respect to this same reference, i.e., antenna
yaw is θ_1. The object is to keep the antenna at a fixed relationship to the
spatial reference regardless of the yaw angle of the aircraft, i.e., to keep
θ_1 small irrespective of θ_2. It has been determined by test that a satisfactory performance is obtained if

$$\theta_1 \cong \frac{1}{300}\theta_2$$

in steady-state. In addition, as transient specifications it is desirable
that the damping factor for yaw disturbances have the range $0.6 \leqslant \zeta \leqslant$

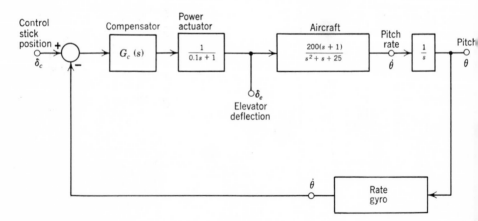

Figure P8.9

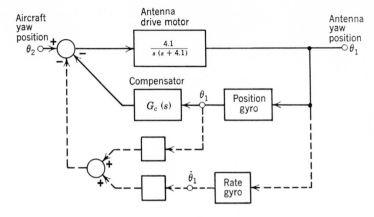

Figure P8.10

0.8 and that a disturbance in aircraft yaw cause a disturbance in antenna yaw, which should settle down in no more than 0.1 sec.

Using only antenna position feedback (shown in heavy lines) design a compensator

$$G_c(s) = A \frac{(s + 1/\tau)}{(s + 2/\tau)}$$

which will go as shown in Fig. 8P.10 and which will result in a satisfactory antenna position response.

9

CLASSICAL COMPENSATION OF DISCRETE-TIME CONTROL SYSTEMS

9.1 INTRODUCTION

The methods of improvement in the performance of continuous-control systems using some form of compensation within the control loop have been investigated in some detail in Chapter 8. The criteria of such compensation have been based on speeding up the step response and increasing the steady-state error constant. These criteria can also be applied to the improvements of performance of sampled-data or discrete-time systems. The same criteria can be used for compensation since the design objectives are generally the same whether or not a system happens to have a sampler in the signal path. We have seen that the mathematical tool that is used to analyze the SDS is the Z-transform. This makes compensation design more complex at times; however, the approach, the basic tools, and the design objectives remain the same as in the continuous case. To obtain insight into the problem of compensation for SDS, let us examine the types of compensation for such systems.

Forms of Compensation

As in the continuous case, the compensator can either be in the forward path or the feedback path of the loop. In the SDS case, however, the compensator can be digital, continuous, or analog-digital. Because of these variations and combinations, the possibilities for designing the controller or compensator are numerous.

In our study here we restrict ourselves to a limited number of possibilities, just enough to show the techniques of design. The other cases usually can be handled by simple extension of these basic methods. Specific designs for various situations are extensively covered in the SDS literature and the reader is referred to this if he has a SDS problem that does not fit any of the basic methods covered here.

1. FORWARD-PATH COMPENSATION

Figure 9.1-1 shows the continuous and digital compensation in the forward path for a general SDS. The forward-path compensators are in series with the controlled systems, i.e., the plant. The discrete compensation in Fig. 9.1-1b may be performed by a digital computer. Hold circuits may or may not be involved in the compensation. In the case of continuous compensation the hold circuit may be part of the compensator or the plant. For digital compensation the hold circuit goes in series with the plant. It should be noted that there are possible positions that the sampler might appear besides those shown in Fig. 9.1-1.

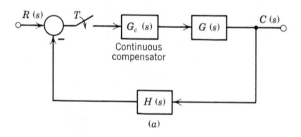

(a)

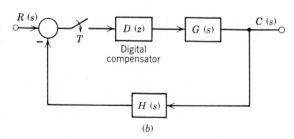

(b)

Figure 9.1-1 (a) Forward-path continuous compensated system. (b) Forward-path digital compensated system.

427

2. FEEDBACK COMPENSATION

Figure 9.1-2 shows two typical examples of feedback compensation. Figure 9.1-2*a* shows compensation by a continuous compensator while Fig. 9.1-2*b* shows digital feedback compensation. Again, there are other possibilities of feedback compensation.

Techniques of Compensation

The basic techniques of compensation are the same as those for the continuous case, because the goals are primarily the same. However, the basic mathematical tool is slightly different. We use the following techniques to design compensators for sampled-data systems:
1. Bode plots and Nichols charts.
2. Root locus method.
3. z-plane synthesis.
4. Deadbeat performance.
5. Mitrovic's method.

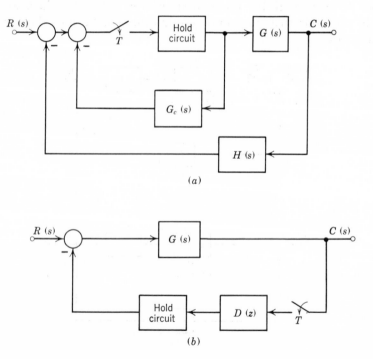

Figure 9.1-2   Feedback compensation. (*a*) Continuous feedback compensator. (*b*) Discrete feedback compensator.

9.2 FORWARD-PATH CONTINUOUS COMPENSATION

In this section we wish to develop techniques to design the forward-path continuous compensator shown in Fig. 9.1-1a. Specifically, let us consider the system shown in Fig. 9.2-1. Here we wish to obtain a suitable compensator $G_c(s)$ which will improve the stability, i.e., it will give larger phase and gain margins than the uncompensated system has. This requires the Bode plots of the frequency responses of both the uncompensated and the compensated systems for comparison. The sampler in the system, which requires the use of the Z-transform for analysis, makes finding frequency responses somewhat more laborious than in the continuous case, especially if hand methods are used (as in Section 6.4). This additional labor can be partially avoided by making a simple approximation, which replaces the sampler and the hold circuit by a delay element.

Approximating the Sampler and Hold by a Delay Element

The open-loop transfer function of the system shown in Fig. 9.2-1 in the z-domain, using (5.2-47), is given by

$$Z\{G_h(s)G_c(s)G(s)\} = G_hG_cG(z)$$
$$= \frac{1}{T}\sum_{k=-\infty}^{\infty} G_h(s+j\omega_rk)G_c(s+j\omega_rk)G(s+j\omega_rk)$$

$$(9.2\text{-}1)$$

where $\qquad \omega_r = \dfrac{2\pi}{T}$

For the case of a zero-order hold,

$$G_h(s) = \frac{1-e^{-sT}}{s} = \frac{e^{-sT/2}\sinh\,(sT/2)}{s/2} \qquad (9.2\text{-}2)$$

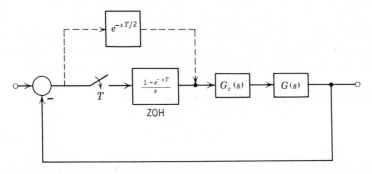

Figure 9.2-1   Continuous-compensator sampled-data system.

The frequency response is obtained by letting $s = j\omega$. We have

$$G_h G_c G(e^{j\omega T}) = \frac{1}{T} \sum_{k=-\infty}^{\infty} \frac{\sin\left[(\omega + k\omega_r)T/2\right]}{(\omega + k\omega_r)/2} \exp\left\{-\left[\frac{j(\omega + k\omega_r)T}{2}\right]\right\}$$
$$\times G_c(j\omega + jk\omega_r) G(j\omega + jk\omega_r) \qquad (9.2\text{-}3)$$

Assuming that the system $G(s)$ acts as a lowpass filter, which is true for the usual control system, we can assume that only low frequencies are important. This means

$$\frac{\sin\left[(\omega + k\omega_r)T/2\right]}{(\omega + k\omega_r)/2} \cong T \qquad (9.2\text{-}4)$$

and hence

$$G_h G_c G(e^{j\omega T}) = G_h G_c G^*(j\omega) \cong e^{-j\omega T/2} G_c(j\omega) G(j\omega) \qquad (9.2\text{-}5)$$

which is the open-loop frequency response of a continuous-time system where the sampler and hold have been replaced by a delay of $T/2$. This is also shown in Fig. 9.2-1. Why the preceding approximation is a good one and when it can be expected to be good can be seen in (9.2-4,5). The foregoing result shows that we can make the Bode and Nyquist plots very easily with and without the compensator because the sampler and hold only add an additional phase shift (of $-\omega T/2$) to the frequency response of the system without a sampler. However, we note that the design will be valid only if the approximation is good. Details will become obvious when we consider an example later.

If the approximation made above is not valid, then we have to consider a more exact method for compensator design. A method first given by Pastel and Thaler[1] is based on the bilinear w-transformation which was discussed in Section 6.4 and which transforms the transfer functions of interest from the z- to the w-domain. The inside of the unit circle in the z-plane transforms into the left half w-plane. The standard frequency-domain techniques can be applied to design in the w-plane and once the compensator is designed in the w-plane, it can be transformed back into the z-plane; thus $G_c(s)$ is obtained.

Design Based on *w*-Plane Transformation

Let

$$w = \frac{z-1}{z+1} \qquad (9.2\text{-}6)$$

---

[1] M. P. Pastel and G. J. Thaler, "Sampled-Data Design by Log-Gain Diagrams," *IRE Transactions on Automatic Control*, 192–197 Nov. 1959, **AC-4**.

Then from Fig. 9.2-1*a* we have

$$Z\{G_h(s)G_c(s)G(s)\} = G_hG_cG(z)$$

with the compensator and

$$Z\{G_h(s)G(s)\} = G_hG(z)$$

without the compensator. The approach can be summarized in the following steps.

Step 1. Since the behavior of the system can be determined from the position of the poles and zeros of the open-loop transfer function with respect to the unit circle in the z-plane, we can use the bilinear transformation $z = (1+w)/(1-w)$ to transfer the open-loop transfer function from the z- to the w-plane. This transforms the unit circle into the imaginary axis and the inside of the unit circle into the left half of w-plane. We transform the open-loop transfer function $G_hG(z)$ without the compensation from the z- to the w-plane, giving us $G_hG(w)$.

Step 2. The behavior of the system can be determined from the location of the poles and zeros of $G_hG(w)$ with respect to the imaginary axis of the w-plane. Standard Bode plot and Nichols chart techniques can be used to determine whether compensation is needed. The compensation in the w-plane may be lag, lead, or lag-lead, whichever is required. The compensation can be written as $G_c(w)$ so that the compensated open-loop transfer function in the w-plane becomes $G_c(w)G_hG(w)$.

Step 3. We now go back to the z-plane by letting $w = (z-1)/(z+1)$. Thus the compensated open-loop transfer function in the z-plane with compensation becomes $G_c(z)G_hG(z)$.

Step 4. To find $G_c(s)$ we use

$$G_c(s) = \frac{Z^{-1}\{G_c(z)G_hG(z)\}}{G_h(s)G(s)}$$

For the preceding equation to give a suitable compensator in the s-plane, we must impose realizability conditions on $G_c(s)$. It is generally desirable (for realizability reasons) to have a $G_c(s)$ that has all its poles on the negative real axis. This condition on $G_c(s)$ means that the poles of $G_c(w)$ in the w-plane must lie on the negative real semi-axis between the $-1+j0$ point and the origin and must be simple (the reader is encouraged to argue this statement by his own logic). There is complete freedom in the choice of zeros.

To illustrate the compensation obtained using the foregoing

approximate and exact procedures, let us consider a simple feedback SDS whose response we wish to compensate suitably.

Example 9.2-1

Consider the system of Fig. 9.2-2a with $G(s) = 8/(s+1)(s+2)$ and $T = 1$ sec. We look for the compensator $G_c(s)$ such that phase margin and other requirements as detailed below are met.

Approximate Method

In this case the sampler and the hold circuit are replaced by a delay element $e^{-j\omega T/2}$ as shown in Fig. 9.2-2b. Thus the open-loop frequency response becomes

$$8 \frac{e^{-j\omega T/2} G_c(j\omega)}{(j\omega+1)(j\omega+2)}, \qquad T = 1 \text{ sec}$$

First let there be no compensator; then $G_c(j\omega) = 1$. In this case we plot the gain and phase plots on a Bode diagram (Fig. 9.2-3) from the open-loop frequency response,

$$\frac{4e^{-j\omega/2}}{(j\omega+1)(j\omega/2+1)}.$$

The magnitude plot is quite straightforward since $|e^{-j\omega/2}| = 1$ for all ω. The phase plot is obtained from the addition of the phase of $1/(j\omega+1)(j\omega/2+1)$ to the phase of $e^{-j\omega/2}$ which is just $-\omega/2$. Note

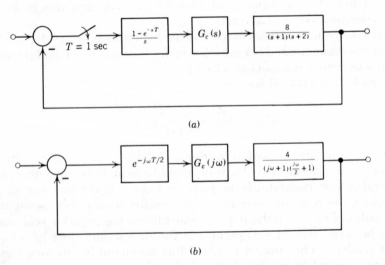

(a)

(b)

Figure 9.2-2   (a) An example of a feedback SDS to be compensated. (b) Delay element approximation to the sampler and hold of (a).

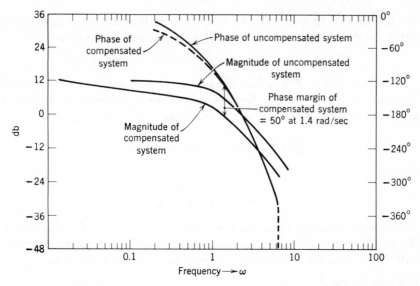

Figure 9.2-3 Bode plots for uncompensated and compensated system of Example 9.2-1.

that at $\omega = 2\pi$, there will be a discontinuity in the phase diagram of the actual sampled systems because of the change of sign of $\sin(\omega T/2)/(\omega T/2)$ at $\omega = 2\pi$. Our approximate method here, however, is based upon the assumption that the significant signal frequencies are all much lower than $\omega = 2\pi$.

From Fig. 9.2-3, phase margin $\approx 0°$ and gain margin ≈ 0 db. Since we are approximating the system here, it may or may not be stable. In any case, if we are going to operate this system, phase and gain margin must be increased. We could simply reduce the loop gain, which would increase both phase and gain margins. However, if we did this, we would decrease the magnitude of the steady-state error constant (K_p here). Let us assume that a decrease in K_p is not tolerable for the application we have in mind. Therefore we shall design a compensator that gives a phase margin of about 50° without decreasing K_p. This can be done by a phase-lag compensator, which will reduce the gain crossover frequency and provides a better phase margin. Let us assume the reduction in response time is acceptable.

A phase margin of about 48° can be achieved if the gain crossover frequency drops to about 1.5 rad/sec provided the phase curve changes very little. Let the lag compensator be

$$G_c(s) = A\alpha\frac{(1+\tau s)}{(1+\alpha\tau s)}, \qquad \alpha > 1, A, \tau > 0$$

We let $A\alpha = 1$. From the Bode plot, we note that the gain curve has to drop about 6 db to give the desired phase margin. Therefore

$$20\log_{10}\tau\omega - 20\log_{10}\alpha\tau\omega = -6\,\text{db} \qquad \text{or} \qquad \alpha \cong 2$$

We also wish a negligible change in phase at 1.5 rad/sec. This is accomplished by placing the upper corner frequency $(\omega = 1/\tau)$ at about 0.1 times the gain crossover frequency. This implies

$$0.1 \times 1.5 = \frac{1}{\tau} \qquad \text{or} \qquad \tau = 6.66$$

Hence

$$G_c(s) = \frac{1 + 6.66s}{1 + 13.33s}$$

The compensated Bode plot is shown in Fig. 9.2-3. On the graph, the compensated phase margin is about 50° and gain margin is about 6 db; K_p will be the same in both cases since we chose $A\alpha = 1$.

w-Plane Method

The open-loop transfer function without compensation is given by

$$G_h(s)G(s) = \frac{8(1 - e^{-sT})}{s(s+1)(s+2)}, \qquad T = 1\,\text{sec}$$

In the z-plane, we have

$$G_hG(z) = \frac{1.6\,(z + 0.368)}{(z - 0.368)(z - 0.135)}$$

We now obtain $G_hG(w)$ by letting $z = (1+w)/(1-w)$:

$$G_hG(w) = \frac{4(1-w)(w/2.165 + 1)}{(w/0.462 + 1)(w/0.761 + 1)}$$

In order to make a Bode plot, we let $w = jv$, where v is frequency in the w-plane. We get

$$G_hG(jv) = \frac{4(1 - jv)(jv/2.165 + 1)}{(jv/0.462 + 1)(jv/0.761 + 1)}$$

The magnitude and phase plot are shown in Fig. 9.2-4. We note that the phase margin is about 32° at a gain crossover frequency of 1.4 rad/sec. Gain margin is about 4.5 db.

We can transfer the Bode plot information to a Nichols chart to determine peak resonance (Fig. 9.2-5). We note in the figure that $M_p = 6$ db (or 2) at a frequency $v \cong 1.5$. An M_p of 2 implies 100% overshoot. This is normally too much. Moreover, the phase margin is too small.

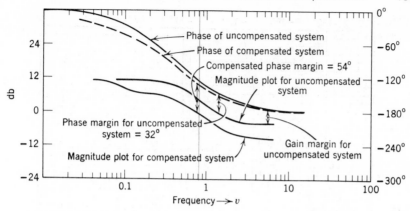

Figure 9.2-4 Bode plots for uncompensated and compensated system by *w*-plane method of Example 9.2-1.

Once again we let the requirement be that phase margin be about 50° and the frequency-response overshoot be about 25%. Hence $M_p \leq 1.25$ or $M_p \approx 2.0$ db or less. In order to increase phase margin, we use a lag compensator of the form

$$G_c(w) = A\alpha \frac{(1+\tau w)}{(1+\alpha\tau w)}, \qquad \alpha > 1, A, \tau > 0$$

If we examine the Bode plot of Fig. 9.2-4, we note that if the gain crossover point moves to $v = 0.8$, we will have the required phase margin of over 50°. In order to achieve this, the magnitude curve must drop about 6 db. Therefore we again let $A\alpha = 1$ and we have

$$20 \log_{10} \tau v - 20 \log_{10} \alpha\tau v = -6 \text{ db} \quad \text{or} \quad \alpha = 2$$

We again choose the upper corner frequency of the compensator at $v = 1/\tau$, about 0.1 times the uncompensated gain crossover frequency:

$$\frac{1}{\tau} = 0.14 \quad \text{or} \quad \tau = 7.15$$

Hence

$$G_c(w) = \frac{1+7.15w}{1+14.3w}$$

The Bode plot for the compensated system is also shown in Fig. 9.2-4. This frequency response is based on the open-loop transfer function:

$$G_c(w)G_hG(w) = \frac{4(w/2.165+1)(1-w)(1+w/0.14)}{(w/0.462+1)(w/0.761+1)(1+w/0.07)}$$

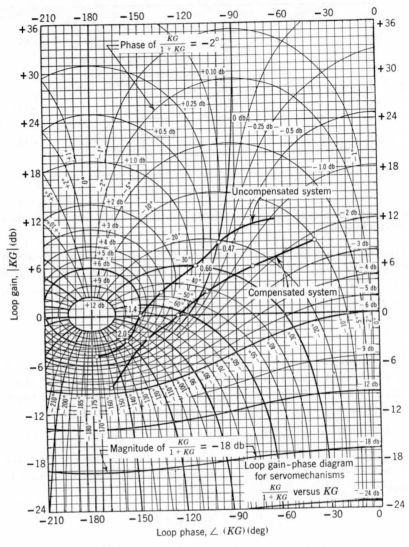

Figure 9.2-5   Plot of Fig. 9.2-4 on Nichols chart.

We note that the compensated phase margin is about 54° and compensated gain margin is 9.5 db. In order to check the frequency-response overshoot M_p, we plot the new curve on the Nichols chart in Fig. 9.2-5. We get

$$M_p \cong 2 \text{ db}$$

or about 26% frequency-response overshoot in the v-domain. This is what we wanted. Thus we have satisfactory compensation at least in the w-plane. To get the compensator in the s-domain we let $w = (z-1)/(z+1)$. This gives

$$G_c(z)G_hG(z) = \frac{0.855(z+0.368)(z-0.755)}{(z-0.368)(z-0.135)(z-0.87)}$$

$$= \frac{z-1}{z}\left[\frac{0.855z(z+0.368)(z-0.755)}{(z-1)(z-0.368)(z-0.135)(z-0.87)}\right]$$

Using the partial fraction expansion we get

$$G_c(s)G_h(s)G(s) = Z^{-1}\{G_c(z)G_hG(z)\}$$

$$Z^{-1}\left\{0.855(1-z^{-1})\left(\frac{4.7z}{z-1}-\frac{3.85z}{z-0.368}+\frac{2.11z}{z-0.135}-\frac{2.96z}{z-0.87}\right)\right\}$$

Note that we factored out the $(1-z^{-1})$ term since we know we have a hold circuit following the sampler. Tables can now be used to go to the s-domain; we have

$$G_c(s)G_h(s)G(s) = 0.855\ (1-e^{-sT})\left[\frac{4.7}{s}-\frac{3.85}{s+1}+\frac{2.11}{s+2}-\frac{2.96}{s+0.14}\right]$$

$$= 1.71\times10^{-2}\frac{(1-e^{-sT})}{s}\cdot\frac{(s+0.3)(s+239.7)}{(s+1)(s+2)(s+0.14)}$$

By eliminating $G_h(s)G(s)$, this gives us

$$G_c(s) = 0.214\times10^{-2}\frac{(s+0.3)(s+239.7)}{(s+0.14)}$$

$G_c(s)$ can be made *RC*-realizable by adding a pole at say $s=-300$. This is far into the LHP relative to the other poles and zero. This pole will have negligible effect on the system response. We write

$$G_c(s) = 0.214\times10^{-2}\frac{(s+0.3)(s+239.7)}{(s+0.14)}\times\frac{300}{(s+300)}$$

$$= 0.642\frac{(s+0.3)(s+239.7)}{(s+0.14)(s+300)}$$

This completes the design.

As an alternate route, we could have neglected the zero at $s = -239.7$. In that case we take

$$G_c(s) = 0.214\times10^{-2}\times239.7\frac{(s+0.3)}{(s+0.14)}$$

$$= 1.1\frac{(1+3.33s)}{(1+7.1s)}$$

This compensator does not compare too badly with the one obtained by the approximate method. We can say that the approximate method worked quite well for this example.

9.3 FORWARD-PATH DIGITAL COMPENSATION

The performance of a SDS can also be improved by using a digital compensator in the forward path as shown in Fig. 9.1-1*b*. Design of such compensation has the advantage that it is accomplished directly in the *z*-plane. Further, the discrete compensator can be synthesized easily by digital computer. It is sometimes required to have a digital computer in the loop of the control system. In such cases, it is logical to use discrete compensation since the same computer can also be utilized to perform the function of the compensator.

In order to show some techniques available to design discrete controllers, we will specifically consider the system shown in Fig. 9.3-1. Two popular design procedures will be developed.

Design Based on w-Plane Transformation

First, we extend the bilinear transformation technique developed in the last section for the design of digital controllers. Because of the use of a digital controller, the procedure is actually less cumbersome. The method is summarized in the following steps.

Step 1. The open-loop transfer function without the digital compensator in the *z*-plane is $G_hG(z)$. We transfer this to the *w*-plane by using the bilinear transformation $z = (w+1)/(1-w)$. This gives $G_hG(w)$.

Step 2. Bode plots and Nichols charts now can be used to determine the phase margin, gain margin, resonant frequency, bandwidth, etc., to see if the uncompensated system is satisfactory.

Step 3. If compensation is needed, we add a factor $G_c(w)$ to $G_hG(w)$, making it $G_c(w)G_hG(w)$. This will improve the gain margin, phase

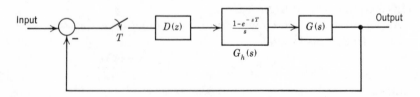

Figure 9.3-1 Sampled-data system with forward-path digital compensation. Design based on *w*-plane transformation.

margin, etc. The forms of $G_c(w)$ to make such improvements are the standard lag, lead, and lag-lead forms of continuous-time compensation. Once $G_c(w)$ is determined, we have the open-loop transfer function in the w-plane as $G_c(w)G_hG(w)$.

Step 4. We come back to the z-plane by letting $w = (z-1)/(z+1)$. Thus we get $G_c(z)G_hG(z)$ as the open-loop transfer function.

Step 5. We have

$$D(z) = G_c(z)$$

where $D(z)$ is the required compensation.

It is desirable to have a $D(z)$ with poles inside the unit circle and with no more zeros than poles.

It is obvious from this discussion that the bilinear transformation technique is simpler for the discrete controller since the compensator can be separated in both the w- and the z-plane. An example illustrating the method will be considered later in this section.

Design Based on the Use of the Root Locus Method

The root locus method essentially determines the location of the closed-loop poles of the system in the z-plane for the SDS as a function of the frequency-independent gain. The location of the poles determines the transient behavior and the stability of the system. The root contour can be plotted if some parameter other than gain is open to choice. The method of designing a suitable digital compensator using the root locus can be summarized in the following steps.

Step 1. After locating open-loop poles and zeros of the system without compensation, the root locus diagram is obtained using the standard procedures developed in Section 5.5.

Step 2. The simplest compensation scheme is to adjust the gain so that the closed-loop poles are in a configuration insuring proper dynamic and steady-state response. Usually, simply adjusting the gain will not give the response desired, so a suitable root locus reshaping is necessary. This requires adding poles and zeros to the open-loop transfer function. These added poles and zeros make up the digital compensation.

Step 3. Adding poles and zeros for suitable performance is a trial and error procedure. Lead, lag, or lag-lead compensation can be used. The dynamic performance may be judged by superimposing constant damping curves on the root locus plots. These curves aid in the proper choice of closed-loop poles.

We now consider a specific example of the design of a digital controller by the w-plane and root-locus methods.

Example 9.3-1

Consider the system of Example 9.2-1 as shown in Fig. 9.3-2. This will give us a comparison with previous continuous-compensator design. The fixed part of the plant is given by

$$G_h(s)G(s) = \frac{8(1-e^{-sT})}{s(s+1)(s+2)}$$

We wish to find a $D(z)$. The specifications for performance will be discussed as we proceed with the design.

w-Plane Method

The w-plane method works as in Example 9.2-1. We first obtained $G_hG(w)$ in the w-plane and found that the w-plane frequency-response phase margin was about 32° with a peak frequency-response overshoot of 100%. We designed a lag compensator $G_c(w)$ which increased phase margin to 54° and reduced overshoot to 26%. Assuming the same specifications here, we can use the same $G_c(w)$. In Example 9.2-1, from the open-loop compensated transfer function in the z-plane we obtained

$$G_c(z)G_hG(z) = \frac{0.855(z+0.368)(z-0.755)}{(z-0.368)(z-0.135)(z-0.87)}$$

Here we do not go to the s-plane to obtain the compensator. The open-loop compensated transfer function is $D(z)G_hG(z)$.

Now we have from Example 9.2-1

$$G_hG(z) = \frac{1.6(z+0.368)}{(z-0.368)(z-0.135)}$$

Hence we get the discrete compensator

$$D(z) = G_c(z) = 0.534\frac{(z-0.755)}{(z-0.87)}$$

This is an easily realizable, stable, discrete controller. We note that

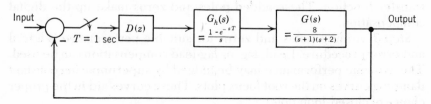

Figure 9.3-2   Example of a system to be compensated by a discrete compensator.

the w-plane method is more direct in the case of the design of discrete compensators as compared to continuous compensators.

Root Locus Method

The design of discrete compensators can be accomplished directly in the z-plane using the root locus method. The design is again a trial and error procedure and we proceed in steps as outlined above.

First we plot the root locus from the open-loop transfer function of the uncompensated system, which is

$$G_h G(z) = \frac{K(z + 0.368)}{(z - 0.368)(z - 0.135)}, \qquad K = 1.6$$

This is a type 0 system; therefore we must consider the position error constant. Let the position error constant of the uncompensated system be

$$K_p = \lim_{z \to 1} G_h G(z).$$

We assume

$$D(z) = \alpha \frac{z - a}{z - b}$$

Then the position error constant with compensation becomes

$$K_p' = \lim_{z \to 1} D(z) G_h G(z) = \alpha K_p \frac{1 - a}{1 - b}$$

This relationship gives a constraint on α, a and b. Note that other error constants can be obtained similarly for higher type of systems.

The constants a and b in $D(z)$ can be chosen to *cancel out* an undesirable pole or zero of the open-loop transfer function $G_h G(z)$ and to insert a new pole in order to get a more suitably shaped root locus diagram. We classify $D(z)$ as a lead or lag compensator if $b < a$ or $b > a$, respectively.

In order to see the effects of lead or lag compensation on our example, we superimpose some constant ζ and ω_n lines of Fig. 9.3-4 on the root locus plot of Fig. 9.3-3. This is shown in Fig. 9.3-5. From Fig. 9.3-5 we note that the uncompensated system is underdamped ($\zeta < 0.1$). The natural frequency here is $\omega_n = 2.35$ rad/sec. Let us assume that this is too high for the application intended. Let us first try adjusting the gain K and see what response can be obtained.

CHANGING GAIN K

If we reduce K, we get a higher damping ratio ζ and of course we lower K_p correspondingly. This is not what is wanted.

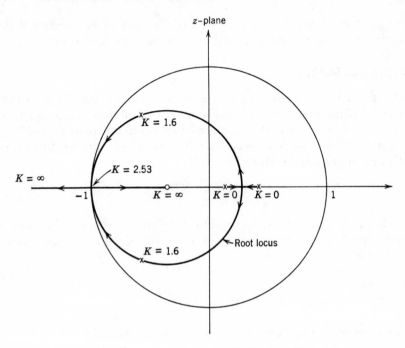

Figure 9.3-3 Root locus plot for Example 9.3-1.

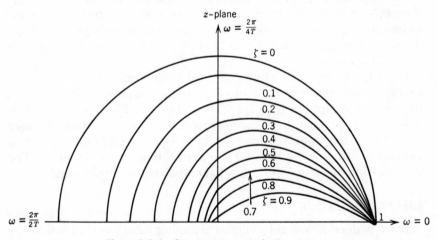

Figure 9.3-4 Constant-ζ curves in the z-plane.

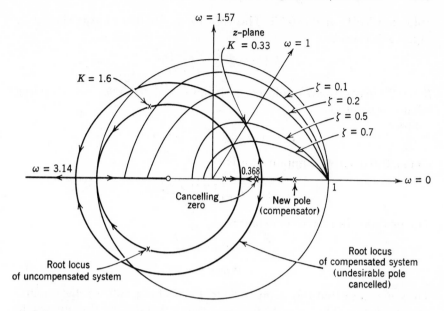

Figure 9.3-5 Root locus plots of uncompensated and compensated system for Example 9.3-1.

DESIGN REQUIREMENTS

Let us now set the specifications for designing the discrete compensator $D(z)$ and require that $\zeta = 0.5$ and $\omega_n \cong 1$. With $D(z)$ thus specified a lead or lag network may be required.

Lead Compensation

Figure 9.3-5 shows that we can cancel one of the poles (either at 0.135 or at 0.368) by the zero of $D(z)$ and add the pole anywhere to the left of the zero. The effect on the root locus will be to put the compensated root locus circle inside the uncompensated circle. This will give higher values of damping constant (with equivalent values of K_p), but in general ω_n will be increased. Since the objective here is a narrower bandwidth, we see that we are going in the wrong direction. Hence let us consider lag compensation.

Lag Compensation

From Fig. 9.3-5 we note that we can cancel out the pole at 0.368 by a zero in $D(z)$ and then insert a new pole to the right such that the new root locus circle passes through the point $\zeta = 0.5$ and $\omega_n \approx 1$. The new pole should be at $b = 0.75$. The root locus for this case is shown in Fig. 9.3-5. Hence

$$D(z) = \alpha \frac{z - 0.368}{z - 0.75}$$

$K = 0.33$ gives a pair of closed-loop poles at $\zeta = 0.5$, $\omega_n = 1$, which is what was wanted.

Since

$$K' = 1.6 \times \alpha, \qquad \alpha \cong 0.2$$

Hence the discrete compensator is

$$D(z) = 0.2 \left\{ \frac{z - 0.368}{z - 0.75} \right\}$$

The new position error constant is

$$K'_p = \lim_{z \to 1} D(z) K_p$$
$$= 0.51 K_p$$

Thus we note that the compensated K_p is about 50% of the uncompensated one. This is not what was wanted. Perhaps the answer to this is that we should not have cancelled out the pole at $z = 0.368$ but should cancel the pole at 0.135. This will increase K'_p. We might also relax the requirement of $\omega_n \cong 1$ to increase K'_p. In order to illustrate this, let us reposition the pole and zero of the lag compensator so that $\zeta = 0.5$ and the position error constant remains at least the same.

In Fig. 9.3-6, we note that $K = 0.4$ for the uncompensated system gives $\zeta = 0.5$. In order to increase K'_p, we put the pole of $D(z)$ closer to the $1 + j0$ point and the zero further from it. Let us put the pole at 0.9 and zero at 0.3. We see that in order to stay on the $\zeta = 0.5$ curve, the new gain K' must be 0.33. The discrete filter is

$$D(z) = \alpha \frac{z - 0.3}{z - 0.9}$$

Since $K' = 0.33$,

$$1.6\alpha = 0.33 \qquad \text{or} \qquad \alpha \cong 0.2$$

and

$$K'_p = \lim_{z \to 1} D(z) K_p$$
$$= \frac{0.2 \times 0.7}{0.1} K_p$$
$$= 1.4 K_p$$

Hence we have been able to increase K_p 40% by moving the compensator pole and zero appropriately. The natural frequency is

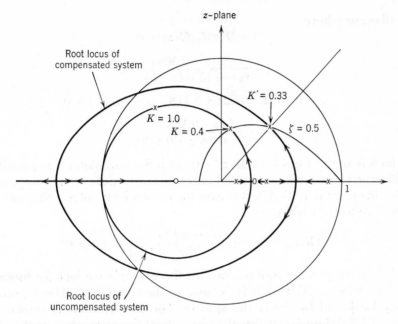

Figure 9.3-6   Another compensated root locus plot of Example 9.3-1.

$\omega_n \approx 0.8$. Since the object was to narrow the bandwidth of the system, we see that we have done a good job.

We have now seen how the lag compensator can be effectively utilized to control ζ, ω_n and K_p of the closed-loop system. These in turn control the overshoot and peak time as discussed in Section 2.3. It is clear that location of the compensating pole and zero can be improved by trial and error. With very little trial it should be possible to meet the requirement of $\zeta = 0.5$, $\omega_n = 1$, and K_p not decreased.

Root Contour Method

We can shape the root locus by varying pole or zero while keeping the gain fixed. As mentioned in Section 8.3, this is known as the root-contour method.

We again illustrate this with a single-pole, single-zero compensator. Let the pole location be the parameter and let the zero cancel out the pole at 0.368. The gain of the compensator is chosen as 0.2. Then

$$D(z) = 0.2\frac{z - 0.368}{z - b}$$

Let us consider b in the range $0.368 \leq b \leq 1$. The closed-loop poles will occur where

$$1 + D(z)G_h G(z) = 0$$

or
$$1 + \frac{0.32(z + 0.368)}{(z - 0.135)(z - b)} = 0$$

or
$$z^2 - 0.135z + 0.32z + 0.118 - b(z - 0.35) = 0$$

or
$$1 - \frac{b(z - 0.135)}{(z^2 + 0.185z + 0.118)} = 0$$

which is in the form $1 + GH(z) = 0$, with the parameter b as a multiplicative factor, so that standard root locus procedures can be applied. However, we note that in this case the closed-loop poles will occur at those values of z where

$$\angle GH(z) = \frac{\angle \{b(z - 0.135)\}}{\angle (z^2 + 0.185z + 0.118)} = \angle 1 = 0°$$

When we plot the root contour for this example we look for those z values where $\angle GH(z)$ is $0°$ instead of the usual $180°$. These points may be found by using the spirule. The root contour is shown in Fig. 9.3-7. When $b = 0.368$, the roots are at the same place as the uncompensated system. Since b is restricted to values between 0.368 and 1 the contour is very short. We can obtain K'_p for each value of b. Here

$$K'_p = \frac{0.2K_p(1 - 0.368)}{1 - b} = \frac{0.126K_p}{1 - b}$$

Hence the compensated K_p can be plotted as an additional parameter along the root contour. We thus will be able to see at a glance the closed-loop pole position and steady-state error constant as a function of the compensator pole location, b.

9.4 DESIGN OF COMPENSATORS BY z-PLANE SYNTHESIS

The w-plane method and the root locus method developed in the last section are trial and error procedures which start with an open-loop transfer function and a compensator and work toward a desired closed-loop response. We start by choosing a compensator form from past experience and then determine whether it gives satisfactory response; if it does not, then it is redesigned to give a better performance. The idea is basically the modification of the open-loop transfer function in order to meet the closed-loop system requirements. A more direct procedure is to first fix the transfer function of

z-plane

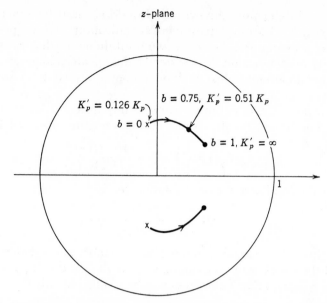

Figure 9.3-7 Root contour plot of Example 9.3-1 with pole of the compensator as variable.

the closed-loop system to meet a set of given specifications, as in the case of s-plane synthesis in Section 8.6. Once this is done, the digital compensator can be found by simple algebra. There are, however, two difficulties with this straightforward procedure. First, we must determine a way of fixing the overall closed-loop transfer function from a given set of specifications and, second, the digital compensator obtained must be realizable. This second requirement will of course affect the choice of closed-loop transfer function.

The procedure for such design was first introduced for continuous systems in Section 8.6 and will be extended here to the case of a SDS.

Basic Technique

Let the system be as shown in Fig. 9.4-1.

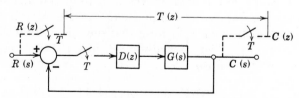

Figure 9.4-1 A feedback sampled-data system.

We will determine the overall closed-loop transfer function in the z-domain, $T(z)$, from a given set of specifications. These specifications can be quite general; however, we limit them, for discussion here, to velocity error constant, K_v, for steady-state performance, and damping ratio for dominant poles, peak overshoot and peak time for transient performance.

In order to develop the restrictions imposed on $T(z)$ by the realizability conditions on the compensator, consider Fig. 9.4-1. We see

$$T(z) = \frac{C(z)}{R(z)} = \frac{D(z)G(z)}{1+D(z)G(z)} \tag{9.4-1}$$

where $G(z)$ is fixed. We now obtain directly

$$D(z) = \frac{T(z)}{G(z)[1-T(z)]} \tag{9.4-2}$$

For $D(z)$ to be realizable, it must have no more zeros than poles.

Let the number of poles and zeros for $T(z)$ and $G(z)$, respectively, be p_T, z_T, and p_G, z_G. Then $D(z)$ will be realizable only if

{Excess number of poles of $T(z)$} $\geq$ {Excess number of poles of $G(z)$}

or $\qquad\qquad\qquad p_T - z_T \geq p_G - z_G \tag{9.4-3}$

Furthermore, if $D(z)$ is to give a stable compensator, it must not have more than one simple pole on the unit circle. However, we can synthesize an unstable compensator, but this will lead to an open-loop unstable system, which usually should be avoided. As in the case mentioned in Section 8.6 on s-plane synthesis, if $G(z)$ has a pole outside the unit circle, $1-T(z)$ must have a zero at the same point. Also, if $G(z)$ has a zero outside the unit circle, $T(z)$ must have a zero at the same point. These conditions are necessary to avoid pole-zero cancellation outside the unit circle.

The pole-zero restrictions on the closed-loop transfer function $T(z)$ affect the transient and steady-state responses of our system. The relationships for the two are developed below. We assume that the transient response of the overall system is to be dominated by a pair of complex poles and a zero within the unit circle. All other poles and zeros are placed in such a fashion that they contribute little to the transient response. The closed-loop transfer function, quadratic model, is then of the form

$$T(z) = \frac{K(z-z_1)}{(z-p_1)(z-p_2)}, \qquad K = \frac{(1-p_1)(1-p_2)}{1-z_1} \tag{9.4-4}$$

The gain K is determined from the steady-state requirements that there be no steady-state error at sampling instants for a unit step input.

In order to satisfy (9.4-3), we must add more poles to $T(z)$ if $p_G - z_G > 1$. We place these additional poles in the z-plane in such a manner that they do not influence the transient response significantly.

Assuming that $p_G - z_G \gg 1$ and that $G(z)$ has no poles or zeros outside the unit circle and using (9.4-2) and (9.4-4), we write the compensator as

$$D(z) = \frac{1}{G(z)}\left[\frac{K(z-z_1)}{(z-p_1)(z-p_2)-K(z-z_1)}\right] \tag{9.4-5}$$

or

$$D(z) = \frac{1}{G(z)}\left[\frac{K(z-z_1)}{z^2-(p_1+p_2+K)z+p_1 p_2+K z_1}\right] \tag{9.4-6}$$

Equation (9.4-6) will give us the required stable compensator if z_1, p_1, and p_2 are chosen to satisfy the required transient and velocity error constant constraints, and the requirement of open-loop stability.

Transient Performance and Pole-Zero Locations

Let us consider the system with overall transfer function

$$T(z) = \frac{C(z)}{R(z)} = \frac{K(z-z_1)}{(z-p_1)(z-p_2)} \tag{9.4-7}$$

where p_1 and p_2 are complex conjugate poles and z_1 is the zero as shown in Figs. 9.4-2a and 9.4-2b and K has been defined in (9.4-4). Note that z_1 can lie on the real axis in such a manner that the perpendicular on the p_1 line can be to the left or right of p_1. For a step input, we have

$$c(nT) = \frac{1}{2\pi j}\oint \frac{K(z-z_1)}{(z-p_1)(z-p_2)} \cdot \frac{z}{z-1} \cdot z^{n-1}dz \tag{9.4-8}$$

Let

$$p_1 = |p|e^{j\phi} \quad \text{and} \quad p_2 = |p|e^{-j\phi} \tag{9.4-9}$$

The integral of (9.4-8) can now be evaluated as

$$c(nT) = 1 - \frac{|p|^{-n}}{\cos\alpha}\cos(n\phi+\alpha) \tag{9.4-10}$$

where α is as shown in Fig. 9.4-2a and 9.4-2b. It can also be shown that

$$\cos\alpha = \frac{|1-z_1|\cdot|p|\sin\phi}{|1-p_2|\cdot|p_1-z_1|} \tag{9.4-11}$$

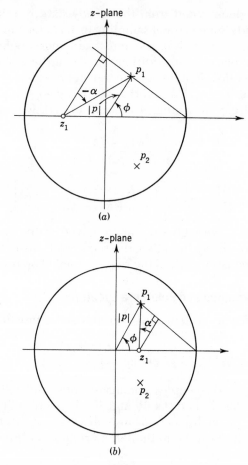

Figure 9.4-2 (a) Pole-zero locations for the quadratic model with zero on negative real axis. (b) Pole-zero locations for the quadratic model with zero on the positive real axis.

Note that α is negative or positive if the perpendicular from z_1 to line p_1 is to the left or the right of p_1 respectively.

Let us now construct a time function $c(t)$, such that at $t = nT$, $c(nT)$ has the same values as those given by (9.4-10). We write $nT = t$ in this equation and define

$$|p| \triangleq e^{-\sigma T}, \qquad \sigma > 0 \qquad (9.4\text{-}12)$$

which gives

$$c(t) = 1 - \frac{e^{-\sigma t}}{\cos \alpha} \cos\left(\frac{t}{T}\phi + \alpha\right) \qquad (9.4\text{-}13)$$

In terms of damping ratio and the resonant frequency of a second-order, continuous-time system,

$$\sigma = \zeta \omega_n \qquad (9.4\text{-}14)$$

and

$$\phi = \angle\, p_1 = \omega T = \omega_n \sqrt{1 - \zeta^2}\, T \qquad (9.4\text{-}15)$$

Hence (9.4-13) reduces to

$$c(t) = 1 - \frac{e^{-\zeta \omega_n t}}{\cos \alpha} \cos\left(\omega_n \sqrt{1 - \zeta^2}\, t + \alpha\right) \qquad (9.4\text{-}16)$$

It will be assumed that the peak overshoot and peak time for a step input as given by this equivalent continuous output $c(t)$ are the same as the peak time and peak overshoot of the SDS with transfer function of (9.4-4). The peak overshoot and peak time are obtained from (9.4-16) using simple differentiation:

$$\text{overshoot} = \frac{\sqrt{1 - \zeta^2}}{\cos \alpha} \exp\left\{ \frac{\zeta}{\sqrt{1 - \zeta^2}} [\cos^{-1}(\sqrt{1 - \zeta^2}) + \alpha - \pi] \right\}$$

$$(9.4\text{-}17)$$

and

$$t_p = \text{peak time} = \frac{1}{\omega_n \sqrt{1 - \zeta^2}} [\pi - \cos^{-1}(\sqrt{1 - \zeta^2}) - \alpha] \qquad (9.4\text{-}18)$$

Figure 9.4-3 gives the overshoot as a function of α for various values of ζ. In order to utilize this graph, we have to know ζ. This is obtained from (9.4-12), (9.4-14), and (9.4-15) as

$$|p| = \exp\left[-\phi\left(\frac{\zeta}{\sqrt{1 - \zeta^2}} \right) \right] \qquad (9.4\text{-}19)$$

In order to relate peak time to poles and zero locations, let us define

$$\gamma = \frac{\pi - \alpha - \cos^{-1}(\sqrt{1 - \zeta^2})}{\pi} \qquad (9.4\text{-}20)$$

Then using (9.4-15) and (9.4-20), we get

$$t_p = \gamma T \frac{\pi}{\phi} \qquad (9.4\text{-}21)$$

Gamma (γ) versus α lines for various values of ζ are shown in Fig. 9.4-4.

Determination of Velocity Error Constant K_v

Finally we consider the velocity error constant K_v. In this design based on the quadratic model the poles and zero are specified. This

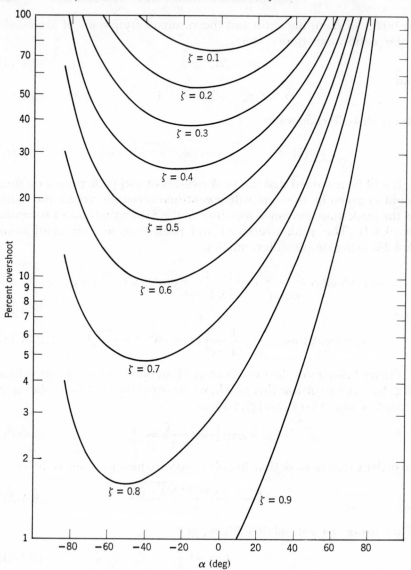

Figure 9.4-3 Overshoot versus α for various values of ζ.

makes it convenient to express K_v in terms of these poles and zero as shown in Section 6.3. Using (6.3-20), we get

$$\frac{1}{K_v T} = \frac{1}{1 - p_1} + \frac{1}{1 - p_2} - \frac{1}{1 - z_1} \qquad (9.4\text{-}22)$$

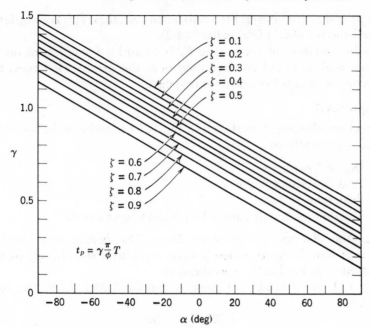

Figure 9.4-4   Peak time determination for various values of α and ζ.

Note that we have only one zero and two poles.

Utilizing the preceding equation, we may find that K_v is not large enough and is below specifications. How do we increase K_v without affecting any other requirement?

This can be done by adding a pole and a zero (called a dipole) to the overall transfer function very near the $z = 1$ point. This is analogous to adding a pole-zero pair near the origin in the s-plane. The new overall transfer function becomes

$$T(z) = \frac{K(z - z_1)}{(z - p_1)(z - p_2)} \times \frac{(1 - p_v)(z - z_v)}{(1 - z_v)(z - p_v)} \qquad (9.4\text{-}23)$$

The constant factor $(1 - p_v)/(1 - z_v)$ is necessary in order to maintain the same steady-state value of the output to a unit step input. The zero is placed to the right of the pole sufficiently close to the $1 + j0$ point to ensure that the transient response is not adversely affected.

The ratio $(1 - p_v)/(1 - z_v)$ is very close to unity (usually between 1.01 to 1.05), so that the peak overshoot is not adversely increased. In this case

$$\frac{1}{K_v T} = \frac{1}{1 - p_1} + \frac{1}{1 - p_2} + \frac{1}{1 - p_v} - \frac{1}{1 - z_1} - \frac{1}{1 - z_v} \qquad (9.4\text{-}24)$$

Once p_v and z_v are chosen for a satisfactory K_v, then $T(z)$ is completely defined and we obtain $D(z)$ using (9.4-2).

Another method of increasing K_v is to add a second zero on the positive real axis and another pole near the origin. However, this arrangement affects the transient response.

Example 9.4-1

The closed-loop system shown in Fig. 9.3-2 must have the following overall specifications:

(a) $|K_v| \geqslant 5 \text{ sec}^{-1}$.
(b) Peak overshoot $\leqslant 20\%$.
(c) Peak time $\leqslant 4$ sec.
(d) The dominant poles are to be placed to give $\zeta = 0.6$.

Design the discrete compensator $D(z)$. The design given in this section is based on zero steady-state error to a step input, so this condition is implied in the specifications.

Step 1. Using $\zeta = 0.6$ and peak overshoot $\leqslant 20\%$, from Fig. 9.4-3 we have

$$-76° \leqslant \alpha \leqslant 30°$$

Step 2. We can use Fig. 9.4-4 to obtain limits on γ for $\zeta = 0.6$:

$$0.64 \leqslant \gamma \leqslant 1.23$$

This leads to

$$\frac{0.64 T\pi}{\phi} \leqslant t_p \leqslant \frac{1.23 T\pi}{\phi}$$

or

$$\frac{115°}{\phi} \leqslant t_p \leqslant \frac{221°}{\phi}$$

We would now like to choose a ϕ in such a manner that the restriction on t_p is satisfied and ω_n is not too high. A reasonable choice is $\phi = 60°$. Then $t_p < 4$ sec; choosing $\alpha = 0$, we have

$$\zeta = 0.6$$
$$\text{overshoot} = 13\%$$
$$\phi = 60°$$
$$t_p = 0.8 \times \frac{\pi}{\pi/3} \times 1 = 2.4 < 4 \text{ sec}$$

Step 3. Using (9.4-19), we have

$$|p| = \exp\left[-\frac{\pi}{3}\left(\frac{0.6}{0.8}\right)\right] = 0.456$$

Using $|p|$, ϕ, and α, as chosen above, we can calculate z_1 from (9.4-11) or obtain it geometrically. We get $z_1 = 0.03$.

Step 4. Noting that $p_1 + p_2 = 2|p| \cos \phi$, $p_1 p_2 = |p|^2$, we have

$$K = \frac{1 - 2|p| \cos \phi + |p|^2}{1 - z_1}$$

$$= 0.775.$$

Step 5. Knowing z_1, K, $|p|$, and ϕ, we now have

$$T(z) = \frac{K(z - z_1)}{(z - p_1)(z - p_2)}$$

$$= \frac{0.775(z - 0.03)}{z^2 - 0.456z + 0.208}$$

Step 6. Before we proceed to obtain $D(z)$, we must see if the velocity error constant K_v requirement is met. We have

$$\frac{1}{K_v} = \frac{1}{1 - p_1} + \frac{1}{1 - p_2} - \frac{1}{1 - z_1}, \qquad T = 1 \text{ sec}$$

$$= \frac{2 - 2|p| \cos \phi}{1 - 2|p| \cos \phi + |p|^2} - \frac{1}{1 - z_1}$$

$$= \frac{2 - 0.456}{0.75} - \frac{1}{0.97} = 1.27$$

or

$$K_v = 0.8$$

This is considerably below 5. In order to increase K_v to this value and and not to affect the other specifications significantly, we add a dipole. We proceed as follows:

(a) The peak overshoot is 13%. We assume that as much as 20% is tolerable. Thus we let it be about 18%. Then

$$\frac{1 - p_v}{1 - z_v} = 1 + \frac{5}{100} = 1.05 \tag{i}$$

(b) Now

$$\frac{1}{K_v} = \frac{1}{1 - p_1} + \frac{1}{1 - p_2} + \frac{1}{1 - p_v} - \frac{1}{1 - z_1} - \frac{1}{1 - z_v}$$

or

$$\frac{1}{5} = 1.27 + \frac{1}{1 - p_v} - \frac{1}{1 - z_v} \tag{ii}$$

We can solve (i) and (ii) to obtain

$$p_v = 0.954, \qquad z_v = 0.956$$

With this addition of a pole and zero, the overall transfer function becomes

$$T(z) = \frac{0.775(z-0.03)}{z^2 - 0.456z + 0.208} \times \frac{(1-p_v)}{(1-z_v)} \frac{(z-z_v)}{(z-p_v)}$$

$$= \frac{0.815(z-0.03)(z-0.956)}{(z-0.954)(z^2-0.456z+0.208)}$$

Now

$$G(z) = \frac{1.6(z+0.368)}{(z-0.368)(z-0.135)}$$

The discrete compensator is then given by (9.4-2) as

$$D(z) = \frac{T(z)}{G(z)[1-T(z)]}$$

$$= \frac{0.51(z-0.03)(z-0.956)(z-0.368)(z-0.135)}{(z+0.368)(z-1)^2(z-0.25)}$$

This is a realizable discrete compensator, though perhaps a complicated one. Note that we could have increased K_v by making a different choice of α and ϕ. However, this would be a trial and error procedure and we might be unsuccessful in increasing K_v to the specified value.

9.5 COMPENSATION FOR DEADBEAT PERFORMANCE

In the last few sections we have developed design techniques for sampled-data systems which have been the extensions of and parallel to the techniques of continuous systems. In this section, we will design a digital controller for specific test input based on what is known as deadbeat performance. There is no comparable technique for continuous-time systems.

A *deadbeat performance implies zero steady-state error at sampling instants and fastest response. Fastest response can be defined as rise time in minimum number of sampling periods and a minimum finite settling time.*

Let us consider the system of Fig. 9.4-1. The closed-loop transfer function is given by

$$T(z) = \frac{C(z)}{R(z)} = \frac{D(z)G(z)}{1+D(z)G(z)} \tag{9.5-1}$$

If we now define

$$V(z) \triangleq \frac{1}{1+D(z)G(z)} \tag{9.5-2}$$

Then
$$T(z) = D(z)G(z)V(z) \tag{9.5-3}$$
Also, the error
$$E(z) = \frac{R(z)}{1 + D(z)G(z)} = R(z)V(z) \tag{9.5-4}$$

Method of Design

For any test signal, we want an $E(z)$ which is a polynomial in powers of z^{-1} with as few terms as possible. From (5.2-4) we see that this means $e(nT) = 0$ for all n higher than the highest power (N) of z^{-1} in $E(z)$. This will insure zero steady-state error and if N is a minimum this will mean a minimum settling time. However, we do want a realizable $D(z)$, which, we shall see, places some restrictions on just how small an N we can choose. The usual test signals are the step and the ramp. However, to be more general let us choose

$$u(t) = t^n, \qquad t > 0 \tag{9.5-5}$$

where $n = 0$ and $n = 1$ give the step and ramp, respectively. Then

$$U(z) = \frac{Q_n(z)}{(1 - z^{-1})^{n+1}} \tag{9.5-6}$$

where $Q_n(z)$ is a finite polynomial in z^{-1} with no zero at $z = 1$. From (9.5-4) we get

$$E(z) = \frac{Q_n(z)V(z)}{(1 - z^{-1})^{n+1}} \tag{9.5-7}$$

No steady-state error at sampling instants means, by the final-value theorem, that

$$\lim_{z \to 1} (1 - z^{-1})E(z) \to 0 \tag{9.5-8}$$

Therefore $V(z)$ must have the form

$$V(z) = (1 - z^{-1})^{n+1}F(z) \tag{9.5-9}$$

where $F(z)$ is a polynomial in z^{-1} with no zero at $z = 1$. This gives

$$E(z) = Q_n(z)F(z) \tag{9.5-10}$$

$Q_n(z)$ is a finite polynomial in powers of z^{-1}. $F(z)$ must then be a finite polynomial in z^{-1} if $E(z)$ is to have only a finite number of terms, thus insuring a finite settling time. If $F(z)$ is unity, then $E(z)$ will have minimum number of terms in powers of z^{-1}. However, this may result in an unrealizable $D(z)$. Thus we choose $F(z)$ as short a polynomial in powers of z^{-1} as possible with the conditions of realizability satisfied. Furthermore, $G(z)$ usually has a factor z^{-1} in the numerator. If we examine (9.5-3), we note that $T(z)$, the closed-loop transfer function, must also have a factor z^{-1} to give a realizable $D(z)$. This is possible only if $1-V(z)$ has this factor as well. This implies that $F(z)$ must be chosen so that $V(z)$ has a constant term, 1, in its expansion. Note also that if $G(z)$ has a transport lag (delay factor), then $T(z)$ must have the same transport lag so that cancellation can occur [see (9.5-3)] to give a realizable $D(z)$. This must also be considered in the choice of $F(z)$.

If $G(z)$ has poles and/or zeros outside the unit circle, we must exercise care in the choice of $V(z)$ so as not to obtain a $D(z)$ that tries to cancel these poles and/or zeros. If $G(z)$ has zeros and poles outside the unit circle, the following conditions must be satisfied:

1. Any poles of $G(z)$ on or outside the unit circle must be zeros of $V(z)$.

2. Any zeros of $G(z)$ on or outside the unit circle must also be zeros of $T(z) = 1 - V(z)$.

These conditions are the same that appear in the case of z-plane synthesis.

We now apply this discussion to two specific examples.

Example 9.5-1

The fixed part of a SDS is given by the transfer function

$$G(s) = \frac{2(1-e^{-sT})}{s(s+1)(s+2)}$$

Find the digital compensator which will give deadbeat response for a step input.

We have

$$G(z) = \frac{0.4z^{-1}(1+0.368z^{-1})}{(1-0.368z^{-1})(1-0.135z^{-1})}$$

We note that $G(z)$ has a factor z^{-1}, has no transport lag, and has no pole or zero outside the unit circle. Also

$$U(z) = Z\{1(t)\} = \frac{1}{(1-z^{-1})}$$

Therefore let

$$E(z) = (1-z^{-1})^{-1}V(z)$$

Now

$$V(z) = (1-z^{-1})F(z).$$

In this case $F(z)$ can be chosen as unity. This will give minimum settling time of one period as well as a $T(z) = 1 - V(z) = z^{-1}$ with a factor z^{-1} in the expansion. Therefore the compensator is

$$D(z) = \frac{T(z)}{G(z)V(z)}$$

$$= \frac{2.5(1 - 0.368z^{-1})(1 - 0.135z^{-1})}{(1 + 0.368z^{-1})(1 - z^{-1})}$$

In this case $E(z) = 1$, therefore

$$e(0) = 1, \qquad e(T) = 0, \qquad e(2T) = 0, \cdots$$

$D(z)$ is a realizable compensator. If this system were to be used for a ramp input, then $n = 2$ in (9.5-6), which gives,

$$E(z) = \frac{Tz^{-1}}{(1 - z^{-1})^2}(1 - z^{-1})$$

$$= Tz^{-1}(1 + z^{-1} + z^{-2} + z^{-3} + \cdots)$$

$$= T(z^{-1} + z^{-2} + \cdots)$$

Then

$$e(0) = 0, \qquad e(T) = e(2T) = \cdots = e(nT) \cdots = T$$

Hence the ramp gives a constant error T after the initial sampling.

This shows that deadbeat response is obtained only for a specific test signal.

Example 9.5-2

Consider a SDS with the transfer function of the fixed plant including the hold circuit as

$$G(s) = \frac{20(1 - e^{-sT})}{s^2(s + 10)(s + 20)}$$

If $T = 0.1$ sec, find the digital compensator which insures deadbeat response for a step input. We have

$$G(z) = Z\left\{\frac{200(1 - e^{-Ts})}{s^2(s + 10)(s + 20)}\right\}\Bigg|_{T = 0.1}$$

$$= \frac{1.64 \times 10^{-2}z^{-1}(1 + 0.12z^{-1})(1 + 1.932z^{-1})}{(1 - z^{-1})(1 - 0.368z^{-1})(1 - 0.135z^{-1})}$$

$G(z)$ has a factor of z^{-1}, has no transport lag, has a pole on the unit circle and a zero outside the unit circle. Since $T(z) = D(z)G(z)V(z)$, the preceding observations imply that in order to have a realizable $D(z)$ and deadbeat response;

1. $V(z)$ must have a zero at $z = 1$.
2. $T(z)$ must have a zero at $z = -1.932$ as well as a factor z^{-1}.

Since the input is a step

$$U(z) = \frac{1}{(1 - z^{-1})}$$

Therefore

$$V(z) = (1 - z^{-1})F(z)$$

Now $T(z)$ must have factors z^{-1} and $(1 + 1.932z^{-1})$ and since $T(z) = 1 - V(z)$, $F(z)$ must be of form $(1 + bz^{-1})$. We choose

$$T(z) = Kz^{-1}(1 + 1.932z^{-1})$$

and

$$V(z) = (1 - z^{-1})(1 + bz^{-1})$$

K and b are found by comparing coefficients since

$$T(z) = 1 - V(z)$$

We get

$$K = 1 - b \quad \text{and} \quad 1.932K = b$$

which give

$$b = 0.66 \quad \text{and} \quad K = 0.34$$

Thus

$$E(z) = 1 + 0.66z^{-1}, \quad e(0) = 1, \quad e(T) = 0.66, \quad e(2T) = 0, \cdots$$

and the compensator is

$$D(z) = \frac{T(z)}{V(z)G(z)}$$

$$= \frac{20.7(1 - 0.135z^{-1})(1 - 0.368z^{-1})}{(1 + 0.12z^{-1})(1 + 0.66z^{-1})}$$

This is a stable realizable compensator.

Note that again the response is deadbeat only for a step input. For a ramp input the error function would be

$$E(z) = U(z)V(z)$$
$$= 0.1z^{-1}(1 + 0.66z^{-1})(1 - z^{-1})^{-1} = 0.1z^{-1} + 0.166z^{-2} + 0.166z^{-3} \cdots$$

which means that the steady-state error is 0.166.

9.6 COMPENSATION BY MITROVIC'S METHOD

We have developed the basic Mitrovic's method for the analysis of SDS based on the characteristic equation in chapter 6. Stability is determined from the basic equations by plotting the region of stability which these imply.

In the upper half of the unit circle shown in Fig. 9.6-1 we let

$$z = -\rho v + j v \sqrt{1 - \rho^2} \qquad (9.6\text{-}1)$$

The characteristic equation is given by

$$F(z) = a_n z^n + a_{n-1} z^{n-1} + \cdots + a_1 z + a_0 = 0 \qquad (9.6\text{-}2)$$

or

$$a_1 z + a_0 = -a_n z^n - a_{n-1}^{n-1} - \cdots - a_2 z^2 \qquad (9.6\text{-}3)$$

Stability is insured by constraining the roots of (9.6-2) to be within the unit circle. We get three basic equations:

1. We put z from (9.6-1) into (9.6-3) and get

$$a_1 = a_2 \phi_2(\rho) v + a_3 \phi_3(\rho) v^2 + \cdots + a_n \phi_n(\rho) v^{n-1}$$

and $$\qquad (9.6\text{-}4)$$

$$a_0 = -v^2 \left[a_2 \phi_1(\rho) + a_3 \phi_2(\rho) v + \cdots + a_n \phi_{n-1}(\rho) v^{n-2} \right]$$

We have $v = 1$ on the unit-circle, so we substitute that in (9.6-4).

2. We get two more stability constraints at $z = 1$ and $z = -1$ from (6.2-11) $[F(1) > 0, (-1)^n F(-1) > 0]$. These applied to (9.6-2) give

$$a_0 > -a_1 - a_2 - a_3 - \cdots - a_n \qquad (9.6\text{-}5)$$

$$a_0 \lessgtr a_1 - a_2 + a_3 - \cdots - (-1)^n a_n, \quad \begin{cases} n \text{ odd} \\ n \text{ even} \end{cases} \qquad (9.6\text{-}6)$$

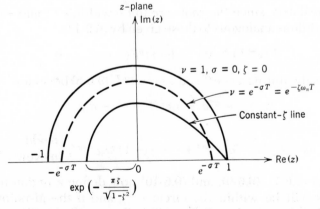

Figure 9.6-1   Nu (v), ζ contours in the z-plane.

Equations (9.6-4), (9.6-5), and (9.6-6) define the region of stability. The M-point (a_1, a_0) must be within this region in the (a_1, a_0)-plane for the system to be stable.

Method of Design

Mitrovic's method for the design of compensators is utilized by choosing lead or lag type compensators. The coefficients of the characteristic equation are then found as functions of the compensator parameters. These parameters are then adjusted to put the roots of the characteristic equation inside a desirable region of the unit circle.

Two basic considerations from the transient analysis viewpoint are the settling time and relative damping ratio.

SETTLING-TIME CONSIDERATION

The quantity that determines the settling time is the negative real part of the root. In the s-plane it is the distance σ or $\zeta\omega_n$. In the z-plane it is the quantity $\nu = e^{-\sigma T}$. If settling time is specified, the roots of the characteristic equation must be within a specified circle of radius $\nu = e^{-\sigma T}$ (Fig. 9.6-1).

If we put $\nu = e^{-\sigma T}$ into (9.6-4), we obtain

$$a_1 = a_2 \phi_2(\rho) e^{-\sigma T} + a_3 \phi_3(\rho) e^{-2\sigma T} + \cdots + a_n \phi_n(\rho) e^{-(n-1)\sigma T}$$

and   (9.6-7)

$$a_0 = -e^{-2\sigma T} [a_2 \phi_1(\rho) + a_3 \phi_2(\rho) e^{-\sigma T} + \cdots + a_n \phi_{n-1}(\rho) e^{-(n-2)\sigma T}]$$

On the real axis, since the roots are to be within $e^{-\sigma T}$ and $-e^{-\sigma T}$, we have conditions analogous to those given by (6.2-11):

$$F(e^{-\sigma T}) > 0, \qquad (-1)^n F(-e^{-\sigma T}) > 0 \qquad (9.6\text{-}8)$$

Then by substituting $z = e^{\sigma T}$ and $z = -e^{-\sigma T}$, (9.6-8) becomes

$$a_0 > -e^{-\sigma T} a_1 - a_2 e^{-2\sigma T} - \cdots - a_n e^{-n\sigma T} \qquad (9.6\text{-}9)$$

and

$$a_0 \lessgtr e^{-\sigma T} a_1 - a_2 e^{-2\sigma T} + \cdots - (-1)^n a_n e^{-n\sigma T}, \begin{cases} n \text{ odd} \\ n \text{ even} \end{cases} \qquad (9.6\text{-}10)$$

Equations (9.6-7), (9.6-9), and (9.6-10) now define a region in which the roots will be within the circle $e^{-\sigma T}$, and if the M-point (a_1, a_0) lies within this region we will be within the specified settling time.

RELATIVE DAMPING RATIO CONSIDERATION

In this case the minimum damping ratio ζ is specified for the dominant poles. A constant-ζ contour can be drawn in the z-plane for this specified value (Fig. 9.6-1). The roots must be within this curve to meet specifications. We know from (6.6-8) and (6.6-9) that

$$\nu = e^{-\zeta \omega_n T} \tag{9.6-11}$$

and

$$\rho = -\cos \omega_n T \sqrt{1 - \zeta^2} \tag{9.6-12}$$

On the constant-ζ curve, ρ will be given by (9.6-12). On the upper half of the unit circle ω varies from 0 to π where

$$\omega = \omega_n T \sqrt{1 - \zeta^2} \tag{9.6-13}$$

Using (9.6-13) for a fixed ζ, from (9.6-4) we have

$$a_1 = a_2 \phi_2(-\cos \omega_n T \sqrt{1 - \zeta^2}) e^{-\zeta \omega_n T} + \cdots$$

$$+ a_n \phi_n(-\cos \omega_n T \sqrt{1 - \zeta^2}) e^{-(n-1)\zeta \omega_n T}$$

and

$$\tag{9.6-14}$$

$$a_0 = -e^{-2\zeta \omega_n T}[a_2 \phi_1(-\cos \omega_n T \sqrt{1 - \zeta^2}) + \cdots$$

$$+ a_n \phi_{n-1}(-\cos \omega_n T \sqrt{1 - \zeta^2}) e^{-(n-2)\zeta \omega_n T}]$$

We must also note the fact that the constant ζ curve passes through $z = 1$ point on the positive real axis while on the negative real axis it passes through the point

$$z = -e^{-\zeta \omega_n T}$$

where

$$-\cos \omega_n T \sqrt{1 - \zeta^2} = \rho = 1$$

Therefore

$$z = -\exp \frac{-\pi \zeta}{\sqrt{1 - \zeta^2}} \tag{9.6-15}$$

The two relations implied by (6.2-11) therefore are

$$F(1) > 0 \quad \text{and} \quad (-1)^n F\left(-\exp \frac{-\pi \zeta}{\sqrt{1 - \zeta^2}}\right) > 0 \tag{9.6-16}$$

Equation (9.6-3) reduces to

$$a_0 > -a_1 - a_2 - a_3 - \cdots - a_n \tag{9.6-17}$$

and

$$a_0 \lessgtr a_1 r - a_2 r^2 + a_3 r^3 - \cdots - (-1)^n a_n r^n, \quad \begin{cases} n \text{ odd} \\ n \text{ even} \end{cases} \tag{9.6-18}$$

where

$$r = \exp \frac{-\pi\zeta}{\sqrt{1-\zeta^2}}$$

Equations (9.6-14), (9.6-17), and (9.6-18) define the region where ζ is equal to or greater than the specified value. If the M-point (a_1, a_0) is within this region, the specifications on ζ will be met.

Note that by the same method, we also could develop regions on the (a_1, a_0)-plane where the roots are specified between two values of σ, σ_1, and σ_2, or between two damping ratios ζ_1 and ζ_2.

We will now demonstrate the compensator design procedure by considering an example.

Example 9.6-1

Consider the system shown in Fig. 9.6-2. Use Mitrovic's method to design the continuous compensator $G_c(s)$ that does not increase gain in the loop but assures at least a 20-fold increase in the velocity error constant and gives for the closed-loop poles $\zeta = 0.5$.

Without the Compensator

$$G(z) = Z\left\{\frac{1}{s(s+1)}\right\} = \frac{0.865z}{(z-1)(z-0.135)}$$

Here $K_v = \frac{1}{2}$ and

$$C(z) = \frac{0.865z^2}{(z-1)(z^2-0.27z+0.135)}$$

$$= 0.865z^{-1} + 1.1z^{-2} + 1.05z^{-3} + 0.885z^{-4} + \cdots$$

The output $c(nT)$ is shown in Fig. 9.6-4. The output at sampling instants has been joined by a smooth curve.

With the Compensator

Since we wish to increase the steady-state error constant K_v, the compensator can be of the form

$$G_c(s) = A\alpha\frac{1+\tau s}{1+\alpha\tau s}, \qquad \alpha > 1$$

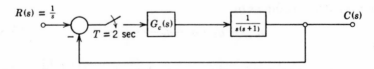

Figure 9.6-2   An example of a SDS.

which is a lag compensator. With this compensator,

$$K_v = A\alpha$$

Hence $A\alpha \geqslant 10$ for K_v to increase at least 20-fold.

In lag compensation the pole is chosen very near the origin in the s-plane. Thus let $\alpha\tau = 50$ and $A = 1$. Now the open-loop transfer function becomes

$$G(s)G_c(s) = \frac{(s+0.02\alpha)}{s(s+0.02)(s+1)}$$

and

$$GG_c(z) = Z\{G(s)G_c(s)\}$$

$$= \frac{(0.023\alpha+0.843)z^2 + (0.011\alpha-0.843)z}{z^3 - 2.096z^2 + 1.226z - 0.13}$$

The characteristic equation becomes

$$F(z) = z^3 + (0.023\alpha - 1.253)z^2$$
$$+ (0.011\alpha + 0.383)z - 0.13 = 0$$

Here

$$a_0 = -0.13$$
$$a_1 = 0.011\alpha + 0.383$$
$$a_2 = 0.023\alpha - 1.253$$
$$a_3 = 1$$

Since we want the roots restricted to $\zeta = 0.5$, we use (9.6-14), (9.6-17), and (9.6-18):

$$a_1 = a_2\phi_2(-\cos 1.732\omega_n)e^{-\omega_n} + \phi_3(-\cos 1.732\omega_n)e^{-2\omega_n}$$

$$a_0 = a_2 e^{-2\omega_n} - \phi_2(-\cos 1.732\omega_n)e^{-3\omega_n} \qquad \text{(I)}$$

and

$$a_0 > -a_1 - a_2 - 1 \qquad \text{(II)}$$

$$a_0 < 0.164a_1 - 0.027a_2 + 0.0044 \qquad \text{(III)}$$

since

$$r = \exp\frac{-0.5\pi}{\sqrt{0.75}} = 0.164$$

Here

$$\omega_n = \frac{\omega}{2\sqrt{1-\zeta^2}} = \frac{\omega}{1.732}, \qquad 0 \leqslant \omega \leqslant \pi$$

and

$$\phi_2(-\cos 1.732\omega_n) = -2\cos 1.732\omega_n$$
$$\phi_3(-\cos 1.732\omega_n) = 1 - 4\cos^2 1.732\omega_n$$

We now plot I, II, and III and the point $M(a_1, a_0)$, for the different values of α. Since we want $\alpha \geq 10$, we have varied α from 10 to 14 in steps of 2. The results are shown in Fig. 9.6-3.

In Fig. 9.6-3a we note that the M-point is inside the region, but ζ of the closed-loop poles is > 0.5.

In Fig. 9.6-3c we note that M-point has gone outside the region and therefore $\zeta < 0.5$.

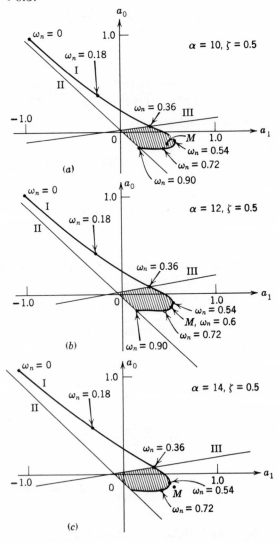

Figure 9.6-3   Plot of equations I, II, and III of Example 9.6-1.

In Fig. 9.6-3*b*, we note that the *M*-point is approximately on the curve of I giving ζ of closed-loop poles = 0.5. Here $\alpha = 12$. Therefore all requirements are met. Hence the compensator is

$$G_c(s) = \frac{s+0.24}{s+0.02}$$

The velocity error constant $K_v = 12$. From the curve we pick

$$\zeta = 0.5$$
$$\omega_n = 0.6$$

Therefore the poles are

$$z = -\rho\nu \pm j\nu \sqrt{1-\rho^2}$$

where

$$\nu = e^{-\zeta\omega_n T} = 0.55$$
$$\rho = -\cos{(\omega_n T\sqrt{1-\zeta^2})} = -0.51$$

The third pole is at

$$z = 0.428$$

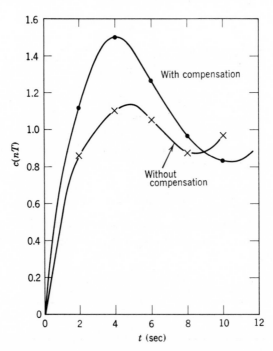

Figure 9.6-4 Step response of the system of Fig. 9.6-2.

The closed-loop transfer function therefore becomes

$$T(z) = \frac{C(z)}{R(z)} = \frac{z(1.12z - 0.71)}{(z^2 - 0.562z + 0.304)(z - 0.428)}$$

For a step input, we get

$$C(z) = 1.12z^{-1} + 1.505z^{-2} + 1.275z^{-3} + 0.980z^{-4} + \cdots$$

$c(nT)$ given by this $C(z)$ is also plotted in Fig. 9.6-4. Again the sampling points are joined by a smooth curve.

We now have designed directly a continuous compensator for a SDS without the use of any approximation technique.

PROBLEMS

9.1. Consider the system shown in Fig. P9.1.
   (a) Design the compensator $G_c(s)$ by using the approximate method insuring a phase margin of 50°.
   (b) Design the compensator $G_c(s)$ by using the w-plane method insuring a minimum phase margin of 50° and $M_p \leq 1.3$ in the w-plane.

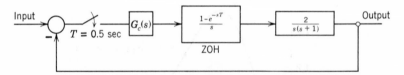

Figure P9.1

9.2. Use the w-plane method to design a compensator $G_c(s)$ if the fixed plant transfer function is given by

$$G(s) = \frac{2.5}{s(s+1)(s+2)}, \qquad T = 1 \text{ sec}$$

The overshoot in the w-plane is to be less than 25%.

9.3. A system is to be compensated by a discrete compensator $D(z)$ as shown in Fig. P9.3.

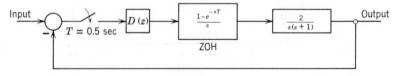

Figure P9.3

(a) Use the w-plane method to find $D(z)$ insuring an overshoot of less than 20% and a phase margin of about 45° in the w-plane.

(b) Use the root locus method to insure the same K_v and $\zeta = 0.7$.

9.4. Use the root locus method to design a digital compensator to insure a 50% gain in velocity error constant and closed-loop poles having $\zeta = 0.6$. The open-loop transfer function of the system in the z-domain is given by

$$G(z) = \frac{K(z+0.8)}{(z-1)(z-0.6)(z-0.5)}$$

9.5. The overall transfer function of a closed-loop sampled-data system is to be determined by ζ (damping of dominant poles) $= 0.7$, $t_p \leqslant 2$ sec, overshoot $\leqslant 10\%$ and $K_v \geqslant 1$ sec^{-1}. Design the discrete compensator $D(z)$ in the forward path using the z-plane synthesis if the fixed part of the system is given by

$$G(s) = \frac{5}{s(s+1)} \quad , \quad T = 0.5 \text{ sec}$$

9.6. The open-loop transfer function of an uncompensated sampled-data system is given by

$$\frac{1-e^{-sT}}{s^2(s+2)} \quad , \quad T = 0.2 \text{ sec}$$

Find the discrete compensator $D(z)$ in the forward path to insure an overall performance based on (a) overshoot $\leqslant 15\%$; (b) $\zeta = 0.6$, (c) $t_p \leqslant 5$ sec, and (d) $K_v \geqslant 10$ sec^{-1}.

9.7. Design a deadbeat performance compensator if the fixed part of the system has a transfer function

$$G(s) = \frac{1-e^{-sT}}{s^2(s+2)}, \quad T = 0.2 \text{ sec}$$

and the input is (a) a step and (b) a ramp.

Now use the modified Z-transform to plot the output as a function of time.

9.8. Consider the system shown in Fig. P9.8. Design $D(z)$ for deadbeat performance for (a) a step input and (b) a ramp input.

9.9. Consider the system shown in Fig. P9.9. Design (a) a continuous compensator $G_c(s)$ and (b) a discrete compensator $D(z)$ such that the ζ for

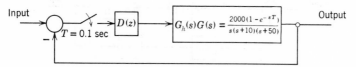

Figure P9.8

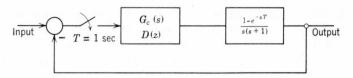

Figure P9.9

the closed-loop poles is $= 0.6$. The error constant is to be increased fivefold. Use Mitrovic's method.

9.10. Design a discrete compensator for a sampled-data system with open-loop transfer function $G(s) = 1/s(s+1)$, $T = 0.1$ sec such that the system settles within five sampling intervals and the velocity constant is increased five times.

9.11. Design the feedback compensator $G_c(s)$ as shown in Fig. P9.11. The following specifications have to be met:
(a) $K_v \geq 25 \text{ sec}^{-1}$
(b) Phase margin $= 45°$
Use the bilinear transformation method.

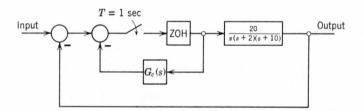

Figure P9.11

9.12. Use the z-plane synthesis to obtain the discrete compensator $D(z)$ shown in Fig. P9.12 if the following requirements are set on the overall response.
(a) $\zeta = 0.6$
(b) $t_p \leq 3$ sec
(c) overshoot $< 20\%$
(d) $K_v \geq 0.5 \text{ sec}^{-1}$

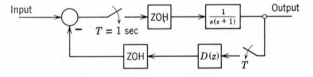

Figure P9.12

9.13 The open-loop transfer function of a sampled-data feedback system is given by

$$G(s) = \frac{(1 - e^{-sT})}{s(s+1)} \ , \quad T = 1 \text{ sec}$$

Design the discrete compensator for deadbeat performance for step input.

9.14 Design $D(z)$ for deadbeat performance for system shown in Fig P9.14. Discuss the effects of zero on the response time.

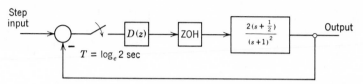

Step input → ⊗ → ⟋ $T = \log_e 2$ sec → $D(z)$ → ZOH → $\dfrac{2(s + \frac{1}{2})}{(s+1)^2}$ → Output

Figure P9.14

10

STATE VARIABLE
FEEDBACK COMPENSATION

10.1 INTRODUCTION

In discussing compensation, we have assumed that the plant is fixed and that the only quantity available for measurement and for feedback into the input is the plant output. Here we relax this by assuming that although the plant is fixed, we can reach into the internal mechanism and bring out for measurement those quantities that determine the plant's dynamic state. Specifically, we assume that a set of state variables for the plant is accessible for measurement and feedback into the input. In practice, this is probably assuming too much, since some of the state variables may be inaccessible. If this is the case, we can either reconstruct the inaccessible state variables from the accessible ones (discussed further in Section 10.2) or try to get sufficient compensation using only the accessible state variables. Thus the methods to be discussed here are those with practical merit. Moreover, the assumption that only the plant output is available for measurement usually is too restrictive. In the practical case the output and several dynamic quantities, though usually not all the state variables, are available for use in feedback. So the cases we consider here are really the extreme cases. Understanding and ability to compensate in these cases should lead to a satisfactory job of compensation in those problems that fall somewhere between the extremes. We consider the continuous-time case first.

10.2 STATE VARIABLE FEEDBACK COMPENSATION OF CONTINUOUS-TIME SYSTEMS

The Basic State Variable Feedback (SVF) Configuration

The plant we consider for compensation by state variable feedback (SVF) is assumed in a state variable formulation as given in Fig. 10.2-1. The simple input to the plant is u, which is related to the state vector $\mathbf{x}$ of the system by the equation

$$\dot{\mathbf{x}} = \mathbf{A}\mathbf{x} + \mathbf{b}u \qquad (10.2\text{-}1)$$

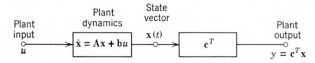

Figure 10.2-1 Block diagram for the state variable formulation of the plant.

The plant output is assumed given by

$$y = \mathbf{c}^T \mathbf{x} \qquad (10.2\text{-}2)$$

Now $\mathbf{x}$ is an n-vector (for an n-order system), $\mathbf{A}$ is an $n \times n$ constant matrix, $\mathbf{b}$ and $\mathbf{c}$ are constant n-vectors. This is the standard state variable formulation for a linear, lumped-parameter, time-invariant system as was discussed in Section 7.2. For this system, it has been shown in (7.2-42) that the transfer function is

$$\frac{Y(s)}{U(s)} = G_f(s) = \mathbf{c}^T [s\mathbf{I} - \mathbf{A}]^{-1} \mathbf{b} \qquad (10.2\text{-}3)$$

The fixed-plant transfer function will be referred to as $G_f(s)$ here.

In state variable feedback we assume that the entire state vector $\mathbf{x}$ is available for feedback. The feedback system is shown in Fig. 10.2-2. We feed back the quantity $\mathbf{k}^T \mathbf{x}$ and subtract it from the system input r, to get an input to the plant which is given by

$$u = r - \mathbf{k}^T \mathbf{x} \qquad (10.2\text{-}4)$$

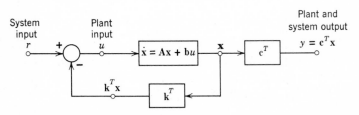

Figure 10.2-2 State variable plant formulation with state variable feedback.

In expanded form this is

$$u = r - [k_1 \; k_2 \cdots k_n] \begin{bmatrix} x_1 \\ x_2 \\ \cdot \\ \cdot \\ \cdot \\ x_n \end{bmatrix}$$

$$= r - \sum_{i=1}^{n} k_i x_i \qquad (10.2\text{-}5)$$

Our assumption now is that the constant feedback gains k_i, $i = 1$, $2, \ldots,\ n$ are available for adjustment by the designer and the compensation problem is essentially reduced to specifying the values of these constants. Although we shall alter this simple scheme slightly, in principle this is what is to be done. It will be seen that this scheme has some real advantages, so far as system compensation is concerned, over the previous method of feeding back the output only and then using a forward-path cascade compensator.

Let us now consider what effect SVF has on the closed-loop transfer function of the system. Let the plant transfer function be of the usual rational form,

$$G_f(s) = \frac{Y(s)}{U(s)} = \frac{a_m s^m + a_{m-1} s^{m-1} + \cdots + a_0}{s^n + b_{n-1} s^{n-1} + \cdots + b_0}, \qquad m < n \qquad (10.2\text{-}6)$$

A plant with this form of transfer function is block diagramed in Fig. 7.2-4 and is suitable for rendition into a state variable formulation. The resulting state variable formulation using $\mathbf{A}$, $\mathbf{b}$, and $\mathbf{c}$ from (7.2-28) and (7.2-29) in (10.2-3) gives

$$G_f(s) = \frac{Y(s)}{U(s)} = [a_0 \; a_1 \quad \cdots \quad a_m \; 0 \; \cdots \; 0]$$

$$\begin{bmatrix} s & -1 & 0 & 0 & \cdot & \cdot & & 0 \\ 0 & s & -1 & 0 & \cdot & \cdot & & 0 \\ 0 & 0 & s & -1 & & & & \\ \cdot & \cdot & & & & & & \cdot \\ \cdot & \cdot & & & & & & \cdot \\ 0 & 0 & \cdot & \cdot & \cdot & 0 & s & -1 \\ b_0 & b_1 & \cdot & \cdot & \cdot & & b_{n-2} & (s+b_{n-1}) \end{bmatrix}^{-1} \begin{bmatrix} 0 \\ 0 \\ \cdot \\ \cdot \\ \cdot \\ 0 \\ 1 \end{bmatrix}$$

$$= \frac{a_m s^m + a_{m-1} s^{m-1} + \cdots + a_0}{s^n + b_{n-1} s^{n-1} + \cdots + b_0} \qquad (10.2\text{-}7)$$

by use of (10.2-6).

Now let us consider the SVF system of Fig. 10.2-2. Using (10.2-4) in (10.2-1) gives

$$\dot{x} = Ax + b(r - k^T x)$$

$$= [A - bk^T]x + br \qquad (10.2\text{-}8)$$

and again from Fig. 10.2-2,

$$y = c^T x \qquad (10.2\text{-}9)$$

Now comparing (10.2-8) and (10.2-9) with (10.2-1) and (10.2-2) we see that they are alike except that

$$A \rightarrow [A - bk^T] \qquad (10.2\text{-}10)$$

$$u \rightarrow r$$

Using this in (10.2-3) we have immediately that

$$\frac{Y(s)}{R(s)} = T(s) = c^T[sI - A + bk^T]^{-1}b \qquad (10.2\text{-}11)$$

with A and b as given in (7.2-28) and k as in (10.2-5). By simple calculation

$$[A - bk^T] = \begin{bmatrix} 0 & 1 & 0 & \cdot & \cdot & \cdot & \cdot & 0 \\ 0 & 0 & 1 & \cdot & 0 & \cdot & \cdot & 0 \\ \cdot & & & & & & & \\ \cdot & & & & & & & 0 \\ \cdot & & & & & & & \\ 0 & 0 & \cdot & \cdot & \cdot & 0 & \cdot & 1 \\ (-b_0 - k_1) & (-b_1 - k_2) & \cdot & \cdot & \cdot & \cdot & \cdot & (-b_{n-1} - k_n) \end{bmatrix}$$

$$(10.2\text{-}12)$$

Using (10.2-12) in (10.2-11) and c and b from (7.2-28) and (7.2-29) gives

$$T(s) = [a_0 \, a_1 \ldots a_m \, 0 \ldots 0]$$

$$\begin{bmatrix} s & -1 & 0 & \cdot & \cdot & \cdot & & 0 \\ 0 & s & -1 & & & & & 0 \\ \cdot & \cdot & \cdot & & & & & \\ \cdot & & \cdot & \cdot & & & & \cdot \\ \cdot & & & \cdot & \cdot & & & 0 \\ 0 & 0 & \cdot & & s & & -1 \\ (b_0 + k_1) & (b_1 + k_2) & \cdot & \cdot & \cdot & \cdot & (b_{n-2} + k_{n-1}) & (s + b_{n-1} + k_n) \end{bmatrix}^{-1} \begin{bmatrix} 0 \\ 0 \\ \cdot \\ \cdot \\ \cdot \\ 0 \\ 1 \end{bmatrix}$$

$$(10.2\text{-}13)$$

Now comparing (10.2-13) with (10.2-7), we see that

$$T(s) = \frac{Y(s)}{R(s)} = \frac{a_m s^m + a_{m-1} s^{m-1} + \cdots + a_0}{s^n + (b_{n-1} + k_n) s^{n-1} + \cdots + (b_0 + k_1)} \quad (10.2\text{-}14)$$

which is the result we seek.

Looking at (10.2-14), we see the closed-loop transfer function for the system of Fig. 10.2-2 where the open-loop transfer function is given by (10.2-6). The effect of the SVF has been to leave the numerator polynomial of the closed-loop transfer function as it was in the open-loop transfer function, i.e., the open-loop and closed-loop zeros are the same. Looking further at (10.2-14), we see that all of the denominator polynomial (the characteristic polynomial) coefficients are linear functions of the feedback coefficients k_i, $i = 1, 2, \ldots, n$. Thus if we have the freedom to choose the feedback coefficients, i.e., the vector k, we can completely control the location of the closed-loop poles. This is just what is desired if we wish to compensate the transient response of a system, for now we can arrange to put the closed-loop poles in a position to give a desired ζ and ω_n for a pair of complex poles which can then be made to dominate the time response. If there are zeros that are near the $j\omega$-axis, they may be cancelled with a pole at the same location if the zeros are in the LHP. If there are RHP zeros, we cannot cancel them. Hence ζ and ω_n for the dominant LHP poles must be chosen so that the effect of the RHP zero or zeros is taken into account. It should also be noted that with SVF a RHP open-loop pole gives no particular trouble. Further, since the k_i's enter linearly into the coefficients of the characteristic equation, the problem of actually calculating values for the k_i is, in general, straightforward once the desired closed-loop pole configuration has been decided upon.

The conclusions which we draw from (10.2-14) are:

1. With state variable feedback the open-loop and closed-loop zeros are the same.

2. The coefficients of the characteristic polynomial are linear functions of the feedback coefficients, the k_i, $i = 1, 2, \ldots, n$; hence the closed-loop poles may be located at will through proper choice of the feedback coefficients.

It should be appreciated that the closed-loop transfer function of (10.2-14) has been shown to hold for only one particular state variable formulation, a formulation which is by no means unique for the system we are considering. However, the transfer function for a system has to be invariant under any choice of state variables. Thus conclusions (1) and (2) which we have drawn from (10.2-14) are valid for any choice of state variables. However, for a different choice of state variables, the feedback coefficients will differ and in fact will enter the characteristic polynomial coefficients in a different, but always linear, manner. This may be seen from (10.2-14) if we simply renumber the state variables with a corresponding renumbering of the feedback coefficients. The order in which the feedback co-efficients appear in the characteristic polynomial coefficients is destroyed of course.

Now that the notion of state variable feedback and its possible uses has been introduced, we must realize that the feedback configuration of Fig. 10.2-2, which is adequate for transient response compensation, generally is inadequate for achieving a desirable steady-state error response simultaneously with a satisfactory transient response. To achieve this it is usually necessary to have some control over the forward-path gain of the system. This can be obtained most easily by adding a frequency-independent gain K, an amplifier, directly at the input of the plant. This is as shown in Fig. 10.2-3a. K is assumed adjustable by the system designer. To see what the closed-loop trans-fer function for this system is, we note that its response, so far as the input and output are concerned, will be the same as for the system of Fig. 10.2-3b.

We now need simply note that Fig. 10.2-3b and Fig. 10.2-2 are of the same form except that Kk and $KR(s)$ replace k and $R(s)$, respectively.

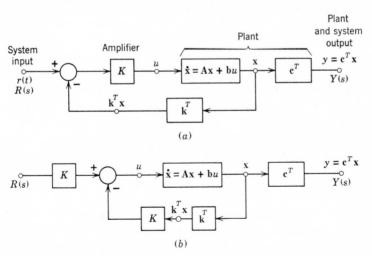

Figure 10.2-3 (a) State variable feedback system with amplifier in forward path. (b) State variable feedback system equivalent to that of (a).

Thus if the plant transfer function is given by (10.2-6) and the closed-loop transfer function for the system of Fig. 10.2-2 is given by (10.2-14), then the closed-loop transfer function of the system of Fig. 10.2-3 must be given by

$$T(s) = \frac{Y(s)}{R(s)} = \frac{K(a_m s^m + a_{m-1}s^{m-1} + \cdots + a_0)}{s^n + (b_{n-1} + Kk_n)s^{n-1} + \cdots + (b_0 + Kk_1)}$$

$$(10.2\text{-}15)$$

Again the closed-loop and open-loop zeros can be seen to be the same from (10.2-6) and (10.2-15). The coefficients of the characteristic polynomial are seen to be linear functions of terms of the form Kk_i, $i = 1, 2, \ldots, n$, instead of just k_i. The system of Fig. 10.2-3 and the corresponding closed-loop transfer function of (10.2-15) shall be used subsequently in the study of SVF compensation of linear time-invariant systems.

Choice of State Variables

Although the compensation methods that are to be described do not depend upon the particular choice of state variables, in order to describe the methods a choice of state variables must be made. Moreover, in order to make a choice of state variables some additional assumptions about the plant must be made. We hence assume that the plant is made up of a series connection of blocks, each of which has a first-order transfer function of the form

$$G_{f_i}(s) = \frac{c_i s + d_i}{s + f_i}$$

The choice of the state variables is now simply the output of each of these first-order blocks. Each of these outputs is assumed to be fed back to the input through a variable gain k_i (Fig. 10.2-4). This form for the plant is perhaps naive in the general case; however, this often is close to what is encountered. Again, the methods described in general work even though the plant may not take on this particular form. What is critical to our methods is the feedback of the state variables and the presence of the amplifier of gain K at the input to the plant.

To apply the compensation methods it will be necessary to determine the closed-loop transfer function for the system of Fig. 10.2-4. To make this job easier, the system has been shown in the form of nested blocks numbered n to 1 from the inside out. For each the output is x_n and the input is r_n. We note that the transfer function for

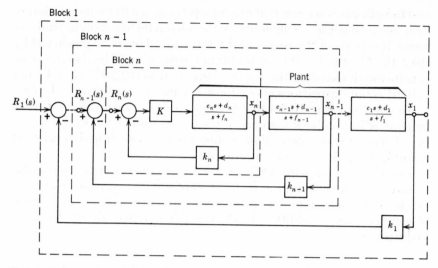

Figure 10.2-4 The assumed form of the plant and the choice of state variable feedback.

block n is found by applying (2.2-37). The system in block n is of the general form seen in Fig. 2.2-8. Thus we have

$$\frac{X_n(s)}{R_n(s)} = \frac{G(s)}{1 + G(s)H(s)} = \frac{K(c_n s + d_n)/(s + f_n)}{1 + Kk_n(c_n s + d_n)/(s + f_n)} \qquad (10.2\text{-}16)$$

Equation (2.2-37) now may be applied iteratively to give

$$\frac{X_i(s)}{R_i(s)} = \frac{[X_{i+1}(s)/R_{i+1}(s)][(c_i s + d_i)/(s + f_i)]}{1 + k_i[X_{i+1}(s)/R_{i+1}(s)][(c_i s + d_i)(s + f_i)]} \qquad (10.2\text{-}17)$$

To find the closed-loop transfer function we work down from $X_n(s)/R_n(s)$ using (10.2-17), until we have

$$\frac{X_1(s)}{R_1(s)} = T(s)$$

which is the overall closed-loop transfer function. For $n = 1$ we have

$$T(s) = \frac{X_1(s)}{R_1(s)} = \frac{K(c_1 s + d_1)}{(1 + Kk_1)s + f_1 + Kk_1 d_1} \qquad (10.2\text{-}18)$$

For $n = 2$ we get

$$T(s) = \frac{K(c_2 s + d_2)(c_1 s + d_1)}{\left\{ \begin{matrix} (1 + Kk_1 c_1 c_2 + Kk_2)s^2 + (Kk_1 c_1 d_2 + Kk_1 c_2 d_1 + Kk_2 f_1 + Kk_2 d_2 \\ + f_1 + f_2)s + (f_1 f_2 + Kk_1 d_1 d_2 + Kk_2 d_2 f_1) \end{matrix} \right\}}$$

$$(10.2\text{-}19)$$

and in both cases we see that open-loop zeros are the same as closed-loop zeros and the coefficients of the characteristic polynomial are linear functions of the terms of the form Kk_i. This is as predicted by (10.2-15). Continuing this on for higher order n's will lead to the same conclusion. Expanding the expressions used in (10.2-17) is really not as hard as the expressions above for $n = 1$ and $n = 2$ seem to indicate.

We see from Fig. 10.2-4 that the output is x_1. Since the interest in general is in feedback systems in which the output tracks the input, $k_1 = 1$. Thus the compensation problem reduces itself to determining K and k_i, $i = 2, 3, \ldots, n$, i.e., n constants.

It may also be seen from Fig. 10.2-4 that if a lead or lag compensator is added to the system, the basic form of the system is not changed. We have simply added one more series first-order block, the output of which is now an additional state variable that may also be fed back to the input.

Transient Response Compensation with State Variable Feedback

Our approach here is to put the system into the block diagram form as shown in Fig. 10.2-4. The closed-loop transfer function is then determined using (10.2-17), which gives us the characteristic polynomial with coefficients that are functions of the gain K and the feedback coefficients k_i, $i = 1, 2, \ldots, n$. We then have to determine K and the k_i so that we get the desired closed-loop pole configuration. What is proposed is basically just a form of s-plane synthesis, only now we have fixed the form of the compensation beforehand. The result is that now we need not be so careful in choosing the closed-loop transfer function so as to come up with a reasonable compensator. The quantities we work with are in general linearly related to the coefficients of the characteristic polynomial so the relationship between what we can control (K and the k_i) and what we want to control (the closed-loop pole configuration) is much more direct. The disadvantage of working with state variable feedback is that we cannot control the location of any of the closed-loop zeros. We can cancel them out with a pole at the same location if they happen to be in the LHP, but we can do no more. The zero locations play an important role in the steady-state error response of a system. We can see this by realizing from (3.3-24) that the velocity error constant K_v, satisfies

$$\frac{1}{K_v} = \sum_{i=1}^{m} \frac{1}{z_i} - \sum_{i=1}^{n} \frac{1}{p_i}$$

for a system that has closed-loop poles at p_i, $i = 1, 2, \ldots, n$ and closed-loop zeros at z_i, $i = 1, 2, \ldots, m$. Since we cannot control the location of any closed-loop zeros by state variable feedback and since we assume a fixed plant, we must add a series compensator of the lead or lag type in order to be able to control the location of a zero. This alternative will be discussed later.

From (10.2-15) we see that the general form of the closed-loop transfer function for the system of Fig. 10.2-4 will be

$$T(s) = \frac{K(a_m s^m + a_{m-1} s^{m-1} + \cdots + a_0)}{b_n(\text{Kk}) s^n + b_{n-1}(\text{Kk}) s^{n-1} + \cdots b_0(\text{Kk})} \qquad (10.2\text{-}20)$$

The characteristic polynomial coefficients are written as functions of Kk, by which we mean they are functions of terms $Kk_1, Kk_2, \ldots, Kk_n$. Since we have $n + 1$ constants at our disposal and since there are $n + 1$ coefficients $b_i(\text{Kk})$, $i = 0, 1, \ldots, n$, we can, in principle at least, choose the location of the n poles of the system, determine what the characteristic equation has to be to give us this pole configuration, and then choose the Kk_i, $i = 1, 2, \ldots, n$ to give us this exact characteristic equation and hence the corresponding pole configuration.

It is usually not necessary to fix the location of every closed-loop pole of the system to obtain a satisfactory response. We have seen that by properly placing two complex poles and by having other poles and zeros either near one another or far into the LHP a very satisfactory system response may be obtained. Hence the approach here will be to locate two dominant poles with an appropriate damping factor ζ and undamped natural frequency ω_n to give the desired transient step response. If there are any zeros near the $j\omega$-axis relative to these two poles and if they are in the LHP, we cancel them out with poles at the same location. We cannot cancel out RHP zeros so we must take these into account when choosing ζ and ω_n for the two dominant complex poles. Any other freedom that we have in determining the constants K and k_i, $i = 1, 2, \ldots, n$ we may use to get the best steady-state error response possible and to get all other poles far into the LHP.

Our basic objectives then in determining the gain K and the feedback coefficients k_i, $i = 1, 2, \ldots, n$ are:

1. To get a pair of dominant poles near the $j\omega$-axis with desired ζ and ω_n and any LHP zeros near the $j\omega$-axis relative to those two poles canceled by poles.
2. To get a desirable steady-state response.
3. To get the remaining poles far into the LHP.

We here accomplish these objectives by forming a factor whose roots are the desired two complex poles and any other poles required to cancel out LHP zeros near the $j\omega$-axis. This factor is then divided into the characteristic polynomial [the denominator of (10.2-20)] by synthetic division. Setting the remainder terms in this division to zero insures that the characteristic polynomial will have roots where desired (i.e., at the location of the desired two complex poles and at the $j\omega$-axis LHP zeros). This gives the conditions on the coefficients of the characteristic equation that have to be satisfied to give the desired location for the closed-loop poles. A suitable steady-state error response is then obtained by adjusting the steady-state error constant K_p, K_v, or K_a, whichever is appropriate, by use of (3.3-18), (3.3-25), and (3.3-35), respectively. Any remaining freedom in the choice of feedback coefficients is used to insure that all the remaining poles fall far into the LHP. Some trial and error may be required here.

Let us now consider the design step by step.

THE STEPS IN THE DESIGN OF STATE-VARIABLE FEEDBACK COMPENSATION.

Step 1. For the plant imbedded into a feedback system as shown in Fig. 10.2-2 the closed-loop transfer function is determined as in (10.2-20).

Step 2. A ζ and an ω_n are determined for a pair of dominant complex poles which will result in a step response satisfactory for the desired use of the system. This ζ and ω_n have to be chosen in the light of any plant RHP zeros which will also be closed-loop zeros, i.e., ζ and ω_n must be chosen with the RHP plant zeros taken into account.

Step 3. Construct a factor which has its roots at the location of the two desired dominant poles (with ζ and ω_n as determined in Step 2) and any plant LHP zeros which are near the $j\omega$-axis (relative to the desired dominant two poles). This will give a factor of the form

$$(s^2 + 2\zeta\omega_n + \omega_n{}^2)(s - z_1)(s - z_2) \cdots \qquad (10.2\text{-}21)$$

After this has been multiplied out, it may be divided into the characteristic polynomial (as determined in Step 1) by synthetic division. Setting the remainder terms to zero insures that the factor of (10.2-21) is a factor of the characteristic equation and gives us the conditions that the coefficients of the characteristic equation, and hence the terms of the form Kk_i, must satisfy to give the desired pole location.

Step 4. The feedback coefficients, k_i, $i = 1, 2, \ldots, n$, and the gain K (see Fig. 10.2-4) are determined so that

(a) The conditions as determined from the synthetic division in Step 3 are satisfied.

(b) The steady-state error constant K_p, K_v, or K_a, depending upon whether the system is type 0, 1, or 2, respectively, is satisfactory. This may be determined by use of (3.3-18), (3.3-25), or (3.3-35), depending upon which is appropriate.

(c) All uncancelled poles other than the desired dominant complex pair are far into the LHP.

Step 5. Determine the step response and steady-state error constant to see that the desired response has actually been achieved.

To see that we can actually accomplish all these things for a given case let us consider an example.

Example 10.2-1

Let us consider the complex we considered in Chapter 8. A d-c motor as shown in Fig. 8.3-6 has transfer function

$$G_f(s) = \frac{1}{s(s+1)(s+4)} \qquad (10.2\text{-}22)$$

which we desire to operate as a feedback position control device that gives a shaft position that tracks an input voltage $r(t)$. Instead of using only the feedback from the output shaft position, as in Fig. 8.3-6, let us assume that we now can measure and feed back all three state variables. This is shown in Fig. 10.2-5 in block diagram form. We note here that if x_1, as shown in Fig. 10.2-5, is the output shaft position of the motor of Fig. 8.3-6, then x_2 will be the angular velocity of the output shaft and hence could be measured with a tachometer on the output shaft. Then x_3 might be the armature current and this too could be measured reasonably for feedback into the input.

Since the desire is to have the output shaft track the input voltage, $r(t)$, we see immediately that we should have $k_1 = 1$. Since the plant

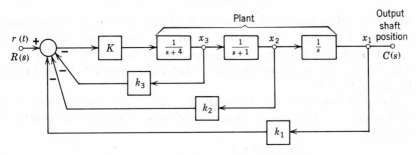

Figure 10.2-5 The plant of (10.2-22) in a state variable feedback system.

has one pole at the origin, it is a type 1 system and hence with $k_1 = 1$ we will have zero steady-state error to a step input and a finite, non-zero velocity error constant.

Let us now use the steps as outlined above to determine the constants, K, k_2, and k_3 (we already know $k_1 = 1$).

Step 1. The system is shown in Fig. 10.2-5; we see that we may use (10.2-16) and (10.2-17) to determine the closed-loop transfer function. We have, in order,

$$\frac{X_3(s)}{R_3(s)} = \frac{K}{s+4+Kk_3}$$

$$\frac{X_2(s)}{R_2(s)} = \frac{[K/(s+4+Kk_3)][1/(s+1)]}{1+[Kk_2/(s+4+Kk_3)][1/(s+1)]}$$

$$= \frac{K}{s^2 + (5+Kk_3)s + 4 + Kk_2 + Kk_3}$$

$$\frac{X_1(s)}{R(s)} = \frac{\{K/[s^2+(5+Kk_3)s+4+Kk_2+Kk_3]\}(1/s)}{1+\{K/[s^2+(5+Kk_3)s+4+Kk_2+Kk_3]\}(1/s)}$$

$$T(s) = \frac{X_1(s)}{R(s)} = \frac{K}{s^3 + (5+Kk_3)s^2 + (4+Kk_2+Kk_3)s + K} \quad (10.2\text{-}23)$$

where we have used the fact that $k_1 = 1$ in (10.2-23).

Step 2. As we have done before, let us take as our desired step-response objective an overshoot of approximately 15% and a settling time of 4 sec. Fifteen percent overshoot (see Fig. 2.3-3) implies $\zeta = 0.5$ and assuming a settling time of four time constants we have $4\tau = 4/\zeta\omega_n = 4$ sec, which implies $\omega_n = 2$.

Step 3. We here construct a factor which has zeros at the pole locations implied by the ζ and ω_n determined in Step 2. In this case the factor is

$$s^2 + 2\zeta\omega_n s + \omega_n^2 = s^2 + 2s + 4 \quad (10.2\text{-}24)$$

Since the plant has no zeros, from (10.2-22), there is no cancellation of zeros required. Hence the factor as determined in (10.2-24) specifies the location of all the poles whose location we wish to fix. The remaining pole (the system is 3rd-order) need only be placed far into the LHP.

We now proceed by dividing the factor of (10.2-24) into the characteristic equation as determined from (10.2-23) by synthetic division:

$$
\begin{array}{r}
s + (3+Kk_3) \\
s^2+2s+4 \overline{\big)\, s^3 + (5+Kk_3)s^2 + (4+Kk_2+Kk_3)s + K}
\end{array}
$$

$$\frac{s^3 \qquad\qquad +2s^2 \qquad\qquad\qquad\qquad +4s}{(3+Kk_3)s^2 \;+\; (Kk_2+Kk_3)s+K}$$

$$\frac{(3+Kk_3)s^2 \;+\; (6+2Kk_3)s+(12+4Kk_3)}{(-6+Kk_2-Kk_3)s+(K-12 \;-4Kk_3)}$$

Now if s^2+2s+4 is to be a factor of the characteristic polynomial, we must have

$$-6+Kk_2-Kk_3=0$$
$$K-12-4Kk_3=0 \qquad\qquad\qquad (10.2\text{-}25)$$

If (10.2-25) is satisfied, we see, also from the synthetic division, that the characteristic equation will have the factor $[s+(3+Kk_3)]$. Thus the characteristic polynomial will have roots (and hence the system will have closed-loop poles) at $s=-1\pm j\sqrt{3}$ and $s=-3-Kk_3$.

Step 4. We now determine the constants K, k_2, and k_3, so that (10.2-25) is satisfied and so that the pole at $s=-3-Kk_3$ is far into the LHP relative to the two complex poles at $s=-1\pm j\sqrt{3}$, which we wish to have dominate the response of the system. We know from Section 2.3 and Fig. 2.3-8 that a LHP pole on the real axis with magnitude six times the real part of the complex poles — six times farther into the LHP — will have negligible effect upon the transient response. Thus in this case we put the pole at $s=-3-Kk_3$ at $s=-6$, which implies $Kk_3=3$, which, used in (10.2-25), gives

$$\begin{aligned} K &= 24 \\ k_2 &= \tfrac{3}{8} \\ k_3 &= \tfrac{1}{8} \end{aligned} \qquad\qquad (10.2\text{-}26)$$

Step 5. Using the values of (10.2-26) in (10.2-23) gives

$$T(s)=\frac{24}{(s^2+2s+4)(s+6)} \qquad\qquad (10.2\text{-}27)$$

The step response for a system with this closed-loop transfer function has already been calculated and is shown as the lead compensated response in Fig. 8.5-2. We see a 16% overshoot and settling time of very nearly 4 sec have been obtained. The step response is very close to that of a two-pole system with $\zeta=0.5$ as shown in Fig. 2.3-3.

Since the plant is type 1 we may apply (3.3-25) to the coefficients of the closed-loop transfer function of (10.2-23) to get

$$K_v=\frac{K}{4+Kk_2+Kk_3}$$

which with the values given in (10.2-26) gives $K_v=1.5$. This is not too interesting in itself since it is the same value of K_v that was obtained by lead compensation by Mitrovic's method in Example 8.5-1. However, if (10.2-25) is solved simultaneously for Kk_2 and Kk_3 in terms of K and the results are substituted into the preceding equation for K_v, we find

$$Kk_3 = -3 + \frac{K}{4}$$

$$K_v = \frac{2K}{K+8}$$

(10.2-28)

And now if we recall from the synthetic division that we have a pole at $s = -3 - Kk_3$, the first equation of (10.2-28) gives the pole to be at

$$s = -\frac{K}{4}$$

(10.2-29)

Now as K is increased we see from (10.2-28) and (10.2-29) that K_v increases and approaches a limiting value of $K_v = 2$, and the third closed-loop pole goes farther into the LHP. Thus we can design our system compensation using as high a value of K as we wish. If the value $K_v = 1.5$ which we have chosen by the compensation of (10.2-26) is not high enough, we can increase it a little by increasing K and changing k_2 and k_3 correspondingly. However, no K_v higher than 2 is to be had, so we may accept $K_v = 1.5$. If a much higher K_v is required, we either have to change the location of the two dominant complex poles—i.e., choose a different ζ and ω_n—or we can add a lag compensator in series-cascade with the plant.

Before moving on to the problem of lag compensation with state variable feedback, let us for the moment consider what would happen if we had no amplifier in series with the plant, i.e., no amplifier with gain K as in Fig. 10.2-3a and 10.2-4. To do this we can simply take $K = 1$ in (10.2-23). This gives

$$T(s) = \frac{1}{s^3 + (5+k_3)s^2 + (4+k_2+k_3)s + 1}$$

If we now wish a 15% overshoot and 4-sec settling time as before, the characteristic polynomial must have a pair of complex poles with $\zeta = 0.5$ and $\omega_n = 2.0$; i.e., it must have a factor $s^2 + 2s + 4$. This implies that (10.2-28) and (10.2-29) must apply with $K = 1$. We have then that $K_v = \frac{2}{9}$ and a closed-loop pole (which before was at $s = -6$) at $s = -\frac{1}{4}$. We thus have a very low value of K_v and a troublesome pole on the real axis very near the origin. This pole will dominate the response of the system so that the whole system response will be very sluggish. The conclusion is that state variable feedback without the series amplifier is not very effective.

Lag Compensation with State Variable Feedback

We have seen that with state variable feedback there is no control over the location of the closed-loop zeros. This means that in general we have inadequate control over the steady-state error response of the system, or, looked at an other way, we have little control over the magnitude of the steady-state error constant. This may be seen by considering the velocity error constant K_v which we know from (3.3-24) to be given by

$$\frac{1}{K_v} = \sum_{i=1}^{m} \frac{1}{z_i} - \sum_{i=1}^{n} \frac{1}{p_i} \tag{10.2-30}$$

for a system with closed-loop poles and zeros at p_i, $i = 1, 2, \ldots, n$ and z_i, $i = 1, 2, \ldots, m$, respectively. The closed-loop pole positions are more or less dominated by transient response requirements, at least the ones which are near the $j\omega$-axis and hence have the major effect upon K_v. Thus to make a large magnitude change in the steady-state error constant some control over a zero is required.

The most straightforward way of accomplishing this is simply to add a series compensator which has a zero location that can be put where desired. As we have seen before, to increase the magnitude of the steady-state error constant, a lag compensator is required. Our approach here is to increase the magnitude of the steady-state error constant by simply adding a lag compensator of standard form in series with the plant as shown in Fig. 10.2-6.

We see in the figure a lag compensator of standard configuration (the gain constant is here taken as K) in series with the plant, which

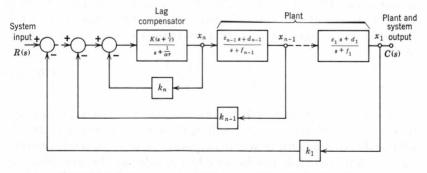

Figure 10.2-6   Lag compensated state variable feedback system.

we have assumed to be in the form of a series of first-order blocks. The assignment of state variables has been made in Fig. 10.2-6 and the output of the compensator (the input to the plant) has been labeled as state variable x_n. This state variable has been fed back into the input through the feedback gain k_n. It should be appreciated that an equivalent series lag compensator may be found which will have the same effect upon the system response as the proposed compensation scheme, but without the necessity of feedback. The system of Fig. 10.2-6, besides having stability advantages which will be seen later, is of exactly the same form as the system of Fig. 10.2-4. Thus after the pole and the zero (i.e., α and τ) of the compensator are specified, the design of the feedback coefficients k_1, $i = 1, 2, \ldots, n$ and the gain of the amplifier K proceed exactly as has been outlined earlier in this section.

Now let us look at the problem of determining the pole and zero location of the compensator. To do this we first have to decide where we want the closed-loop pole and zero, which result from adding the compensator pole and zero, to fall. We have a well developed technique for doing just this from the s-plane synthesis method of designing cascade compensators as discussed in Section 8.6. Briefly, what we do is compensate the system, in this case by SVF, to get a desired transient response and correspondingly closed-loop poles and zeros p_i, $i = 1, 2, \ldots, n$ and z_i, $i = 1, 2, \ldots m$, respectively. Now we can use (10.2-30) to determine a K_v for the system. We assume that our plant is type 1: if it is not, we can use (3.3-17) or (3.3-33) and argue with K_p or K_a, whichever is appropriate. If K_v as determined by (10.2-30) is slightly smaller than the desired value of velocity error constant $(K_v{}^d)$, then we can consider rearranging the closed-loop pole locations somewhat so as to get the K_v we want, $K_v{}^d$. If this fails or if $K_v{}^d$ is much larger than the K_v we get from (10.2-30), lag compensation is called for. This means the addition of an open-loop pole and zero which in turn will add a closed-loop pole and zero which we may call p_j and z_j, respectively. Using (10.2-30) here, we want to set p_j and z_j so that

$$\frac{1}{K_v{}^d} = \frac{1}{K_v} + \frac{1}{z_j} - \frac{1}{p_j} \tag{10.2-31}$$

In general the pole at p_j will be very close to the origin on negative real axis. Though the zero at z_j will be close to the pole at p_j, this pole will tend to degrade the transient response. How much it will degrade the transient response may be seen by considering the step-response overshoot. For a pole and zero, p_j and z_j, added to the closed-loop transfer function, the step-response overshoot will be increased by $X\%$. In Section 8.6, (8.6-24), it was seen that, to good approximation,

$$\frac{p_j}{z_j} \cong 1 + \frac{X}{100} \tag{10.2-32}$$

Now if we decide on the increase in step-response overshoot and the value of K_v^d, (10.2-31) and (10.2-32) may be used to determine where the additional closed-loop pole and zero, i.e. p_j and z_j should go.

Referring back to the lag compensator of Fig. 10.2-6 (since we know that with SVF open-loop zeros fall in the same location as closed-loop zeros), we have

$$-\frac{1}{\tau} = z_j \tag{10.2-33}$$

which determines τ. Now all that is required to design the compensator is the location of the pole at $-(1/\alpha\tau)$. We find that this pole is not at all critical. We simply choose a convenient value of α and take the open-loop pole wherever it happens to fall with this value since we know that the closed-loop poles may be placed as desired with the feedback coefficients k_i. Knowing the pole and zero location of the lag compensator, we proceed to determine the feedback coefficients, k_i, $i = 1, 2, \ldots, n$ and the gain K so that we have the same pole locations as before, adding the lag compensation and with the additional pole at p_j as determined by (10.2-31) and (10.2-32).

STEPS IN THE DESIGN OF LAG COMPENSATION WITH STATE VARIABLE FEEDBACK

Step 1. Design state variable feedback compensation to obtain the desired transient response as outlined previously.

Step 2. Determine the velocity error constant K_v using (3.3-25) or (10.2-30), whichever is most convenient. Also determine what the velocity error constant (K_v^d) will have to be to get the desired steady-state error response. If K_v [as determined by (3.3-25) or (10.2-30)] satisfies

$$K_v \geqslant K_v^d$$

then nothing more is required. If K_v is less than K_v^d, we can try to rearrange the closed-loop pole configuration, guided by (10.2-30), to get K_v up to the desired level, K_v^d. If this is impossible without degrading the transient response intolerably, lag compensation is required, so go to the next step.

Step 3. We add a lag compensator to the system as shown in Fig.

10.2-6. The additional closed-loop pole and zero (at p_j and z_j, respectively) that this implies are found by determining what percentage, X, increase in step-response overshoot can be reasonably tolerated. Then p_j and z_j are determined by

$$\frac{1}{K_v^d} = \frac{1}{K_v} + \frac{1}{z_j} - \frac{1}{p_j}$$

$$\frac{p_j}{z_j} \cong 1 + \frac{X}{100} \tag{10.2-34}$$

where K_v^d and K_v are as determined in Step 2. We then choose α and τ of the compensator (see Fig. 10.2-6) so that

$$z_j = -\frac{1}{\tau}$$

and any convenient value of α. Since the steady-state error constant is usually increased by approximately the factor α with lag compensation (see Section 8.1) we can make

$$\alpha \cong \frac{K_v^d}{K_v} \tag{10.2-35}$$

Step 4. Having determined α and τ of the compensator, we determine the feedback coefficients, k_i, and the gain K (see Fig. 10.2-6) so that the system has the same closed-loop pole configuration as in Step 1, plus an additional pole at p_j as determined by (10.2-34) of Step 3.

Step 5. Determine the transient and steady-state response of the resulting lag compensated SVF system to see if all requirements have been met.

Example 10.2-2

Let us continue with Example 10.2-1, with the plant as given in (10.2-22) and the compensation for transient response with the SVF configuration as shown in Fig. 10.2-5. The resulting closed-loop transfer function is given in (10.2-27). Step 1 has been covered in Example 10.2-1.

Step 2. K_v for this system was determined using (10.2-28) to be $K_v = 1.5$. We saw in the argument following (10.2-28) that this value could possibly be increased up to a limiting value of $K_v = 2$ if the gain K were increased. This much can be done with negligible effect upon the transient response. As an example let us consider increasing the velocity error constant by a factor of four:

$$K_v^d = 4 \times 1.5 = 6$$

This will give $e_{ss}(\text{ramp}) = \frac{1}{6}$, which we assume will be adequate. This is much too large a change in K_v to be realized by relocating the closed-loop poles, i.e., it is too large a change to be realized without materially altering the transient response. Hence a lag compensator is required. We add this to the system as shown in Fig. 10.2-7.

Step 3. The step response for the closed-loop transfer function of (10.2-27), the closed-loop system resulting from the transient compensation achieved in Step 1, is shown as the lead compensated response in Fig. 8.5-2. We see an overshoot of approximately 16%. Let us assume here, as was done in the s-plane synthesis of Example 8.6-2, that an additional step-response overshoot of 5% can be tolerated reasonably in the use for which this system is intended. The total overshoot will then be 21%. This gives $X = 5$ to be used in (10.2-34). In Step 2, $K_v = 1.5$, $K_v{}^d = 6$ were determined. Using these values in (10.2-34) gives

$$\frac{1}{6} = \frac{1}{1.5} + \frac{1}{z_j} - \frac{1}{p_j}$$

$$\frac{p_j}{z_j} = 1.05$$

Hence

$$p_j = -0.1$$

$$z_j = \frac{-0.1}{1.05}$$

(10.2-36)

These values used in (10.2-33) with α as chosen by (10.2-35) give

$$\tau = 10.5$$

$$\alpha = 4$$

(10.2-37)

These specify the pole and the zero of the lag compensator of Fig. 10.2-7.

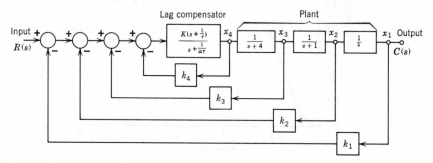

Figure 10.2-7   Lag compensated state variable feedback system of Example 10.2-2.

Step 4. Using the values of (10.2-37) and determining the closed-loop transfer function of the system of Fig. 10.2-7, we have

$$T(s) = \cfrac{K(s+1/10.5)}{\left[\begin{array}{l}(1+Kk_4)s^4 + (5.023+Kk_3+5.095Kk_4)s^3 \\ + (4.119+Kk_2+1.095Kk_3+4.476Kk_4)s^2 \\ + [0.095+K+0.095(Kk_2+k_3+4Kk_4)]s \\ + K/10.5\end{array}\right]} \quad (10.2\text{-}38)$$

We wish to add a pole and zero as given in (10.2-36) to the closed-loop transfer function of (10.2-27). The result is a desired closed-loop transfer function

$$T(s) = \frac{25.2(s+1/10.5)}{(s^2+2s+4)(s+6)(s+0.1)} \quad (10.2\text{-}39)$$

where the multiplying constant has been adjusted to give zero steady-state error to a step input. By multiplying out the denominator in (10.2-39) and comparing coefficients with (10.2-38), we determine, after some routine calculations, that

$$\begin{aligned} K &= 6.28 \\ k_2 &= 0.403 \\ k_3 &= 0.131 \\ k_4 &= -0.120 \end{aligned} \quad (10.2\text{-}40)$$

These values used in the system of Fig. 10.2-7, with α and τ as given in (10.2-37), give the desired closed-loop transfer function of (10.2-39).

Step 5. The step response for the system was calculated by using the closed-loop transfer function of (10.2-39); the result is shown in Fig. 10.2-8. We see that the step-response overshoot has been

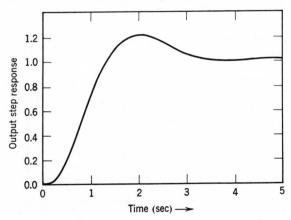

Figure 10.2-8 Lag compensated state variable feedback step response.

increased by 5% as assumed in Step 3. This is seen by comparing this response with the lead compensated response as seen in Fig. 8.5-2. This is the response for the system compensated as in Example 10.2-1. We know from our choice of closed-loop transfer function, (10.2-39), that $K_v = 6$, so this need not be checked.

Compensation in the Feedback Path

An alternative method of controlling the position of a closed-loop zero and thereby controlling the steady-state error constant is to put a compensator into one of the feedback paths (Fig. 10.2-9). The system here is the same SVF one seen in Fig. 10.2-3a except that one state varible, x_i, has been brought out and is fed back to the input through the compensator with transfer function

$$\frac{\beta}{s+\gamma} \qquad (10.2\text{-}41)$$

This system will now have a closed-loop zero at $s = -\gamma$. That this is true may be seen by realizing that the system differs from the system of Fig. 10.2-3 only in that the ith state variable, x_i, is fed back through $k_i + \beta/(s+\gamma)$ instead of just k_i The form of the closed-loop transfer function for this system hence may be found by replacing k_i by $k_i + \beta/(s+\gamma)$ in (10.2-15), which is the transfer function of the closed-loop system of Fig. 10.2-3. The result is a closed-loop transfer function of order one higher and with a zero at $s = -\gamma$.

To compensate this system we choose γ so as to get a closed-loop zero where it is wanted and then determine the constants K, k_i, and β to get the desired closed-loop pole configuration. This follows by the same steps as in designing a lag compensated SVF system with the compensator in the forward path.

Example 10.2-3

Here let us consider increasing the steady-state error constant for the SVF system of Fig. 10.2-5. We take the same problem considered

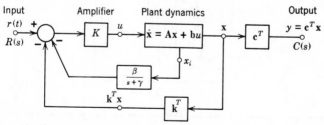

Figure 10.2-9 State variable feedback system with a compensator in a feedback path.

in Example 10.2-2. The design requirements then were met by the closed-loop transfer function as given in (10.2-39). Now let us attempt to achieve this closed-loop transfer function for the system of Fig. 10.2-5 by adding a compensator in the feedback path coming from state variable x_3. This is shown in Fig. 10.2-10.

The closed-loop transfer function for this system is given by

$$T(s) = \frac{K(s+\gamma)}{\begin{cases} s^4 + (5 + Kk_3 + \gamma)s^3 \\ + [4 + 5\gamma + Kk_3(1+\gamma) + Kk_2 + K\beta]s^2 \\ + [4\gamma + K + Kk_3\gamma + Kk_2\gamma + K\beta]s + K\gamma \end{cases}} \quad (10.2\text{-}42)$$

We wish to achieve the closed-loop transfer function of (10.2-39). By comparing coefficients in (10.2-39) and (10.2-42), we have

$$K = 25.2$$
$$\gamma = 1/10.5$$
$$\beta = 0.049$$
$$k_2 = 0.408$$
$$k_3 = 0.119$$

With these values used in the system of Fig. 10.2-10 the closed-loop transfer function will be given by (10.2-39), the step response will be as shown in Fig. 10.2-8, and we will have $K_v = 6$, which was desired.

Inaccessible State Variables

Thus far in this section we have assumed that all state variables are available for feedback to the input. In practice, this is not usually the case; therefore means of handling the situation when inaccessible

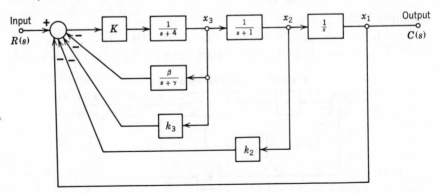

Figure 10.2-10 Feedback compensated system of Example 10.2-3.

state variables appear should be considered. Basically, there are two approaches to this problem. First, we can work with the state variables that are accessible. This means that in the compensation methods that have been considered we set the feedback coefficient $k_i = 0$ if state variable x_i is inaccessible, and then try to obtain a closed-loop pole configuration that is satisfactory by properly adjusting the remaining feedback coefficients (which correspond to the accessible state variables). This generally makes the compensation problem more difficult, so more trial and error work is involved and hence more patience is required to achieve an adequate compensation. However, if the system is of high order, i.e., if n is high, then we may very well have more freedom than can be effectively used if there are n feedback coefficients to be chosen. Hence letting $k_i = 0$ for a few state variables may take care of a lot of unusable freedom in the choice of the feedback coefficients. In any case, if there are inaccessible state variables, the possibility of just doing without the feedback of these variables should be thoroughly investigated before turning to the second possibility, the reconstruction of state variables.

THE RECONSTRUCTION OF STATE VARIABLES

To see the basic idea here, let us consider the system of Fig. 10.2-4 and let us look at a typical block, as shown in Fig. 10.2-11. In Fig. 10.2-11*b* we assume that the output of the block under consideration, state variable x_i, is inaccessible, while the input (state variable x_{i-1}) is accessible. Hence a compensator block is added, tapped into the accessible state variable x_{i-1}, which has the same transfer function as the plant block. The output of this compensator, which will be accessible, will be a reconstruction of state variable x_i. This output may be fed back to the input in the same manner in which state variable x_i would be fed back if it were accessible.

The second alternative for state variable reconstruction is shown in Fig. 10.2-11*c*, where the input state variable x_{i-1} to the block is inaccessible. Here we add a compensator block with its input connected to the accessible state variable x_i, whose transfer function is the inverse of the transfer function of the block. The output of the compensator will be the reconstruction of the inaccessible state variable x_{i-1}. This reconstruction of x_{i-1} will be accessible.

In reconstructing state variables, it should be borne in mind that there is a great deal of freedom available to the designer in the choice of state variables. What this means in the case of the reconstruction of state variables as in Fig. 10.2-11 is that the zero of the reconstructing

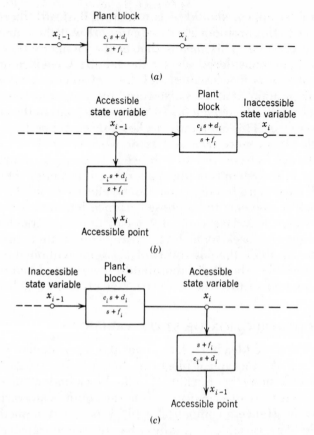

Figure 10.2-11   Two possibilities in the reconstruction of state variables. (*a*) A typical first-order block. (*b*) and (*c*) generating accessible point.

compensator may be chosen arbitrarily, and the output (of the compensator) used as the missing state variables. Care must be used in this reconstruction when a pole is introduced into a feedback path. The order of the system will be increased by one and the closed-loop transfer function will have a zero at the location of the feedback pole. We will simply have manufactured a new state variable by the introduction of this feedback pole unless the open-loop forward-path transfer function has a pole at this same location to cancel out the feedback pole. This cancellation is insured in Fig. 10.2-11*b* by an open-loop pole at the same location but ahead of the feedback pole in the signal path of the system. In Fig. 10.2-11*c* this cancellation is insured by the open-loop zero at the same location but behind the feedback pole in the signal path of the system. If either of the compensator blocks of Fig. 10.2-11 is inserted into the feedback system of Fig. 10.2-4, the result will be a closed-loop transfer function of order n and with no zeros at the location of the feedback poles. In the reconstruction of state variables, what is wanted is not to increase the order of the characteristic polynomial and to have each coefficient (or at least as many coefficients as necessary) of the characteristic polynomial a function of a variable that may be chosen by the system designer. Any scheme that does this should be looked upon as a practical possibility.

Another method for producing an inaccessible state variable which is simple in concept but may be difficult in practice is differentiation of the state variables that are available. This is as shown in Fig. 10.2-12. Though this is simple in principle, it is usually difficult in practice to realize a differentiator, so this scheme has its drawbacks. However, an important case in practice is when a state variable appears as the position of a shaft. Usually for this case a tachometer may be connected to the shaft so the information on the derivative is fairly easy to obtain here.

The Root Locus for a State Variable Feedback System

It is possible to draw a root locus very easily for a SVF system. It is with the root locus that we begin to see the inherent advantages in

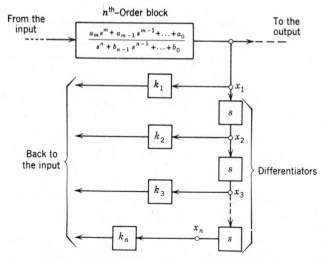

Figure 10.2-12 Construction of state variables by differentiation.

using SVF. To draw up a root locus for a SVF system as in Fig. 10.2-3 consider the closed-loop transfer function for this system in (10.2-15). Dividing the numerator and denominator of this transfer function by the open-loop characteristic polynomial $s^n + b_{n-1}s^{n-1} + \cdots + b_0$ gives

$$T(s) = \frac{K\left(\dfrac{a_m s^m + a_{m-1}s^{m-1} + \cdots + a_0}{s^n + b_{n-1}s^{n-1} + \cdots + b_0}\right)}{1 + K\left(\dfrac{k_n s^{n-1} + k_{n-1}s^{n-2} + \cdots + k_1}{s^n + b_{n-1}s^{n-1} + \cdots + b_0}\right)} \tag{10.2-43}$$

This transfer function is of the form

$$T(s) = \frac{G(s)}{1 + G(s)H(s)} \tag{10.2-44}$$

where

$$G(s) = K\frac{a_m s^m + a_{m-1}s^{m-1} + \cdots + a_0}{s^n + b_{n-1}s^{n-1} + \cdots + b_0}$$

$$\tag{10.2-45}$$

$$H(s) = \frac{k_n s^{n-1} + k_{n-1}s^{n-2} + \cdots + k_1}{a_m s^m + a_{m-1}s^{m-1} + \cdots + a_0}$$

This, it may be recalled, is the transfer function of the closed-loop system of the configuration of Fig. 2.2-8 for which the transfer function is given in (2.2-37). By looking at the $G(s)$ of (10.2-45) and (10.2-6) with Fig. 10.2-1, we see that $G(s)$ here is just the plant transfer function in series with the amplifier, i.e., it is just the open-loop transfer function for the feedback system of Fig. 10.2-3a. Hence we see by comparing Fig. 2.2-8 with Fig. 10.2-3a, that SVF may be thought of as having added a feedback compensator with transfer function $H(s)$ as given in (10.2-45). We see from (10.2-45) that the compensator $H(s)$ will usually be of much higher order than the usual first-order lead or lag type compensation we have been considering. In general, then, much more should be possible with SVF, as has been the case in the compensation that has been achieved so far in this section. However, to really see the inherent advantages of SVF let us consider the root locus for a SVF system, as in Fig. 10.2-3a.

The transfer function for this system is in (10.2-43), which tells us that the root locus for the system will be given by the root locus of the transfer function

$$G(s)H(s) = K\frac{k_n s^{n-1} + k_{n-1}s^{n-2} + \cdots + k_1}{s^n + b_{n-1}s^{n-1} + \cdots + b_0} \tag{10.2-46}$$

Now looking at the open-loop transfer function in (10.2-46) we can see some general features of the root locus. First, we have $n-1$ open-loop zeros if all the feedback coefficients are nonzero (i.e., if we feed back all the state variables) and we have n open-loop poles. By rule 1 of Section 4.2 we know that the root locus thus has n branches. By rule 2 of the same section these n branches begin on the poles and end on the zeros. We now have $n-1$ of these branches ending on the $n-1$ open-loop zeros which are the roots of the polynomial in the numerator of (10.2-46). We note that the loation of these zeros is at the disposal of the system designer if he can adjust the magnitude of the feedback coefficients. Generally, except for one or two coefficients which may be fixed by steady-state error considerations, this is usually the case. Hence he can exert a considerable influence on the shape of the root locus. Furthermore, since with $n-1$ open-loop zeros and n poles there is an excess of only one pole by rule 4 of Section 4.2 there will be one branch of the root locus going off to the zero at $s = \infty$, and this branch goes off at an angle of $-180°$, i.e., it goes off along the negative real axis. This is exactly where one would want this branch to be, since it means that as the loop gain [K as defined by (4.2-5b)] increases the closed-loop pole on this branch will always be in the LHP, i.e., it will always be a stable pole. Furthermore, if the designer chooses the feedback coefficients properly, all the open-loop zeros [the roots of the numerator of (10.2-46)] will also be in the LHP. Hence for high values of loop gain the system will have all its closed-loop poles in the LHP and hence will certainly be stable. This is decidedly different from most of the cases we have been considering heretofore in that we have usually had an excess of at least three poles, which means (see Fig. 4.2-4) that for high values of loop gain the system is certain to have a RHP closed-loop pole and be unstable. In fact, if the plant has poles (the open-loop poles) all in the LHP, the possibility exists for the designer to locate the zeros so that all branches of the root locus lie completely in the LHP. For this case, the system will be stable for all values of loop gain.

Note that the particularly simple form of the numerator of (10.2-46) is a function of the choice of state variables. For a general choice of state variables it may be expected that the coefficients of the numerator polynomial in (10.2-46) will be linear combinations of the feedback coefficients instead of being identically equal to the feedback coefficients. However, since the closed-loop transfer function is invariant under any choice of state variables, the form of $G(s)H(s)$ given in (10.2-46) will remain the same, as will the conclusions we have drawn from this particular form of open-loop transfer function.

Example 10.2-4

As an example of a root locus for a SVF system let us consider Example 10.2-1 (Fig. 10.2-5). Let us take the values for the feedback coefficients k_2 and k_3 as determined in (10.2-26):

$$k_2 = \tfrac{3}{8}$$
$$k_3 = \tfrac{1}{8} \tag{10.2-47}$$

and let us leave the gain of the amplifier, K, as the free parameter on the root locus. Now the closed-loop transfer function for this system is given in (10.2-23). Dividing the numerator and denominator of this equation by $s^3 + 5s^2 + 4s$, the characteristic polynomial for the open-loop system, we have

$$T(s) = \cfrac{\cfrac{K}{s(s+1)(s+4)}}{1 + K \cfrac{k_3 s^2 + (k_2 + k_3)s + 1}{s(s+1)(s+4)}} \tag{10.2-48}$$

from which we have the root locus for our system given by the root locus of the open-loop transfer function

$$G(s)H(s) = \frac{K}{8} \frac{s^2 + 4s + 8}{s(s+1)(s+4)}$$

$$= \frac{K(s+2-2j)(s+2+2j)}{8s(s+1)(s+4)} \tag{10.2-49}$$

The root locus for $G(s)H(s)$ as given by the $G(s)H(s)$ of (10.2-49), the root locus for the system under consideration, is shown in Fig. 10.2-13. The parameter used on the root locus here is the gain K of the amplifier in series with the plant.

The root locus of Fig. 10.2-13 has a lot to recommend it as the locus for a closed-loop system. We see first that all branches are in the LHP. Thus the closed-loop system is stable for all $K > 0$. Moreover, we see that $\zeta = 0.5$ is approximately the minimum value of damping factor that can be obtained. $K = 24$, which was found to give the desired response in Example 10.2-1, gives this minimum value of damping. We see from the root locus that as K is increased, the damping factor approaches a maximum value of $\zeta = 0.7$, which still gives a very desirable form of step response. Also as K is increased, ω_n (the distance of the closed-loop pole from the origin) increases and the response of the system will be faster. On the other hand, it can also be seen from the root locus that the gain K may be decreased down to $K = 2.5$ and the system will still have a damping factor $\zeta = 0.7$,

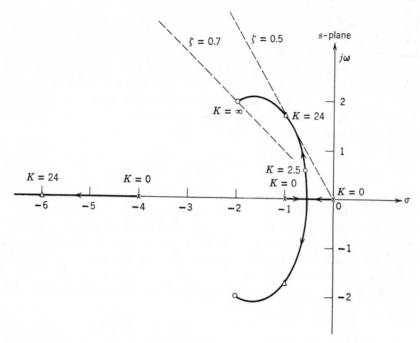

Figure 10.2-13 Root locus for the state variable feedback system of Fig. 10.2-5. $k_1 = 1, k_2 = 3/8, k_3 = 1/8$.

although ω_n will be smaller and hence the system response will be slower. Thus the system may be expected to give a usable response over a remarkable range of the gain K. Over all the values of K considered, it may be noted that the third closed-loop pole remains relatively far into the LHP, so the response will in all cases be dominated by the two complex poles, which is as desired.

Example 10.2-5

One way of understanding the effect of inaccessible state variables is by setting some of the feedback coefficients (the k_i's in the numerator of (10.2-46) to zero. The effect on the root locus then can be seen from the fact that the designer no longer has complete control over the location of the open-loop zeros, hence the branches' terminations are no longer at the disposal of the designer. In fact, lacking feedback from some inaccessible state variables may be expected to lead to RHP open-loop zeros. If this is the case, the system will be unstable for high values of gain in the forward path.

We illustrate this discussion with a simple example. Let us consider the system of Fig. 10.2-5, but now let the state variable x_3 be inaccessible. This means that we are restricted (unless we wish to go to the trouble of reconstructing state variable x_3) to $k_3 = 0$. The root locus for this same system with x_3 assumed accessible is shown in Fig. 10.2-13, where $k_2 = \frac{3}{8}$, and $k_3 = \frac{1}{8}$. Let us use this value of k_2. The transfer function for the system under consideration is given in (10.2-48). Using $k_3 = 0$, $k_2 = \frac{3}{8}$ in this equation, we see that the root locus of the system will now be given by the root locus of

$$G(s)H(s) = \frac{3K}{8} \frac{s + \frac{8}{3}}{s(s+1)(s+4)}$$

The root locus for this open-loop transfer function is shown in Fig. 10.2-14, which is drawn with K as the parameter. We now see that the lone finite zero is at $s = -\frac{8}{3}$. Two open-loop zeros occur at $s = \infty$ and these two we cannot control by adjustment of the remaining feedback coefficient, k_2. The result is that two branches of the root locus go off to infinity at an angle of $\pm 90°$. Now for high values of gain K, the two poles dominating the response of the system will

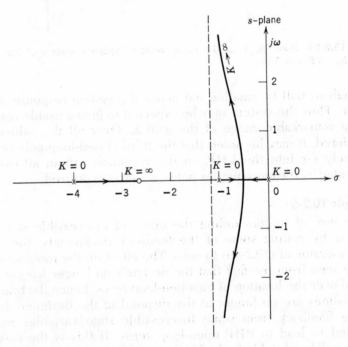

Figure 10.2-14 Root locus for the state variable feedback system of Fig. 10.2-5. $k_1 = 1, k_2 = 3/8, k_3 = 0$.

have a damping factor that approaches zero. A very oscillatory type of response may be expected for high values of K, even though the system will, as in the case of Fig. 10.2-13, be stable for all $K > 0$. The stability properties here may be improved somewhat by changing the location of the zero (it is at $s = -1/k_2$) i.e., changing k_2, but the general form of the root locus will remain the same.

We may carry this analysis one step further by considering the case where both x_2 and x_3 are inaccessible and hence $k_2 = k_3 = 0$. This is the case where only the output is fed back to the input. The root locus for this case was considered in Section 8.3 and is shown in Fig. 8.3-7. Here we see that there are no finite zeros and hence we have two branches of the root locus which go off into the RHP for $K \rightarrow \infty$.

By looking at Figs. 10.2-13, 10.2-14, and 8.3-7, we see that with each loss of feedback of another state variable, more desirable stability properties are lost. The message is clear in that the more state variables that are fed back, the more stable the resulting system. When all the state variables can be fed back and the amount of each is properly adjusted, a truly stable system may be expected.

10.3. STATE VARIABLE FEEDBACK COMPENSATION OF DISCRETE-TIME SYSTEMS

We have discussed the state variable feedback sampled-data system (SVF SDS) briefly in Section 7.3. Let us now consider compensating such a system by adjusting the amount of feedback of each state variable. The general objectives and methods of approach are exactly the same as for the case of continuous-time SVF systems, so these will not be repeated here. In general we shall limit the discussion to showing a particular SVF SDS with emphasis upon those points by which SVF compensation for SDS differs from SVF compensation in the continuous case. We shall consider two examples which illustrate how similar compensation by SVF works in the SDS and continuous-time cases.

The State Variable Feedback Sampled-Data System

The SVF SDS that we shall consider here is the same one that was considered for the continuous case in Section 10.2 (Fig. 10.2-3*a*) except that we assume that a sampler followed by a zero-order hold appears immediately after the input summer (Fig. 10.3-1). This basic system is the same that was considered in Section 7.3 except

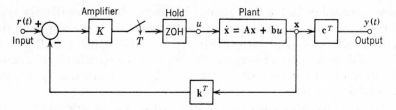

Figure 10.3-1   State variable feedback sampled-data system.

that the amplifier of gain K does not appear in the system (Fig. 7.3-5) considered in that section. This amplifier appears here (Fig. 10.3-1) for the same reasons as it appears in the continuous-time case system of Fig. 10.2-3: basically, it is required to give adequate control over the steady-state characteristics of the system (see Section 10.2). The z-transfer function for the system of Fig. 7.3-5 has been given in (7.3-36) and (7.3-42). By an argument that is exactly analogous to the one used on the system of Fig. 10.2-3, we may show [using (7.3-36) and (7.3-42)] that the z-transfer function for the system of Fig. 10.3-1 is

$$T(z) = \frac{Y(z)}{R(z)} = \frac{K\mathbf{c}^T[z\mathbf{I} - \mathbf{A}_d]^{-1}\mathbf{h}(T)}{1 + K\mathbf{k}^T[z\mathbf{I} - \mathbf{A}_d]^{-1}\mathbf{h}(T)} \qquad (10.3\text{-}1)$$

$$T(z) = \frac{Y(z)}{R(z)} = K\mathbf{c}^T[z\mathbf{I} - \mathbf{A}_d + K\mathbf{h}(T)\mathbf{k}^T]^{-1}\mathbf{h}(T) \qquad (10.3\text{-}2)$$

where now

$$\mathbf{A}_d = \mathbf{e}^{\mathbf{A}T}$$

$$\mathbf{h}(T) = \int_0^T \mathbf{e}^{\mathbf{A}\tau}\, \mathbf{b}\, d\tau \qquad (10.3\text{-}3)$$

$$= \mathbf{A}^{-1}[\mathbf{e}^{\mathbf{A}T} - \mathbf{I}]\mathbf{b}$$

where the second expression for $\mathbf{h}(T)$ applies if $\mathbf{A}$ is nonsingular. Since our compensation method involves expanding $T(z)$ [using (10.3-1) or (10.3-2)] in terms of $K\mathbf{k}$, we can save labor by applying (7.2-106) to (10.3-1) to get

$$T(z) = \frac{Y(z)}{R(z)} = \frac{\det\,[z\mathbf{I} - \mathbf{A}_d + K\mathbf{h}(T)\mathbf{c}^T] - \det\,[z\mathbf{I} - \mathbf{A}_d]}{\det\,[z\mathbf{I} - \mathbf{A}_d + K\mathbf{h}(T)\mathbf{k}^T]} \qquad (10.3\text{-}4)$$

Equation (10.3-4) is usually easier to evaluate since $[z\mathbf{I} - \mathbf{A}_d]$ need not be inverted. Also, to evaluate (10.3-4) we need only evaluate $\det\,[z\mathbf{I} - \mathbf{A}_d + K\mathbf{h}(T)\mathbf{k}^T]$, the denominator. The numerator is then found by letting $\mathbf{k} = \mathbf{c}$ in this evaluation and then disregarding all terms that do not involve K. We note that the open-loop z-transfer function,

(10.3-6), may also be found simply be evaluating (10.3-4) first and then setting $\mathbf{k} = \mathbf{0}$.

We note from (10.3-1) that open-loop zeros are also closed-loop zeros, and that closed-loop poles occur where

$$1 + K\mathbf{k}^T[z\mathbf{I} - \mathbf{A}_d]^{-1}\mathbf{h}(T) = 0 \qquad (10.3\text{-}5)$$

If we consider our SVF SDS as a nonunity feedback system as in Fig. 5.2-4c for which $T(z)$ is given in (5.2-34), we have immediately [comparing (5.2-34) with (10.3-1)] that

$$G(z) = K\mathbf{c}^T[z\mathbf{I} - \mathbf{A}_d]^{-1}\mathbf{h}(T) \qquad (10.3\text{-}6)$$

$$GH(z) = K\mathbf{k}^T[z\mathbf{I} - \mathbf{A}_d]^{-1}\mathbf{h}(T) \qquad (10.3\text{-}7)$$

If we wish a root locus for the SVF SDS of Fig. 10.3-1, then we plot the root locus for $GH(z)$ as given (10.3-7). A frequency response for this system is found by finding $GH(e^{j\omega T})$ for the expression in (10.3-7).

Transient Response Compensation of the State Variable Feedback Sampled Data System

Our basic approach to transient response compensation of SVF SDS is the same as that for continuous-time SVF systems as discussed in Section 10.2.

Step 1. We choose a closed-loop transfer function which gives the desired transient response. This may be done by using the methods of Section 9.4 on z-plane synthesis.

Step 2. We choose K and $\mathbf{k}$ so that the transfer function found in Step 1 is realized. The only restriction on this straightforward procedure comes from the fact that with SVF the closed-loop zeros are the same as open-loop zeros. If the open-loop zeros are inside the unit circle in the z-plane, we may arrange to cancel them with a closed-loop pole at the same point, but we cannot add a closed-loop zero without adding some form of compensator. Thus the transfer function chosen in Step 1 can have zeros only where the open-loop transfer function [as given by (10.3-6)] has its zeros. Also, if the open-loop transfer function has a zero outside the unit circle, the closed-loop transfer function chosen in Step 1 must have a zero at the same point. Since we cannot cancel it, simply accept it. The closed-loop poles chosen in Step 1 must take these zeros (outside the unit circle) into account. Let us consider an example.

Example 10.3-1

Let us consider a fixed plant with transfer function given by

$$G(s) = \frac{1}{s(s+1)} \qquad (10.3\text{-}8)$$

which is to be operated in series with a sampler for which $T = 1$ sec, which is followed by a zero-order hold. We wish to design SVF compensation for this plant so that overshoot to a step input will be less than 20%, the damping factor ζ of the dominant poles will be approximately $\zeta = 0.6$, and the time to the first peak $t_p \leqslant 4$ sec. The system with the chosen state variables is shown in Fig. 10.3-2. For this system

$$sX_1 = X_2$$
$$(s+1)X_2 = U(s)$$
$$Y(s) = X_1$$

which, after applying the inverse $\mathscr{L}$-transform, give

$$\dot{x}_1 = x_2$$
$$\dot{x}_2 = -x_2 + u(nT), \qquad nT \leqslant t < (n+1)T \qquad (10.3\text{-}9)$$
$$y = x_1$$

Thus

$$\mathbf{A} = \begin{bmatrix} 0 & 1 \\ 0 & -1 \end{bmatrix}, \qquad \mathbf{b} = \begin{bmatrix} 0 \\ 1 \end{bmatrix}, \qquad \mathbf{c} = \begin{bmatrix} 1 \\ 0 \end{bmatrix} \qquad (10.3\text{-}10)$$

Using (7.3-54),

$$e^{\mathbf{A}t} = \mathscr{L}^{-1}[s\mathbf{I} - \mathbf{A}]^{-1} = \begin{bmatrix} 1 & (1-e^{-t}) \\ 0 & e^{-t} \end{bmatrix} \qquad (10.3\text{-}11)$$

Using (10.3-3) with $T = 1$ and the other values from (10.3-10), we have

$$\mathbf{A}_d = e^{\mathbf{A}T} = \begin{bmatrix} 1 & 0.632 \\ 0 & 0.368 \end{bmatrix}$$

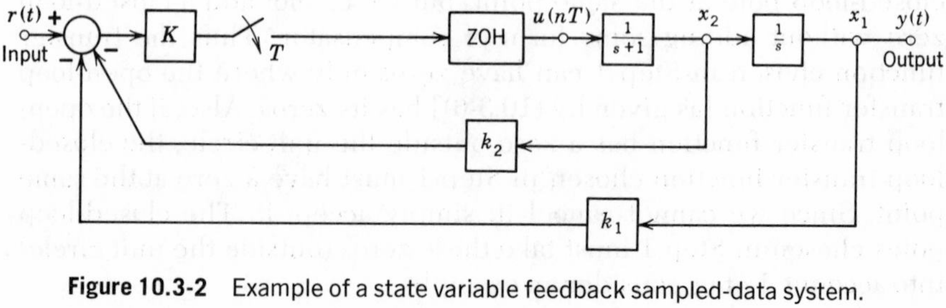

Figure 10.3-2   Example of a state variable feedback sampled-data system.

$$h(T) = \int_0^T e^{A\tau}b\, d\tau = \begin{bmatrix} 0.368 \\ 0.632 \end{bmatrix}. \qquad \text{for } T = 1 \text{ sec} \qquad (10.3\text{-}12)$$

$$c = \begin{bmatrix} 1 \\ 0 \end{bmatrix}$$

We may use the quantities of (10.3-12) in (10.3-4) to get

$$T(z) = \frac{\det\begin{bmatrix} (z-1+0.368K) & -0.632 \\ 0.632K & (z-0.368) \end{bmatrix} - \det\begin{bmatrix} (z-1) & -0.632 \\ 0 & (z-0.368) \end{bmatrix}}{\det\begin{bmatrix} (z-1+0.368Kk_1) & (-0.632+0.368Kk_2) \\ 0.632Kk_1 & (z-0.368+0.632Kk_2) \end{bmatrix}}$$

$$= \frac{0.368K(z+0.72)}{z^2+(-1.368+0.368Kk_1+0.632Kk_2)z+(0.368+0.265Kk_1-0.632Kk_2)}$$

$$(10.3\text{-}13)$$

We note that the numerator in (10.3-13) is just the denominator with $k = c$, i.e., with $k_1 = 1$, $k_2 = 0$ and with terms not involving K discarded. We may find the open-loop z-transfer function from (10.3-13) by letting $k = 0$ to get

$$G(z) = \frac{0.368K(z+0.72)}{z^2-1.368z+0.368} \qquad (10.3\text{-}14)$$

Let us now design SVF compensation for this system to realize the transient response specifications given.

Step 1. We see from (10.3-13) and (10.3-14) that there is a closed-loop and hence an open-loop zero at $z = -0.72$. In choosing a desired closed-loop transfer function for this system, we can have either no zero (cancel out the zero with a pole) or a zero at $z = -0.72$. If we keep the zero at $z = -0.72$, it may be found by the z-plane synthesis methods (discussed in Section 9.4) that

$$T(z) = \frac{0.438(z+0.72)}{z^2-0.456z+0.208}$$

$$= \frac{0.438(z+0.72)}{(z-0.228-j0.395)(z-0.228+j0.395)} \qquad (10.3\text{-}15)$$

has an overshoot of approximately 10% with time to the first peak $t_p \cong 3$ sec and damping factor $\zeta \cong 0.6$, which meet the design specifications. We also note that $T(1) = 1$ for the $T(z)$ of (10.3-15), so steady-state error to a step input will be zero. Thus we may choose $T(z)$ of (10.3-15) as the desired closed-loop transfer function for our system, which completes Step 1.

Step 2. We must now choose K and $k^T = [k_1 \, k_2]$ so that the closed-loop transfer function of our SVF system (of Fig. 10.3-2), which is given in (10.3-13), will be the same as the desired closed-loop transfer function as given in (10.3-15). Comparing (10.3-13) and (10.3-15), we must have

$$0.368\,K = 0.438$$
$$-1.368 + 0.368\,Kk_1 + 0.632\,Kk_2 = -0.456$$
$$0.368 + 0.265\,Kk_1 - 0.632\,Kk_2 = 0.208$$

Solving these simultaneous equations gives

$$K = 1.19$$
$$k_1 = 1.0 \hspace{2cm} (10.3\text{-}16)$$
$$k_2 = 0.631$$

K and k as in (10.3-16) will give the closed-loop transfer function of (10.3-15). The corresponding step response is shown in Fig. 10.3-3, where it is labeled the SVF compensated response (the solid curve). We see $t_p \cong 3$ sec and an overshoot of 10% since the closed-loop transfer function was chosen to give these transient response specifications.

We may check the steady-state response by considering the velocity

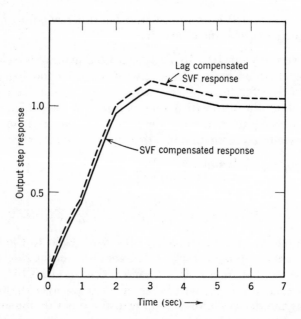

Figure 10.3-3   State variable feedback step responses for Example 10.3-1.

error constant K_v for the system. From (6.3-20) we have

$$\frac{1}{K_v T} = -\sum_{i=1}^{m} \frac{1}{1-z_i} + \sum_{i=1}^{n} \frac{1}{1-p_j} \qquad (10.3\text{-}17)$$

where z_i, $i = 1, 2, \ldots m$ and p_i, $i = 1, 2, \ldots, n$ are the closed-loop zeros and poles, respectively, and T is the sampling period. For the system here, with compensation as given by (10.3-16), the closed-loop zero and poles are given in (10.3-15). We have then, using (10.3-17), (recall that $T = 1$)

$$\frac{1}{K_v} = \frac{1}{1+0.72} - \frac{1}{1+0.228+j0.395} - \frac{1}{1+0.228-j0.395}$$

$$= \frac{1}{0.675}$$

and therefore

$$K_v = 0.675 \qquad (10.3\text{-}18)$$

If this is a sufficiently high K_v for the purposes for which the system is intended, the compensation as given in (10.3-16) may be used. If this is too low a value but fairly close to what is desired, some relocation of the closed-loop poles can be expected to increase it while still satisfying the transient response specifications. However, if a significant increase in K_v is required, lag compensation of the system is called for. This may also be done in a manner very similar to the lag compensation of a continuous-time SVF system discussed in Section 10.2.

Lag Compensation of State Variable Feedback Sampled-Data Systems

To increase the steady-state error constant, we add a compensator to the system to give us an additional closed-loop zero and pole which we may adjust to give the desired error constant. The additional zero and pole are placed near the $1 + j0$ point in the z-plane so that their effect upon the transient response will be small. This is exactly analogous to adding a closed-loop pole and zero near the origin in the case of continuous-time SVF systems when we wish to increase the error constant.

Consider then a system for which we have a satisfactory transient response but with a velocity error constant K_v which is too small and for which a closed-loop pole p_j and zero z_j are to be added so as to realize a desired velocity error constant K_v^d. Using (10.3-17), we have

$$\frac{1}{K_v^d T} = \frac{1}{K_v T} - \frac{1}{1 - z_j} + \frac{1}{1 - p_j} \tag{10.3-19}$$

From (10.3-19) it can be seen that if we wish $K_v^d > K_v$ then $|1 - z_j| < |1 - p_j|$ must apply; i.e., the zero at z_j must be nearer the $z = 1 + j0$ point than the pole. Of course, a stable closed-loop system is desired, so p_j must be chosen inside the unit circle. Equation (10.3-19) gives us one condition on p_j and z_j; we require another to determine the two free parameters z_j and p_j.

We may, as in the case of continuous-time SVF systems, consider the effect of the additional pole and zero on the step-response overshoot. It may be shown by a procedure exactly analogous to the one by which (8.6-24) was obtained and also discussed in Section 9.4 that for an additional pole-zero pair near the $1 + j0$ point

$$\frac{1 - p_j}{1 - z_j} = 1 + \frac{X}{100} \tag{10.3-20}$$

where X is the percent increase in step-response overshoot. Now, given the desired velocity error constant K_v^d and the increase in step-response overshoot that can be tolerated, X, (10.3-19) and (10.3-20) may be used to determine p_j and z_j.

Once we have determined the location of the additional closed-loop pole and zero, the next question is where to put the compensator to get the additional closed-loop pole and zero. We must also choose the free parameters available to us so that the desired transfer function with the additional closed-loop pole and zero at p_j and z_j is realized. There are several possibilities; however, a straightforward means is to add a discrete compensator in series with the plant (Fig. 10.3-4). This is analogous to adding the lag compensator to the SVF continuous-time system in Fig. 10.2-6, only a feedback path has not been added around the compensator. This is done for the sake of convenience and is no limitation (see Problem 10.12). A feedback path around the compensator may be added if so desired.

Assuming a system of the general form of the one in Fig. 10.3-4, the

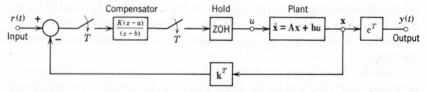

Figure 10.3-4 Lag compensated state variable feedback system.

compensation problem reduces itself to determining K, a, b, and k so that the desired closed-loop transfer function [which we now assume has the additional pole and zero, p_j and z_j, as determined by (10.3-19) and (10.3-20)] is realized. To do this we require the closed-loop transfer function for the system of Fig. 10.3-4. This can be found directly by comparing Fig 10.3-1 with Fig. 10.3-4. We see that they are the same except that the amplifier of gain K is replaced by the compensator with transfer function $K(z-a)/(z-b)$. Thus the transfer function for the system of Fig. 10.3-4 can be found by replacing K by $K(z-a)/(z-b)$ in $T(z)$ of (10.3-1) or (10.3-2), which is the closed-loop transfer function for the system of Fig. 10.3-1. The result is

$$T(z) = \frac{K[(z-a)/(z-b)]\mathbf{c}^T[z\mathbf{I}-\mathbf{A}_d]^{-1}\mathbf{h}(T)}{1+K[(z-a)/(z-b)]\mathbf{k}^T[z\mathbf{I}-\mathbf{A}_d]^{-1}\mathbf{h}(T)} \quad (10.3\text{-}21)$$

or

$$T(z) = K\frac{(z-a)}{(z-b)}\mathbf{c}^T\left[z\mathbf{I}-\mathbf{A}_d+K\frac{(z-a)}{(z-b)}\mathbf{h}(T)\mathbf{k}^T\right]^{-1}\mathbf{h}(T) \quad (10.3\text{-}22)$$

The basic method of compensation then is to determine the closed-loop transfer function that is wanted. For the given system, (10.3-21) or (10.3-22) is expanded and K, a, b, and k are determined to give the desired transfer function. The steps in the lag compensation of a SVF SDS are as follows.

Step 1. A closed-loop transfer function is determined which meets the transient response specifications for the system. This may be done as described earlier. By using (10.3-17), K_v may be determined for this choice of closed-loop transfer function. [This assumes $K_p = \infty$, i.e., $e_{ss}(\text{step}) = 0$, and $K_a = 0$ — if this is not the case, work with either K_p or K_a, whichever is appropriate.] If K_v is satisfactory, then compensation may be completed as discussed earlier. If K_v is not quite satisfactory, a new choice of closed-loop poles may be attempted [guided by (10.3-17)]. If a satisfactory K_v and transient response can be obtained by this means, compensation should be completed, again as shown earlier. If a satisfactory K_v cannot be obtained by this means, lag compensation is in order. Thus a transfer function which gives a satisfactory transient response should be found before proceeding to Step 2.

Step 2. An additional closed-loop pole-zero pair is found for the transfer function found in Step 1 by using (10.3-19) and (10.3-20). The desired transfer function has poles and zeros as determined in Step 1 with an additional pole p_j and zero z_j as determined by (10.3-19) and (10.3-20). The multiplying constant should be adjusted so that e_{ss} (step) $= 0$, i.e., so that $T(1) = 1$.

Example 10.3-2

Let us consider the system and the same transient specifications as given in Example 10.3-1, but let K_v be 10 as the least acceptable value for the velocity error constant, i.e., we have now $K_v{}^d \geqq 10$. $G(s)$ is given in (10.3-8). A satisfactory transient response was realized using the SVF compensation of Fig. 10.3-2 with K and k as given by (10.3-16). The transfer function obtained is given in (10.3-15) with the corresponding K_v given by (10.3-17), $K_v = 0.675$. This is considerably below the $K_v{}^d = 10$, so lag compensation is required. Let us design such a compensation.

Step 1. This step has been completed earlier. The closed-loop transfer function that gives a satisfactory transient response is given (10.3.15). The corresponding time response is shown as the SVF compensated response in Fig. 10.3-3. The response has 10% step overshoot with the time to first peak $t_p \cong 3$ sec.

Step 2. We require a lag compensator to increase K_v to obtain $K_v{}^d \geqq 10$. Adding a lag compensator to the SVF compensated system of Fig. 10.3-2 gives the system with state variables assigned as in Fig. 10.3-5. We wish to add a pole and a zero to the closed-loop transfer function of (10.3-15) so that $K_v{}^d$ will be obtained and so that the transient response will remain essentially unchanged. To do this, first note that for $T(z)$ of (10.3-15) $K_v = 0.675$ from (10.3-18). If p_j and z_j are the locations of the additional pole and zero, then (recalling that $T = 1$) by (10.3-19)

$$\frac{1}{K_v{}^d} = \frac{1}{10} = \frac{1}{0.675} - \frac{1}{1 - z_j} + \frac{1}{1 - p_j} \tag{10.3-23}$$

$T(z)$ of (10.3-15) gives 10% step-response overshoot. The original specification in Example 10.3-1 was overshoot $\leqq 20\%$. Additional

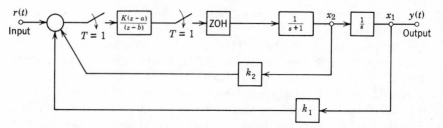

Figure 10.3-5   Example of a lag compensated state variable feedback system.

overshoot of 5% can thus be tolerated; this allows an additional 5% margin for error. Thus we choose $X = 5$ in (10.3-20) and obtain

$$\frac{1-p_j}{1-z_j} = 1.05 \qquad (10.3\text{-}24)$$

Solving (10.3-23) and (10.3-24) simultaneously gives $p_j = 0.9638$ and $z_j = 0.9655$. Hence the desired closed-loop transfer function is

$$T(z) = \frac{0.438(z+0.72)}{(z^2-0.456z+0.208)} \frac{(1-p_j)}{(1-z_j)} \frac{(z-z_j)}{(z-p_j)}$$

$$= \frac{0.46(z+0.72)(z-0.966)}{z^3-1.42z^2+0.648z-0.2} \qquad (10.3\text{-}25)$$

where the constant multiplier has been adjusted so that $T(1) = 1$, i.e., so that $e_{ss}(\text{step}) = 0$. This completes Step 2.

Step 3. The closed-loop transfer function for our lag compensated SVF SDS as shown in Fig. 10.3-5 may be found by comparing this system with the one of Fig. 10.3-2. The difference in the two systems is solely that the compensator $K(z-a)/(z-b)$ has replaced the amplifier of gain K. Thus the transfer function for the system of Fig. 10.3-5 may be obtained by replacing K by $K(z-a)/(z-b)$ in the transfer function of the system in Fig. 10.3-2, which is given in (10.3-13). This gives

$$T(z) = \frac{0.368K(z+0.72)(z-a)}{\left\{ \begin{array}{l} z^3 + (-1.368 - b + 0.368Kk_1 + 0.632Kk_2)z^2 \\ + [0.368 + 1.368b - a(0.368Kk_1 + 0.632Kk_2) + 0.265Kk_1 \\ -0.632\,Kk_2]z - 0.368b - a(0.265Kk_1 - 0.632Kk_2) \end{array} \right\}}$$

$$(10.3\text{-}26)$$

Now we wish to determine K, a, b, and k so that the transfer function in (10.3-25) will be realized as the transfer function for the system of Fig. 10.3-5; this means that (10.3-25) and (10.3-26) are the same transfer function. We may note that open-loop zeros are the same as closed-loop zeros; so we must have $a = z_j = 0.966$; also, directly from the numerators of (10.3-25) and (10.3-26), $0.368K = 0.46$ so $K = 1.25$. Comparing like coefficients in the denominators of (10.3-25) and (10.3-26), after solving the resulting set of simultaneous linear equations we find

$$\begin{array}{l} K = 1.25 \\ a = 0.966 \\ b = 0.998 \\ k_1 = 1.0 \\ k_2 = 0.623 \end{array} \qquad (10.3\text{-}27)$$

which completes the compensation. The step response for the closed-loop transfer function in (10.3-25) with compensation as in (10.3-27) is shown as the lag compensated SVF response in Fig. 10.3-3. The 15% overshoot that was predicted in Step 2 has been realized. Time to first peak is still 3 sec.

The effect of the added closed-loop pole at $p_j = 0.9638$ can be seen clearly in Fig. 10.3-3. The response has been lifted by approximately 5% over the whole range of time shown. This additional 5% will decay out eventually, in fact, as $(p_j)^n$. Since p_j is nearly 1.0, this will be very slow.

We know from the choice of the additional pole and zero in the closed-loop transfer function in Step 2 that $K_v = 10$ for the lag compensated system. Thus $K_v{}^d \geq 10$ has been met.

By comparing the compensation obtained in Example 10.3-1, (10.3-16), with the compensation obtained in example 10.3-2, (10.3-27), we see that K, k_1, and k_2 have not changed appreciably. This suggests that as a practical approach to designing lag compensation for these SVF SDS we may choose K and k to meet the transient response requirements as discussed earlier in this section. The lag-compensator parameters a and b (see Fig. 10.3-4) can then be found by letting $a = z_j$ as determined by (10.3-19) and (10.3-20). If the transient response compensated error constant is K_v and the desired error constant is $K_v{}^d$, then b may be determined from (6.3-12):

$$K_v{}^d = \lim_{z \to 1} K \frac{(z-a)}{(z-b)} \frac{(z-1)}{T} G(z) = \frac{(1-a)}{(1-b)} K_v$$

With a and b so determined and with K and k as determined for the transient response compensation, the resulting response should be very nearly as desired. It should be at least close enough so that the desired results can be achieved by simple system tuning, which usually has to be done anyway.

The Root Locus of a State Variable Feedback Sampled-Data System

The root locus for the SVF SDS shown in Fig. 10.3-1 may be found by plotting the root locus for the $GH(z)$ in (10.3-7). This was discussed in the text following (10.3-7). Let us consider plotting the root locus for the SVF SDS considered in Example 10.3-1 (Fig. 10.3-2).

$GH(z)$ for this system may be evaluated by evaluating (10.3-7) with $\mathbf{A}_d$, $\mathbf{h}(T)$, and $\mathbf{c}$ as given (10.3-12). An alternate method is to consider the closed-loop transfer function for this system as given in (10.3-13). If the numerator and denominator of this transfer function are divided by the terms in the denominator not involving K, we have

$$T(z) = \frac{K\dfrac{0.368(z+0.72)}{z^2-1.368z+0.368}}{1+\dfrac{K[(0.368k_1+0.632k_2)z+0.265k_1-0.632k_2]}{z^2-1.368z+0.368}} \qquad (10.3\text{-}28)$$

Comparing (10.3-28) with (10.3-1), we have

$$G(z) = K\frac{0.368(z+0.72)}{z^2-1.368z+0.368}$$

$$(10.3\text{-}29)$$

$$GH(z) = \frac{K[(0.368k_1+0.632k_2)z+0.265k_1-0.632k_2]}{z^2-1.368z+0.368}$$

$GH(z)$ as given in (10.3-29) may now be used to plot the root locus for the SVF SDS of Fig. 10.3-2. If we use the values for k_1 and k_2 found in the transient response compensation in Example 10.3-1, $k_1 = 1.0$,

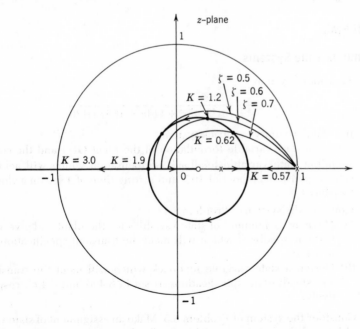

Figure 10.3-6 Root locus for the state variable feedback system of Fig. 10.3-2. $k_1 = 1, k_2 = 0.631$.

$k_2 = 0.631$ from (10.3-16), then (10.3-29) gives

$$GH(z) = K\frac{0.767z - 0.134}{z^2 - 1.368z + 0.368}$$

$$= \frac{0.767K(z - 0.175)}{(z - 1)(z - 0.368)} \tag{10.3-30}$$

The root locus plot for the $GH(z)$ of (10.3-30) is shown in Fig. 10.3-6. The root locus circle is seen to be well within the unit circle. The locus, however, has a branch that goes off to minus infinity along the negative real axis. So the system will be unstable for $K \geq 3.0$. This value of K is considerably higher than the value [$K = 1.19$ from (10.3-16)] for which the desired transient response was achieved in Example 10.3-1. Thus the system as compensated in Example 10.3-1 has an adequate stability margin. Also we see from Fig. 10.3-6 that the root locus is in the range of desirable damping factors (ζ) over a large portion of its length inside the unit circle. In fact, any value of K in the range $1.9 \leq K \leq 0.62$ will give a reasonable damping factor. The transient response hence should not be sensitive to small changes in the gain K. This is a desirable characteristic for a control system.

PROBLEMS

Continuous-Time Systems

10.1. Consider a system

$$G(s) = \frac{1}{s^5 + 10s^4 + 45s^3 + 114s^2 + 166s + 136}$$

in a standard unity-feedback system.

Give a state variable formulation of the plant $G(s)$ and the corresponding state variable feedback compensation which will achieve step-response overshoot of 16% and setting time of 4 sec on a closed-loop basis.

10.2. Consider the system of Problem 8.4.
   (a) Make an assignment of state variables for the plant and give state variable feedback which will meet the transient specifications as in 8.4(a).
   (b) Design a state variable feedback which will meet the transient and steady-state specifications as set in 8.4(a) and 8.4(c), respectively.

10.3. Consider the system of Problem 8.5. Make an assignment of state variables and design state variable feedback compensation to meet the transient and steady-state specifications as in 8.5(c).

10.4. For a system whose open loop plant transfer function is

$$G(s) = \frac{10(s+5)(s+13)}{s(s+4)(s+9)(s^2 - 22s + 122)}$$

design a state variable (after making proper choice) feedback compensation scheme such that overshoot is 10%, $t_p = 0.7$ sec, and $K_v \geqslant$ 15.

10.5. Consider the antenna positioning system of Problem 8.8 shown in Fig. P8-8.

(a) Assuming that antenna shaft position θ_a antenna shaft angular velocity θ_α (from a tachometer), d-c generator voltage e_g, and generator field voltage e_f are available for feedback, design a state variable feedback which will result in a system response which will meet the response specifications of 8.8(d).

(b) Design a state variable feedback compensation which will meet the response specifications of (a), but will in addition have $K_v \geqslant$ 5.

(c) Determine and plot the step response for the system with compensation as determined in (a). What is K_v for this case?

(d) Determine and plot the step response for the system with compensation as determined in (b).

10.6. Consider the radar antenna system of Problem 8.10. The state variable feedback is shown in dotted lines in Fig. P8.10.

(a) Make a suitable state variable diagram.

(b) Find k_1 and k_2 [note that now $G_c(s) = 0$] such that the specifications given in Problem 8.10 are met.

Discrete-Time Systems

10.7. An open-loop sampled-data system is shown in Fig. P10.7. Find the forward gain K as well as state variable feedback gains k_1 and k_2 such that the closed-loop system has $\zeta = 0.7$, $t_p \leqslant 3$ sec, and overshoot $\leqslant 18\%$. What is the value of position error constant?

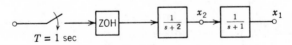

T = 1 sec

Figure P10.7

10.8. A sampled-data system is shown in Fig. P10.8.

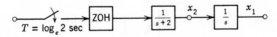

$T = \log_e 2$ sec

Figure P10.8

(a) Find forward gain K as well as state-variable feedback gains k_1 and k_2 such that the closed loop system has $\zeta = 0.65$, $t_p \leq 3.5$ sec, and overshoot $\leq 20\%$.

(b) Using k_1, k_2 as obtained in (a) and varying K, draw the root locus of the system and discuss the various limitations of the system.

10.9. It is desired to increase tenfold K_v of system of Problem 10.8 by a digital lag compensator, as shown in Fig. P10.9, keeping everything else approximately the same. Find K, k_1, k_2, k_3, a, and b.

10.10. It is now desired to increase tenfold K_v of the system of Problem 10.8 by a continuous type lag compensator, as shown in Fig. P10.10, keeping everything approximately the same. Find K, k_1, k_2, k_3, α, and τ.

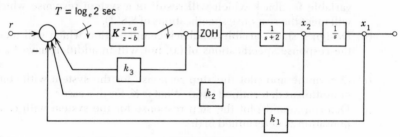

Figure P10.9

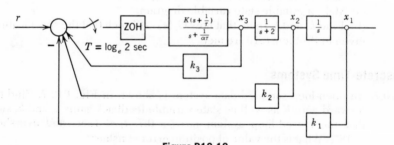

Figure P10.10

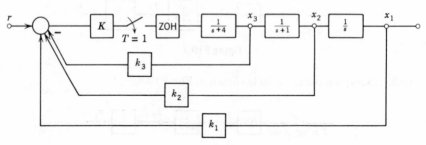

Figure P10.11

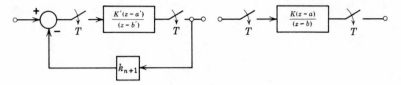

Figure P10.12

10.11. For the system shown in Fig. P10.11 find K, k_1, k_2, k_3 such that the dominant poles of the closed-loop system assure us of $\zeta = 0.707$, $t_p \leq$ 2.5 sec, and overshoot of $< 15\%$. Under these conditions what maximum K_v can you assure? Using k_1, k_2, and k_3 obtained above, draw the root locus of the system and determine whether you can improve on design of the system.

10.12. Find K', a', and b' as functions of k_{n+1}, K, a, and b so that the two compensators of Fig. P10.12 are equivalent.

11

INTEGRAL SQUARE ERROR COMPENSATION

11.1 INTRODUCTION

In this chapter the problem of compensation is viewed from a different point of view. We consider the actual output of the given system, $c(t)$, and hypothesize that we wish this output to be a certain desired function of time, $i(t)$. Ideally we require that

$$c(t) = i(t), \qquad \forall t \tag{11.1-1}$$

After the requisite mathematical tools have been evolved, our efforts will be directed toward the design of networks (filters) that help match the output to a preset ideal, subject to some predetermined criteria. This notion seems quite different from the criterion for design of filters previously discussed, but the two approaches indeed lead to the design of stable systems. As an example, consider the transient response of an uncompensated system. Ideally we want the output to be a step function of time. What we can actually expect to get is not a step function but something like the functions that take a finite time to rise up to the desired value and may overshoot and then oscillate about the desired final value. Thus we know that we will in most cases have to live with something less than the ideal.

Since the output generally will not track the ideal output function $i(t)$, we define the error, $e(t)$, as

$$e(t) = i(t) - c(t) \tag{11.1-2}$$

We now may look upon the problem of system compensation as that of keeping the error tolerably small over all time t. This is more or less what we have been doing all along by properly choosing the damping factor ζ and undamped natural frequency ω_n of the dominant poles or by arranging to get a high enough value of steady-state error constant or by other means.

We shall take a more recent approach to compensation by defining a measure (or criterion) on the error and then seeking the compensation which minimizes this measure (or criterion). The measure we shall

524

use is the integral square error (ISE), defined by

$$\text{ISE} = \int_{-\infty}^{\infty} e^2(t)\,dt = \int_{-\infty}^{\infty} [i(t) - c(t)]^2 dt \qquad (11.1\text{-}3)$$

We choose this particular measure because we can find a linear, time-invariant compensator which produces the minimum value for this particular measure, which we cannot do for practically any other measure. Looking at (11.1-3), we see that ideally ISE = 0 and any departure of the output from the ideal will produce ISE > 0. In general, the larger the divergence from the ideal, the larger will be the corresponding ISE. Conversely, if ISE is kept small for a given system, it may be expected that the actual output will approximate satisfactorily the ideal output. A compensator that produces minimum ISE can be expected to be a good one in this sense. Thus ISE is a very good measure for the general compensation problem, as is borne out in practice. Our discussion dwells on deterministic (known) signal inputs. The criteria developed here have been successfully extended to stochastic situations.

The ISE criterion can be effectively utilized for the design of compensators for both continuous and discrete-time systems. The basic models that we consider are shown in Fig. 11.1-1 for continuous- and discrete-time systems. We emphasize that the techniques developed are applicable to the design of compensators for configurations other than those in Fig. 11.1-1 both for continuous- and discrete-time systems, but we will restrict ourselves here to the models shown. The problem then is to design $G_c(s)$ or $D(z)$ such that ISE is minimized.

The compensator can also be designed (while minimizing ISE) when we have some constraint on another variable in the system, such as saturation. The effects of external disturbances can also be minimized while designing a compensator. However, with every constraint we put, the difficulty of designing the compensator analytically increases. The problem of design of the systems discussed here

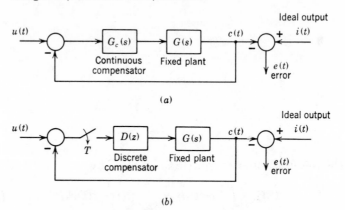

Figure 11.1-1 (*a*) Continuous-time compensated control system. (*b*) Discrete-time compensated control system.

as well as other systems can be approached in two ways:

1. The form of the compensator can be given. The design problem then is to choose the parameters of the compensator in such a manner that ISE with or without constraints is minimized. The solution to this problem is at least conceptually straightforward. We first obtain the integral-square error with or without some constraint in terms of these parameters. The error is then minimized with respect to the parameters. The parameters then give the required compensator.

2. We find the form of the compensator subject to the condition that integral-square error is minimized with or without some constraint.

We see from the preceding discussion that it is very difficult to consider all types of configurations. This is primarily due to the large number of possible combinations. It should be noted further that there may be elements in the feedback path of the given system. Before we proceed with the compensator design, we develop the basic mathematical tools used for analytical representation of ISE.

11.2   BASIC TOOLS

Translation Functions of Continuous-Time Functions and Their Use

The basic tool in ISE analysis is the translation function. Given time functions $x(t)$ and $y(t)$ defined for $-\infty < t < \infty$, the translation function of $x(t)$ with $y(t)$ $[I_{xy}(\tau)]$ is defined by

$$I_{xy}(\tau) \triangleq \int_{-\infty}^{\infty} x(t)\, y(t+\tau)\, dt \qquad (11.2\text{-}1)$$

It can be seen that the translation function for two functions can be defined only for those cases where the integral in (11.2-1) exists. In the special case where

$$x(t) = y(t)$$

the resulting translation function $I_{xx}(\tau)$ is termed an *autotranslation function.*

USEFUL PROPERTIES OF TRANSLATION FUNCTIONS

1. By a simple change of variable of integration it may be shown that for $I_{xy}(\tau)$ as defined in (11.2-1)

$$I_{xy}(-\tau) = I_{yx}(\tau) \tag{11.2-2}$$

Correspondingly, for autotranslation functions

$$I_{xx}(\tau) = I_{xx}(-\tau) \tag{11.2-3}$$

i.e., autotranslation functions are even functions.

2. Consider

$$\int_{-\infty}^{\infty} \left(\frac{x(t)}{\sqrt{I_{xx}(0)}} \pm \frac{y(t+\tau)}{\sqrt{I_{yy}(0)}} \right)^2 dt \geqslant 0 \tag{11.2-4}$$

where

$$I_{xx}(0) = \int_{-\infty}^{\infty} x^2(t)\, dt \tag{11.2-5}$$

and similarly for $I_{yy}(0)$. Expanding the argument in (11.2-4) and recognizing that

$$I_{yy}(0) = \int_{-\infty}^{\infty} y^2(t+\tau)\, dt \tag{11.2-6}$$

we have

$$1 \pm \frac{I_{xy}(\tau)}{\sqrt{I_{xx}(0)}\,\sqrt{I_{yy}(0)}} \geqslant 0 \tag{11.2-7}$$

From (11.2-7) we have immediately that

$$|I_{xy}(\tau)| \leqslant \sqrt{I_{xx}(0)}\,\sqrt{I_{yy}(0)} \tag{11.2-8}$$

and for the case of $x = y$

$$|I_{xx}(\tau)| \leqslant I_{xx}(0) \tag{11.2-9}$$

From (11.2-8) and (11.2-9) we may conclude that if functions $x(t)$ and $y(t)$ are square integrable over the whole real axis, then the corresponding translation and autotranslation functions exist.

3. By definition, we note

$$I_{xx}(0) \stackrel{\Delta}{=} \int_{-\infty}^{\infty} x^2(t)\, dt \tag{11.2-10}$$

In words, the integral square value of a function is the value of its autotranslation function at the origin.

THE $\mathcal{L}_B$-TRANSFORM OF TRANSLATION FUNCTIONS

We have

$$I_{xy}(\tau) = \int_{-\infty}^{\infty} x(t)y(t+\tau)\,dt \tag{11.2-11}$$

Taking the Laplace transform of both sides (note that we have the bilateral Laplace ($\mathcal{L}_B$)-transform here because the variable τ runs from $-\infty$ to ∞), we have

$$I_{XY}(s) = \mathcal{L}_B\{I_{xy}(\tau)\} = \int_{-\infty}^{\infty}\left\{\int_{-\infty}^{\infty} x(t)y(t+\tau)\,dt\right\} e^{-s\tau}\,d\tau \tag{11.2-12}$$

$$= \int_{-\infty}^{\infty} x(t)e^{st}\,dt \int_{-\infty}^{\infty} y(t+\tau)e^{-s(t+\tau)}\,d\tau$$

Now

$$\int_{-\infty}^{\infty} y(t+\tau)e^{-s(t+\tau)}\,d\tau = \mathcal{L}_B\{y(\tau)\} = Y(s) \tag{11.2-13}$$

Also

$$\int_{-\infty}^{\infty} x(t)e^{st}\,dt = \mathcal{L}_B\{x(t)\}\big|_{s\to -s} = X(-s) \tag{11.2-14}$$

Therefore

$$I_{XY}(s) = X(-s)Y(s) \tag{11.2-15}$$

We will assume that (11.2-15) is valid here.

It may be seen that only a very small class of functions has corresponding well-defined translation functions. Excluded from this class are some very important functions in linear control system analysis such as the step, the ramp, and the sine functions. The situation is complicated by the fact that these functions all have well defined $\mathcal{L}_B$-transforms. Hence in these cases the straightforward application of (11.2-15) leads quickly to nonsense if one does not take care to note that the *functions involved must be square integrable on the infinite interval*.

To extend slightly the domain of definition of the translation function, instead of considering the function $f(t)$, consider weighting it with an exponential, i.e., consider $e^{-\epsilon|t|}f(t)$ where ϵ is a small, non-zero, positive number. With this weighting, it is obvious that for the step, the ramp, and the sine functions the corresponding weighted autotranslation function will exist. Hence the corresponding translation functions will also exist. The effect of the weighting function is to shift those poles and zeros corresponding to negative time to the right by ϵ and those poles and zeros corresponding to positive time to the left by ϵ. The functions that do not have corresponding auto-translation functions such as the step, it may be noted, all have poles on the $j\omega$-axis. The weighting function simply moves these poles off the $j\omega$-axis and into the RHP or LHP depending on whether the poles correspond to negative time or positive time, respectively.

In the sequel, (11.2-15) will be used as given. When the result is an $I_{XY}(s)$ which has a pole or poles on the $j\omega$-axis, it will be understood that it will be shifted by ϵ into the LHP if it is a positive time pole and into the RHP it if is a negative time pole; i.e., the weighting factor $e^{-\epsilon|t|}$ will be applied as required. The final results obtained are then considered in the limit as $\epsilon \rightarrow 0$.

APPLICATION TO A LINEAR SYSTEM

Consider the system of Fig. 11.2-1, a simple, single-input, single-output, time-invariant linear system with transfer function $W(s)$. Given $I_{UU}(s)$ (the $\mathscr{L}_B$-transform of the input autotranslation function) and $W(s)$, what is $I_{CC}(s)$?

If $U(s)$ is the $\mathscr{L}_B$-transform of the input, then by (11.2-15),

$$I_{UU}(s) = U(-s)U(s) = I_{UU}(-s) \qquad (11.2\text{-}16)$$

and

$$I_{CC}(s) = C(-s)\,C(s) \qquad (11.2\text{-}17)$$

However,

$$C(s) = U(s)\,W(s) \qquad (11.2\text{-}18)$$

Therefore, using (11.2-16),

$$I_{CC}(s) = U(-s)\,W(-s)\,U(s)\,W(s)$$
$$= W(-s)\,I_{UU}(s)\,W(s) \qquad (11.2\text{-}19)$$

Similarly it may be shown that

$$I_{UC}(s) = W(s)I_{UU}(s) \qquad (11.2\text{-}20)$$

$$I_{CU}(s) = W(-s)I_{UU}(s) = I_{UC}(-s) \qquad (11.2\text{-}21)$$

Thus it may be seen how the translation function of the output of a system may be determined knowing the autotranslation function of the input and the system transfer function.

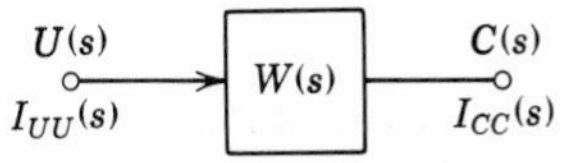

Figure 11.2-1   Linear time-invariant system.

EVALUATION OF INTEGRAL SQUARE VALUES (ISV)

Consider now an autotranslation function $I_{xx}(\tau)$ as the inverse of the $\mathscr{L}_B$-transform of $I_{XX}(s)$:

$$I_{xx}(\tau) \triangleq \frac{1}{2\pi j} \int_{-j\infty}^{+j\infty} I_{XX}(s)\, e^{s\tau} ds \qquad (11.2\text{-}22)$$

By letting $\tau = 0$ and using (11.2-10), we immediately get

$$\int_{-\infty}^{\infty} x^2(t)\, dt = I_{xx}(0) = \frac{1}{2\pi j} \int_{-j\infty}^{j\infty} I_{XX}(s)\, ds \qquad (11.2\text{-}23)$$

From (11.2-23) it may be seen that the square integral of a function in the time domain may be evaluated by an integration in the complex-frequency domain of the corresponding transformed autotranslation function. In general it is much easier to evaluate an integral in the complex plane since it usually reduces itself to the evaluation of residues of the poles. Thus the reason for the introduction of translation functions is clear.

In the case $I_{XX}(s)$ is a rational function such that

$$I_n = I_{XX}(s) = \frac{c(s)}{d(s)} \cdot \frac{c(-s)}{d(-s)} \qquad (11.2\text{-}24)$$

where

$$c(s) = c_{n-1}s^{n-1} + \cdots + c_0$$

$$\qquad\qquad\qquad\qquad\qquad (11.2\text{-}25)$$

$$d(s) = d_n s^n + \cdots + d_0$$

The integral is evaluated in terms of coefficients of the polynomials of $c(s)$ and $d(s)$. Results up to sixth order are given in Table 11.2-1. One can also obtain I_n for the general case.[1]

Example 11.2-1

Let us find $\int_{-\infty}^{\infty} x^2(t)\, dt$ where $x(t) = e^{-|t|}$. We have

$$\mathscr{L}_B\{x(t)\} = X(s) = \mathscr{L}\{x(t)\} + \mathscr{L}\{x(-t)\}\Big|_{s\to -s} = \frac{1}{s+1} + \frac{1}{-s+1}$$

or $$X(s) = \frac{2}{-s^2+1}$$

[1]E. I. Jury and A. G. Dewey, A General Formulation of the Total Square Integrals for Continuous Systems, *IEEE Transactions on Automatic Control*, January 1965, **AC-10**, 119–120.

Table 11.2-1

$$I_n = \frac{1}{2\pi j} \int_{-j\infty}^{j\infty} \frac{c(s)c(-s)}{d(s)d(-s)} \, ds$$

$$c(s) = c_{n-1}s^{n-1} + \cdots + c_0, \qquad d(s) = d_n s^n + \cdots + d_0$$

$d(s)$ *must have all LHP roots.*

$$I_1 = \frac{c_0^2}{2d_0 d_1}$$

$$I_2 = \frac{c_1^2 d_0 + c_0^2 d_2}{2d_0 d_1 d_2}$$

$$I_3 = \frac{c_2^2 d_0 d_1 + (c_1^2 - 2c_0 c_2) d_0 d_3 + c_0^2 d_2 d_3}{2 d_0 d_3 (-d_0 d_3 + d_1 d_2)}$$

$$I_4 = \frac{c_3^2(-d_0^2 d_3 + d_0 d_1 d_2) + (c_2^2 - 2c_1 c_3) d_0 d_1 d_4 + (c_1^2 - 2c_0 c_2) d_0 d_3 d_4 + c_0^2(-d_1 d_4^2 + d_2 d_3 d_4)}{2d_0 d_4 (-d_0 d_3^2 - d_1^2 d_4 + d_1 d_2 d_3)}$$

$$I_5 = \frac{1}{2\Delta_5} [c_4^2 m_0 + (c_3^2 - 2c_2 c_4) m_1 + (c_2^2 - 2c_1 c_3 + 2c_0 c_4) m_2 + (c_1^2 - 2c_0 c_2) m_3 + c_0^2 m_4]$$

where

$$m_0 = \frac{1}{d_5}(d_3 m_1 - d_1 m_2) \qquad m_3 = \frac{1}{d_0}(d_2 m_2 - d_4 m_1)$$

$$m_1 = -d_0 d_3 + d_1 d_2 \qquad m_4 = \frac{1}{d_0}(d_2 m_3 - d_4 m_2)$$

$$m_2 = -d_0 d_5 + d_1 d_4 \qquad \Delta_5 = d_0(d_1 m_4 - d_5 m_3 + d_5 m_2)$$

$$I_6 = \frac{1}{2\Delta_6}[c_5^2 m_0 + (c_4^2 - 2c_3 c_5) m_1 + (c_3^2 - 2c_2 c_4 + 2c_1 c_5) m_2 + (c_2^2 - 2c_1 c_3 + 2c_0 c_4) m_3 + (c_1^2 - 2c_0 c_2) m_4 + c_0^2 m_5]$$

where

$$m_0 = \frac{1}{d_6}(d_4 m_1 - d_2 m_2 + d_0 m_3) \qquad m_4 = \frac{1}{d_0}(d_2 m_3 - d_4 m_2 + d_6 m_1)$$

$$m_1 = -d_0 d_1 d_5 + d_0 d_3^2 + d_1^2 d_4 - d_1 d_2 d_3 \qquad m_5 = \frac{1}{d_0}(d_2 m_4 - d_4 m_3 + d_6 m_2)$$

$$m_2 = d_0 d_3 d_5 + d_1^2 d_6 - d_1 d_2 d_5 \qquad \Delta_6 = d_0(d_1 m_5 - d_3 m_4 + d_5 m_3)$$

$$m_3 = d_0 d_5^2 + d_1 d_3 d_6 - d_1 d_4 d_5$$

From H. M. James, N. B. Nichols, and R. S. Phillips, *Theory of Servo-Mechanisms*, McGraw-Hill, New York, 1947.

$$I_{xx}(0) = \int_{-\infty}^{\infty} x^2(t)\,dt = \frac{1}{2\pi j} \int_{-j\infty}^{j\infty} X(-s)X(s)\,ds$$

$$= \frac{1}{2\pi j} \int_{-j\infty}^{j\infty} \frac{2 \cdot 2}{(s+1)^2(-s+1)^2}\,ds$$

and thus

$$c(s) = 2, \qquad d(s) = s^2 + 2s + 1$$

Using Table 11.2-1, we have

$$I_2 = \frac{c_1{}^2 d_0 + c_0{}^2 d_2}{2d_0 d_1 d_2} = \frac{4}{4} = 1$$

We also may evaluate $\int_{-\infty}^{\infty} x^2(t)\,dt$ directly in this case by

$$\int_{-\infty}^{\infty} (e^{-|t|})^2 dt = \int_{-\infty}^{\infty} e^{-2|t|} dt = 2\int_{0}^{\infty} e^{-2t}\,dt = -e^{-2t}\Big|_{0}^{\infty} = 1$$

which checks.

Translation Functions of Sampled-Time Functions and Their Use

The autotranslation function for a continuous-time function $x(t)$ has been defined earlier as

$$I_{xx}(\tau) \triangleq \int_{-\infty}^{\infty} x(t)x(t+\tau)\,dt \tag{11.2-26}$$

We wish to find out what happens when $x(t)$ is passed through a sampler as shown in Fig. 11.2-2. Now

$$x^*(t) = \sum_{n=-\infty}^{\infty} x(nT)\delta(t-nT) \tag{11.2-27}$$

We therefore can write

$$I_{x^*x^*}(\tau) = \int_{-\infty}^{\infty} \sum_{k=-\infty}^{\infty} x(kT)\delta(t-kT) \sum_{\rho=-\infty}^{\infty} x(\rho T)\delta(t+\tau-\rho T)\,dt$$

$$= \int_{-\infty}^{\infty} \sum_{k=-\infty}^{\infty} \sum_{\rho=-\infty}^{\infty} x(kT)x(\rho T)\delta(t-kT)\delta(t+\tau-\rho T)\,dt$$

$$= \sum_{k=-\infty}^{\infty} \sum_{\rho=-\infty}^{\infty} x(kT)x(\rho T)\delta(\tau-\rho-kT) \tag{11.2-28}$$

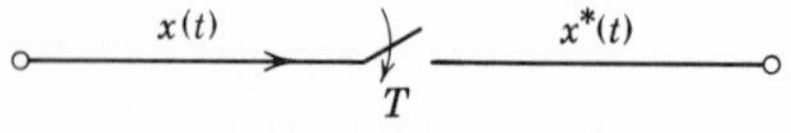

Figure 11.2-2   Sampler.

utilizing the sifting property of the δ-function. Now, letting $\rho - k = n$, we obtain

$$I_{x^{\bullet}x^{\bullet}}(\tau) = \sum_{k=-\infty}^{\infty} \sum_{n=-\infty}^{\infty} x(kT)x(\overline{n+kT})\delta(\tau - nT) \qquad (11.2\text{-}29)$$

From (11.2-26), we have for $\tau = nT$

$$I_{xx}(nT) = \int_{-\infty}^{\infty} x(t)x(t+nT)dt \qquad (11.2\text{-}30)$$

If t has values at sampling instants only, then (11.2-30) becomes

$$I_{xx}(nT) = \sum_{k=-\infty}^{\infty} x(kT)x(\overline{n+kT}) \qquad (11.2\text{-}31)$$

Substituting (11.2-31) in (11.2-29), we obtain

$$I_{x^{\bullet}x^{\bullet}}(\tau) = \sum_{n=-\infty}^{\infty} I_{xx}(nT)\delta(t-nT) \qquad (11.2\text{-}32)$$

We know, however, that

$$\sum_{n=-\infty}^{\infty} I_{xx}(nT)\delta(t-nT) \overset{\Delta}{=} I_{xx}^{*}(\tau) \qquad (11.2\text{-}33)$$

Hence we can say that autotranslation function of the sampled signal can be obtained by impulse modulating the autotranslation function of the nonsampled signal. We can therefore summarize the results obtained as

$$I_{x^{\bullet}x^{\bullet}}(\tau) = I_{xx}^{*}(\tau)$$

$$= \sum_{n=-\infty}^{\infty} I_{xx}(nT)\delta(t-nT) \qquad (11.2\text{-}34)$$

where

$$I_{xx}(nT) \overset{\Delta}{=} \sum_{k=-\infty}^{\infty} x(kT)x(\overline{n+kT}) \qquad (11.2\text{-}35)$$

These results are equally applicable to the cross-translations function provided both the variables are sampled. We wish to point out that the translation functions for the sampled case have the same properties mentioned earlier for the continuous case.

Z-TRANSFORMS OF SAMPLED TRANSLATION FUNCTIONS

We have shown that the translation function of the sampled signals is obtained by impulse modulating the translation function of the pre-sampled signals. This means that we have to use the Z-transform analysis for treating the sampled translation functions. We can write

$$I_{XX}(z) = \sum_{n=-\infty}^{\infty} I_{xx}(nT)z^{-n}$$

$$= Z\left\{ \sum_{n=-\infty}^{\infty} I_{xx}(nT)\delta(t-nT) \right\}$$

$$= Z\left\{ I_{xx}^*(\tau) \right\} \tag{11.2-36}$$

We can rewrite

$$I_{XX}(z) = \sum_{n=0}^{\infty} I_{xx}(nT)z^{-n} + \sum_{n=0}^{\infty} I_{xx}(nT)z^{n} - I_{xx}(0)$$

$$= I'_{XX}(z) + I'_{XX}(z^{-1}) - I_{xx}(0) \tag{11.2-37}$$

where $I'_{XX}(z)$ represents the one-sided Z-transform of $I_{xx}(\tau)$. We can also write

$$I_{XX}(z) \triangleq Z\left\{ \sum_{k=-\infty}^{\infty} x(kT)x(\overline{n+kT}) \right\}$$

$$= \sum_{k=-\infty}^{\infty} \sum_{n=-\infty}^{\infty} x(kT)x(\overline{n+kT})z^{-n}$$

$$= \sum_{k=-\infty}^{\infty} x(kT)z^{k} \sum_{n=-\infty}^{\infty} x(\overline{n+kT})z^{-(n+k)}$$

$$= \sum_{k=-\infty}^{\infty} x(kT)z^{k} \sum_{p=-\infty}^{\infty} x(pT)z^{-p}, \qquad p = n+k$$

$$= X(z^{-1})X(z) \tag{11.2-38}$$

It can similarly be shown that

$$I_{XY}(z) = X(z^{-1})Y(z) \tag{11.2-39}$$

It should be noted that $X(z)$ and $Y(z)$ are the bilateral or two-sided Z-transforms (limits of summations running from $-\infty$ to ∞).

APPLICATION TO A LINEAR SYSTEM

Consider a linear system $G(s)$, which has a sampled input as shown in Fig. 11.2-3. We find, using modified Z-transform theory, that

$$C(z, m) = G(z, m)U(z) \tag{11.2-40}$$

Now we can define

$$I_{CC}(z, m) = C(z, m)C(z^{-1}, m) \tag{11.2-41}$$

Substituting (11.2-40) leads to

$$I_{CC}(z, m) = G(z, m)G(z^{-1}, m)I_{UU}(z) \tag{11.2-42}$$

Figure 11.2-3   A linear, sampled-data, open-loop system.

We can also develop the relationships

$$I_{CU}(z, m) = G(z^{-1}, m) I_{UU}(z) \qquad (11.2\text{-}43)$$

$$I_{UC}(z, m) = G(z, m) I_{UU}(z) \qquad (11.2\text{-}44)$$

The foregoing relationships clearly indicate that if the autotranslation function of the sampled input to a system is known, then the translation function at the output of the system can be easily obtained.

EVALUATION OF THE INTEGRAL SQUARE VALUES (ISV)

Consider the system shown in Fig. 11.2-3. The input to the system is sampled, while the output of the system also exists between-sampling instants. We must be able to evaluate the total integral square value including between-sampling instants, especially when such values do exist.

We know that in order to get information on between-sampling instants we use the modified Z-transform approach. Equation (11.2-42) gives the autotranslation function, which takes into account the information between-sampling instants. We can write

$$I_{CC}(z, m) = \sum_{n=-\infty}^{\infty} I_{cc}(nT, m) z^{-n}, \qquad 0 \leqslant m < 1 \qquad (11.2\text{-}45)$$

where

$$I_{cc}(nT, m) = \sum_{k=-\infty}^{\infty} c(kT, m) c(\overline{k+n}T, m), \qquad 0 \leqslant m < 1 \qquad (11.2\text{-}46)$$

We know from the modified Z-transform theory that

$$c(kT, m) = c(\overline{k+m-1}T) \qquad 0 \leqslant m < 1 \qquad (11.2\text{-}47)$$

and

$$c(\overline{k+n}T, m) = c(\overline{k+n+m-1}T) \qquad (11.2\text{-}48)$$

Hence

$$I_{cc}(nT, m) = \sum_{k=-\infty}^{\infty} c(\overline{k+m-1}T) c(\overline{k+n+m-1}T), \qquad 0 \leqslant m < 1$$
$$(11.2\text{-}49)$$

Substituting $n = 0$ on both sides of (11.2-49), we have

$$I_{cc}(0, m) = \sum_{k=-\infty}^{\infty} c^2(\overline{k+m-1}T), \qquad 0 \leqslant m < 1 \qquad (11.2\text{-}50)$$

Since m is a variable, we can define the integral square value

$$\text{ISV}_c = \overline{c^2(t)} = \int_0^1 \sum_{k=-\infty}^{\infty} c^2(\overline{k+m-1}T)\, d(mT)$$

$$= \int_0^1 T \cdot I_{cc}(0, m)\, dm \qquad (11.2\text{-}51)$$

From the inversion integral, we have

$$I_{cc}(nT, m) = \frac{1}{2\pi j} \oint_{\text{unit circle}} I_{CC}(z, m) z^{n-1} dz \qquad (11.2\text{-}52)$$

Putting $n = 0$ in (11.2-52), we get

$$I_{cc}(0, m) = \frac{1}{2\pi j} \oint_{\text{unit circle}} I_{CC}(z, m) \frac{dz}{z} \qquad (11.2\text{-}53)$$

Hence

$$\text{ISV}_c = \overline{c^2(t)} = \frac{T}{2\pi j} \oint_{\text{unit circle}} \frac{dz}{z} \int_0^1 I_{CC}(z, m)\, dm \qquad (11.2\text{-}54)$$

Equation (11.2-54) gives the average integral square value and can be utilized whenever the variable has values between the sampling instants.

In order to evaluate the ISV integral, we write (11.2-54) in the form

$$I_n = \frac{1}{2\pi j} \oint_{\text{unit circle}} F(z)F(z^{-1}) \frac{dz}{z} \qquad (11.2\text{-}55)$$

where

$$F(z) = \frac{b_0 z^n + b_1 z^{n-1} + \cdots + b_n}{a_0 z^n + a_1 z^{n-1} + \cdots + a_n} \qquad (11.2\text{-}56)$$

Results of I_n up to $n = 3$ are given in Table 11.2-2. The general result for any n is also available.[2]

Example 11.2-2

Evaluate $\overline{c^2(t)}$ for the system shown in Fig. 11.2-4.
We have

$$I_{UU}(z) = U(z^{-1})U(z) = \frac{4}{(2z^{-1} - 1)(2z - 1)}$$

Now

$$G(s) = \frac{1 - e^{-sT}}{s(s+1)}$$

---

[2] E. I. Jury, A Note on the Evaluation of the Total Square Integral, *IEEE Transactions on Automatic Control*, AC-10 January, 1965, pp. 110–111. Also see reference B.43.

Table 11.2-2

$$I_n = \frac{1}{2\pi j} \oint_{\text{unit circle}} F(z)F(z^{-1}) \frac{dz}{z}, \qquad F(z) = \frac{b_0 z^n + b_1 z^{n-1} + \cdots + b_n}{a_0 z^n + a_1 z^{n-1} + \cdots + a_n}$$

Evaluation of I_n for $n = 1, 2, 3$

n	$F(z)$	I_n
1	$\dfrac{b_0 z + b_1}{a_0 z + a_1}$	$I_1 = \dfrac{(b_0{}^2 + b_1{}^2)a_0 - 2b_0 b_1 a_1}{a_0(a_0{}^2 - a_1{}^2)}$
2	$\dfrac{b_0 z^2 + b_1 z + b_2}{a_0 z^2 + a_1 z + a_2}$	$I_2 = \dfrac{B_0 a_0 e_1 - B_1 a_0 a_1 + B_2(a_1{}^2 - a_2 e_1)}{a_0[(a_0{}^2 - a_2{}^2)e_1 - (a_0 a_1 - a_1 a_2)a_1]}$
		where $$\begin{aligned} B_0 &= b_0{}^2 + b_1{}^2 + b_2{}^2 \\ B_1 &= 2(b_0 b_1 + b_1 b_2) \\ B_2 &= 2b_0 b_2 \\ e_1 &= a_0 + a_2 \end{aligned}$$
3	$\dfrac{b_0 z^3 + b_1 z^2 + b_2 z + b_3}{a_0 z^3 + a_1 z^2 + a_2 z + a_3}$	$I_3 = \dfrac{a_0 B_0 Q_0 - a_0 B_1 Q_1 + a_0 B_2 Q_2 - B_3 Q_3}{[(a_0{}^2 - a_3{}^2)Q_0 - (a_0 a_1 - a_2 a_3)Q_1 + (a_0 a_2 - a_1 a_3)Q_2]a_0}$
		where $$\begin{aligned} B_0{}^2 &= b_0{}^2 + b_1{}^2 + b_2{}^2 + b_3{}^2 \\ B_1 &= 2(b_0 b_1 + b_1 b_2 + b_2 b_3) \\ B_2 &= 2(b_0 b_2 + b_1 b_3), \qquad B_3 = 2b_0 b_3 \\ Q_0 &= (a_0 e_1 - a_3 a_2), \qquad Q_1 = (a_0 a_1 - a_1 a_3) \\ Q_2 &= (a_1 e_2 - a_2 e_1) \\ Q_3 &= (a_1 - a_3)(e_2{}^2 - e_1{}^2) + a_0(a_0 e_2 - a_3 e_1) \\ e_1 &= a_0 + a_2, \; e_2 = a_1 + a_3 \end{aligned}$$

From E. I. Jury, *Theory and Application of the Z-Transform Method*, Wiley, New York, 1964.

Hence

$$G(z, m) = \frac{z-1}{z}\left(\frac{1}{z-1} - \frac{2^{-m}}{z-1/2}\right)$$

$$= \frac{z - 1/2 - (z-1)2^{-m}}{z(z-1/2)}$$

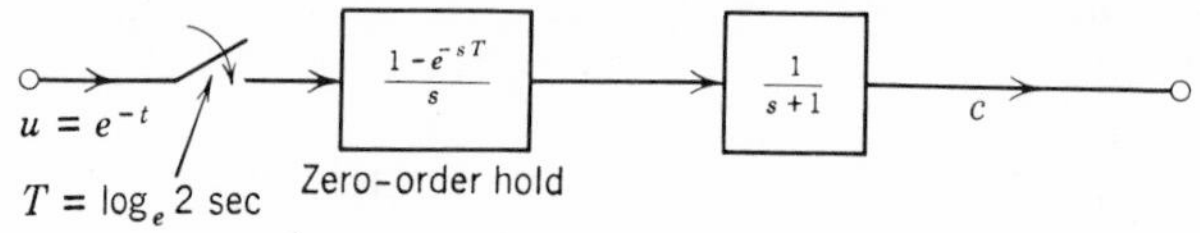

Figure 11.2-4 System of Example 11.2-2.

Utilizing (11.2-54),

$$\overline{c^2(t)} = \frac{T}{2\pi j} \oint \frac{dz}{z} \int_0^1 I_{CC}(z, m)\, dm, \qquad T = \log_e 2$$

where

$$I_{CC}(z, m) = \frac{4}{(2z^{-1}-1)(2z-1)} \cdot \frac{z^{-1}-1/2-(z^{-1}-1)2^{-m}}{z^{-1}(z^{-1}-1/2)}$$

$$\cdot \frac{(z-1/2)-(z-1)2^{-m}}{z(z-1/2)}$$

The increased difficulty in handling the algebraic work when dealing with integral square values for sampled-data systems is apparent. We can first utilize Table 11.2-2 to obtain

$$\overline{c^2(nT, m)} = \frac{T}{2\pi j} \oint I_{CC}(z, m)\, \frac{dz}{z}$$

$$= \frac{20 \cdot 4 \cdot 5(1-2^{-m})^2 + 16(1-2^{-m})^2 \cdot -16}{4 \cdot 27} \cdot T$$

$$= \tfrac{4}{3}(1 - 2^{-m+1} + 2^{-2m}) \cdot \log_e 2$$

Therefore

$$\overline{c^2(t)} = \int_0^1 \tfrac{4}{3}(1 - 2^{-m+1} + 2^{-2m})\log_e 2\, dm$$

$$= \tfrac{4}{3}\log_e 2 - \tfrac{5}{6}$$

11.3 PARAMETER OPTIMIZATION USING INTEGRAL SQUARE ERROR

Continuous-Time Systems

In this case we assume the form of $G_c(s)$. The only thing we need to know are the unknown parameters. The system shown in Fig. 11.1-1a can be redrawn as shown in Fig. 11.3-1. We note that

$$T(s) = \frac{G_c(s)G(s)}{1 + G_c(s)G(s)} \tag{11.3-1}$$

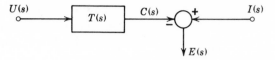

Figure 11.3-1   Equivalent system for Fig. 11.1-1*a*: the basic system.

and therefore

$$G_c(s) = \frac{T(s)}{\{1 - T(s)\} G(s)} \tag{11.3-2}$$

Since we want to obtain the ISE, we must evaluate $I_{EE}(s)$. From Fig. 11.3-1,

$$E(s) = I(s) - T(s)U(s) \tag{11.3-3}$$

Hence

$$
\begin{aligned}
I_{EE}(s) &= E(-s)E(s) \\
&= [I(-s) - T(-s)U(-s)][I(s) - T(s)U(s)] \\
&= I(-s)I(s) - T(-s)U(-s)I(s) - I(-s)U(s)T(s) \\
&\quad + T(-s)U(-s)U(s)T(s) \\
&= I_{II}(s) - T(-s)I_{UI}(s) - I_{IU}(s)T(s) + T(-s)I_{UU}(s)T(s) \tag{11.3-4}
\end{aligned}
$$

And now by use of (11.2-23) the ISE for this system may be evaluated as

$$I_e = I_{ee}(0) = \int_{-\infty}^{\infty} e^2(t)\, dt = \frac{1}{2\pi j} \int_{-j\infty}^{j\infty} I_{EE}(s)\, ds \tag{11.3-5}$$

where $I_{EE}(s)$ under the far right integral sign is given by (11.3-4).

Consider the case of the system in Fig. 11.3-1 where now the transfer function $T(s)$ is fixed to within one free parameter, K, i.e., $T(s) = T(K, s)$. What we are really saying is that $G_c(s)$ has one free parameter K. Let the problem be to adjust K so that the ISE for the system is at a minimum. $T(K, s)$ may be used in (11.3-4) and the result in (11.3-5). The final result will be the integral square error (I_e) expressed as a function of K, i.e., $I_e(K)$. Now, the minimizing value of K will occur where

$$\frac{dI_e(K)}{dK} = 0 \tag{11.3-6}$$

which may be used to solve for K. If there is more than one free parameter $(K_1, K_2, \ldots, K_n,$ say), we may use the same procedure and express the ISE as a function of the free parameters. The minimizing values for these free parameters will occur when

$$\frac{\partial I_e(K_1, K_2, \ldots K_n)}{\partial K_i} = 0, \qquad i = 1, 2, \ldots, n \tag{11.3-7}$$

and we thus have a set of n simultaneous equations in n unknowns, which may be solved to give the desired minimizing values for the parameters. Though in principle this may be done, computational difficulties increase rapidly with increasing n.

Example 11.3-1

As an example of the use of this procedure, consider the system of Fig. 11.3-1 where

$$T(s) = \frac{\omega_n^2}{s^2 + 2\zeta\omega_n s + \omega_n^2}$$

This is a typical transfer function for a second-order, positional servo system. Let $U(s) = I(s) = 1/s$, i.e., for a step input, it is desired to have the output follow the input exactly. For a servo system this is usually desired. Now assume that ζ is free to be adjusted as desired. We would like to determine what value of ζ should be used to produce minimum ISE.

We proceed as follows. From Fig. 11.3-1,

$$E(s) = I(s) - T(s)U(s)$$

$$= \frac{1}{s} - \frac{\omega_n^2}{s(s^2 + 2\zeta\omega_n s + \omega_n^2)}$$

$$= \frac{s + 2\zeta\omega_n}{s^2 + 2\zeta\omega_n s + \omega_n^2}$$

Therefore

$$I_{EE}(s) = E(-s)E(s) = \frac{-s + 2\zeta\omega_n}{(-s)^2 - 2\zeta\omega_n s + \omega_n^2} \frac{s + 2\zeta\omega_n}{s^2 + 2\zeta\omega_n s + \omega_n^2}$$

$$I_e(\zeta) = \frac{1}{2\pi j}\int_{-j\infty}^{+j\infty} I_{EE}(s)\, ds$$

In this case the integral to be evaluated is of the form

$$\frac{1}{2\pi j}\int_{-j\infty}^{j\infty} \frac{c(s)c(-s)}{d(s)d(-s)}\, ds$$

where $c(s)$ and $d(s)$ are polynomials in s. Integrals of this form are evaluated in Table 11.2-1 in terms of the coefficients of $c(s)$ and $d(s)$. For the case at hand $d(s)$ is a second-order polynomial, and the value of the integral is

$$I_2 = \frac{c_1^2 d_0 + c_0^2 d_2}{2 d_0 d_1 d_2}$$

where c_i, d_i are the coefficients of s^i in the polynomials $c(s)$ and $d(s)$, respectively. Substituting in the values of the coefficients from $E(s)$

we have

$$I_e(\zeta) = \frac{1 + 4\zeta^2}{4\zeta\omega_n}$$

Taking the derivative of $I_e(\zeta)$ with respect to ζ and setting the result to zero gives

$$\frac{dI_e(\zeta)}{d\zeta} = -\frac{1}{4\zeta^2\omega_n} + \frac{1}{\omega_n} = 0$$

From this it is determined immediately that

$$\zeta = \tfrac{1}{2}$$

is the value of ζ that produces minimum integral square error here. This value of ζ is in the middle of the range in which control system designers usually try to hold the damping factor.

Using the minimizing value of ζ, we have

$$I_{e(\min)} = \frac{1}{\omega_n}$$

from which we see that to minimize ISE for this problem, ω_n should be made as large as possible, not a surprising result!

With the knowledge of $T(s)$, we can now obtain $G_c(s)$ for a given $G(s)$ using (11.3-2).

Discrete-Time or Sampled-Data Systems

In this case we assume that the form of $D(z)$ is known. We consider the system configuration of Fig. 11.1-1b. We can redraw this system as shown in Fig. 11.3-2. Even if there are elements in the feedback path, the system can be reduced to the one shown in this figure.

From Fig. 11.3-2, we write

$$\begin{aligned} E(z, m) &= I(z, m) - C(z, m) \\ &= I(z, m) - T(z, m)U(z) \end{aligned} \tag{11.3-8}$$

Therefore

$$\begin{aligned} I_{EE}(z, m) &= I_{II}(z, m) + T(z, m)T(z^{-1}, m)I_{UU}(z) \\ &\quad - T(z^{-1}, m)I(z, m)U(z^{-1}) - T(z, m)I(z^{-1}, m)U(z) \end{aligned} \tag{11.3-9}$$

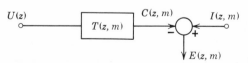

Figure 11.3-2 Equivalent system of Fig. 11.1-1b.

Assuming $D(z)$ has unknown parameters $K_1, \ldots, K_n$, by using (11.3-9), we can obtain

$$\overline{e^2(t)} = \frac{T}{2\pi j} \oint \frac{dz}{z} \int_0^1 I_{EE}(z, m)\,dm = f(K_1, \ldots, K_n) \tag{11.3-10}$$

Since we want to minimize $\overline{e^2(t)}$, we must obtain $K_1, \ldots, K_n$ from

$$\frac{\partial \overline{e^2(t)}}{\partial K_1} = \cdots = \frac{\partial \overline{e^2(t)}}{\partial K_n} = 0 \tag{11.3-11}$$

Example 11.3-2

Given a system shown in Fig. 11.3-3. The desired output is the same as the input. Find the parameters K and a such that ISE is minimized.

We have

$$I_{UU}(z) = U(z)U(z^{-1}) = \frac{z}{z-1} \cdot \frac{z^{-1}}{z^{-1}-1}$$

and

$$G(z, m) = Z_m\left\{\frac{1-e^{-sT}}{s^2}\right\} = \frac{1}{z}\left[m + \frac{1}{z-1}\right]$$

Therefore

$$T(z, m) = \frac{K[m(z-1)+1]}{z^2 + (K-a-1)z + a}$$

Also

$$I(z, m) = Z_m\{U(z)\} = \frac{1}{z-1}$$

We now write

$$E(z, m) = I(z, m) - T(z, m)U(z)$$

$$= \frac{z(1-Km)-a}{z^2 + (K-a-1)z + a}$$

We can again utilize Table 11.2-2 to obtain

$$\overline{e^2(nT, m)} = \frac{T}{2\pi j} \oint E(z, m)\,E(z^{-1}, m)\,\frac{dz}{z}$$

$$= \frac{\left\{\begin{array}{l}(1+a^2)(1+a)+2a(K-a-1)\\ +(K^2m^2-2Km)(1+a)-2aKm(K-a-1)\end{array}\right\}}{(1-a^2)(1+a)-(K-a-1)(K-1+a^2-Ka)}$$

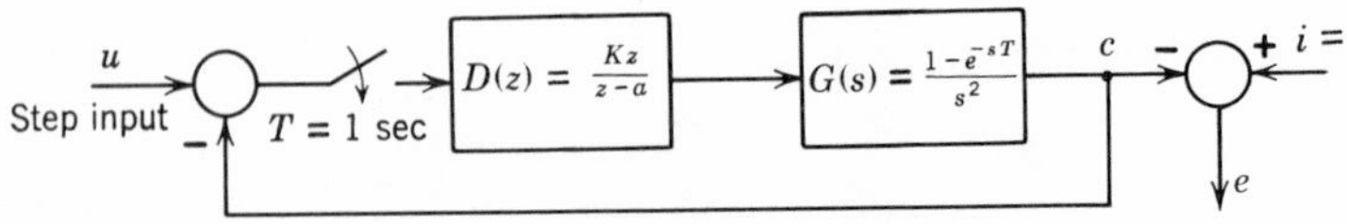

Figure 11.3-3   System of Example 11.3-2.

We can now obtain $\overline{e^2(t)}$ as

$$\text{ISE} = \overline{e^2(t)} = \int_0^1 \overline{e^2(nT, m)}\ dm$$

$$= \frac{(1+a^2)(1+a) + 2a(K-a-1) + (K^2/3 - K)(1+a) - aK(K-a-1)}{(1-a^2)(1+a) - (K-a-1)(K-1+a^2-Ka)}$$

In order to minimize $\overline{e^2(t)}$, we must differentiate it with respect to a and K. This is a very tedious procedure. We can, however, search for the minimum of $\overline{e^2(t)}$ utilizing the digital computer as a and K are varied. The variation of a and K is limited because of stability requirements. For stability, we must have the roots of the characteristic equation within the unit circle. The characteristic equation is

$$z^2 + (K - a - 1)z + a = 0$$

For stability (Jury's criterion),

$$-1 < a < 1$$
$$K > 0$$

$$a > -1 + \frac{K}{2}$$

The optimum values of K and a are obtained approximately, while maintaining stability, from Fig. 11.3-4:

$$K = 1.27 \quad \text{and} \quad a = 0$$

Further,

$$\overline{e^2(t)} = 0.29$$

Hence the compensator is $D(z) = 1.27$, just a gain.

11.4 COMPENSATOR DESIGN WITHOUT CONSTRAINTS

In this section we develop the technique of finding the compensator under the ISE criterion. In the next section we will consider the problem of putting constraints on some variable while minimizing ISE. We start here with continuous-time systems.

Continuous-Time Systems

Let the given system be as shown in Figure 11.1-1a. We draw the system and an equivalent as shown in Fig. 11.4-1. Note that the equivalent drawn here is different from the one in the preceding section.

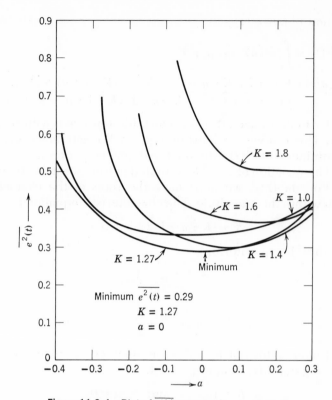

Figure 11.3-4 Plot of $\overline{e^2(t)}$ versus a for constant K.

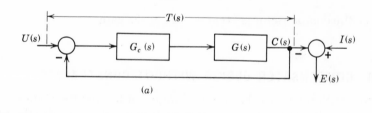

Figure 11.4-1 Continuous-time system and an equivalent model.

We have to design a realizable $G_c(s)$ such that $\int_{-\infty}^{\infty} e^2(t)\,dt$ is minimized. From Fig. 11.4-1, we have

$$T(s) = \frac{G_c(s)G(s)}{1+G_c(s)G(s)} = G(s)W(s) \tag{11.4-1}$$

Therefore

$$W(s) = \frac{G_c(s)}{1+G_c(s)G(s)} \tag{11.4-2}$$

and

$$G_c(s) = \frac{W(s)}{1-W(s)G(s)} \tag{11.4-3}$$

We proceed to obtain $W(s)$. Then we can obtain the required compensator $G_c(s)$ from (11.4-3). From Fig. 11.4-1b we have

$$E(s) = I(s) - G(s)W(s)U(s) \tag{11.4-4}$$

Hence

$$I_{EE}(s) = I_{II}(s) - W(s)G(s)I_{IU}(s) - W(-s)G(-s)I_{UI}(s)$$
$$+ W(s)W(-s)G(s)G(-s)I_{UU}(s) \tag{11.4-5}$$

We can now obtain

$$\text{ISE} = I_e = \int_{-\infty}^{\infty} e^2(t)\,dt = \frac{1}{2\pi j} \int_{-j\infty}^{j\infty} I_{EE}(s)\,ds$$

$$= \frac{1}{2\pi j} \int_{-j\infty}^{j\infty} [I_{II}(s) - I_{IU}(s)W(s)G(s) - I_{UI}(s)W(-s)G(-s)$$
$$+ W(s)W(-s)G(s)G(-s)I_{UU}(s)]\,ds \tag{11.4-6}$$

We want to find the $W(s)$ that minimizes I_e. The method used to do this is a technique of *calculus of variations*. We assume that $W(s)$ is subject to a small variation such that

$$W(s) = \overline{W}(s) + \epsilon W_0(s)$$

Then I_e changes also to give us

$$I_e + \epsilon \delta I_e \tag{11.4-8}$$

Substituting (11.4-7) and (11.4-8) in (11.4-6), we obtain

$$I_e(\epsilon) = I_e + \epsilon\delta I_e = \frac{1}{2\pi j} \int_{-j\infty}^{j\infty} \{I_{II}(s) - I_{IU}(s)G(s)[\overline{W}(s) + \epsilon W_0(s)]$$

$$- I_{UI}(s)G(-s)[\overline{W}(-s) + \epsilon W_0(-s)] + I_{UU}(s)G(s)G(-s)$$
$$\times [\overline{W}(s) + \epsilon W_0(s)][\overline{W}(-s) + \epsilon W_0(-s)]\}\,ds \tag{11.4-9}$$

In order to minimize I_e, we minimize $I_e(\epsilon)$ as $\epsilon \rightarrow 0$. This will be the case if

$$\left.\frac{\partial I_e(\epsilon)}{\partial \epsilon}\right|_{\epsilon \rightarrow 0} = 0 \qquad (11.4\text{-}10)$$

Applying (11.4-10) to (11.4-9), we obtain

$$\frac{1}{2\pi j} \int_{-j\infty}^{j\infty} \{-I_{IV}(s)G(s)W_0(s) - I_{VI}(s)G(-s)W_0(-s) + I_{VV}(s)G(s)G(-s)$$

$$[\overline{W}(s)W_0(-s) + \overline{W}(-s)W_0(s)]\} \, ds = 0 \qquad (11.4\text{-}11)$$

A realizable $W(s)$ which satisfies (11.4-11) gives the optimum $\overline{W}(s)$, which will minimize I_e.

It should be pointed out here that $\partial^2 I_e(\epsilon)/\partial \epsilon^2$ as $\epsilon \rightarrow 0$ is positive, insuring that I_e or $I_e(\epsilon)$ is indeed a minimum.

Equation (11.4-11) can be rewritten as

$$\frac{1}{2\pi j} \int_{-j\infty}^{j\infty} W_0(s) [I_{VV}(s)G(s)G(-s)\overline{W}(-s) - I_{IV}(s)G(s)] \, ds$$

$$+ \frac{1}{2\pi j} \int_{-j\infty}^{j\infty} W_0(-s) [I_{VV}(s)G(s)G(-s)\overline{W}(s) - I_{VI}(s)G(-s)] \, ds = 0$$

$$(11.4\text{-}12)$$

Letting $s = -s$ in first integral of (11.4-12), we note that (11.4-12) reduces to

$$\frac{1}{2\pi j} \int_{-j\infty}^{j\infty} W_0(-s) [I_{VV}(s)G(s)G(-s)\overline{W}(s) - I_{VI}(s)G(-s)] \, ds = 0$$

$$(11.4\text{-}13)$$

Now we have to find a *realizable* $\overline{W}(s)$ which is a solution of (11.4-13). Let

$$I_{VV}(s) = [I_{VV}(s)]^+ \cdot [I_{VV}(s)]^- = I_{VV}^+(s)I_{VV}^-(s) \qquad (11.4\text{-}14)$$

and

$$G(s)G(-s) = G^+(s)G^-(s)$$

where the $+$ (plus) part and the $-$ (minus) part contain zeros and poles in the left half and right half s-plane, respectively. Equation (11.4-13) then becomes

$$\frac{1}{2\pi j} \int_{-j\infty}^{j\infty} W_0(-s)G^-(s)I_{VV}^-(s) \left[\overline{W}(s)G^+(s)I_{VV}^+(s) - \frac{G(-s)I_{VI}(s)}{G^-(s)I_{VV}^-(s)} \right] ds$$

$$(11.4\text{-}15)$$

We note the following:

1. $\overline{W}(s)$ and $W_0(s)$ can have no poles in the right-hand side of the s-plane for stability and realizability reasons.

2. $W_0(-s)G^-(s)I_{\overline{UU}}(s)$ can have no poles in the left-hand side of the s-plane.

3. We can express

$$\frac{G(-s)I_{UI}(s)}{G^-(s)I_{\overline{UU}}(s)} = \left[\frac{G(-s)I_{UI}(s)}{G^-(s)I_{\overline{UU}}(s)}\right]_+ + \left[\frac{G(-s)I_{UI}(s)}{G^-(s)I_{\overline{UU}}(s)}\right]_- \quad (11.4\text{-}16)$$

where $[\ ]_+$ and $[\ ]_-$ represent the parts having poles in the left-hand side and the right-hand side plane of the s-plane, respectively.

In view of these conditions and since we are only looking for a realizable $\overline{W}(s)$, (11.4-15) can be written as

$$\frac{1}{2\pi j}\int_{-j\infty}^{j\infty} W_0(-s)G^-(s)I_{\overline{UU}}(s)\left\{\overline{W}(s)G^+(s)I_{UU}^+(s) - \left[\frac{G(-s)I_{UI}(s)}{G^-(s)I_{\overline{UU}}(s)}\right]_+\right\} ds = 0$$

$$(11.4\text{-}17)$$

Equation (11.4-17) is satisfied if

$$\overline{W}(s) = \frac{\left[\dfrac{G(-s)I_{UI}(s)}{G^-(s)I_{\overline{UU}}(s)}\right]_+}{G^+(s)I_{UU}^+(s)} \quad (11.4\text{-}18)$$

This is then the transfer function $W(s)$, that gives minimum ISE for the system of Fig. 11.4-1. From this we can obtain $G_c(s)$.

Example 11.4-1

Consider the feedback control system of Fig. 11.4-2. For a unit step input, find $G_c(s)$ which minimizes $\int_{-\infty}^{\infty} e^2(t)\, dt$.

As a first step in the solution we must make an equivalent model that matches the model and for which we have a solution. This is shown in Fig. 11.4-3. Note that the examination of Fig. 11.4-2 will reveal that our ideal output is actually just the input. $G_c(s)$ will be given by (11.4-3). We see that $G(s)$ is a nonminimum phase transfer function because it has a zero in RHP.

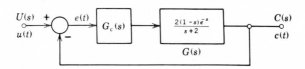

Figure 11.4-2 A unity-feedback system with fixed plant.

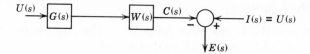

Figure 11.4-3 Equivalent model for Fig. 11.4-2.

Here

$$I(s) = U(s) = \frac{1}{s}$$

Therefore

$$I_{UU}(s) = I_{UI}(s) = \frac{1}{-s^2}$$

Also

$$G(s)G(-s) = \frac{2(1-s)e^{-s}}{(s+2)} \cdot \frac{2(1+s)e^{s}}{(-s+2)} = G^+(s)G^-(s)$$

Therefore

$$G^+(s) = \frac{2(1+s)}{s+2}, \qquad G^-(s) = \frac{2(1-s)}{(-s+2)}$$

Further,

$$I_{UU}^+(s) = \frac{1}{s} \quad \text{and} \quad I_{UU}^-(s) = \frac{1}{-s}$$

Note how we have left $1/+s$ in [ ]$^+$ and $1/-s$ in [ ]$^-$ of the $1/-s^2$ factor. If we keep track of these poles in this manner we do not have to use ϵ and then let $\epsilon \to 0$ at the end. Now we use (11.4-18) to obtain

$$\overline{W}(s) = \frac{\left[\dfrac{2(1+s)e^{s}/(-s+2)\,[1/(-s^2)]}{2(1-s)/(-s+2)\,[1/(-s)]}\right]_+}{\dfrac{2(1+s)}{(s+2)} \cdot \dfrac{1}{s}} = \frac{\left[\dfrac{(1+s)e^{s}}{s(1-s)}\right]_+}{\dfrac{2(1+s)}{s(s+2)}} = \frac{s+2}{2(1+s)}$$

Note that [ ]$_+$ has only one LHP pole at $s = 0$. Therefore

$$\left[\frac{(1+s)e^{s}}{s(1-s)}\right]_+ = \frac{1}{s}$$

Using the preceding $\overline{W}(s)$, the given $G(s)$, we have from (11.4-3)

$$G_c(s) = \frac{1}{2} \frac{s+2}{(s+1)+(s-1)e^{-s}}$$

This is not an easy compensator to synthesize, but it can be done approximately or by delay devices.

From this example we see that although we had a nonminimum phase plant with a delay element in it, we were able to design a compensator analytically without great difficulty. This is not easily done by previously discussed methods. Hence we can see that ISE compensation does offer advantages.

Sampled-Data Systems

In this case, the actual system and the corresponding model are shown in Figs. 11.4-4. Comparing Figs. 11.4-4a and 11.4-4b, we obtain[3]

$$T(z, m) = \frac{D(z)G(z, m)}{1 + D(z)G(z)} = W(z)G(z, m) \qquad (11.4\text{-}19)$$

Therefore

$$D(z) = \frac{W(z)}{1 - W(z)G(z)} \qquad (11.4\text{-}20)$$

Equation (11.4-19) implies that if $T(z, m)$ can be obtained, thus minimizing $e^2(t)$, we can easily obtain the required compensator $D(z)$.

We want to find a realizable $W(z)$ [i.e., $D(z)$] such that $\overline{e^2(t)}$ is minimized. We proceed to get

$$E(z, m) = I(z, m) - T(z, m)U(z)$$
$$= I(z, m) - G(z, m)W(z)U(z) \qquad (11.4\text{-}21)$$

[3]If $G(z, m)$ contains a factor of the form z^{-k}, $T(z, m)$ must also have this factor, otherwise cancellation of this transportation lag by $D(z)$ would require a physically unrealizable form for $D(z)$. Moreover, if $G(z, m)$ has a zero outside the unit circle, $T(z, m)$ must also have this zero. This follows from stability considerations. It is therefore convenient to express $G(z, m)$ and $T(z, m)$ in powers of z^{-1}, except for zeros outside the unit circle.

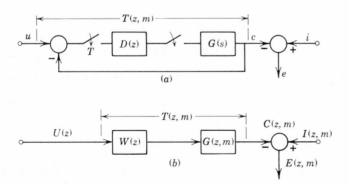

Figure 11.4-4 (*a*) Actual unity-feedback sampled-data system. $\overline{e^2(t)}$ minimization. (*b*) Equivalent model.

We now have to evaluate

$$\overline{e^2(t)} = \frac{T}{2\pi j} \int_0^1 dm \oint I_{EE}(z, m) \frac{dz}{z} \qquad (11.4\text{-}22)$$

where $\bigcirc$ is the unit circle and

$$I_{EE}(z, m) = E(z, m)E(z^{-1}, m) \qquad (11.4\text{-}23)$$

Let us define

$$A_1(z) = \int_0^1 G(z, m)G(z^{-1}, m)dm \qquad (11.4\text{-}24)$$

$$A_2(z) = \int_0^1 G(z, m)I(z^{-1}, m)dm \qquad (11.4\text{-}25)$$

$$A_3(z) = \int_0^1 I(z, m)I(z^{-1}, m)dm \qquad (11.4\text{-}26)$$

Now substituting (11.4-21) in (11.4-22) and utilizing (11.4-24), (11.4-25) and (11.4-26),

$$\overline{e^2(t)} = \frac{T}{2\pi j} \oint [A_3(z) - A_2(z)W(z)U(z) - A_2(z^{-1})W(z^{-1})U(z^{-1})$$
$$+ A_1(z)W(z)W(z^{-1})I_{UU}(z)]\frac{dz}{z} \qquad (11.4\text{-}27)$$

To find the condition for minimum $\overline{e^2(t)}$, we assume that $W(z)$ is subject to a small variation such that

$$W(z) = \overline{W}(z) + \epsilon W_0(z) \qquad (11.4\text{-}28)$$

Replacing this in (11.4-27), we obtain

$$I_e(\epsilon) = \overline{e^2(t)} + \epsilon \delta \overline{e^2(t)} = \frac{1}{2\pi j} \oint \{A_3(z) - U(z)A_2(z)[\overline{W}(z) + \epsilon W_0(z)]$$
$$- U(z^{-1})A_2(z^{-1})[\overline{W}(z^{-1}) + \epsilon W_0(z^{-1})]$$
$$+ I_{UU}(z)A_1(z)[\overline{W}(z) + \epsilon W_0(z)][\overline{W}(z^{-1}) + \epsilon W_0(z^{-1})]\}\frac{dz}{z} \qquad (11.4\text{-}29)$$

In order to minimize $\overline{e^2(t)}$, we minimize $I_e(\epsilon)$ as $\epsilon \to 0$. This implies

$$\left.\frac{\partial I_e}{\partial \epsilon}\right|_{\epsilon \to 0} = 0 \qquad (11.4\text{-}30)$$

Going through this step in (11.4-29) gives

$$\frac{1}{2\pi j} \oint W_0(z)[A_1(z)\overline{W}(z^{-1})I_{UU}(z) - U(z)A_2(z)]\frac{dz}{z} + \frac{1}{2\pi j} \oint W_0(z^{-1})$$
$$\times [A_1(z)\overline{W}(z)I_{UU}(z) - U(z^{-1})A_2(z^{-1})]\frac{dz}{z} = 0 \qquad (11.4\text{-}31)$$

Let $z = z^{-1}$ in first integral of (11.4-31). We then note that (11.4-31) reduces to

$$\frac{1}{2\pi j} \oint W_0(z^{-1})[A_1(z)\overline{W}(z)I_{UU}(z) - U(z^{-1})A_2(z^{-1})]\frac{dz}{z} = 0 \qquad (11.4\text{-}32)$$

In order to solve this equation for a realizable $\overline{W}(z)$, we let

$$A_1(z) = A_1^+(z)A_1^-(z) = [A_1(z)]^+[A_1(z)]^- \qquad (11.4\text{-}33)$$

and

$$I_{UU}(z) = I_{UU}^+(z)I_{UU}^-(z) = [I_{UU}(z)]^+[I_{UU}(z)]^- \qquad (11.4\text{-}34)$$

where the + (plus) part contains zeros and poles inside the unit circle and the − (minus) part contains zeros and poles outside the unit circle. We can write (11.4-32) as

$$\frac{1}{2\pi j} \oint W_0(z^{-1})A_1^-(z)I_{UU}^-(z)\left[\frac{I_{UU}^+(z)A_1^+(z)\overline{W}(z)}{z} - \frac{U(z^{-1})A_2(z^{-1})}{zA_1^-(z)I_{UU}^-(z)}\right]dz = 0$$
$$(11.4\text{-}35)$$

In (11.4-35), we note the following:

1. $\overline{W}(z)$ and $W_0(z)$ can have no poles outside the unit circle for stability and realizability reasons.

2. $W_0(z^{-1})A_1^-(z)I_{UU}^-(z)$ have no poles inside the unit circle.

3. We can express

$$\left[\frac{U(z^{-1})A_2(z^{-1})}{zA_1^-(z)I_{UU}^-(z)}\right] = \left[\frac{U(z^{-1})A_2(z^{-1})}{zA_1^-(z)I_{UU}^-(z)}\right]_+ + \left[\frac{U(z^{-1})A_2(z^{-1})}{zA_1^-(z)I_{UU}^-(z)}\right]_- \qquad (11.4\text{-}36)$$

where $[\ ]_+$ and $[\ ]_-$ represent the parts that have poles inside and outside the unit circle, respectively.

In view of these conditions, (11.4-35) reduces to

$$\frac{1}{2\pi j} \oint W_0(z^{-1})A_1^-(z)I_{UU}^-(z)\left\{\frac{I_{UU}^+(z)A_1^+(z)\overline{W}(z)}{z} - \left[\frac{U(z^{-1})A_2(z^{-1})}{zA_1^-(z)I_{UU}^-(z)}\right]_+\right\}$$
$$\times\, dz = 0 \qquad (11.4\text{-}37)$$

Equation (11.4-37) is satisfied only if

$$\overline{W}(z) = \frac{z[U(z^{-1})A_2(z^{-1})/zA_1^-(z)I_{UU}^-(z)]_+}{I_{UU}^+(z)A_1^+(z)} \qquad (11.4\text{-}38)$$

Once $\overline{W}(z)$ is obtained, $D(z)$ is calculated by (11.4-20).

Example 11.4-2

Consider the system shown in Fig. 11.4-4a. Let $G(s) = (1 - e^{-sT})/s^2$. For a step input, design $D(z)$. Let the ideal output be the same as the input.

Here again we can use the formula developed in (11.4-38) directly. We need to calculate $I_{UU}(z)$, $A_1(z)$, and $A_2(z)$. Now

$$I_{UU}(z) = U(z)U(z^{-1}) = \frac{z}{z-1} \cdot \frac{z^{-1}}{z^{-1}-1}$$

In order to avoid factoring difficulties, we can either keep track of the $(z-1)$ factor or introduce the factor $z-(1-\epsilon)$ and let $\epsilon \to 0$ at the end. Here we will just keep track of the $(z-1)$ factor by putting one inside and one outside the circle. We can, therefore, write

$$I^+_{UU}(z) = \frac{z}{z-1}, \qquad I^-_{UU}(z) = \frac{1}{1-z}$$

Now

$$G(z, m) = Z_m\{G(s)\} = Z_m\left\{\frac{1-e^{-sT}}{s^2}\right\} = \frac{m(z-1)+1}{z(z-1)}$$

We can now obtain

$$A_1(z) = \int_0^1 G(z, m)G(z^{-1}, m)\, dm = \frac{1}{6}\frac{z^2+4z+1}{(z-1)(1-z)}$$

Therefore we can choose

$$A_1^+(z) = \frac{1}{6}\frac{z+2-\sqrt{3}}{z-1}, \qquad A_1^-(z) = \frac{z+2+\sqrt{3}}{1-z}$$

Similarly,

$$A_2(z) = \int_0^1 G(z, m)I(z^{-1}, m)\, dm, \qquad \text{where } I(z, m) = Z_m\left\{\frac{1}{s}\right\} = \frac{1}{z-1}$$

$$= \int_0^1 \frac{m(z-1)+1}{z(z-1)} \cdot \frac{z}{1-z} \cdot dm$$

$$= \frac{z+1}{2(z-1)(1-z)}$$

Therefore

$$\left[\frac{A_2(z^{-1})U(z^{-1})}{zA_1^-(z)I^-_{UU}(z)}\right]_+ = \left[\frac{\dfrac{z^{-1}+1}{2(z^{-1}-1)(1-z^{-1})} \cdot \dfrac{z^{-1}}{z^{-1}-1}}{z\dfrac{z+2+\sqrt{3}}{1-z} \cdot \dfrac{1}{1-z}}\right]_+$$

$$= \left[\frac{(z+1)}{2(z-1)(z+2+\sqrt{3})}\right]_+ = \frac{1}{(3+\sqrt{3})(z-1)}$$

Hence (using 11.4-38)

$$W(z) = \frac{\dfrac{z}{(3+\sqrt{3})(z-1)}}{\dfrac{1}{6}\dfrac{(z+2-\sqrt{3})}{(z-1)}\cdot\dfrac{z}{z-1}} = \frac{1.27(z-1)}{z+0.27}$$

or

$$D(z) = \frac{W(z)}{1-W(z)G(z)} = 1.27$$

In this case, the compensation is just a gain.

11.5 COMPENSATOR DESIGN SUBJECT TO CONSTRAINTS

The problems considered thus far in this chapter have been mainly of academic interest. It is very seldom in engineering practice that one finds real problems clear enough to permit the use of the methods we have discussed. Moreover, when the methods are applicable, the solution obtained is usually trivially obvious to one who is familiar with the problems. It is with the introduction of the notion of constraints that we may begin to solve problems with more of an engineering flavor to them—this is where the methods have practical worth. Here, too, the compensators obtained are no longer obvious.

In using constraints here, the basic idea is that the compensator should be designed so that it not only produces the minimum integral square error but also causes the system to respond to an input (which may or may not be the input for which minimum ISE is desired) in such a way that the integral square of some system quantity is held equal to or less than some specified amount. In using a constraint such as this we realize that we will usually have to accept a larger ISE than if no such constraint were applied.

An example of the use of such a constraint is the design of a compensator for a d-c motor so chosen that it produces minimum tracking error to a step-input subject to the constraint that the integral square of the armature current will be held equal to or less than a specified amount. In this case the constraint on armature current makes sense because the integral square of armature current is a measure of the power consumed in the resistance of the armature circuit.

In general we do not always want to constrain the integral square of a quantity, but we may want to constrain the instantaneous or peak magnitude, or integral of the absolute value of the quantity, or some such quantity. So the question of why such importance is given to integral square constraints arises. The answer is that it is only in the case of integral square constraints that the problem has something approaching a general solution. Of course, there are many instances where the integral square of a quantity is what one wishes to constrain. Furthermore, useful compensators for other types of constraints can often be found through the appropriate use of integral square constraints.

The Compensation for the Case of Minimum ISE with an Integral Square Constraint for Continuous-Time Systems

Consider the system block diagrams of Fig. 11.5-1. Two separate systems are diagrammed: Fig. 11.5-1*a* is the block diagram that has been considered earlier in this chapter in finding the ISE minimizing compensator for a fixed plant; Fig. 11.5-1*b* is a simple system with a fixed plant $G_s(s)$ and compensator $W(s)$. Note that $W(s)$ is common to both systems. For the system of Fig. 11.5-1, we want to find that compensator $\overline{W}(s)$ that produces the minimum $\int_{-\infty}^{\infty} e^2(t)\,dt$ subject to the constraint

$$\int_{-\infty}^{\infty} v^2(t)\,dt \leq \alpha \tag{11.5-1}$$

where α is some real, positive quantity, and $u_1(t)$, $u_2(t)$, $i(t)$, $G(s)$, and $G_s(s)$ are assumed given.

There are many ways of introducing the constraint. The method just discussed has been found by experience to be the simplest one which still has general applicability in control problems. Thus in order to minimize $\int_{-\infty}^{\infty} e^2(t)\,dt$ subject to the constraint on $\int_{-\infty}^{\infty} v^2(t)\,dt$, we use a method of calculus of variations. We form a functional

$$J_1[W(s),\lambda] = \int_{-\infty}^{\infty} e^2(t)\,dt + \lambda \int_{-\infty}^{\infty} v^2(t)\,dt \tag{11.5-2}$$

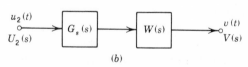

Figure 11.5-1 The basic system considered in the problem of minimum ISE with an integral square constraint for continuous-time systems.

where λ is the Lagrangian multipler. It can be shown that we can minimize $\int_{-\infty}^{\infty} e^2(t)\, dt$ by minimizing J_1. We want to find λ and $W(s)$ which will minimize J_1 and satisfy the constraint. Therefore we specify J_1 as a function of these two quantities. We now apply the calculus of variations.

For the system of Fig. 11.5-1 we have

$$E(s) = I(s) - W(s)G(s)U_1(s) \qquad (11.5\text{-}3)$$

$$V(s) = W(s)G_s(s)U_2(s) \qquad (11.5\text{-}4)$$

Now

$$\int_{-\infty}^{\infty} e^2(t)\, dt + \lambda \int_{-\infty}^{\infty} v^2(t)\, dt = \frac{1}{2\pi j} \int_{-j\infty}^{j\infty} E(s)E(-s)\, ds$$

$$+ \frac{\lambda}{2\pi j} \int_{-j\infty}^{j\infty} V(s)V(-s)\, ds \qquad (11.5\text{-}5)$$

Therefore substituting (11.5-3) and (11.5-4) in (11.5-5), we obtain from (11.5-2)

$$J_1[W(s), \lambda] = \frac{1}{2\pi j} \int_{-j\infty}^{j\infty} \{I_{II}(s) - W(-s)G(-s)I_{U_1I}(s) - I_{IU_1}(s)G(s)W(s)$$

$$+ W(-s)W(s)[G(s)G(-s)I_{U_1U_1}(s) + \lambda G_s(s)G_s(-s)I_{U_2U_2}(s)]\} \, ds \qquad (11.5\text{-}6)$$

Next we seek a $W(\lambda, s)$ that minimizes $J_1[W(s), \lambda]$. We follow the same procedure used in Section 11.4. We note that (11.5-6) is similar to (11.4-6) and hence the solution will be similar to (11.4-18). It is given by

$$\overline{W}(\lambda, s) = \frac{\left[\dfrac{G(-s)I_{U_1I}(s)}{[G(-s)G(s)I_{U_1U_1}(s) + \lambda G_s(s)G_s(-s)I_{U_2U_2}(s)]^-} \right]_+}{[G(-s)G(s)I_{U_1U_1}(s) + \lambda G_s(s)G_s(-s)I_{U_2U_2}(s)]^+} \qquad (11.5\text{-}7)$$

We note that $\overline{W}(s)$ determined in (11.5-7) is a function of λ, so we have designated it as $\overline{W}(\lambda, s)$. Now we show how to determine λ.

THE DETERMINATION OF THE LAGRANGIAN MULTIPLIER λ

The determination of the Lagrangian multiplier factor λ proceeds directly by the consideration of the constraint equation:

$$\int_{-\infty}^{\infty} v^2(t)\, dt \leq \alpha \qquad (11.5\text{-}8)$$

Since the compensator $W(s)$ to be used in the system of Fig. 11.5-1 is now known up to the factor λ, $\int_{-\infty}^{\infty} v^2(t)\, dt$ may now be determined as a function of λ. From Fig. 11.5-1b [using $\overline{W}(\lambda, s)$ of (11.5-7) as the compensator $W(s)$], we have

$$V(\lambda, s) = W(\lambda, s)G_s(s)U_2(s) \tag{11.5-9}$$

Now letting

$$I_v(\lambda) = \int_{-\infty}^{\infty} v^2(\lambda, t)\, dt \tag{11.5-10}$$

we have

$$I_v(\lambda) = \frac{1}{2\pi j} \int_{-j\infty}^{j\infty} \overline{W}(\lambda, -s)G_s(-s)U_2(-s)U_2(s)\overline{W}(\lambda, s)G_s(s)\, ds \tag{11.5-11}$$

Using Table 11.2-1, $I_v(\lambda)$ can, in the usual cases at least, be evaluated explicitly as a function of λ. The determination of the proper value of λ is reduced to the determination of λ such that

$$I_v(\lambda) \leq \alpha \tag{11.5-12}$$

Since this is an inequality instead of an equality, we proceed in the following fashion. Consider $I_v(0)$, the value of $\int_{-\infty}^{\infty} v^2(t)\, dt$ for $\lambda = 0$. This will be the value of $\int_{-\infty}^{\infty} v^2(t)\, dt$ for the ISE minimizing compensator which is determined without any consideration of the constraint on v. If $I_v(0) \leq \alpha$ (which means that the constraint is satisfied without considering it in constructing the ISE minimizing compensator), then $\lambda = 0$ is the proper value to use to determine the desired compensator. If, however, $I_v(0) > \alpha$ (which means that the compensator which minimizes ISE without consideration of the constraint leads to a violation of the constraint), then we determine the desired value of the multiplier λ, λ^*, such that $I_v(\lambda^*) = \alpha$:

$$\frac{1}{2\pi j} \int_{-j\infty}^{j\infty} \overline{W}(\lambda^*, -s)G_s(-s)U_2(-s)U_2(s)G_s(s)\overline{W}(\lambda^*, s)\, ds = \alpha \tag{11.5-13}$$

Once λ^* is determined from (11.5-13) and substituted in (11.5-7), $\overline{W}(s)$ is known and hence $G_c(s)$ can be obtained.

Here we have imposed only a single integral square constraint. Multiple integral square constraints can be introduced and resolved by amplifying the techniques indicated. The procedure is algebraically cumbersome but easy to visualize.

Example 11.5-1 *Phase-Locked Loop*

A simplified diagram of a phase-locked loop is shown in Fig. 11.5-2. Basically the function of the loop is to control the oscillator so that the output of the oscillator is locked in phase with the input signal. The phase of the input is some function of time. The problem in the design of phase-locked loops is to choose the filter so that there is minimum phase difference between the input and output subject to the con-

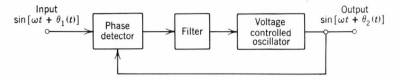

Figure 11.5-2   A phase-locked loop system.

straint that the bandwidth of the overall system is equal to or less than some fixed number.

The system block diagram may be put into the form shown in Fig. 11.5-3. In Fig. 11.5-3 only the phase of the input $[\theta_1(t)]$ and the phase of the output $[\theta_2(t)]$ are considered. The phase detector is assumed to be linear over the range of phase differences to be encountered. The constraint is now on the bandwidth of the system. To handle this constraint, it must be translated into a constraint on the integral square of some system element. As it turns out, the value of the integral square of the impulse response of a system is a good measure of the band width of the system.

If $T(s)$ is the overall transfer function, then we can write

$$\text{Bandwidth, } B = \int_{-\infty}^{\infty} T^2(t)\, dt = \frac{1}{2\pi j} \int_{-j\infty}^{j\infty} T(s)T(-s)\, ds$$

With this in mind, the problem may be seen to be as diagrammed in Fig. 11.5-4 (where the closed loop system has been converted to the equivalent open loop one). The problem is to find the compensator, $W(s)$, which produces minimum ISE subject to the constraint

$$\int_{-\infty}^{\infty} v^2(t)\, dt = \int_{-\infty}^{\infty} T^2(t)\, dt \leqslant \alpha,$$

where now α is chosen to give the desired bandwidth of the system.

Before proceeding to the solution, $\theta_1(t)$ must be decided upon. We will examine the case when there is a step change in phase. We have

$$\theta_1(t) = 1(t)$$

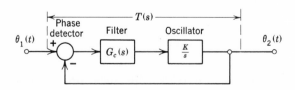

Figure 11.5-3   The phase-locked loop system in standard block diagram form.

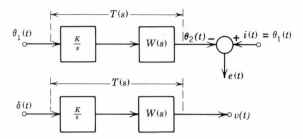

Figure 11.5-4 The equivalent open-loop problem with bandwidth constraint for the phase-locked loop system.

Comparing with Fig. 11.5-1,

$$u_1(t) = \theta_1(t) = i(t) = 1(t)$$

$$u_2(t) = \delta(t)$$

$$G(s) = G_s(s) = \frac{K}{s}$$

Here

$$I_{U_1 U_1}(s) = -\frac{1}{s^2} = I_{U_1 I}(s)$$

and

$$I_{U_2 U_2}(s) = 1$$

The desired compensator is now given by (11.5-7):

$$\overline{W}(\lambda, s) = \frac{\left[\dfrac{(K/-s)(1/-s^2)}{[(K^2/s^4) - \lambda(K^2/s^2)]^{-}} \right]_{+}}{[(K^2/s^4) - \lambda(K^2/s^2)]^{+}}$$

Now

$$\frac{K^2}{s^4} - \lambda \frac{K^2}{s^2} = \underbrace{\frac{K}{s}\left[\frac{1}{s} + \sqrt{\lambda}\right]}_{[\ ]^{+}} \cdot \underbrace{\frac{K}{-s}\left[\frac{1}{-s} + \sqrt{\lambda}\right]}_{[\ ]^{-}}$$

Therefore

$$\left[\frac{(K/-s)(1/-s^2)}{[K^2/s^4 - \lambda(K^2/s^2)]^{-}} \right]_{+} = \left[\frac{1/(-s^2)}{1/(-s) + \sqrt{\lambda}} \right]_{+} = \left[\frac{1}{s(1 - \sqrt{\lambda}s)} \right]_{+} = \frac{1}{s}$$

Hence

$$\overline{W}(\lambda, s) = \frac{s}{K(1 + \sqrt{\lambda}s)}$$

To determine λ, we use the bandwidth constraint

$$\int_{-\infty}^{\infty} v^2(t) \, dt = \int_{-\infty}^{\infty} T^2(t) \, dt \leq \alpha$$

where $T(s) = \overline{W}(\lambda, s)(K/s)$. Using Table 11.2-1, we get

$$I_v(\lambda) = \frac{1}{2\pi j} \int_{-j\infty}^{j\infty} \overline{W}(\lambda, s)\overline{W}(\lambda, -s)\frac{K^2}{-s^2} \, ds$$

$$= \frac{1}{2\pi j} \int_{-j\infty}^{j\infty} \frac{1}{1+\sqrt{\lambda}s} \cdot \frac{1}{1-\sqrt{\lambda}s} \, ds$$

$$= \frac{1}{2\sqrt{\lambda}}$$

From this we see that $I_v(0) \to \infty$, thus as explained earlier the constraint as given must be satisfied with equality. Hence λ must be so determined that

$$I_v(\lambda) = \int_{-\infty}^{\infty} v^2(t) \, dt = \alpha$$

and so

$$\sqrt{\lambda} = \frac{1}{2\alpha}$$

Hence

$$\overline{W}(s) = \frac{s}{K[1+(1/2\alpha)s]}$$

What is really desired is the equivalent closed-loop compensator $G_c(s)$ which is given by

$$G_c(s) = \frac{\overline{W}(s)}{1-\overline{W}(s)G(s)}$$

$$= \frac{2\alpha}{K}$$

Hence $G_c(s)$ is just a constant gain. We may also note that by increasing bandwidth (α), we may use a larger gain in the loop and thus tracking should be better.

The Compensation for the Case of Minimum ISE with an Integral Square Constraint for Sampled-Data Systems

In this case the actual model and the constraint model are shown in Figs. 11.5-5a and 11.5-5b, respectively.

Here again we have to minimize $\overline{e^2(t)}$ subject to the constraint

$$\overline{v^2(t)} \leqslant \alpha \tag{11.5-14}$$

The problem is to find $\overline{W}(z)$ such that

$$J_1[W(z), \lambda] = \overline{e^2(t)} + \lambda\overline{v^2(t)} \tag{11.5-15}$$

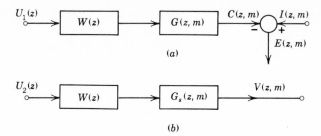

Figure 11.5-5   (*a*) Actual model. (*b*) Constraint model.

is minimized. Lambda (λ) is the Lagrangian multiplier to be found from the constraint relationship.

We have, from the figures,

$$E(z, m) = I(z, m) - G(z, m)W(z)U_1(z) \qquad (11.5\text{-}16)$$

and

$$V(z, m) = G(z, m)W(z)U_2(z) \qquad (11.5\text{-}17)$$

We first form

$$
\begin{aligned}
I_{EE}(z, m) + \lambda I_{VV}(z, m) = {}& I_{II}(z, m) - G(z, m)I(z^{-1}, m)W(z)U_1(z) \\
& - G(z^{-1}, m)I(z, m)W(z^{-1})U_1(z^{-1}) \\
& + W(z)W(z^{-1}) \left[G(z^{-1}, m)G(z, m)I_{U_1U_1}(z) \right. \\
& \left. + \lambda G_s(z, m)G_s(z^{-1}, m)I_{U_2U_2}(z) \right] \qquad (11.5\text{-}18)
\end{aligned}
$$

Now we can define

$$
\begin{aligned}
A_1(\lambda, z) = \int_0^1 {}& [G(z^{-1}, m)G(z, m)I_{U_1U_1}(z) \\
& + \lambda G_s(z, m)G_s(z^{-1}, m)I_{U_2U_2}(z)]\, dm \qquad (11.5\text{-}19)
\end{aligned}
$$

$$A_2(z) = \int_0^1 G(z, m)I(z^{-1}, m)\, dm \qquad (11.5\text{-}20)$$

$$A_3(z) = \int_0^1 I_{II}(z, m)\, dm$$

We then obtain

$$
\begin{aligned}
J_1[W(z), \lambda] = \frac{1}{2\pi j} \oint {}& [A_3(z) - A_2(z)W(z)U_1(z) - A_2(z^{-1})W(z^{-1})U_1(z^{-1}) \\
& + W(z)W(z^{-1})A_1(\lambda, z)]\, \frac{dz}{z} \qquad (11.5\text{-}21)
\end{aligned}
$$

To minimize this, we follow the same procedure as in Section 11.4

and get optimum $W(z)$ as a function of λ:

$$\overline{W}(\lambda, z) = \frac{z\left[\dfrac{A_2(z^{-1})U_1(z^{-1})}{zA_1^{-}(\lambda, z)}\right]_+}{A_1^{+}(\lambda, z)} \tag{11.5-22}$$

We can now obtain

$$V(z, m) = \overline{W}(\lambda, z)G_s(z, m)U_2(z) \tag{11.5-23}$$

Hence we can obtain $\overline{v^2(t)}$ and then obtain λ from (11.5-14). Thus we get $\overline{W}(z)$ and hence $D(z)$.

Example 11.5-2

Consider the system shown in Fig. 11.5-6. Design the compensator $D(z)$ such that $\overline{e^2(t)}$ is minimized with $\overline{v^2(t)} \leqslant \alpha$.

We note that

$$G_s(z, m) = Z_m\left\{\frac{1 - e^{-sT}}{s}\right\} = \frac{1}{z}\left(m + \frac{1}{z - 1}\right)$$

Also

$$V(z, m) = \frac{D(z)}{1 + D(z)G(z)} \cdot G_s(z, m) \cdot U(z)$$

Therefore according to our models

$$U_1(z) = U_2(z) = U(z) = \frac{z}{z - 1}$$

and

$$G_s(z, m) = Z_m\left\{\frac{1 - e^{-sT}}{s}\right\} = \frac{1}{z}$$

Further,

$$I(z, m) = U(z, m) = \frac{1}{z - 1}$$

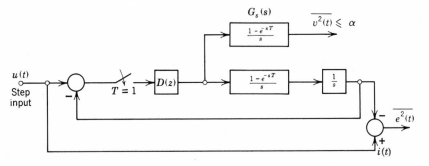

Figure 11.5-6 System of Example 11.5-2.

We now compute

$$A_1(\lambda, z) = \int_0^1 [G(z^{-1}, m)G(z, m)I_{U_1 U_1}(z)$$

$$+ \lambda G_s(z, m)G_s^{\,j}(z^{-1}, m)I_{U_2 U_2}(z)]\, dz$$

$$= \left[\frac{z^2 + 4z + 1}{6(z-1)(1-z)} + \lambda \right] \frac{z}{z-1} \cdot \frac{z^{-1}}{z^{-1} - 1}$$

or

$$A_1(\lambda, z) = \frac{6\lambda - 1}{6} \cdot \frac{z^2 - [(12\lambda + 4)/(6\lambda - 1)]z + 1}{(z-1)^2} \cdot \frac{-z}{(1-z)^2}$$

$$= \frac{6\lambda - 1}{6} \frac{(z - \lambda_1)(z - \lambda_2)}{(z-1)^2} \cdot \frac{-z}{(1-z)^2}$$

where

$$\lambda_1 + \lambda_2 = \frac{12\lambda + 4}{6\lambda - 1}$$

and   (I)

$$\lambda_1 \lambda_2 = 1$$

and we let $\lambda_1 < 1$ and $\lambda_2 > 1$. We can therefore write

$$A_1^+(\lambda, z) = \frac{6\lambda - 1}{6} \frac{z(z - \lambda_1)}{(z-1)^2}$$

$$A_1^-(\lambda, z) = -\frac{(z - \lambda_2)}{(1-z)^2}$$

Also

$$A_2(z) = \int_0^1 G(z, m)I(z^{-1}, m)\, dz$$

$$= \frac{z+1}{2(z-1)} \cdot \frac{z^{-1}}{z^{-1} - 1}$$

Now we have

$$\left[\frac{A_2(z^{-1})U_1(z^{-1})}{zA_1^-(\lambda, z)} \right] = \left[\frac{\dfrac{z^{-1}+1}{2(z^{-1}-1)} \cdot \dfrac{z}{z-1} \cdot \dfrac{z^{-1}}{z^{-1}-1}}{z \cdot \dfrac{-(z - \lambda_2)}{(1-z)^2}} \right]_+$$

$$= \left[\frac{-(z+1)}{2(z - \lambda_2)(z-1)} \right]_+$$

$$= \frac{1}{(\lambda_2 - 1)(z-1)}$$

Hence we get

$$\tilde{W}(\lambda, z) = \frac{\dfrac{z}{(\lambda_2 - 1)(z - 1)}}{\dfrac{6\lambda - 1}{6} \dfrac{z(z - \lambda_1)}{(z - 1)^2}}$$

$$= \frac{6(z - 1)}{(\lambda_2 - 1)(6\lambda - 1)(z - \lambda_1)}$$

We now obtain

$$V(z, m) = \overline{W}(\lambda, z) G_s(z, m) U(z)$$

$$= \frac{6(z - 1)}{(\lambda_2 - 1)(6\lambda - 1)(z - \lambda_1)} \cdot \frac{1}{z} \cdot \frac{z}{(z - 1)}$$

$$= \frac{6}{(\lambda_2 - 1)(6\lambda - 1)(z - \lambda_1)}$$

Using Table 11.2-2,

$$\overline{v^2(t)} = \frac{T}{2\pi j} \oint \frac{dz}{z} \int_0^1 V(z, m) V(z^{-1}, m)\, dm$$

$$= \frac{36}{(\lambda_2 - 1)^2 (6\lambda - 1)^2 (1 - \lambda_1^2)}$$

$$= \frac{3}{\sqrt{36\lambda + 3}}, \qquad \text{using relations (I)}$$

Since $\overline{v^2(t)} \le \alpha$, we can obtain

$$\lambda = \frac{3 - \alpha^2}{12\alpha^2}$$

Therefore

$$\overline{W}(z) = \frac{4}{(\alpha^{-2} - 1)(\lambda_2 - 1)} \cdot \frac{(z - 1)}{(z - \lambda_1)}$$

$D(z)$ can now be easily obtained from

$$D(z) = \frac{\overline{W}(z)}{1 - \overline{W}(z) G(z)}$$

$$= \frac{4(z - 1)}{(\alpha^{-2} - 1)(\lambda_2 - 1)(z - \lambda_1) - 4}$$

If we choose $\alpha = \frac{1}{2}$, then $\lambda = \frac{11}{12}$, $\lambda_1 = \frac{1}{3}$, $\lambda_2 = 3$, and $D(z) = \frac{2}{3}$, just a gain.

A more practical application of the sampled-data material developed here would be a digital phase-locked loop.[4]

11.6 SOME COMMENTS ON THE DESIGN OF COMPENSATORS FOR DISCRETE-TIME SYSTEMS USING SUM SQUARE ERROR CRITERION

In the last few sections we have used the ISE criterion for the design of digital compensators for sampled-data systems. In many discrete-time system applications we may not even have the value of some variables except at sampling instants. In such cases integral square values do not have much meaning. However, the design can be carried out based on the minimization of sum square error. In fact, even when the values are available for between-sampling instants, it may be desirable to design the compensators based on sum square error criterion with or without the use of sum square constraints.

In order to use the sum square criterion we can easily deduce from Section 11.2 that

$$\sum_{n=-\infty}^{\infty} x^2(nT) = \frac{1}{2\pi j} \oint_{\text{unit circle}} I_{XX}(z) \frac{dz}{z} \qquad (11.6\text{-}1)$$

In fact we need only the Z-transform and not the modified Z-transform since we need to concern ourselves only with information at sampling instants.

The relationship in (11.6-1) can be easily exploited to design compensators following a procedure similar to that shown in Sections 11.3, 11.4, and 11.5 based on sum square error criterion with or without constraints.

PROBLEMS

Continuous-Time Systems

11.1. Given the feedback system of Fig. P11.1 for a unit step input. The desired output is also a unit step, i.e., $U(s) = I(s) = 1/s$.
(a) If $a > 0$, what value of K minimizes ISE?
(b) If $K = 8$, what value of a minimizes ISE?

11.2. Given the system of Fig. P11.2. What is $I_{ED}(s) = \mathscr{L}_B\{I_{ed}(\tau)\}$ in terms of $U(s)$, $G(s)$, and $H(s)$?

---

[4]S. C. Gupta, On Optimum Digital Phase Locked Loops, *IEEE Transactions on Communication Technology*, April 1968, **COM-16**, 340–344.

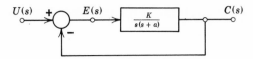

Figure P11.1

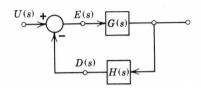

Figure P11.2

11.3. Consider the system of Fig. P11.3 with $U(s) = \mathscr{L}_B\{u(t)\} = 1/s$; $G(s) = 1/s$.
(a) Find $G_c(s)$ which minimizes $\text{ISE} = \int_{-\infty}^{\infty} e^2(t)\,dt$.
(b) What is this minimum value of ISE?
(c) How would you approximate $G_c(s)$ using physically realizable components?

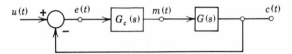

Figure P11.3

11.4. Given the open-loop system configuration as shown in Fig. P11.4. Find the integral equation which a realizable $w(t)\,[w(t) = 0, \forall t < 0]$ must satisfy which minimizes

$$\int_{-\infty}^{\infty} a(t)e^2(t)\,dt, \qquad a(t) \geq 0\ \forall t$$

This equation should be in terms of $a(t)$, $w(t)$, $u(t)$, and $i(t)$ only.

11.5. Given the system configuration of Fig. P11.5.
(a) Determine b and c (in terms of a) such that ISE is a minimum.
(b) For what value of a is this minimum value of ISE a minimum?

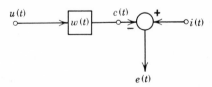

Figure P11.4

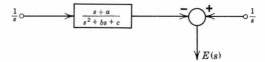

Figure P11.5

11.6. Given the system of Fig. P11.6; $\int_{-\infty}^{\infty} m^2(t)\,dt$ is to be minimized. T is fixed and gains K_1 and K_2 are variable. For $\int_{-\infty}^{\infty} c^2(t)\,dt \leq C_{\max}$, roughly plot $\int_{-\infty}^{\infty} m^2(t)\,dt$ versus T and comment upon the stability of the system.

11.7. For the system shown in Fig. P11.7a with $u(t) = t1(t)$ (ramp) and $i(t) = 1(t)$ (step) and $g(t) = \mathcal{L}^{-1}\{G(s)\}$ as shown in Fig. P11.7b find $G_c(s)$ which gives minimum ISE.

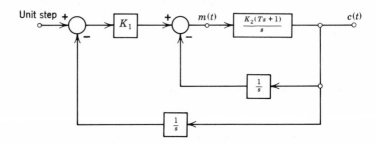

Figure P11.6

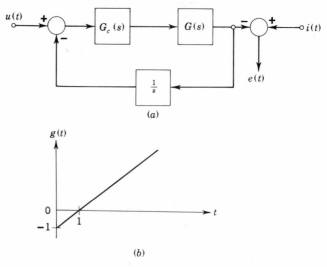

(a)

(b)

Figure P11.7

11.8. For the system of Fig. P11.3 with $u(t) = 1(t)$,

$$G(s) = \frac{2(1-s)e^{-s}}{s+2}$$

(a) Find $G_c(s)$ which minimizes $\int_0^\infty e^2(t)\,dt$.
(b) What is the minimum $\int_0^\infty e^2(t)\,dt$?

11.9. The system of Fig. P11.9 is an example of a d-c motor feedback controller. The object here is to design the compensator $G_c(s)$ so that the output shaft's angular velocity will track the input step with minimum ISE subject to the constraint that the angular acceleration of the output shaft will not exceed 10 rad/sec^2 in magnitude at any instant of time.

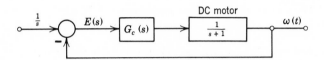

Figure P11.9

11.10. Given the feedback system of Fig. P11.3 with

$$G(s) = \frac{10}{s(s+1)}, \qquad u(t) = 50t1(t)$$

(a) Find $G_c(s)$ which minimizes $\int_0^\infty m^2(t)\,dt$ subject to the constraint $\int_0^\infty e^2(t)\,dt \leqslant 2$.
(b) What is $\int_0^\infty m^2(t)\,dt$ for the $G_c(s)$ found in (a)?
(c) Repeat the problem of (a) with

$$G(s) = \frac{10}{s^2(s+1)}$$

(d) What is $\int_0^\infty m^2(t)\,dt$ for the $G_c(s)$ found in (c)?
(e) For $G(s)$ as in (c), find the compensator $G_c(s)$ which minimizes $\int_0^\infty e^2(t)\,dt$ subject to $\int_0^\infty m^2(t)\,dt \leqslant 4$.

11.11. The pitch axis control system for an aircraft is shown in Fig. P.11.11a. The input is the control stick deflection δ_c and the output is the pitch of the aircraft θ. Through tests with pilots on ground trainers a model as shown in Fig. P11.11b was found to have pitch axis characteristics which gave the most satisfactory all-round performance of the pilot-aircraft combination. The desire now is to make the response of the aircraft pitch axis look like the response of the model so far as the pilot who produces the control stick deflection δ_c is concerned. However, it was also found that if the airframe was not to be unduly strained by the resulting aircraft motions, a severe constraint on the elevator

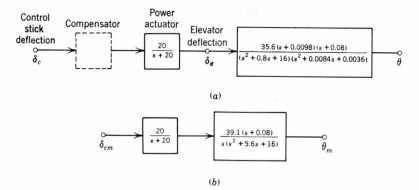

(a)

(b)

Figure P11.11 (a) Aircraft pitch axis control system. (b) Model system for pitch axis of aircraft.

deflection had to be imposed. It was hence decided that a compensator would be designed to go between the control stick and the elevator power actuator (as shown in the figure); this would make the pitch impulse-response as near as possible like the impulse response of the model under the constraint on the elevator deflection. Through trial on a mock-up it was found that for this aircraft $\int_0^\infty \delta_e^2(t)\,dt \leq 1.0$ gave in general a tolerable aircraft response for unit step control stick deflections.

Design a compensator to go between the control stick and the elevator actuator which will minimize $\int_0^\infty [\theta(t) - \theta_m(t)]^2 dt$ subject to the constraint on elevator deflection.

11.12. For the phase-locked loop as shown in Figs. 11.5-2, 11.5-3, and 11.5-4 find the closed-loop compensator, $G_c(s)$, which produces minimum ISE for a unit ramp change in input phase $[\theta_1(s) = 1/s^2]$ subject to the bandwidth constraint.

11.13. Given a system in Fig. P11.13 determine K and T so as to minimize ISE.

11.14. A unity-feedback control system has a fixed plant as given by

$$G(s) = \frac{1-s}{1+s}$$

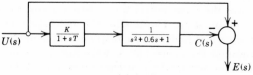

Figure P11.13

Assuming that $U(s) = I(s) = 1/s$, find $G_c(s)$ based on ISE criterion. What is the value of ISE?

Discrete-Time Systems

11.15. In Fig. P11.15 assume

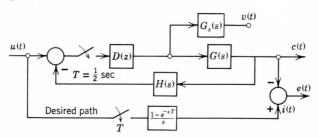

Figure P11.15

$$D(z) = \frac{z-a}{z-b}, \qquad G(s) = \frac{1-e^{-sT}}{s(s+1)}, \qquad H(s) = 1,$$

$$u = e^{-0.1t}, \qquad G_s(s) = \frac{1-e^{-sT}}{s}$$

Find the unknown parameters a and b such that
(a) $\overline{e^2(t)}$ is minimized.
(b) $\overline{e^2(t)}$ is minimized subject to $\overline{v^2(t)} \leqslant 2$.

11.16. (a) In Problem 11.15 assume that $D(z)$ is unknown and other transfer functions are the same. Design $D(z)$ such that
   (i) $\overline{e^2(t)}$ is minimized.
   (ii) $\overline{e^2(t)}$ is minimized such that $\overline{v^2(t)} \leqslant 2$.
(b) Repeat if $H(s) = 1/s$.

11.17. In Fig. P11.15, if the desired path sample and hold are replaced by an integrator, design a $D(z)$ such that $\overline{e^2(t)}$ is minimized subject to $\overline{v^2(t)} \leqslant \frac{1}{2}$. Assume $G(s)$ and $G_s(s)$ and $U(s)$ of Problem 11.15 and (a) $H(s) = 1$, (b) $H(s) = 1/s$.

11.18. Consider the system shown in Fig. P11.18.
(a) Design the compensator $D(z)$ such that $\overline{e^2(t)}$ is minimized. Assume step input.
(b) Design the compensator if the delay factor is removed.

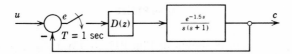

Figure P11.18

11.19. In Fig. P11.19, find K and a such that $\overline{e^2(t)}$ is minimized.

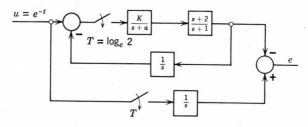

Figure P11.19

11.20. Find the compensator $D(z)$ such that $\overline{e^2(t)}$ in Fig. P11.20 is minimized.

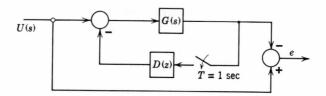

Figure P11.20

11.21. In Fig. P11.20, if

$$G(s) = \frac{1}{s}, \qquad U(s) = \frac{1}{s+1}$$

Find $D(z)$ such that $\overline{e^2(t)}$ is minimized.

11.22. Consider the system shown in Fig. P11.22. Minimize $\overline{e^2(t)}$ such that
(a) $F(s)$ is fixed and $G(s)$ is to be optimized.
(b) $G(s)$ is fixed and $F(s)$ is to be optimized.
(c) $F(s)$ and $G(s)$ are to be optimized simultaneously.

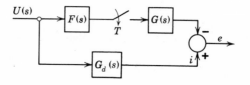

Figure P11.22

11.23. A simple system is shown in Fig. P11.23. Find $\overline{e^2(t)}$ and plot it as a function of aT. Under what conditions is a delay of $T/2$ a good substitute for a sampler of period T and hold for the given $u(t)$?

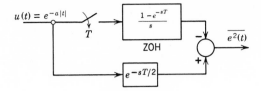

Figure P11.23

11.24. An analog-digital phase-locked loop is shown in Fig. P11.24. Find $D(z)$ such that $\overline{e^2(t)}$ is minimized subject to the constraint that bandwidth $B \le \alpha$. It is given that

$$B = \frac{T}{2\pi j} \int_0^1 dm \oint_{\text{unit circle}} T(z, m)T(z^{-1}, m) \frac{dz}{z}$$

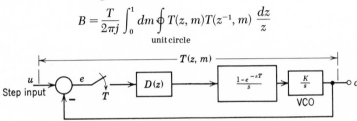

Figure P11.24

SELECTED BIBLIOGRAPHY

A. Continuous-time Control Systems

BOOKS

A.1 Bode, H. W., *Network Analysis and Feedback Amplifier Design*, Van Nostrand, 1945.

A.2 Guillemin, E. A., *The Mathematics of Circuit Analysis*, Wiley, 1949.

A.3 Van Valkenburg, M. E., *Network Analysis*, 2nd Ed., Prentice-Hall, 1964.

A.4 Evans, W. R., *Control System Dynamics*, McGraw-Hill, 1954.

A.5 James, H. M., N. B. Nichols, and R. S. Phillips, *Theory of Servo-Mechanisms*, McGraw-Hill, 1947.

A.6 Newton, G. C., Jr., L. A. Gould, and J. F. Kaiser, *Analytical Design of Linear Feedback Controls*, Wiley, 1957.

A.7 Bruns, R. A., and R. M. Saunders, *Analysis of Feedback Control Systems*, McGraw-Hill, 1955.

A.8 Chestnut, H., and R. W. Mayer, *Servo Mechanisms and Regulating System Design*, Wiley, 1951.

A.9 Truxal, J. G., *Automatic Feedback Control System Synthesis*, McGraw-Hill, 1955.

A.10 Aseltine, J. A., *Transform Method in Linear System Analysis*, McGraw-Hill, 1953.

A.11 Smith, O. J. M., *Feedback Control Systems*, McGraw-Hill, 1958.

A.12 Savant, C. J., Jr., *Basic Feedback Control System Design*, McGraw-Hill, 1958.

A.13 Wilts, C. H., *Principles of Feedback Control*, Addison-Wesley, 1960.

A.14 Brown, G. S., and D. P. Campbell, *Principles of Servo Mechanisms*, Wiley, 1948.

A.15 Fuchs, B. A., and B. V. Shabat, *Functions of a Complex Variable and Some of Their Applications*, Vol. I, Addison-Wesley, 1964.

A.16 Coddington, E. A., and N. Levinson, *Theory of Ordinary Differential Equations*, McGraw-Hill, 1955.

A.17 Schultz, M. A., *Control of Nuclear Reactors and Power Plants*, McGraw-Hill, 1955.

A.18 Horowitz, I. M., *Synthesis of Feedback Systems*, Academic Press, 1963.

A.19 Robertson, B. L., and L. J. Black, *Electrical Circuits and Machines*, 2nd Ed., Van Nostrand, 1957.

A.20 Gardner, M. F., and J. L. Barnes, *Transients in Linear Systems*, Vol. 1, Wiley, 1942.

A.21 Thaler, G. J., and R. G. Brown, *Analysis and Design of Feedback Control Systems*, 2nd Ed., McGraw-Hill, 1960.

A.22 D'Azzo, J. J., and C. H. Houpis, *Feedback Control System Analysis and Synthesis*, McGraw-Hill, 1960.

A.23 Murphy, G. J., *Basic Automatic Control Theory*, 2nd Ed., Van Nostrand, 1966.

A.24 Clark, R. N., *Introduction to Automatic Control Systems*, Wiley, 1962.

A.25 Tou, J. T., *Modern Control Theory*, McGraw-Hill, 1964.

A.26 Langill, A. W., Jr., *Automatic Control System Engineering*, Vol. I and II Prentice-Hall, 1965.

A.27 Kuo, B. C., *Automatic Control Systems*, 2nd Ed., Prentice-Hall, 1967.

A.28 Friedman, B., *Principles and Techniques of Applied Mathematics*, Wiley, 1956.

A.29 Zadeh, L. A., and C. A. Desoer, *Linear System Theory: The State Space Approach*, McGraw-Hill, 1963.

A.30 Gupta, S. C., *Transform and State Variable Methods in Linear Systems*, Wiley, 1966.

A.31 Dorf, R. C., *Modern Control Systems*, Addison-Wesley, 1967.

A.32 Elgerd, O. I., *Control Systems Theory*, McGraw-Hill, 1967.

A.33 Shinners, S. M., *Control System Design*, Wiley, 1964.

A.34 Saucedo, R., and E. E. Schiring, *Introduction to Continuous and Digital Control Systems*, Macmillan, 1968.

A.35 Blackwell, W. A., *Mathematical Modeling of Physical Networks*, Macmillan, 1968.

A.36 Schultz, D. G., and J. L. Melsa, *State Functions and Linear Control Systems*, McGraw-Hill, 1967.

A.37 Ogata, K., *State Space Analysis of Control Systems*, Prentice-Hall, 1967.

A.38 DeRusso, P. M., R. J. Roy, and C. M. Close, *State Variables for Engineers*, Wiley, 1965.

A.39 Huelsman, L. P., *Circuits Matrices and Linear Vector Spaces*, McGraw-Hill, 1963.

A.40 Bellman, B., *Introduction to Matrix Analysis*, McGraw-Hill, 1960.

A.41 Nering, E. D., *Linear Algebra and Matrix Theory*, Wiley, 1963.

A.42 Melsa, J. L., and D. G. Schultz, *Linear Control Systems*, McGraw-Hill, 1969.

A.43 Dorf, R. C., *Time-Domain Analysis and Design of Control Systems*, Addison-Wesley, 1965.

A.44 Schwarz, R. J., and B. Friedland, *Linear Systems*, McGraw-Hill, 1965.

A.45 Gantmacher, F. B., *The Theory of Matrices*, Chelsea, 1959.

575

A.46 Coddington, E. A., and N. Levinson, *Theory of Ordinary Differential Equations*, McGraw-Hill, 1955.

A.47 Timothy, L. K., and B. E. Bona, *State Space Analysis: An Introduction*, McGraw-Hill, 1968.

PAPERS

A.48 Kalman, R. E., "Mathematical Description of Linear Dynamical Systems," *Jour. Soc. of Ind. & Appl. Math.*, 1963, Ser. A., Control, Vol. 1, No. 2, 1963, pp. 152–192.

A.49 Brockett, B. W., "Poles, Zeros and Feedback: State Space Interpretation," *IEEE Trans. on Auto. Cont.*, Vol. AC-10, No. 2, April 1965, pp. 129–135.

A.50 Desoer, C. A., "An Introduction to State-Space Techniques in Linear Systems," *Proc. Joint Auto. Cont. Conf.*, 10(2), AIEE, 1962, pp. 1–5.

A.51 Evans, W. B., "Graphical Analysis of Control Systems," *Trans. AIEE*, **67**, 1948, pp. 547–551.

A.52 Routh, E. J., "Stability of a Given State of Motion," Adams Prize Essay, Macmillan, London, 1877.

A.53 Nyquist, H., "Regeneration Theory," *Bell System Tech. Jour.*, Vol. II, No. 1, Jan. 1932, pp. 126-147.

A.54 Mitrovic, D., "Graphical Analysis and Synthesis of Feedback Control Systems: Part I, Theory and Analysis; Part II, Synthesis," *Trans. AIEE*, Pt. II, Jan. 1959.

A.55 Siljak, D. D., "Analysis and Synthesis of Feedback Control Systems in the Parameter Plane: Pt. I-Linear Continuous Systems," *IEEE Trans. on Applications and Industry*, Vol. 83, Nov. 1964, pp. 449–458.

A.56 Siljak, D. D., "Generalization of the Parameter Plane Method." *IEEE Trans. on Auto. Cont.*, Vol. AC-11, No. 1, Jan. 1966, pp. 63–70.

B. Discrete-Time Control Systems

BOOKS

B.1 Truxal, J. G., *Control System Synthesis*, McGraw-Hill, 1955.

B.2 Aseltine, J. A., *Transform Methods in Linear System Analysis*, McGraw-Hill, 1958.

B.3 Jury, E. I., *Sampled-Data Control Systems*, Wiley, New York, 1958.

B.4 Ragazzini, J. R., and G. F. Franklin, *Sampled-Data Control Systems*, McGraw-Hill, 1958.

B.5 Tou, J. T., *Digital and Sampled Data Control Systems*, McGraw-Hill, 1959.

B.6 Tschauner, J., *Introduction to Theory of Sampled-Data Systems*, R. Oldenbourgh, Munich, 1960. [In German]

B.7 Chang, S. S. L., *Synthesis of Optimum Control Systems*, McGraw-Hill, 1961.

B.8 Monroe, J., *Digital Processes for Sampled-Data Systems*, Wiley, 1961.

B.9 Brown R. G., and J. W. Nilsson, *Introduction to Linear System Analysis*, Wiley, 1962.

B.10 Tsypkin, Y. Z. *Theory of Linear Sampled-Data Systems*, State Press for Physics and Mathematical Literature, Moscow, 1963. [In Russian]

B.11 Kuo, B., *Analysis and Synthesis of Sampled-Data Control Systems*, Prentice-Hall, 1963.

B.12 Jury, E. I., *Theory and Application of the Z-Transform Method*, Wiley, 1964.

B.13 Shinners, S. M., *Control System Design*, Wiley, 1964.

B.14 Freeman, H., *Discrete-Time Systems*, Wiley, 1965.

B.15 Lindorff, D. P. *Theory of Sampled Data Control Systems*, Wiley, 1965.

B.16 Gupta, S. C., *Transform and State Variable Methods in Linear Systems*, Wiley, 1966.

B.17 Saucedo, R., and E. E. Schiring, *Introduction to Continuous and Digital Control Systems*, Macmillan, 1968.

PAPERS

B.18 Linvill, W. K., "Sampled Data Control Systems Studied Through Comparison of Amplitude Modulation," *Trans. AIEE*, Vol. 70, Part 2, 1951, pp. 1779.

B.19 Ragazzini, J. R., and L. A. Zadeh, "Analysis of Sampled Data Systems," *Trans. AIEE*, Vol. 71, Part 2, 1952, pp. 225–232.

B.20 Barker, R. H., "The Pulse Transfer Function and its Application to Servo Systems," *Proc. IEE*, Vol. 99, Part IV, 1952, pp. 202–217.

B.21 Bergen, A. R., and J. R. Ragazzini, "Sampled Data Processing Techniques for Feedback Control Systems," *Trans. AIEE*, Vol. 73, Part 2, 1954, pp. 236–247.

B.22 Johnson, G. W., D. P. Lindorff, and C. G. A. Nordling, "Extension of Continuous Data System Design Techniques to Sampled Data Control Systems," *Trans. AIEE*, Part 2, 1955, pp. 252–263.

B.23 Franklin, G., "Linear Filtering of Sampled-Data," *IRE Convention Record*, Part 4, 1955, pp. 119–128.

B.24 Maitra, K. K., and P. E. Sarachik, "Digital Compensation of Continuous Data Feedback Control Systems," *Trans. AIEE*, Vol. 75, Part 2, 1956, pp. 107–116.

B.25 Jury, E. I., "Synthesis and Critical Study of Sampled-Data Control Systems," *Trans. AIEE*, Vol. 75, Part 2, 1956, pp. 141–151.

B.26 Mori, M., "Root Locus Method of Pulse Transfer Functions for Sampled Data Control Systems," *IRE Trans. on Auto. Cont.*, Vol. 3, 1957, pp. 13–20.

B.27 Freeman, H., and O. Lowenschuss, "Bibliography of Sampled-Data Control Systems and Z-Transform Applications," *IRE Trans. on Auto. Cont.*, Vol. PG AC-4, 1958, pp. 28–30.

B.28 Chang, S. S. L., "Statistical Design Theory of Digital Controlled Continuous Systems," *Trans. AIEE*, Vol. 77, Part 2, 1958, pp. 191–201.

B.29 Kalman, R., and J. Bertram, "A Unified Approach to the Theory of Sampling Systems," *Journal of Franklin Institute*, Vol. 267, 1959, pp. 405–436.

B.30 Lindorff, D. P., "Application of Pole-Zero Concepts to Design of Sampled Data Systems," *IRE Trans. on Auto. Cont.*, Vol. PG AC-5, 1959, pp. 173–184.

B.31 Jury, E. I., "Recent Advances in the Field of Sampled and Digital Controlled Systems," *Proceedings of 1960 IFAC*, Butterworth, England, 1960. pp. 240–246.

B.32 Jury, E. I., "Contribution to the Modified Z-Transformation Theory," *Journal of Franklin Institute*, Vol. 270, No. 2, 1960, pp. 114–124.

B.33 Jury, E. I., and J. Blanchard, "A Stability Test for Linear Discrete Systems in Table Form," *IRE Proceedings*, Vol. 49, No. 12, 1961, pp. 1947–1948.

B.34 Tou, J. T., "Statistical Design of Linear Discrete Data Control Systems via the Modified Z-Transform Method," *Journal of Franklin Institute*, Vol. 271, No. 4, 1961, pp. 249–262.

B.35 Jury, E. I., "A Simplified Stability Criterion for Linear Discrete Systems," *IEEE Proceedings*, Vol. 50, No. 6, 1962, pp. 1493–1500.

B.36 Gupta, S. C., "Increasing the Sampling Efficiency of a Control System," *IEEE Trans. on Auto. Cont.*, Vol. AC-8, 1963, pp. 262–264.

B.37 Gupta, S. C., and L. Hasdorff, "Method of Minimizing the Effect of External Disturbance in a Sampled Data Control System," *Regelungstechnik*, Vol. 12, 1964, pp. 397–402. [In German]

B.38 Greaves, C. J., and J. A. Cadzow, "The Optimal Discrete Filter Corresponding to a Given Analog Filter," *IEEE Trans. on Auto Cont.*, Vol. AC-12, No. 3, 1967, pp. 304–307.

B.39 Jury, E. I., "A General Z-Transform Formula for Sampled Data Systems," *IEEE Trans. on Auto. Cont.*, Vol. AC-12, No. 5, 1967, pp. 606–608.

B.40 Gupta, S. C., "On Optimum Digital Phase Locked Loops," *IEEE Trans. on Comm. Tech.*, Vol. COM-16, No. 2, 1968, pp. 340–344.

B.41 Proch, G. E., and S. C. Gupta, "Error Constants for Feedback Systems from State Variable Equations," *Proc. National Electronics Conference*, Vol. 22, October 1966, pp. 667–670.

B.42 Siljak, D. D., "Analysis and Synthesis of Feedback Control Systems in the Parameter Plane: II Sampled-Data Case," *IEEE Trans. on Applications and Industry*, Vol. 83, No. 75, November 1964, pp. 458–466.

B.43 Åström, K. J., E. I. Jury, and R. G. Agniel, "A Numerical Method for Evaluation of Complex Integrals," *IEEE Transactions on Automatic Control*, **AC-15**, 1970.

Index